Lecture Notes in Computer Science 15718

The series Lecture Notes in Computer Science (LNCS), including its subseries Lecture Notes in Artificial Intelligence (LNAI) and Lecture Notes in Bioinformatics (LNBI), has established itself as a medium for the publication of new developments in computer science and information technology research, teaching, and education.

LNCS enjoys close cooperation with the computer science R & D community, the series counts many renowned academics among its volume editors and paper authors, and collaborates with prestigious societies. Its mission is to serve this international community by providing an invaluable service, mainly focused on the publication of conference and workshop proceedings and postproceedings. LNCS commenced publication in 1973.

Edward Curry · Maribel Acosta ·
Maria Poveda-Villalón · Marieke van Erp ·
Adegboyega Ojo · Katja Hose · Cogan Shimizu ·
Pasquale Lisena
Editors

The Semantic Web

22nd European Semantic Web Conference, ESWC 2025
Portoroz, Slovenia, June 1–5, 2025
Proceedings, Part I

Editors
Edward Curry
University of Galway
Galway, Ireland

Maribel Acosta
Technical University of Munich
Heilbronn, Germany

Maria Poveda-Villalón
Universidad Politécnica de Madrid
Madrid, Spain

Marieke van Erp
KNAW - Humanities Cluster
Amsterdam, The Netherlands

Adegboyega Ojo
Carleton University
Ottawa, ON, Canada

Katja Hose
Technical University of Vienna
Vienna, Austria

Cogan Shimizu
Wright State University
Dayton, OH, USA

Pasquale Lisena
EURECOM
Biot, France

ISSN 0302-9743 ISSN 1611-3349 (electronic)
Lecture Notes in Computer Science
ISBN 978-3-031-94574-8 ISBN 978-3-031-94575-5 (eBook)
https://doi.org/10.1007/978-3-031-94575-5

This Springer imprint is published by the registered company Springer Nature Switzerland AG
The registered company address is: Gewerbestrasse 11, 6330 Cham, Switzerland

Preface

These two volumes (LNCS 15718-9) contain the main proceedings of the 22nd edition of the European Semantic Web Conference (ESWC 2025). ESWC is a major venue for discussing the latest scientific results and innovations related to the semantic web, knowledge graphs, and web data. This year we acknowledge semantics's critical role in enabling different forms of knowledge with the theme of "Empowering Knowledge through Semantics: From Knowledge Graphs to Neurosemantics".

This year ESWC's Research track addressed the theoretical, analytical, and empirical aspects of the Semantic Web, semantic technologies, knowledge graphs, and semantics on the Web in general. The In-use track focused on contributions that use and apply state-of-the-art semantic technologies or resources in real-world settings. The Resource track welcomed resource contributions that are on the one hand innovative or novel, and on the other hand sharable and reusable (e.g. datasets, knowledge graphs, ontologies, workflows, benchmarks, frameworks), and provide the necessary scaffolding to support the generation of scientific work and advance the state of the art.

The main scientific program of ESWC 2025 contained 45 papers selected out of 155 submissions (98 research, 22 in-use, 41 resource): 26 papers in the Research track, 8 in the In-use track, and 11 in the Resource track. The overall acceptance rate was 29% (26.5% research, 36% in-use, 27% resource). As in previous years, the review process included a rebuttal phase, allowing authors to respond to reviewers' comments and clarify technical contributions. This interaction helped ensure a fair decision process. The program chairs are grateful to the 23 senior PC members, the 277 PC members, and the 44 external reviewers for providing their feedback on the scientific program, and to all other community members who contributed to reviewing. Each paper received an average of 4.0 reviews, with the Research track being dual anonymous, and Resource and In-Use being single anonymous. We adopted ACM's terminology in all calls to improve ESWC's Diversity, Equity, and Inclusion principles in naming review guidelines.

We welcomed invited keynotes from three world-renowned speakers, spanning industry and academia: Raphaël Troncy (EURECOM), Sonja Zillner (Siemens Technology), and Leonid Libkin (University of Edinburgh).

ESWC 2025 featured 15 workshops and two tutorials. The workshops covered a large variety of topics, ranging from established research topics such as knowledge graph construction and neuro-symbolic artificial intelligence to emerging topics such as data spaces, generative agents, and causal artificial intelligence. Some of the workshops focused on specific application domains such as biomedicine, cultural heritage, and scientific publishing, while others addressed cross-cutting topics such as sustainability and responsible artificial intelligence. The tutorials were on nanopublications and graph shapes.

The conference also offered other opportunities to discuss the latest research and innovation work, including a poster and demo session, workshops and tutorials, a PhD symposium, an EU project networking session, and an industry track. We thank Marta

Sabou and Andreas Harth for organising the Workshop and Tutorials track. We are also thankful to Valentina Presutti and John McCrae for successfully running the Posters and Demos track. We are grateful to Mehwish Alam and Pieter Colpaert for coordinating the PhD Symposium, which welcomed 12 PhD students who had the opportunity to present their work and receive feedback in a constructive environment. Thanks go to Josiane Xavier Parreira and Diego Collarana for their management of the Industry Track, which welcomed submissions from several large industry players.

We also thank Irene Celino and Rafiq ul Haque for increasing the networking potential of ESWC by running the Project Networking Session. A special thanks to Sanju Tiwari and Yang Yang for their amazing job as Web and Publicity chairs, and to Pasquale Lisena for preparing this volume with Springer and the conference metadata. We thank JSI for supporting the conference organisation, particularly Marjia Komatar for their invaluable support. We thank our sponsors for supporting ESWC 2025 and our sponsorship chair, Marko Grobelnik, for securing them. We are grateful to John Domingue, the ESWC steering committee, Albert Meroño Peñuela, and the ESWC 2024 organising committee for their invaluable support and advice.

On December 29th, 2024, one of the ESWC founders, Dieter Fensel, passed away in Innsbruck after a long illness. Dieter was a central figure who provided visionary and strategic leadership to the Semantic Web community. His passing is a profound loss to our community. We acknowledge Dieter's significant scientific and community contributions over the last two decades, especially his establishment of this conference series. We introduce the ESWC Dieter Fensel Visionary Award to honour his memory and to recognise the most visionary contribution in each year's conference.

April 2025

Edward Curry
Maribel Acosta
Maria Poveda-Villalón
Marieke van Erp
Adegboyega Ojo
Katja Hose
Cogan Shimizu
Pasquale Lisena

In memory of Dieter Fensel

It is with great sadness that we report that one of the ESWC founders, Dieter Fensel, passed away in Innsbruck on December 29th, 2024 after a long illness. Dieter was born in Nuremberg, Germany, on October 10th, 1960. His early education was conducted in and around Nuremberg after which he went on to study mathematics at the Free University of Berlin where he earned a Master's degree in Social Science. He also earned a Master's degree in Computer Science from the Technical University of Berlin.

Following this Dieter enrolled at the University of Karlsruhe, where he obtained his doctorate in 1993 under the supervision of Rudi Studer. He later completed his habilitation at the same university and was honored with the Carl Adam Petri Award for his work. In 1994 Dieter moved to the University of Amsterdam after gaining a two-year scholarship from the German Research Foundation. Following this, he took on the role of associate professor at the Vrije Universiteit Amsterdam. The 2000s saw the Semantic Web emerge as a new and vibrant research field, and in 2002 Dieter became a full professor at the University of Innsbruck and in 2003 at National University of Ireland, Galway as well. It was at this time that he founded the Digital Enterprise Research Institute (DERI) at both of his new affiliations in Innsbruck and Galway. The Innsbruck institute was later renamed the Semantic Technology Institute (STI) Innsbruck, where Dieter remained serving as the Chair until his passing.

Whilst the above captures the main essentials of Dieter's career it is not easy to put into words his contributions to ESWC and the Semantic Web movement in general. Dieter was a true visionary and he saw at the very beginnings of the Semantic Web movement that a comprehensive and broad international research infrastructure would be required for the field to truly take off and have lasting impact. It was he that advocated for the setting up of the main conferences, ISWC as well as ESWC, and also that a high-quality journal to capture research outputs was required as well as a summer school to train and inculcate the next generation.

As stated on the ESWC history page, this conference series started over 20 years ago as the European Semantic Web Symposium held in Heraklion, Greece in May of 2004. This first event was established and supported through three large European projects that Dieter had a leading role in (SEKT, DIP and Knowledge Web). Since that time, as ESWC grew, and was relocated to Budva in Montenegro, Innsbruck, Tenerife, Montpellier and Portoroz (more than once), before coming back to Heraklion, Dieter was the main driver behind the scenes ensuring that the conference has had sustained success over a two-decade period.

Over that time, he never lost his passion for all things related to semantics and his vision that semantics takes its rightful place at the centre of computer science impacting society globally. This insight, as AI establishes itself at the top of the news agenda, remains as true today as it was back in the last century.

As all who worked with him quickly found out Dieter was a unique individual. He valued honesty, authenticity, clarity and excellence, whilst advocating that the main barriers in work (and life) were self-imposed. He also believed that culture and environment were fundamental aspects of fulfilling work and expended substantial energy engineering the highest quality of both within his research institutes and for colleagues and participants at any event or meeting he organised. Dieter was a very rare individual able to provide visionary and strategic leadership whilst also ensuring that the well-being of staff, colleagues and event participants was comprehensively covered.

We remain hugely grateful for all of Dieter's contributions over the last two decades, especially for establishing this conference series. We will all miss him greatly.

May he rest in peace.

January 2025 ESWC Steering Committee

Organization

General Chair

Edward Curry — University of Galway, Ireland

Research Track Program Chairs

Maribel Acosta — Technical University of Munich, Germany
Maria Poveda-Villalón — Universidad Politécnica de Madrid, Spain

In-Use Track Program Chairs

Marieke van Erp — KNAW Humanities Cluster, Netherlands
Adegboyega Ojo — Carleton University, Canada

Resource Track Program Chairs

Katja Hose — TU Wien, Austria
Cogan Shimizu — Wright State University, USA

Workshops and Tutorials Chairs

Marta Sabou — WU Vienna, Austria
Andreas Harth — Friedrich-Alexander-Universität Erlangen-Nürnberg and Fraunhofer IIS, Germany

Poster and Demo Chairs

Valentina Presutti — University of Bologna, Italy
John McCrae — Insight Centre for Data Analytics, University of Galway, Ireland

PhD Symposium Chairs

Mehwish Alam	Institut Polytechnique de Paris, France
Pieter Colpaert	Ghent University, Belgium

Industry Track Program Chairs

Josiane Xavier Parreira	Siemens AG, Austria
Diego Collarana	Fraunhofer FIT, Germany

Sponsorship

Marko Grobelnik	Jožef Stefan Institute, Slovenia

Project Networking

Irene Celino	Cefriel, Italy
Rafiq ul Haque	University of Galway, Ireland

Web and Publicity

Sanju Tiwari	Sharda University, India
Yang Yang	University of Galway, Ireland

Proceedings and Conference Metadata

Pasquale Lisena	EURECOM, France

Program Committee

Alessandro Adamou	Bibliotheca Hertziana - Max Planck Insitute for Art History, Italy
Shqiponja Ahmetaj	Vienna University of Technology, Austria
Mehwish Alam	Institut Polytechnique de Paris, France
Céline Alec	Université de Caen-Normandie, France

Alsayed Algergawy	University of Jena, Germany
Manzoor Ali	Paderborn University, Germany
Saeed Alsamhi	University of Galway, Ireland
Elvira Amador-Domínguez	Universidad Politécnica de Madrid, Spain
José Luis Ambite	University of Southern California, USA
Rafael Angarita	Paris Nanterre University, France
Kemafor Anyanwu	North Carolina State University, USA
Julián Arenas-Guerrero	Universidad Politécnica de Madrid, Spain
Natanael Arndt	German National Library, Germany
Ghislain Atemezing	European Union Agency for Railways, France
Maurizio Atzori	University of Cagliari, Italy
Nathalie Aussenac-Gilles	IRIT CNRS, France
Booma Sowkarthiga Balasubramani	University of Illinois Chicago, USA
Christian Beecks	University of Hagen, Germany
Moritz Blum	Semalytix GmbH, Germany
Carlos Bobed	University of Zaragoza, Spain
Fernando Bobillo	University of Zaragoza, Spain
Georgeta Bordea	Université de Bordeaux, France
Andreas Both	DATEV eG, Germany
Paolo Bouquet	University of Trento, Italy
Loris Bozzato	Fondazione Bruno Kessler, Italy
Janez Brank	Jožef Stefan Institute, Slovenia
Carlos Buil Aranda	Universidad Técnica Federico Santa María, Chile
Gregoire Burel	Open University, UK
Davide Buscaldi	LIPN, Université Sorbonne Paris Nord, France
Moritz Busch	Fraunhofer FIT, Germany
Jean-Paul Calbimonte	University of Applied Sciences and Arts Western Switzerland HES-SO, Switzerland
Diego Calvanese	Free University of Bozen-Bolzano, Italy
Antonella Carbonaro	University of Bologna, Italy
Alessio Carenini	Cefriel, Italy
Sylvie Cazalens	INSA de Lyon, France
Irene Celino	Cefriel, Italy
Javad Chamanara	Technische Informationsbibliothek, Germany
Pierre-Antoine Champin	Université Claude Bernard Lyon 1, France
William Charles	IRIT, France
David Chaves-Fraga	Universidade de Santiago de Compostela, Spain
Gong Cheng	Nanjing University, China
Antrea Christou	Wright State University, USA
Philipp Cimiano	Bielefeld University, Germany
Andrea Cimmino Arriaga	Universidad Politécnica de Madrid, Spain

Diego Collarana	Fraunhofer FIT, Germany
Pieter Colpaert	Ghent University, Belgium
Oscar Corcho	Universidad Politécnica de Madrid, Spain
Francesco Corcoglioniti	Free University of Bozen-Bolzano, Italy
Julien Corman	Free University of Bozen-Bolzano, Italy
Claudia D'Amato	University of Bari, Italy
Mathieu D'Aquin	University of Lorraine, France, France
Enrico Daga	Open University, UK
Ioannis Dasoulas	KU Leuven, Belgium
Jérôme David	Université Grenoble Alpes, France
Victor de Boer	Vrije Universiteit Amsterdam, Netherlands
Stefano De Giorgis	Institute for Cognitive Sciences and Technologies - National Research Council (ISTC-CNR), Italy
Ben De Meester	Ghent University, Belgium
Daniele Dell'Aglio	Aalborg University, Denmark
Emanuele Della Valle	Politecnico di Milano, Italy
Elena Demidova	University of Bonn, Germany
Gayo Diallo	University of Bordeaux, France
Stefan Dietze	GESIS, Germany
Anastasia Dimou	KU Leuven, Belgium
Christian Dirschl	Wolters Kluwer, Germany
Daniil Dobriy	Vienna University of Economics and Business, Austria
Milan Dojchinovski	Czech Technical University in Prague, Czech Republic
Ivan Donadello	Free University of Bozen-Bolzano, Italy
Mauro Dragoni	Fondazione Bruno Kessler, Italy
Xuemin Duan	KU Leuven, Belgium
Michel Dumontier	Maastricht University, Netherlands
Shusaku Egami	National Institute of Advanced Industrial Science and Technology, Japan
Fajar J. Ekaputra	Vienna University of Economics and Business, Austria
Andreas Ekelhart	SBA Research, Austria
Vadim Ermolayev	Ukrainian Catholic University, Ukraine
Beatriz Esteves	Ghent University, Belgium
Lorena Etcheverry	Universidad de la República, Uruguay
Pavlos Fafalios	Technical University of Crete and FORTH-ICS, Greece
Alessandro Faraotti	IBM, Italy
Catherine Faron	Université Côte d'Azur, France

Anna Fensel	Wageningen University and Research, Netherlands
Alejandro Fernandez	Universidad Nacional de La Plata, Argentina
Jesualdo Tomás Fernández-Breis	Universidad de Murcia, Spain
Sebastián Ferrada	Universidad de Chile, Chile
Erwin Filtz	Siemens AG Österreich, Austria
Manuel Fiorelli	University of Rome Tor Vergata, Italy
Giorgos Flouris	ICS-FORTH, Greece
Flavius Frasincar	Erasmus University Rotterdam, Netherlands
Naoki Fukuta	Shizuoka University, Japan
Michael Färber	ScaDS.AI & TU Dresden, Germany
Mohamed H. Gad-Elrab	Bosch Center for Artificial Intelligence, Germany
Alban Gaignard	CNRS, France
Aldo Gangemi	Università di Bologna & CNR-ISTC, Italy
Raúl García-Castro	Universidad Politécnica de Madrid, Spain
Daniel Garijo	Universidad Politécnica de Madrid, Spain
Anna Lisa Gentile	IBM Research, USA
Pouya Ghiasnezhad Omran	Australian National University, Australia
Martin Giese	University of Oslo, Norway
Birte Glimm	Universität Ulm, Germany
Esteban González Guardia	Universidad Politécnica de Madrid, Spain
Jorge Gracia	University of Zaragoza, Spain
Arianna Graciotti	University of Bologna, Italy
Irlan Grangel González	Bosch Corporate Research, Germany
Alexander Graß	Fraunhofer FIT, Germany
Paul Groth	University of Amsterdam, Germany
Claudio Gutierrez	Universidad de Chile, Chile
José Manuel Gómez Pérez	expert.ai, Spain
Peter Haase	metaphacts, Germany
Mohand-Saïd Hacid	Université Claude Bernard Lyon 1, France
Torsten Hahmann	University of Maine, USA
Lavdim Halilaj	Bosch Corporate Research, Germany
Rafiqul Haque	University of Galway, Ireland
Andreas Harth	Friedrich-Alexander-Universität Erlangen-Nürnberg and Fraunhofer IIS, Germany
Olaf Hartig	Linköping University, Sweden
Oktie Hassanzadeh	IBM, USA
Manfred Hauswirth	TU Berlin, Germany
Ivan Heibi	University of Bologna, Italy
Lars Heling	Stardog Union, USA
Daniel Hernandez	University of Stuttgart, Germany

Julio Hernandez	Trinity College Dublin, Ireland
Nathalie Hernandez	IRIT, France
Ryohei Hisano	ETH Zurich, Switzerland
Rinke Hoekstra	Elsevier, Netherlands
Aidan Hogan	Universidad de Chile, Chile
Andreas Hotho	University of Würzburg, Germany
Wei Hu	Nanjing University, China
Luis Ibanez-Gonzalez	University of Southampton, UK
Ana Iglesias-Molina	Universidad Politécnica de Madrid, Spain
Maxime Jakubowski	TU Wien, Austria
Ernesto Jimenez-Ruiz	City St. George's, University of London, UK
Simon Jupp	SciBite/Elsevier, Netherlands
Jan-Christoph Kalo	University of Amsterdam, Netherlands
Eduard Kamburjan	IT University of Copenhagen, Denmark
Tomi Kauppinen	Aalto University, Finland
Takahiro Kawamura	University of Tokyo, Japan
Mayank Kejriwal	Information Sciences Institute, USA
Natthawut Kertkeidkachorn	Japan Advanced Institute of Science and Technology, Japan
Sabrina Kirrane	WU Wien, Austria
Tomáš Kliegr	Prague University of Economics and Business, Czech Republic
Matthias Klusch	DFKI, Germany
Haridimos Kondylakis	ICS-FORTH, Greece
George Konstantinidis	University of Southampton, UK
Stasinos Konstantopoulos	NCSR Demokritos, Greece
Roman Kontchakov	Birkbeck, University of London, UK
Manolis Koubarakis	National and Kapodistrian University of Athens, Greece
Kouji Kozaki	Osaka Electro-Communication University, Japan
Ralf Krestel	ZBW - Leibniz Information Centre for Economics & Kiel University, Germany
Tobias Käfer	Karlsruhe Institute of Technology, Germanyy
Jose Emilio Labra Gayo	Universidad de Oviedo, Spain
Frederique Laforest	Laboratoire d'Informatique en Image et Système d'Information UMR CNRS, France
Sarasi Lalithsena	IBM Watson, USA
Christoph Lange	Fraunhofer Institute for Applied Information Technology FIT and RWTH Aachen University, Germany
Davide Lanti	Free University of Bozen-Bolzano, Italy
Nicolas Lazzari	University of Bologna, Italy
Danh Le Phuoc	TU Berlin, Germany

Anh Le-Tuan	TU Berlin, Germany
Maxime Lefrançois	MINES Saint-Étienne, France
Huanyu Li	Linköping University, Sweden
Anna Sofia Lippolis	CNR ISTC, Italy
Pasquale Lisena	EURECOM, France
Wenqiang Liu	Xi'an Jiaotong University, China
Maria Maleshkova	Helmut-Schmidt-Universität/Universität der Bundeswehr Hamburg, Germany
Claudia Marinica	Polytech Nantes, France
Miguel A. Martinez-Prieto	University of Valladolid, Spain
Jose L. Martinez-Rodriguez	Autonomous University of Tamaulipas, Mexico
Patricia Martín-Chozas	Universidad Politécnica de Madrid, Spain
John P. McCrae	University of Galway, Ireland
Jamie McCusker	Rensselaer Polytechnic Institute, USA
Lionel Medini	CNRS, France
Albert Meroño-Peñuela	King's College London, UK
Nandana Mihindukulasooriya	IBM, USA
Daniel Miranker	University of Texas at Austin, USA
Pascal Molli	Nantes Université, France
Pierre Monnin	Université Côte d'Azur, France
Gabriela Montoya	Nantes Université, France
Alba Catalina Morales Tirado	Open University, UK
Boris Motik	University of Oxford, UK
Alamgir Munir Qazi	University of Galway, Ireland
Raghava Mutharaju	IIIT-Delhi, India
Tuan-Phong Nguyen	Max Planck Institute for Informatics, Germany
Vinh Nguyen	National Library of Medicine, NIH, USA
Andriy Nikolov	AstraZeneca, UK
Andrea Giovanni Nuzzolese	CNR - Institute of Cognitive Sciences and Technologies, Italy
Cliff O'Reilly	City St. George's, University of London, UK
Atul Kr Ojha	University of Galway, Ireland
Femke Ongenae	Ghent University, Belgium
Francesco Osborne	Open University, UK
Matteo Palmonari	University of Milano-Bicocca, Italy
Harshvardhan J. Pandit	Dublin City University, Ireland
George Papadakis	National Technical University of Athens, Greece
Pierre-Henri Paris	Université Paris-Saclay, Greece
Theodore Patkos	ICS-FORTH, Greece
Heiko Paulheim	University of Mannheim, Germany
Terry Payne	University of Liverpool, UK
Tassilo Pellegrini	University of Applied Sciences St. Pölten, Austria

Bernardo Pereira Nunes	Australian National University, Australia
Sujan Perera	IBM Watson, USA
Romana Pernisch	Vrije Universiteit Amsterdam, Netherlands
Catia Pesquita	Universidade de Lisboa, Portugal
Rafael Peñaloza	University of Milano-Bicocca, Italy
Guangyuan Piao	Dell Technologies, Ireland
Lydia Pintscher	Wikimedia Deutschland, Germany
Giuseppe Pirrò	University of Calabria, Italy
Dimitris Plexousakis	ICS-FORTH, Greece
María Poveda-Villalón	Universidad Politécnica de Madrid, Spain
Nicoleta Preda	Université de Versailles, France
Valentina Presutti	University of Bologna, Italy
Tara Raafat	University of Surrey, UK
Alexandre Rademaker	IBM Research and EMAp/FGV, Brazil
Marco Ratta	Open University, UK
Georg Rehm	DFKI, Germany
Achim Rettinger	Trier University, Germany
Artem Revenko	Semantic Web Company GmbH, Austria
Mariano Rico	Universidad Politécnica de Madrid, Spain
Giuseppe Rizzo	LINKS Foundation, Italy
Sergio José Rodríguez Méndez	Australian National University, Australia
Edelweis Rohrer	Universidad de la República, Uruguay
Julián Rojas	Ghent University, Belgium
Maria Del Mar Roldan-Garcia	Universidad de Málaga, Spain
Miguel Romero	Universidad Católica de Chile, Chile
Oscar Romero	Universitat Politècnica de Catalunya, Spain
Henry Rosales-Méndez	University of Chile, Chile
Marco Rospocher	Università degli Studi di Verona, Italy
Catherine Roussey	INRAE, France
Sebastian Rudolph	TU Dresden, France
Anisa Rula	University of Brescia, Italy
Alessandro Russo	ISTC-CNR, Italy
Marta Sabou	WU Vienna, Austria
Angelo Salatino	Open University, UK
Muhammad Saleem	Leipzig University, Germany
Emanuel Sallinger	TU Wien, Austria
Md Kamruzzaman Sarker	Bowie State University, USA
Bruno Sartini	Ludwig-Maximilians-Universität München, Germany
Fatiha Saïs	Université Paris-Saclay, France
Vittorio Scarano	University of Salerno, Italy
Ralf Schenkel	Trier University, Germany

Andrea Schimmenti	University of Bologna, Italy
Daniel Schwabe	PUC-Rio, Brazil
Umutcan Serles	Onlim GmbH, Austria
Patricia Serrano Alvarado	Nantes Université, France
Bariş Sertkaya	Frankfurt University of Applied Sciences, Germany
Pavel Shvaiko	Informatica Trentina, Italy
Lucia Siciliani	University of Bari, Italy
Leslie Sikos	Edith Cowan University, Australia
Gerardo Simari	Universidad Nacional del Sur and CONICET, Argentina
Hala Skaf-Molli	Nantes Université, France
Blerina Spahiu	University of Milano-Bicocca, Italy
Marc Spaniol	Université de Caen-Normandie, France
Bram Steenwinckel	Ghent University, Belgium
Nadine Steinmetz	University of Applied Sciences Erfurt, Germany
Armando Stellato	University of Rome Tor Vergata, Italy
Simon Steyskal	Siemens AG, Austria
Umberto Straccia	ISTI-CNR, Italy
Chang Sun	Maastricht University, Netherlands
Zequn Sun	Nanjing University, China
Vojtěch Svátek	Prague University of Economics and Business, Czech Republic
Danai Symeonidou	INRAE, France
Ruben Taelman	Ghent University, Belgium
Lionel Tailhardat	Orange, France
Valentina Tamma	University of Liverpool, UK
Andrea Tettamanzi	Université Côte d'Azur, France
Ratan Bahadur Thapa	University of Stuttgart, Germany
Krishnaprasad Thirunarayan	Wright State University, USA
Ilaria Tiddi	Vrije Universiteit Amsterdam, Netherlands
Tabea Tietz	FIZ Karlsruhe, Germany
Mohan Timilsina	University of Galway, Ireland
Konstantin Todorov	University of Montpellier, France
Dominik Tomaszuk	University of Białystok, Poland
Riccardo Tommasini	INSA de Lyon, France
Sebastian Tramp	eccenca GmbH, Germany
Cassia Trojahn	UT2J & IRIT, France
Raphaël Troncy	EURECOM, France
Yannis Tzitzikas	University of Crete and FORTH-ICS, Greece
Jürgen Umbrich	WU Wien, Austria
Chukwudi Uwasomba	Open University, UK

Sahar Vahdati	InfAI, Germany
Ludger Van Elst	DFKI, Germany
Dylan Van-Assche	Ghent University, Belgium
Miel Vander Sande	Meemoo, Belgium
Ruben Verborgh	Ghent University, Belgium
Maria-Esther Vidal	Leibniz University Hannover, Germany
Fabio Vitali	University of Bologna, Italy
Domagoj Vrgoc	Universidad Católica de Chile, Chile
Abdul Wahid	University of Galway, Ireland
Laura Waltersdorfer	TU Wien, Austria
Haofen Wang	Tongji University, China
Kewen Wang	Griffith University, Australia
Ruijie Wang	University of Zurich, Switzerland
Thilini Wijesiriwardene	University of South Carolina, USA
Xander Wilcke	Vrije Universiteit Amsterdam, Netherlands
Peter Winstanley	Semantechs Consulting, UK
Honghan Wu	King's College London, UK
Zhe Wu	eBay, USA
Josiane Xavier Parreira	Siemens AG, Austria
Ran Yu	University of Bonn, Germany
Jicheng Yuan	TU Berlin, Germany
Fouad Zablith	American University of Beirut, Lebanon
Ondřej Zamazal	Prague University of Economics and Business, Czech Republic
Omnia Zayed	University of Galway, Ireland
Xiaowang Zhang	Tianjin University, China
Yizheng Zhao	Nanjing University, China
Antoine Zimmermann	École des Mines de Saint-Étienne, France

Additional Reviewers

Akaichi, Inès
Alsamhi, Saeed
Antakli, André
Barba-Gonzalez, Cristobal
Benítez Hidalgo, Antonio
Boukhers, Zeyd
Braun, Christoph
Ceriani, Miguel
Charles, William
Charpenay, Victor
Conde-Herreros, Diego
Fathallah, Nadeen
Fernández Álvarez, Daniel
Fischer, Elisabeth
Henselmann, Daniel
Hirschbrunn, Joshua
Illich, Moritz
Kharbanda, Mayank
Kugler, Kai
Lippolis, Anna Sofia
Martín Chozas, Patricia
Marx, Edgard

Maynou, Marc
Mohammadi, Hossein
Morales Tirado, Alba Catalina
Mountantonakis, Michalis
Mustaf, Daham
Papantoniou, Katerina
Pons, Gerard
Popovic, Nicholas
Ragazzi, Luca
Raoufi, Ensiyeh
Sagi, Tomer
Schlör, Daniel
Schraudner, Daniel
Schreieder, Tobias
Sierra-Múnera, Alejandro
Smessaert, Ieben
Troullinou, Georgia
Vargas-Rojas, Felipe
Werner, Simon
Wunderle, Julia
Zehe, Albin
Zuppiroli, Sara

Sponsors

Platinum Sponsors

VideoLectures.NET is an award-winning free and open-access educational video lectures repository. The lectures are given by distinguished scholars and scientists at the most important and prominent events like conferences, summer schools, workshops and science promotional events from many fields of Science. The portal is aimed at promoting science, exchanging ideas and fostering knowledge sharing by providing high-quality didactic contents not only to the scientific community but also to the general public. All lectures, accompanying documents, information and links are systematically selected and classified through the editorial process taking into account also users' comments.

Gold Sponsors

The **Jožef Stefan Institute** (Ljubljana, Slovenia) is the leading Slovenian research organisation. It is responsible for a broad spectrum of basic and applied research in the fields of natural sciences and technology. The staff of around 960 specialize in research in physics, chemistry and biochemistry, electronics and information science, nuclear technology, energy utilization and environmental science. The Institute was founded in 1949 at a time when scientific research was expanding rapidly throughout the world. Initially established as an institute for Physics within the Slovenian Academy of Sciences and Arts, it is today involved in a wide variety of fields of both scientific and economic interest. After more than 60 years of scientific achievement, the Institute has become part of the image of Slovenia. The basic goals of the Institute are to provide expert scientific and applied output in the form of processes, products and consultancy, and to produce well-trained young scientists.

Silver Sponsors

Springer is part of Springer Nature, a leading global research, educational and professional publisher, home to an array of respected and trusted brands providing quality content through a range of innovative products and services. Springer Nature is the world's largest academic book publisher, publisher of the world's most influential journals and a pioneer in the field of open research. The company numbers almost 13,000 staff in over 50 countries and has a turnover of approximately €1.5 billion. Springer Nature was formed in 2015 through the merger of Nature Publishing Group, Palgrave Macmillan, Macmillan Education and Springer Science+Business Media.

Visionary Award

Graphwise enables organizations to unlock ROI for enterprise AI by delivering the most comprehensive and trusted industry solution in the field of knowledge graphs and semantic AI technologies. As enterprises pour millions into AI investment, Graphwise delivers the critical knowledge graph infrastructure to ensure enterprises are ready to realize the technology's full potential, is trusted, and can be implemented at scale. Graphwise, which is the result of the merger between tech visionaries Ontotext and Semantic Web Company, has over 200 employees worldwide, with offices located across North America, Europe and APAC.

Best Paper Award

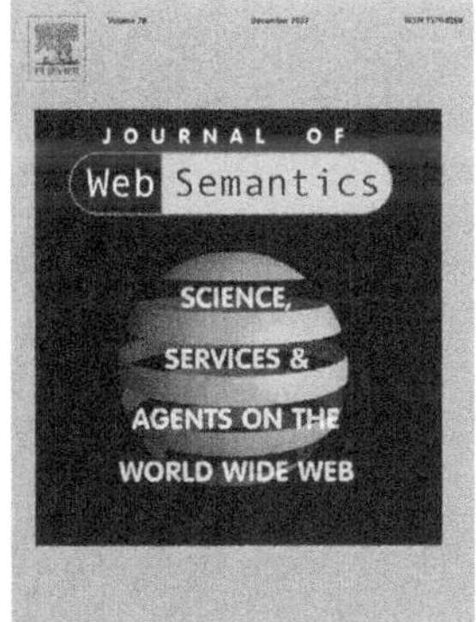

The **Journal of Web Semantics** is an interdisciplinary journal based on research and applications of various subject areas that contribute to the development of a knowledge-intensive and intelligent service Web. These areas include: knowledge technologies, ontology, agents, databases and the semantic grid, obviously disciplines like information retrieval, language technology, human-computer interaction and knowledge discovery are of major relevance as well. The Journal of Web Semantics addresses various prominent application areas including: e-business, e-community, knowledge management, e-learning, digital libraries and e-sciences.

Contents – Part I

Research

Contents – Part II

In-Use

Resource

Research

An Algebraic Foundation for Knowledge Graph Construction

Sitt Min Oo[1(✉)] and Olaf Hartig[2(✉)]

[1] Ghent University - imec, Ghent, Belgium
x.sittminoo@ugent.be
[2] Linköping University, Linköping, Sweden
olaf.hartig@liu.se

Abstract. Although they exist since more than ten years already, have attracted diverse implementations, and have been used successfully in a significant number of applications, declarative mapping languages for constructing knowledge graphs from heterogeneous types of data sources still lack a solid formal foundation. This makes it impossible to introduce implementation and optimization techniques that are provably correct and, in fact, has led to discrepancies between different implementations. Moreover, it precludes studying fundamental properties of different languages (e.g., expressive power). To address this gap, this paper introduces a language-agnostic algebra for capturing mapping definitions. As further contributions, we show that the popular mapping language RML can be translated into our algebra (by which we also provide a formal definition of the semantics of RML) and we prove several algebraic rewriting rules that can be used to optimize mapping plans based on our algebra.

1 Introduction

Knowledge graphs (KGs) are often created or populated by extracting and transforming data from other sources of structured or semi-structured data. One approach to this end is to employ a mapping engine for which the desired mapping from the source data to the KG can be specified using a declarative language. This approach has gained traction in recent years, with applications across diverse domains such as COVID research [17], railway networks [16], and incident management in ICT systems [22]. To provide a basis for this approach, several declarative mapping languages, in particular for RDF-based KGs, have been proposed, including languages that focus only on a single type of input data sources (e.g., R2RML for relational databases [21]) and, more interestingly, also languages designed for multiple types of input data sources [23]. The latter can be classified further into mapping-focused extensions of query languages and dedicated languages designed solely for the declarative description of mappings.

So far, most of these mapping languages—in particular, the dedicated ones—are defined only informally. Yet, such an informal definition can easily lead to discrepancies between different implementations. For instance, for RML [4,5], which has several implementations [1,6,9,18,20], the introduction of conformance tests [8] has revealed various cases in which different implementations

E. Curry et al. (Eds.): ESWC 2025, LNCS 15718, pp. 3–22, 2025.
https://doi.org/10.1007/978-3-031-94575-5_1

fail such a test [7]. It may be argued that the developers of these implementations interpreted the informal definition of the language differently. Without a formal definition it is impossible to introduce implementation and optimization techniques that are provably correct, or toproperly compare languages or language features in terms of fundamental properties, such as their expressive power. Therefore, the goal of our work in this paper is to provide a formal foundation for such declarative mapping languages. Especially, instead of focusing on a single specific language, we aim for a more general formalization approach.

Our main contribution is a language-agnostic algebra to capture definitions of mappings from heterogeneous types of data sources to RDF-based KGs. The basis of this mapping algebra is a variation of the relational data model (Sect. 3), for which we define five types of operators that can be arbitrarily combined into algebra expressions (Sect. 4). Due to its language-agnostic nature, the algebra can serve as a foundation to various mapping languages formally. As our second contribution, we demonstrate this benefit by providing an algorithm that translates mappings defined using RML into our algebra (Sect. 5). Through this algorithm, we show not only that our algebra is at least as expressive as RML, but we also provide a formal definition of the semantics of RML. To the best of our knowledge, this is the first formal approach to capture the RML semantics. Another important value of having an algebra that captures declarative mapping definitions is that it may be used as the basis of a systematic and well-defined approach to plan and to optimize the execution of KG construction processes in mapping engines. Related to this option, we make our third contribution: We show several algebraic equivalences which can be used as rewriting rules to optimize mapping plans that are based on our algebra (Sect. 6).

2 Related Work

As mentioned above, declarative mapping languages for KG construction can be classified into languages that are extensions of other types of languages (mainly, query languages), with XSPARQL [2], SPARQL-Generate [11], and Facade-X [3] being examples of such "repurposed mapping languages" [23], and "dedicated mapping languages" [23] that have been created specifically for the purpose of KG construction, with R2RML [21] and RML [4] as examples.

Given our focus on formal approaches to capture such languages, for the repurposed ones we notice they can rely on the formalization of their underlying language. For instance, XSPARQL combines the semantics of XQuery and SPARQL for mapping between XML and RDF data [2], and to also support relational databases, the authors extend their work by utilizing the semantics of relational algebra [12]. Similarly, the formalization of SPARQL-Generate [11] builds on the algebra-based formalization of SPARQL and, thus, is similar in spirit to our approach, which we see as more relevant for dedicated mapping languages.

When considering dedicated mapping languages, to the best of our knowledge, only R2RML has a formally defined semantics. That is, Sequeda et al. [19] provide such a formalization by means of 57 Datalog rules. As this definition is for R2RML, it is limited to mappings from relational databases to RDF, whereas our formalization in this paper can capture mappings of different types of sources of structured and semi-structured data. Moreover, we argue that an algebraic formalization such as ours has the additional advantage that it can also be utilized to represent a form of logical execution plans in mapping engines.

Besides R2RML, a few studies with formalizations related to RML have been published in recent years [9,14]. Iglesias et al. formalize RML-based mapping processes based on Horn clauses [9], where the focus of this formalization is to define optimization techniques presented by the authors. Lastly, an initial study to capture the semantics of mappings in a language-agnostic manner through algebraic mapping operators [14] has a very similar goal as our work in this paper. Yet, in contrast to our work, the authors do not cover all aspects of popular dedicated mapping language such as RML (e.g., joining data from multiple sources) and several of their definitions remain informal or rely on undefined concepts.

3 Data Model

This section introduces the data model based on which our mapping algebra is defined. Hence, this model captures the types of intermediate results of any mapping process described by our formalism. Informally, the data model is a version of the relational data model, restricted to a special kind of relations, that we call *mapping relations* and that contain RDF terms as possible values.

As a preliminary step for defining the model formally, we introduce relevant concepts of RDF: We let $\mathcal{S}$ be the countably infinite set of strings (sequences of Unicode code points) and $\mathcal{I}$ be the subset of $\mathcal{S}$ that consists of all strings that are valid IRIs. Moreover, $\mathcal{L}$ is a countably infinite set of pairs $(lex, dt) \in \mathcal{S} \times \mathcal{I}$, which we call *literals*. For each such literal $\ell = (lex, dt)$, lex is its *lexical form* and dt is its *datatype IRI*.[1] Furthermore, we assume a countably infinite set $\mathcal{B}$ of *blank nodes*, which is disjoint from both $\mathcal{S}$ and $\mathcal{L}$. The set $\mathcal{T}$ of all *RDF terms* is defined as $\mathcal{T} = \mathcal{I} \cup \mathcal{B} \cup \mathcal{L}$, and we write $\bar{\mathcal{T}}$ to denote the set of all possible sequences of terms; i.e., each element of $\bar{\mathcal{T}}$ is a sequence of elements of $\mathcal{T}$. As usual, an *RDF triple* is a tuple $(s, p, o) \in (\mathcal{I} \cup \mathcal{B}) \times \mathcal{I} \times \mathcal{T}$, an *RDF graph* is a set of RDF triples, and an *RDF dataset* is a set $\{G_{\mathsf{dflt}}, (n_1, G_1), \ldots, (n_m, G_m)\}$ where $G_{\mathsf{dflt}}, G_1, \ldots, G_m$ are RDF graphs, $m \geq 0$, and $n_i \in \mathcal{I} \cup \mathcal{B}$ for all $i \in \{1, \ldots, m\}$.

As additional ingredients for defining mapping relations, we assume a countably infinite set $\mathcal{A}$ of *attributes* and a special value ϵ that is not an RDF term (i.e., $\epsilon \notin \mathcal{T}$) and that shall be used to capture processing errors. The last ingredient are *mapping tuples*, which are the type of tuples used in mapping relations.

Definition 1. A **mapping tuple** is a partial function $t : \mathcal{A} \to \mathcal{T} \cup \{\epsilon\}$.

We say that two mapping tuples t and t' are *compatible* if, for every attribute $a \in \mathrm{dom}(t) \cap \mathrm{dom}(t')$, it holds that $t(a) = t'(a)$. Given two mapping tuples t

[1] To avoid overly complex formulas in this paper, we ignore the option of literals that have a language tag. Yet, the formalism can easily be extended to cover them.

Table 1. Mapping relation r discussed in Examples 1 and 2.

	a_x	a_s	a_p	a_o	a_g
t	("12", xsd:integer)	ex:alice	foaf:knows	ex:bob	rr:defaultGraph
t'	ex:alice	ex:alice	("knows", xsd:string)	ex:charles	ex:g1
t''	ϵ	ex:bob	foaf:name	("Bob", xsd:string)	ex:g2

and t' that are compatible, we write $t \cup t'$ to denote the mapping tuple t'' that is the merge of t and t'; that is, $\text{dom}(t'') = \text{dom}(t) \cup \text{dom}(t')$ and, for every attribute $a \in \text{dom}(t'')$, $t''(a) = t(a)$ if $a \in \text{dom}(t)$, and $t''(a) = t'(a)$ otherwise.

Now we are ready to define our notion of a mapping relation.

Definition 2. A **mapping relation** r is a tuple (A, I), where $A \subset \mathcal{A}$ is a finite, non-empty set of attributes and I is a set of mapping tuples such that, for every such tuple $t \in I$, it holds that $\text{dom}(t) = A$. We call A the **schema** of the mapping relation and I is the **instance** of the mapping relation.

For the sake of conciseness, we sometimes write $t \in r$ to denote that t is a mapping tuple in the instance I of a mapping relation $r = (A, I)$.

Example 1. Consider the mapping relation $r = (A, I)$ that is represented in a tabular form in Table 1. It holds that $A = \{a_x, a_s, a_p, a_o, a_g\}$ and I contains three mapping tuples, t, t', and t'', such that $t(a_x)$ is the literal `("12", xsd:integer)`, $t'(a_x)$ is the IRI `ex:alice`, $t''(a_x) = \epsilon$, etc.

While mapping relations are intermediate results of mapping processes described by our algebra, the ultimate goal of such processes is to produce an RDF dataset. Therefore, we also need to define how mapping relations are mapped into RDF datasets. To this end, we assume four special attributes: $a_s, a_p, a_o, a_g \in \mathcal{A}$. Every mapping relation whose schema contains these attributes is mapped to an RDF dataset by using the values that the tuples in the relation have for these attributes. That is, if the values that a mapping tuple has for a_s, a_p, and a_o form a valid RDF triple, then the graph determined by the a_g-value of the tuple contains this triple. Formally, we define this approach as follows.

Definition 3. Let $r = (A, I)$ be a mapping relation with $a_s, a_p, a_o, a_g \in A$, and

$$I_{\text{valid}} = \Big\{t \in I \,\Big|\, \big(t(a_s), t(a_p), t(a_o)\big) \text{ is an RDF triple and } t(a_g) \in \mathcal{I} \cup \mathcal{B}\Big\} \text{ and}$$
$$N = \Big\{t(a_g) \in \mathcal{I} \cup \mathcal{B} \,\Big|\, t \in I_{\text{valid}} \text{ and } t(a_g) \text{ is not the IRI } \texttt{rr:defaultGraph}\Big\}.$$

The **RDF dataset resulting from** r is the RDF dataset

$$D = \{G_{\text{dflt}}\} \cup \big\{(n, G_n) \,\big|\, n \in N\big\},$$

where, for every $n \in N$,

$$G_n = \Big\{\big(t(a_s), t(a_p), t(a_o)\big) \,\Big|\, t \in I_{\text{valid}} \text{ such that } t(a_g) = n\Big\} \text{ and}$$
$$G_{\text{dflt}} = \Big\{\big(t(a_s), t(a_p), t(a_o)\big) \,\Big|\, t \in I_{\text{valid}} \text{ such that } t(a_g) \text{ is } \texttt{rr:defaultGraph}\Big\}.$$

Example 2. For the mapping relation in Example 1 it holds that $I_{\text{valid}} = \{t, t''\}$ and $N = \{\texttt{ex:g2}\}$. Therefore, the RDF dataset resulting from this mapping relation is $D = \{G, (\texttt{ex:g2}, G')\}$ with $G = \{(\texttt{ex:alice}, \texttt{foaf:knows}, \texttt{ex:bob})\}$ and $G' = \{(\texttt{ex:bob}, \texttt{foaf:name}, (\texttt{"Bob"}, \texttt{xsd:string}))\}$. Notice that $t' \notin I_{\text{valid}}$ because $(t'(a_{\text{s}}), t'(a_{\text{p}}), t'(a_{\text{o}}))$ is not an RDF triple (since $t'(a_{\text{p}})$ is a literal).

4 Mapping Algebra

This section introduces our mapping algebra which consists of five types of operators. The output of each such operator is a mapping relation and the inputs (if any) are mapping relations as well. Hence, the operators can be combined into arbitrarily complex algebra expressions, where each such expression defines a mapping from one or more sources of (semi-)structured data into an RDF dataset.[2]

4.1 Source Operator

The basis of every expression that defines a mapping using our algebra are operators that establish a relational view of each source of input data. From the perspective of the algebra, these *source operators* are nullary operators; they do not have any mapping relation as input. The mapping relation that such an operator creates as output depends on the data of the corresponding data source and can be specified based on a language appropriate to identify and extract relevant pieces of data from the data source. As our mapping algebra should be applicable for various types of data sources, for which different such languages are suitable, we introduce an abstraction of this extraction process and define our notion of source operators based on this abstraction.

Our abstraction considers two infinite sets, $\mathcal{D}$ and $\mathcal{Q}$, which are disjoint. $\mathcal{D}$ is the set of all possible *data objects*. In practice, there are different kinds of data objects, even within the context of the same type of data source, and data objects may be nested within one another. For instance, each line of a CSV file may be seen as a data object, and so may each value within such a line, as well as the CSV file as a whole. In our abstraction, any particular kind of data objects is considered as a specific subset of $\mathcal{D}$.

The set $\mathcal{Q}$ captures possible *query languages* to retrieve data objects from other data objects. We consider each such language $L \in \mathcal{Q}$ as a set, where every element $q \in L$ is one of the expressions admitted by the language. In practice, each such language is designed to be used for a specific kind of data objects (i.e., a specific subset of $\mathcal{D}$). Therefore, we define the notion of the *types of data sources* considered for the source operators of our algebra as a structure that specifies the relevant kinds of data objects and the corresponding evaluation semantics of the languages to be used for data sources of such a type.

[2] We assume here that the mapping relation resulting from the root operator of the expression contains the aforementioned attributes a_{s}, a_{p}, a_{o}, and a_{g} (see Definition 3).

Definition 4. A **source type** is a tuple $(\mathcal{D}^{\text{ds}}, \mathcal{D}^{\text{c1}}, \mathcal{D}^{\text{c2}}, L, L', eval, eval', cast)$ where

- $\mathcal{D}^{\text{ds}} \subseteq \mathcal{D}$, $\mathcal{D}^{\text{c1}} \subseteq \mathcal{D}$, $\mathcal{D}^{\text{c2}} \subseteq \mathcal{D}$,
- L and L' are query languages in $\mathcal{Q}$,
- $eval$ is a function $eval\colon \mathcal{D}^{\text{ds}} \times L \rightarrow \overline{\mathcal{D}^{\text{c1}}}$,
- $eval'$ is a partial function $eval'\colon \mathcal{D}^{\text{ds}} \times \mathcal{D}^{\text{c1}} \times L' \rightarrow \overline{\mathcal{D}^{\text{c2}}}$, and
- $cast$ is a partial function $cast\colon \mathcal{D}^{\text{c2}} \rightarrow \mathcal{L}$,

where $\overline{\mathcal{D}^{\text{c1}}}$, resp. $\overline{\mathcal{D}^{\text{c2}}}$, is the set of all sequences of data objects in $\mathcal{D}^{\text{c1}}$, resp. $\mathcal{D}^{\text{c2}}$.

Definition 5. A **data source** s is a tuple $(type, D)$ in which $type$ is a source type $(\mathcal{D}^{\text{ds}}, \mathcal{D}^{\text{o1}}, \mathcal{D}^{\text{o2}}, L, L', eval, eval', cast)$ and $D \in \mathcal{D}^{\text{ds}}$.

As per Definitions 4 and 5, the set $\mathcal{D}^{\text{ds}}$ of a source type determines the kind of data objects that the data sources of this type may provide access to. For instance, for CSV-based data sources this may be the (infinite) set of all possible CSV files. The functions $eval$ and $eval'$ are meant to define the evaluation semantics of the languages L and L', respectively. Notice that the domain of $eval'$ consists of 3-tuples that contain an additional data object as their second element, which is meant as a context object for the evaluation. The idea of the source operators shall be to use L to specify relevant context objects and, then, to use L' to retrieve values for creating a mapping tuple per context object.

While a formal definition of concrete source types is beyond the scope of this paper, the following two examples outline informally how JSON-based data sources and CSV-based data sources may be captured within our abstraction.

Example 3. To define $type_{\mathsf{json}} = (\mathcal{D}^{\text{ds}}_{\mathsf{json}}, \mathcal{D}^{\text{c1}}_{\mathsf{jp}}, \mathcal{D}^{\text{c2}}_{\mathsf{jp}}, L_{\mathsf{jp}}, L'_{\mathsf{jp}}, eval_{\mathsf{jp}}, eval'_{\mathsf{jp}}, cast_{\mathsf{json}})$ as a source type for JSON-based data sources, we let $\mathcal{D}^{\text{ds}}_{\mathsf{json}}$ be the infinite set of all possible JSON documents, both $\mathcal{D}^{\text{c1}}_{\mathsf{jp}}$ and $\mathcal{D}^{\text{c2}}_{\mathsf{jp}}$ are the infinite set of all possible elements that may occur within JSON documents (i.e., $\mathcal{D}^{\text{c1}}_{\mathsf{jp}} = \mathcal{D}^{\text{c2}}_{\mathsf{jp}}$), L_{jp} is the set of all JSONPath expressions that begin with the selector for the root element (i.e., `$`), and L'_{jp} are all JSONPath expressions that do not begin with `$`. Then, $eval_{\mathsf{jp}}$ and $eval'_{\mathsf{jp}}$ capture the evaluation semantics of JSONPath. That is, for a JSON document $D \in \mathcal{D}^{\text{ds}}_{\mathsf{json}}$ and a query $q \in L_{\mathsf{jp}}$, $eval_{\mathsf{jp}}(D, q)$ returns the elements of D selected by q. If $q' \in L'_{\mathsf{jp}}$, and assume $d \in \mathcal{D}^{\text{c1}}_{\mathsf{jp}}$ is a JSON object within D, then $eval'_{\mathsf{jp}}(D, d, q')$ returns the elements of D that can be reached from d as per the traversal specified by q'. Finally, $cast_{\mathsf{json}}$ maps JSON values to RDF literals. For instance, string values may be mapped to `xsd:string` literals, numeric values that are integers may be mapped to `xsd:integer` literals, etc.

Example 4. As a source type for CSV-based data sources, we may introduce $type_{\mathsf{csv}} = (\mathcal{D}^{\text{ds}}_{\mathsf{csv}}, \mathcal{D}^{\text{c1}}_{\mathsf{csv}}, \mathcal{D}^{\text{c2}}_{\mathsf{csv}}, L_{\mathsf{csv}}, L'_{\mathsf{csv}}, eval_{\mathsf{csv}}, eval'_{\mathsf{csv}}, cast_{\mathsf{csv}})$ as follows. $\mathcal{D}^{\text{ds}}_{\mathsf{csv}}$ is the infinite set of all possible CSV files, $\mathcal{D}^{\text{c1}}_{\mathsf{csv}}$ is the infinite set of all possible rows of CSV files, $\mathcal{D}^{\text{c2}}_{\mathsf{csv}} \subset \mathcal{S}$ is the infinite set of strings that may be values within CSV files, and the $cast_{\mathsf{csv}}$ function maps all these string values in $\mathcal{D}^{\text{c2}}_{\mathsf{csv}}$ to `xsd:string` literals. As the language L_{csv}, we may use the singleton set that contains (only) the empty string ε, and $L'_{\mathsf{csv}} \subset \mathcal{S}$ is the set of all strings that may be used as column names in CSV files. Informally, $eval_{\mathsf{csv}}$ and $eval'_{\mathsf{csv}}$ may then be defined as follows. For every CSV file $D \in \mathcal{D}^{\mathsf{ds}}_{\mathsf{csv}}$, $eval_{\mathsf{csv}}(D, \varepsilon)$ returns all rows of D, and for every such row $d \in eval_{\mathsf{csv}}(D, \varepsilon)$ and every string c that is a column name in D, $eval'_{\mathsf{csv}}(D, d, c)$ is the value that the row d has in column c.

Example 5. As the basis of a running example for the rest of this section, we assume a data source $s_{\mathsf{ex}} = (type_{\mathsf{csv}}, D_{\mathsf{ex}})$ where D_{ex} is a CSV file with four columns, named id, firstname, lastname, and age, and the following two rows:

$$(1, \texttt{Alice}, \texttt{Lee}, 23) \quad \text{and} \quad (2, \texttt{Bob}, \texttt{Malice}, \texttt{unknown}).$$

If we write d_1 to denote the first of these rows, then $eval'_{\mathsf{csv}}(D_{\mathsf{ex}}, d_1, \mathsf{firstname})$ is the string `Alice`, and $cast_{\mathsf{csv}}(\texttt{Alice})$ is the RDF literal $(\texttt{"Alice"}, \texttt{xsd:string})$.

Now we are ready to define our notion of source operators. While they are nullary operators, they are specified using three parameters: the data source from which the data for the produced mapping relation is extracted, a query that selects relevant context objects from the data source, and a map that associates attributes with queries, where these queries are meant to select the values that a mapping tuple produced for a context object has for the attributes. Formally, we define the mapping relation produced by a source operator as follows.

Definition 6. Let $s = (type, D)$ be a data source with $type = (\mathcal{D}^{\mathsf{ds}}, \mathcal{D}^{\mathsf{c1}}, \mathcal{D}^{\mathsf{c2}}, L, L', eval, eval', cast)$, let $q \in L$, and let $\mathbb{P}$ be a partial function $\mathbb{P} \colon \mathcal{A} \to L'$. The **$(q, \mathbb{P})$-specific mapping relation obtained from** s, denoted by $\text{Source}^{(s,q,\mathbb{P})}$, is the mapping relation (A, I) such that $A = \text{dom}(\mathbb{P})$ and

$$I = \big\{\{a_1 \to cast(v_1), \ldots, a_n \to cast(v_n)\} \;\big|\; d \text{ is in } \bar{O} \text{ and } ((a_1, v_1), \ldots, (a_n, v_n)) \in X_d\big\},$$

where $\bar{O} = eval(D, q)$ and

$$X_d = \bigtimes_{a \in \text{dom}(\mathbb{P})} \big\{(a, v) \;\big|\; v \text{ is in } eval'(D, d, q') \text{ with } q' = \mathbb{P}(a)\big\}.$$

Example 6. The result of $\text{Source}^{(s_{\mathsf{ex}}, \varepsilon, \mathbb{P}_{\mathsf{ex}})}$ with source s_{ex} of Example 5 and $\mathbb{P}_{\mathsf{ex}} = \{a_1 \to \mathsf{id}, a_2 \to \mathsf{firstname}, a_3 \to \mathsf{age}\}$ is the relation $\big(\{a_1, a_2, a_3\}, \{t_1, t_2\}\big)$ with

$$t_1 = \big\{a_1 \to (\texttt{"1"}, \texttt{xsd:string}), a_2 \to (\texttt{"Alice"}, \texttt{xsd:string}), a_3 \to (\texttt{"23"}, \texttt{xsd:string})\big\}$$

and

$$t_2 = \big\{a_1 \to (\texttt{"2"}, \texttt{xsd:string}), a_2 \to (\texttt{"Bob"}, \texttt{xsd:string}), a_3 \to (\texttt{"unknown"}, \texttt{xsd:string})\big\}.$$

4.2 Extend Operator

Extend operators are unary operators that can be used to extend a mapping relation with an additional attribute. The values that the tuples of the relation shall have for this attribute are specified via a so-called *extend expression* that can be formed using *extension functions*. In the following, we first introduce the notions of such functions and such expressions, and then define extend operators.

Definition 7. An **extension function** is a function f of the form

$$f\colon (\mathcal{T} \cup \{\epsilon\}) \times \cdots \times (\mathcal{T} \cup \{\epsilon\}) \rightarrow (\mathcal{T} \cup \{\epsilon\}).$$

Example 7. A concrete example of such an extension function may be a unary function called `toInt` that converts `xsd:string` literals representing integers into `xsd:integer` literals. Formally, for every $v \in (\mathcal{T} \cup \epsilon)$, we define:

$$\texttt{toInt}(v) = \begin{cases} (lex, \texttt{xsd:integer}) & \text{if } v \text{ is a literal } (lex, dt) \text{ such that } lex \text{ is in the} \\ & \text{lexical space of the datatype denoted by the} \\ & \text{IRI } \texttt{xsd:integer} \text{ and } dt \text{ is the IRI } \texttt{xsd:string}, \\ \epsilon & \text{else.} \end{cases}$$

Definition 8. Extend expressions are defined recursively as follows:

1. Every RDF term in $\mathcal{T}$ is an extend expression.
2. Every attribute in $\mathcal{A}$ is an extend expression.
3. If $\varphi_1, \ldots, \varphi_n$ are extend expressions and f is an extension function (as per Definition 7), then the tuple $(f, \varphi_1, \ldots, \varphi_n)$ is an extend expression.

For every extend expression φ, we write $\operatorname{attrs}(\varphi)$ to denote the set of all attributes mentioned in φ. Formally, this set is defined recursively as follows.

1. If φ is an RDF term, then $\operatorname{attrs}(\varphi) = \emptyset$.
2. If φ is an attribute $a \in \mathcal{A}$, then $\operatorname{attrs}(\varphi) = \{a\}$.
3. If φ is of the form $(f, \varphi_1, \ldots, \varphi_n)$, then $\operatorname{attrs}(\varphi) = \bigcup_{i \in \{1,\ldots,n\}} \operatorname{attrs}(\varphi_i)$.

While Definition 8 introduces the syntax of extend expressions, their interpretation as a function to obtain values for mappings tuples is defined as follows.

Definition 9. Let φ be an extend expression and t be a mapping tuple. The **evaluation of** φ **over** t, denoted by $eval(\varphi, t)$, is either an RDF term or the error value (ϵ), and is determined recursively as follows:

$$eval(\varphi, t) = \begin{cases} \varphi & \text{if } \varphi \text{ is an RDF term,} \\ t(\varphi) & \text{if } \varphi \in \mathcal{A} \text{ and } \varphi \in \operatorname{dom}(t), \\ f\big(eval(\varphi_1, t), \ldots, eval(\varphi_n, t)\big) & \text{if } \varphi \text{ is of the form } (f, \varphi_1, \ldots, \varphi_n) \\ & \text{such that } f \text{ is an } n\text{-ary function,} \\ \epsilon & \text{else.} \end{cases}$$

Example 8. If φ_{ex} is the extend expression $(\texttt{toInt}, a_3)$, for the tuples t_1 and t_2 of Example 6, $eval(\varphi_{\mathsf{ex}}, t_1)$ is the literal $(\texttt{"23"}, \texttt{xsd:integer})$ and $eval(\varphi_{\mathsf{ex}}, t_2) = \epsilon$.

Now we can define the semantics of extend operators as follows.

Definition 10. Let $r = (A, I)$ be a mapping relation, a be an attribute that is not in A (i.e., $a \in \mathcal{A} \setminus A$), and φ be an extend expression. The (a, φ)**-extension of** r, denoted by $\operatorname{Extend}_{\varphi}^{a}(r)$, is the mapping relation (A', I') such that $A' = A \cup \{a\}$ and $I' = \big\{t \cup \{a \rightarrow eval(\varphi, t)\} \mid t \in I\big\}$.

Example 9. We may extend the mapping relation r of Example 6 with an attribute a_4 for which each of the two tuples of r gets the value obtained by applying the extend expression φ_{ex} of Example 8. This operation can be captured as $\text{Extend}^{a_4}_{\varphi_{\mathsf{ex}}}(r)$ and results in the mapping relation $\big(\{a_1, a_2, a_3, a_4\}, \{t'_1, t'_2\}\big)$ with

$$t'_1 = t_1 \cup \{a_4 \rightarrow (\texttt{"23"}, \texttt{xsd:integer})\} \quad \text{and} \quad t'_2 = t_2 \cup \{a_4 \rightarrow \epsilon\}.$$

While the example shows how extend operators can be used to convert literals from one datatype to another, other use cases include: adding the same constant RDF term to all tuples of a mapping relation, creating IRIs from other attribute values, and applying arbitrary functions to one or more other attribute values.

4.3 Relational Algebra Operators

In addition to the operators introduced above, any operator of the relational algebra can also be used for mapping relations. For instance, the following definition introduces the notion of *projection*, adapted to mapping relations.

Definition 11. Let $r = (A, I)$ be a mapping relation and $P \subseteq A$ be a non-empty subset of the attributes of r. The **P-specific projection of** r, denoted by $\text{Project}^P(r)$, is the mapping relation (P, I') such that $I' = \{t[P] \mid t \in I\}$, where, for every mapping tuple $t \in I$, $t[P]$ denotes the mapping tuple t' that is the restriction of t to P; i.e., $\text{dom}(t') = P$ and $t'(a) = t(a)$ for all $a \in P$.

Example 10. Let r be the mapping relation of Example 9. Assuming that further steps of a potential mapping process that involve this relation use only the attributes a_2 and a_4 of r, we may project away the rest of the attributes. To this end, we use $\text{Project}^{\{a_2,a_4\}}(r)$ and obtain the relation $\big(\{a_2, a_4\}, \{t''_1, t''_2\}\big)$ with

$$t''_1 = \big\{a_2 \rightarrow (\texttt{"Alice"}, \texttt{xsd:string}),\ a_4 \rightarrow (\texttt{"23"}, \texttt{xsd:integer})\big\} \text{ and}$$
$$t''_2 = \big\{a_2 \rightarrow (\texttt{"Bob"}, \texttt{xsd:string}),\ a_4 \rightarrow \epsilon\big\}.$$

As two more examples of adopting traditional relational algebra operators for mapping relations, we present the following two definitions.

Definition 12. Let $r_1 = (A_1, I_1)$ and $r_2 = (A_2, I_2)$ be mapping relations such that $A_1 \cap A_2 = \emptyset$, and $\mathbb{J} \subseteq A_1 \times A_2$. The **$\mathbb{J}$- based equijoin of** r_1 **and** r_2, denoted by $\text{EqJoin}^{\mathbb{J}}(r_1, r_2)$, is the mapping relation (A, I) such that $A = A_1 \cup A_2$ and

$$I = \{t_1 \cup t_2 \mid t_1 \in I_1 \text{ and } t_2 \in I_2 \text{ such that } t_1(a_1) = t_2(a_2) \text{ for all } (a_1, a_2) \in \mathbb{J}\}.$$

Definition 13. The **union** of mapping relations $r_1 = (A_1, I_1)$ and $r_2 = (A_2, I_2)$ with $A_1 = A_2$, denoted by $\text{Union}(r_1, r_2)$, is the mapping relation $(A_1, I_1 \cup I_2)$.

5 Algebra-Based Definition of RML

This section introduces an algorithm to convert mappings defined using the mapping language RML (specifically, version 1.1.2 [5]) into our mapping algebra. This algorithm shows that *our algebra is at least as expressive as RML*. Moreover, through this algorithm, in combination with the algebra, we provide a *formal definition of the semantics of RML mappings.* We emphasize that this formal semantics coincides with the informally-defined semantics of RML v1.1.2 [5], which we have verified by running the official RML test cases [8] on a prototypical implementation of our approach.[3] Hereafter, we describe the algorithm step by step, while a complete pseudocode representation is given as Algorithm 1. For the description, we assume familiarity with the concepts of RML [4,5,10].

Algorithm 1: Translates an RML mapping into our mapping algebra.

Input: G - an RDF graph (assumed to describe an RML mapping)
base - an IRI considered as base IRI
S2B - an injective function that maps every string to a unique blank node
Output: a mapping relation with attributes a_s, a_p, a_o, and a_g

$G_{\mathsf{NF}} \leftarrow$ normalized version of G (apply the update queries of Appendix A to G)
$\omega_{\mathsf{spog}} \leftarrow$ the empty mapping relation $(\{a_s, a_p, a_o, a_g\}, \emptyset)$
foreach $u_{\mathsf{tm}} \in \mathcal{I} \cup \mathcal{B}$ for which there exists $o \in \mathcal{I} \cup \mathcal{B}$ such that G_{NF} contains the triple $(u_{\mathsf{tm}}, \texttt{rr:predicateObjectMap}, o)$ **do**
  $\omega \leftarrow \mathrm{Source}^{(s,q,\mathbb{P})}$, where $(s,q) = \textsc{SrcAndRootQuery}(u_{\mathsf{tm}}, G_{\mathsf{NF}})$ // Definition 14
   and $\mathbb{P} = \textsc{ExtractQueries}(u_{\mathsf{tm}}, G_{\mathsf{NF}})$ // Algorithm 2
  $u_{\mathsf{sm}} \leftarrow o$, where $o \in \mathcal{I} \cup \mathcal{B}$ such that $(u_{\mathsf{tm}}, \texttt{rr:subjectMap}, o) \in G_{\mathsf{NF}}$
  $u_{\mathsf{pom}} \leftarrow o$, where $o \in \mathcal{I} \cup \mathcal{B}$ such that $(u_{\mathsf{tm}}, \texttt{rr:predicateObjectMap}, o) \in G_{\mathsf{NF}}$
  $u_{\mathsf{pm}} \leftarrow o$, where $o \in \mathcal{I} \cup \mathcal{B}$ such that $(u_{\mathsf{pom}}, \texttt{rr:predicateMap}, o) \in G_{\mathsf{NF}}$
  $u_{\mathsf{om}} \leftarrow o$, where $o \in \mathcal{I} \cup \mathcal{B}$ such that $(u_{\mathsf{pom}}, \texttt{rr:objectMap}, o) \in G_{\mathsf{NF}}$
  $\omega \leftarrow \mathrm{Extend}_{\varphi}^{a_s}(\omega)$, where $\varphi = \textsc{CreateExtExpr}(u_{\mathsf{sm}}, G_{\mathsf{NF}}, base, S2B, \mathbb{P})$ // Alg.3
  $\omega \leftarrow \mathrm{Extend}_{\varphi}^{a_p}(\omega)$, where $\varphi = \textsc{CreateExtExpr}(u_{\mathsf{pm}}, G_{\mathsf{NF}}, base, S2B, \mathbb{P})$
  if there is a $u_{\mathsf{ptm}} \in \mathcal{I} \cup \mathcal{B}$ such that $(u_{\mathsf{om}}, \texttt{rr:parentTriplesMap}, u_{\mathsf{ptm}}) \in G_{\mathsf{NF}}$ **then**
    $\omega' \leftarrow \mathrm{Source}^{(s',q',\mathbb{P}')}$, where $(s',q') = \textsc{SrcAndRootQuery}(u_{\mathsf{ptm}}, G_{\mathsf{NF}})$
     and $\mathbb{P}' = \textsc{ExtractQueries}(u_{\mathsf{ptm}}, G_{\mathsf{NF}})$ // Algorithm 2
    $\mathbb{J} \leftarrow \emptyset$
    foreach $u_{\mathsf{jc}} \in \mathcal{I} \cup \mathcal{B}$ for which $(u_{\mathsf{om}}, \texttt{rr:joinCondition}, u_{\mathsf{jc}}) \in G_{\mathsf{NF}}$ **do**
      $a \leftarrow \mathbb{P}^{-1}(lex)$, where $(u_{\mathsf{jc}}, \texttt{rr:child}, o) \in G_{\mathsf{NF}}$ with $o = (lex, dt) \in \mathcal{L}$
      $a' \leftarrow \mathbb{P}'^{-1}(lex)$, where $(u_{\mathsf{jc}}, \texttt{rr:parent}, o) \in G_{\mathsf{NF}}$ with $o = (lex, dt) \in \mathcal{L}$
      $\mathbb{J} \leftarrow \mathbb{J} \cup \{(a, a')\}$
    $\omega \leftarrow \mathrm{EqJoin}^{\mathbb{J}}(\omega, \omega')$
    $u_{\mathsf{sptm}} \leftarrow o$, where $o \in \mathcal{I} \cup \mathcal{B}$ such that $(u_{\mathsf{ptm}}, \texttt{rr:subjectMap}, o) \in G_{\mathsf{NF}}$
    $\omega \leftarrow \mathrm{Extend}_{\varphi}^{a_o}(\omega)$, where $\varphi = \textsc{CreateExtExpr}(u_{\mathsf{sptm}}, G_{\mathsf{NF}}, base, S2B, \mathbb{P}')$
  else
    $\omega \leftarrow \mathrm{Extend}_{\varphi}^{a_o}(\omega)$, where $\varphi = \textsc{CreateExtExpr}(u_{\mathsf{om}}, G_{\mathsf{NF}}, base, S2B, \mathbb{P})$
  if there is a $u_{\mathsf{gm}} \in \mathcal{I} \cup \mathcal{B}$ such that $(u_{\mathsf{sm}}, \texttt{rr:graphMap}, u_{\mathsf{gm}}) \in G_{\mathsf{NF}}$ or $(u_{\mathsf{pom}}, \texttt{rr:graphMap}, u_{\mathsf{gm}}) \in G_{\mathsf{NF}}$ **then**
    $\omega \leftarrow \mathrm{Extend}_{\varphi}^{a_g}(\omega)$, where $\varphi = \textsc{CreateExtExpr}(u_{\mathsf{gm}}, G_{\mathsf{NF}}, base, S2B, \mathbb{P})$
  else
    $\omega \leftarrow \mathrm{Extend}_{\varphi}^{a_g}(\omega)$, where φ is the IRI `rr:defaultGraph`
  $\omega_{\mathsf{spog}} \leftarrow \mathrm{Union}(\omega_{\mathsf{spog}}, \mathrm{Project}^{\{a_s,a_p,a_o,a_g\}}(\omega))$
return ω_{spog}

[3] https://github.com/s-minoo/eswc-2025-poster-algebra-implementation.

Input and Output. Since RML mappings are captured as RDF graphs that use the RML ontology [10], the main input to our algorithm is such an RDF graph (we assume it uses IRIs of the RML ontology only in the way as intended by that ontology). A second input is an IRI considered as a base for resolving relative IRIs [5]. As a third input, we assume an injective function $S2B : \mathcal{S} \rightarrow \mathcal{B}$ that maps every string to a unique blank node (which shall become relevant later). The output is a mapping relation that is obtained by applying operators of our algebra and that can be converted into an RDF dataset as per Definition 3.

Normalization Phase. To minimize the variations of RML descriptions to be considered, the algorithm begins by converting the given RML mapping into a *normal form* in which i) no shortcut properties are used, ii) all referencing object maps have a join condition, and iii) every triples map has only a single predicate-object map with a single predicate map, a single object map, and at most one graph map. Every RML mapping can be converted into this normal form by applying the following sequence of rewriting steps, which we have adopted from Rodríguez-Muro and Rezk [15], with adaptations specific to RML. While we present these steps informally, Appendix A captures them formally in terms of update queries. None of these steps changes the meaning of the defined mapping.

1. Every class IRI definition is expanded into a predicate-object map.
2. All shortcut properties for constant-valued term maps are expanded.
3. Predicate-object maps with multiple predicate maps or multiple object maps are duplicated to have a single predicate map and a single object map each.
4. Referencing object maps without a join condition are replaced by an object map with the subject IRI of the parent triples map as the reference value.
5. Every triples map with multiple predicate-object maps is duplicated such that each of the resulting triples maps has a single predicate-object map only.
6. Every triples map with multiple graph maps is duplicated such that each of the resulting triples maps has one of these graph maps.

Main Loop. After normalization, the algorithm iterates over all triples maps defined by the (normalized) RML mapping (lines 3–27). For each of them, it combines algebra operators, as described below, to define a mapping relation that, in the end, has the attributes a_{s}, a_{p}, a_{o}, and a_{g}. These mapping relations are combined using union (line 27), resulting in a mapping relation (line 28) that captures the complete RDF dataset defined by the given RML mapping.

Translation of the Logical Source. As the first step for each triples map, the algorithm creates a source operator (line 4). The first two parameters for this operator—the data source s and the query expression q for selecting relevant context objects (cf. Definition 6)—depend on the logical source defined for the triples map. Since RML is extensible regarding the types of logical sources, we have to abstract the creation of these two parameters by the following function.

Definition 14. Given an IRI or blank node u_{tm} (assumed to denote an RML triples map) and an RDF graph G (assumed to describe an RML mapping, including u_{tm}), SrcAndRootQuery(u_{tm}, G) is a pair (s, q) of a data source $s = (type, D)$ with $type = (\mathcal{D}^{\mathrm{ds}}, \mathcal{D}^{\mathrm{c1}}, \mathcal{D}^{\mathrm{c2}}, L, L', eval, eval', cast)$ and a query expression $q \in L$, as extracted from the description of the logical source of u_{tm} in G.

Example 11. Given an RDF graph G_{ex} that contains the following RML mapping, SrcAndRootQuery(`ex:tm`, G_{ex}) $= (s'_{\mathsf{ex}}, \varepsilon)$ where $s'_{\mathsf{ex}} = (type_{\mathsf{csv}}, D'_{\mathsf{ex}})$ such that $type_{\mathsf{csv}}$ is the source type for CSV-based data sources (cf. Example 4), D'_{ex} is the CSV file mentioned in the second triple, and ε is the empty string.

```
ex:tm rml:logicalSource [ rml:source "data.csv";  rml:referenceFormulation ql:CSV ];
      rr:subjectMap [ rr:template "http://example.com/person_{ID}";  rr:termType rr:IRI ];
      rr:predicateObjectMap [ rr:predicate rdfs:label;
                              rr:objectMap [rml:reference "Name"] ].
```

The third parameter for the source operator—the partial function $\mathbb{P}$ that maps attributes to query expressions (cf. Definition 6)—is populated by Algorithm 2, which extracts the query expressions from the term maps of the given triples map. In particular, the term maps may use query expressions as references (lines 3–4 of Algorithm 2), in string templates (lines 5–6), and as child and parent queries of join conditions (lines 7–10). To focus only on the term maps of the given triples map, the extraction is restricted to the relevant subgraph of the given RDF graph (line 1 of Algorithm 2), which we define formally as follows.

Algorithm 2: ExtractQueries - extract query expressions from triples maps.

Input: u_{tm} - IRI or blank node of the triples map to extract queries from
G - an RDF graph (assumed to contain a *normalized* RML description)
Output: an injective partial function $\mathbb{P}$ that maps attributes to query expressions

1 $G_{\mathsf{tm}} \leftarrow$ the u_{tm}-rooted subgraph of G // Definition 15
2 $Q \leftarrow \emptyset$ // Initially empty set of query expressions
3 **forall** $s \in \mathcal{I} \cup \mathcal{B}$ and $o \in \mathcal{L}$ for which there is a triple $(s, \texttt{rml:reference}, o) \in G_{\mathsf{tm}}$ **do**
4     $Q \leftarrow Q \cup \{lex\}$, where $o = (lex, dt)$
5 **forall** $s \in \mathcal{I} \cup \mathcal{B}$ and $o \in \mathcal{L}$ for which there is a triple $(s, \texttt{rr:template}, o) \in G_{\mathsf{tm}}$ **do**
6     $Q \leftarrow Q \cup Q'$, where $o = (lex, dt)$ and Q' is the set of all substrings of lex that are enclosed by "{" and "}"
7 $T_{\mathrm{c}} \leftarrow \{(s, p, o) \in G_{\mathsf{tm}} \mid p$ is the IRI `rr:child` and $o \in \mathcal{L}\}$
8 $T_{\mathrm{p}} \leftarrow \{(s, p, o) \in G \mid p$ is the IRI `rr:parent`, $o \in \mathcal{L}$, and there is an $s' \in \mathcal{I} \cup \mathcal{B}$ s.t. $(s', \texttt{rr:joinCondition}, s) \in G$ and $(s', \texttt{rr:parentTriplesMap}, u_{\mathsf{tm}}) \in G\}$
9 **foreach** triple $(s, p, o) \in (T_{\mathrm{c}} \cup T_{\mathrm{p}})$ **do**
10     $Q \leftarrow Q \cup \{lex\}$, where $o = (lex, dt)$
11 $\mathbb{P} \leftarrow$ (initially) empty function from attributes to query expressions
12 **foreach** query expression $q \in Q$ **do**
13     $\mathbb{P} \leftarrow \mathbb{P} \cup \{a \rightarrow q\}$, where a is a fresh attribute from $\mathcal{A} \setminus \{a_{\mathrm{s}}, a_{\mathrm{p}}, a_{\mathrm{o}}, a_{\mathrm{g}}\}$ that has not been used before in the whole translation process
14 **return** $\mathbb{P}$

Definition 15. Let G be an RDF graph and $u \in \mathcal{I} \cup \mathcal{B}$. The ***u*-rooted subgraph of** G is an RDF graph $G' \subseteq G$ that is defined recursively as follows:

1. G' contains every triple $(s, p, o) \in G$ for which it holds that $s = u$.
2. G' contains every triple $(s, p, o) \in G$ for which there is already another triple $(s', p', o') \in G'$ such that $s = o'$ and p' is not the IRI `rr:parentTriplesMap`.

Example 12. For the RDF graph G_{ex} that contains the RML mapping of Example 11, ExtractQueries(ex:tm, G_{ex}) returns $\mathbb{P} = \{a_1 \rightarrow \mathsf{ID}, a_2 \rightarrow \mathsf{Name}\}$, where a_1 and a_2 are arbitrary attributes such that $a_1 \neq a_2$ and $a_1, a_2 \notin \{a_\mathrm{s}, a_\mathrm{p}, a_\mathrm{o}, a_\mathrm{g}\}$.

Translation of the Subject Map and the Predicate Map. The source operator constructed in the previous step creates a mapping relation with values obtained via the extracted query expressions (along the lines of Example 6). This relation can now be extended by using these values to create RDF terms as defined by the term maps of the triples map. To this end, for the subject (term) map and the predicate (term) map, the algorithm adds an extend operator, respectively (lines 9–10 in Algorithm 1). The construction of the extend expressions of these operators (see Definitions 8 and 10) is captured in Algorithm 3 and uses the following extension functions, which are defined formally in Appendix B.

- `toIRI` converts string literals representing IRIs into these IRIs.
- `toBNode`^{S2B}, parameterized by the aforementioned function *S2B* (see the discussion of the input of Algorithm 1), converts string literals into blank nodes.
- `toLiteral` is a binary extension function that converts string literals into literals with a datatype IRI that is given as the second argument.
- `concat` is a binary extension function that concatenates two string literals.

Algorithm 3: CreateExtExpr - creates an extend expression for a term map.

Input: u - IRI or blank node of the term map to create the extend expression for
G - an RDF graph (assumed to describe the term map u)
base - an IRI considered as base IRI
S2B - an injective function that maps every string to a unique blank node
$\mathbb{P}$ - an injective partial function that maps attributes to query expressions

Output: an extend expression (as per Definition 8)

if G contains a triple (u, rr:constant, o) **then return** o
$\varphi \leftarrow$ initially empty extend expression
if G contains a triple (u, rr:reference, o) such that o is a literal (lex, dt) **then**
    $\varphi \leftarrow a$, where a is the attribute in $\mathrm{dom}(\mathbb{P})$ such that $\mathbb{P}(a) = lex$
else if G contains a triple (u, rr:template, o) such that o is a literal (lex, dt) **then**
    $\bar{S} \leftarrow$ Split(lex) //Definition 16
    $\varphi \leftarrow$ (s_1, xsd:string), where s_1 is the first element in $\bar{S}$
    foreach s_i in $\bar{S}$ without the first element **do** //Iterate based on the order in $\bar{S}$
        if s_i is a string that starts with "{" and ends with "}" **then**
            $\varphi \leftarrow$ concat(φ, $\mathbb{P}^{-1}(q)$), where q is s_i without enclosing "{" and "}"
        else
            $\varphi \leftarrow$ concat(φ, (s_i, xsd:string))
if G contains the triple (u, rr:termType, rr:BlankNode) **then return** toBNode^{S2B}(φ)
else if G contains (u, rr:datatype, o) with $o \in \mathcal{I}$ **then return** toLiteral(φ, o)
else if G contains (u, rr:termType, rr:Literal) **then return** toLiteral(φ, xsd:string)
else if G contains (t, rr:objectMap, u) with $t \in \mathcal{I} \cup \mathcal{B}$ **and** G contains (u, rml:reference, o) with $o \in \mathcal{L}$ **then return** toLiteral(φ, xsd:string)
else return toIRI(φ, *base*)

The extend expression created by Algorithm 3 depends on whether the given term map is constant-valued (line 1), reference-valued (lines 3–4), or template-valued (lines 5–12). In the latter case, the corresponding template string is first split into a sequence of substrings (line 6), defined as follows.

Definition 16. For every string *tmpl* $\in \mathcal{S}$, we write $\textsc{Split}(tmpl)$ to denote a sequence $\bar{S}$ of strings that is constructed by the following two steps.

1. Partition *tmpl* into a sequence $\bar{S}$ of *query substrings* and *normal substrings* in the order in which these substrings appear in *tmpl* from left to right, where
   - a *query substring* is a substring starting with an opening curly brace (i.e., "{") and ending with a closing curly brace (i.e., "}"), and
   - a *normal substring* is every substring between two query substrings, as well as the one before the first and the one after the last query substring.
2. If $\bar{S}$ is the empty sequence, insert the empty string as a new substring into $\bar{S}$.

The extend expression is then created to concatenate the resulting substrings (lines 7–12), where every query substring "$\{q\}$" is replaced by the attribute a for which it holds that $\mathbb{P}(a) = q$, because that attribute holds the values that the source operator obtains for the query expression q.

Example 13. Assume b_{sm} and b_{om} are the blank nodes denoting the subject map and the object map in RDF graph G_{ex} in Example 11, respectively, *base* is an arbitrary base IRI, and $\mathbb{P} = \{a_1 \to \mathsf{ID}, a_2 \to \mathsf{Name}\}$ as in Example 12. Then, $\textsc{CreateExtExpr}(b_{\mathsf{om}}, G_{\mathsf{ex}}, base, S2B, \mathbb{P})$ returns the extend expression $\texttt{toLiteral}(a_2, \texttt{xsd:string})$. Moreover, $\textsc{CreateExtExpr}(b_{\mathsf{sm}}, G_{\mathsf{ex}}, base, S2B, \mathbb{P})$ returns the extend expression $\texttt{toIRI}\big(\texttt{concat}(\ell, a_1), base\big)$ where ℓ is the literal $(\texttt{"http://example.com/person\_"}, \texttt{xsd:string})$.

Translation of the Object Map. For object (term) maps, the algorithm also has to consider the case of referencing object maps (lines 11–20 in Algorithm 1), which capture a join with RDF terms defined by the subject map of another triples map. In this case, another source operator needs to be added (for this other triples map) and the mapping relation created by that source operator needs be joined with the mapping relation produced before (line 18). To this end, an equijoin operator is used, for which the join attributes are determined based on the join condition defined for the referencing object map (lines 13–17). Ordinary object maps are handled like subject and predicate maps before (line 22).

Translation of the Graph Map. Finally, if the triples map has a graph map, the algorithm adds an extend operator, with attribute a_{g}, to create IRIs defined by this graph map (lines 23–24). If the triples map has no graph map, an extend operator that assigns a_{g} to the constant IRI `rr:defaultGraph` is added (line 26).

6 Algebraic Equivalences

In addition to providing a basis to formally define the semantics of a mapping language such as RML, our mapping algebra may also be used as a foundation for implementing such languages in mapping tools. More precisely, such tools may use our algebra as a form of language to internally represent the plans that they create for converting data according to a given mapping description, and then they may use algebraic equivalences to rewrite initial mapping plans into more efficient ones (with the provable guarantee that the rewritten plans produce the same result!). While an extensive study of such optimization opportunities is beyond the scope of this paper, in this section we show a number of such equivalences and describe how they may be relevant for a plan optimizer.

6.1 Projection Pushing

The first group of equivalences focuses on pushing project operators deeper into a given plan. This may be useful to reduce the size of intermediate mapping relations in terms of values for attributes that are not used anymore later in the plan. Dropping such attributes from intermediate mapping relations as early as possible reduces the memory space required for executing the mapping plan.

As a typical example of an algebraic equivalence that an optimizer may use for this purpose, consider the following result, which shows that a project operator over an extend-based subexpression may be pushed into this subexpression if the set of projection attributes contains both the extend attribute and the attributes mentioned by the corresponding extend expression (if any).[4]

Proposition 1. *Let $r = (A, I)$ be a mapping relation, $a \in \mathcal{A}$ be an attribute that is not in A (i.e., $a \notin A$), φ be an extend expression, and $P \subseteq A$ be a non-empty subset of the attributes of r. If $(\mathrm{attrs}(\varphi) \cap A) \subseteq P$, then it holds that*

$$\mathit{Project}^{P \cup \{a\}}\big(\mathit{Extend}^{a}_{\varphi}(r)\big) = \mathit{Extend}^{a}_{\varphi}\big(\mathit{Project}^{P}(r)\big).$$

To satisfy the condition for the equivalence in Proposition 1 and, thus, to facilitate projection pushing based on this equivalence, in some cases, it may be necessary to first add another projection with additional projection attributes. The following result shows an equivalence that can be used for this purpose.

Proposition 2. *Let $r = (A, I)$ be a mapping relation and $P \subseteq A$ be a non-empty subset of the attributes of r. For every $P' \subseteq A$, it holds that*

$$\mathit{Project}^{P}(r) = \mathit{Project}^{P}\big(\mathit{Project}^{P \cup P'}(r)\big).$$

While further equivalences for projection pushing can be shown (e.g., into a join), we also want to highlight the existence of cases in which a project operator can be removed completely. For instance, every project operator with a source operator as input may be removed, as shown by the following result.

[4] We provide the proof of Proposition 1 in Appendix C. The following propositions in this section can be shown in the same way.

Proposition 3. *Let* $s = (type, D)$ *be a data source with* $type = (\mathcal{D}^{ds}, \mathcal{D}^{c1}, \mathcal{D}^{c2}, L, L', eval, eval', cast)$, *let* $q \in L$, *let* $\mathbb{P}$ *be a partial function* $\mathbb{P}\colon \mathcal{A} \rightarrow L'$, *and* $P \subseteq \mathrm{dom}(\mathbb{P})$. *It holds that* $Project^{P}\big(Source^{(s,q,\mathbb{P})}\big) = Source^{(s,q,\mathbb{P}')}$, *where* $\mathbb{P}'$ *is the restriction of* $\mathbb{P}$ *to* P*; i.e.,* $\mathbb{P}'(a) = \mathbb{P}(a)$ *for all* $a \in \mathrm{dom}(\mathbb{P}') = P$.

6.2 Pushing or Pulling Extend Operators

The extended version of this paper shows a second group of equivalences which focus on moving extend operators either on top of a join-based subplan or under the corresponding join operator (i.e., into one of the inputs of the join) [13].

7 Concluding Remarks and Future Work

The work presented in this paper opens the door for several new directions of research in the domain of mappings-based KG construction. First and foremost, our approach may be used to formalize other mapping languages besides RML. In this context we emphasize that, while the five types of operators defined in this paper are sufficient to capture the expressive power of RML, the algebra may easily be extended with additional types of operators if needed for a particular language or use case. Similarly, the algebraic equivalences shown in this paper may be complemented with further equivalences. A natural next step then is to develop concrete implementation and optimization techniques based on the algebra and the equivalences. Given the language-agnostic nature of the algebra, it even becomes trivial for resulting engines to support multiple languages.

Acknowledgments. The presented work was supported by the imec.icon project PACSOI (HBC.2023.0752), which was co-financed by imec and VLAIO and brings together the following partners: FAQIR Foundation, FAQIR Institute, MoveUP, Byteflies, AContrario, and Ghent University – IDLab, and by the European Union's Horizon Europe research and innovation programme under grant agreement no. 101058682.

Disclosure of Interests. The authors have no competing interests to declare.

A SPARQL Update Queries to Create Normalized RML

```
DELETE { ?sm rr:class ?sm_class . }
INSERT { ?tm rr:predicateObjectMap [
                rr:predicateMap [
                    rr:termType rr:IRI ;
                    rr:constant rdf:type ] ;
                rr:objectMap [
                    rr:termType rr:IRI ;
                    rr:constant ?sm_class ] ] }
WHERE { ?tm rr:subjectMap ?sm .
        ?sm rr:class ?sm_class . }
```

Query 1. Normalization step 1 (expand shortcuts for class IRIs).

```
DELETE { ?tm rr:predicateObjectMap  ?pom.
         ?pom rr:predicateMap ?pm ;
              rr:objectMap ?om ;  rr:graphMap ?gm }
INSERT { ?tm rr:predicateObjectMap [
                 rr:predicateMap ?pm ;
                 rr:objectMap ?om ;
                 rr:graphMap ?gm ] }
WHERE {    ?tm rr:predicateObjectMap ?pom .
           ?pom rr:predicateMap ?pm ;
           ?pom rr:objectMap ?om
 OPTIONAL {?pom rr:graphMap ?gm} }
```

Query 2. Normalization step 3 (duplicate multi-predicate-object maps into singletons).

```
DELETE {  ?tm   rr:subject ?sm_constant .   ?pompm rr:predicate ?pm_constant .
          ?termMap rr:graph ?gm_constant .  ?pomom rr:object  ?om_constant .   }
INSERT { ?tm       rr:subjectMap    [ rr:constant ?sm_constant ].
         ?pompm    rr:predicateMap  [ rr:constant ?pm_constant ].
         ?pomom    rr:objectMap     [ rr:constant ?om_constant ].
         ?termMap  rr:graphMap      [ rr:constant ?gm_constant ]. }
WHERE { { ?tm       rr:subject   ?sm_constant }
  UNION { ?pompm    rr:predicate ?pm_constant }
  UNION { ?pomom    rr:object    ?om_constant }
  UNION { ?termMap  rr:graph     ?gm_constant } }
```

Query 3. Normalization step 2 (expand shortcuts for constant-valued term maps).

```
DELETE { ?om rr:parentTriplesMap ?ptm }
INSERT { ?om rr:reference  ?ref ;
             rr:template ?template;
             rr:constant ?const ;
             rr:termType rr:IRI . }
WHERE { ?om rr:parentTriplesMap ?ptm .
        ?ptm rr:subjectMap ?sm .
  OPTIONAL{ ?sm rr:reference ?ref }
  OPTIONAL{ ?sm rr:template ?template }
  OPTIONAL{ ?sm rr:constant ?const }
  FILTER NOT EXISTS { ?om rr:joinCondition ?jc } }
```

Query 4. Normalization step 4 (replace referencing object maps without join conditions).

```
DELETE { ?tm rdf:type rr:TriplesMap ;
             rml:logicalSource ?ls ;
             rr:subjectMap ?sm ;
             rr:predicateObjectMap ?pom }
INSERT { [] rml:logicalSource ?ls ;
            rr:subjectMap ?sm ;
            rr:predicateObjectMap ?pom }
WHERE { ?tm rml:logicalSource ?ls ;
            rr:subjectMap ?sm ;
            rr:predicateObjectMap ?pom }
```

Query 5. Normalization step 5 (duplicate triples maps that contain multiple pred.-object maps).

```
DELETE { ?tm rr:predicateObjectMap ?pom .
        ?pom rr:graphMap ?pom_gm .       }
INSERT { [] rml:logicalSource ?ls ;
            rr:subjectMap [
                rr:reference ?ref ;
                rr:template ?template ;
                rr:constant ?const ;
                rr:termType ?ttype ;
                rr:graphMap ?sm_gm ;
                rr:graphMap ?pom_gm ] ;
            rr:predicateObjectMap ?pom .  }
WHERE { ?tm rml:logicalSource ?ls ;
            rr:subjectMap ?sm ;
            rr:predicateObjectMap ?pom .
        ?pom rr:graphMap ?pom_gm .
  OPTIONAL { ?sm rr:graphMap ?sm_gm }
  OPTIONAL { ?sm rr:reference ?ref }
  OPTIONAL { ?sm rr:template ?template }
  OPTIONAL { ?sm rr:constant ?const }
  OPTIONAL { ?sm rr:termType ?ttype }     }
```

Query 6. Normalization step 6a (duplicate triples maps with subject maps where the predicate-object maps contain multiple graph maps).

```
DELETE { ?sm rr:graphMap ?gm1 . }
INSERT { [] rml:logicalSource ?ls ;
            rr:subjectMap [
                rr:reference ?ref ;
                rr:template ?template ;
                rr:constant ?const ;
                rr:termType ?ttype ;
                rr:graphMap ?gm1 ] ;
            rr:predicateObjectMap ?pom }
WHERE { ?tm rml:logicalSource ?ls ;
            rr:subjectMap ?sm ;
            rr:predicateObjectMap ?pom .
        ?sm rr:graphMap ?gm1 ;
            rr:graphMap ?gm2
        FILTER ( ?gm1 != ?gm2 )
  OPTIONAL { ?sm rr:reference ?ref }
  OPTIONAL { ?sm rr:template ?template }
  OPTIONAL { ?sm rr:constant ?const }
  OPTIONAL { ?sm rr:termType ?ttype } }
```

Query 7. Normalization step 6b (duplicate subject maps that contain multiple graph maps).

B Definition of Relevant Extension Functions

For every $v \in (\mathcal{T} \cup \epsilon)$, $dt \in (\mathcal{T} \cup \epsilon)$, $v' \in (\mathcal{T} \cup \epsilon)$, and $base \in \mathcal{I}$, we define:

$$\texttt{toBNode}^{S2B}(v) = \begin{cases} S2B(lex) & \text{if } v \text{ is a literal } (lex, dt) \text{s.t. } dt = \texttt{xsd:string}, \\ \epsilon & \text{else.} \end{cases}$$

$$\texttt{toLiteral}(v, dt) = \begin{cases} (lex, dt) & \text{if } v \text{ is a literal } (lex, dt') \text{s.t. } dt = \texttt{xsd:string}, \\ \epsilon & \text{else.} \end{cases}$$

$$\texttt{toIRI}(v, base) = \begin{cases} lex & \text{if } v \text{ is a literal } (lex, dt) \text{ such that } lex \text{ is} \\ & \text{a valid IRI and } dt \text{ is the IRI } \texttt{xsd:string}, \\ base \circ lex & \text{if } v \text{ is a literal } (lex, dt) \text{ such that } lex \text{ is not} \\ & \text{a valid IRI, but } base \circ lex \text{ is a valid IRI, and} \\ & dt \text{ is the IRI } \texttt{xsd:string}, \\ \epsilon & \text{else.} \end{cases}$$

$$\texttt{concat}(v, v') = \begin{cases} (lex \circ lex', \texttt{xsd:string}) & \text{if } v \text{ and } v' \text{ are literals } (lex, dt) \text{ and} \\ & (lex', dt'), \text{ respectively, such that} \\ & dt = dt' = \texttt{xsd:string}, \\ \epsilon & \text{else,} \end{cases}$$

where $lex \circ lex'$ is the string obtained from concatenating the strings lex and lex'.

C Proof of Proposition 1

We assume that $(\text{attrs}(\varphi) \cap A) \subseteq P$, and let $(A_1, I_1) = \text{Project}^{P \cup \{a\}}(\text{Extend}^a_\varphi(r))$ and $(A_2, I_2) = \text{Extend}^a_\varphi(\text{Project}^P(r))$. To show that $(A_1, I_1) = (A_2, I_2)$ we first note that, by Definition 11, it holds that $A_1 = P \cup \{a\}$, and by Definition 10, $A_2 = P \cup \{a\}$. Hence, $A_1 = A_2$ and, thus, it remains to show that $I_1 = I_2$.

We first show that $I_1 \subseteq I_2$, for which we consider an arbitrary mapping tuple $t \in I_1$ and show that $t \in I_2$. Given that $t \in I_1$, by Definition 11, it holds that i) $\text{dom}(t) = P \cup \{a\}$ and ii) there exists a mapping tuple $t' \in \text{Extend}^a_\varphi(r)$ such that $t(a') = t'(a')$ for all $a' \in P \cup \{a\}$. Since $t' \in \text{Extend}^a_\varphi(r)$, by Definition 10, it holds that i) $\text{dom}(t') = A \cup \{a\}$ and ii) there exists a mapping tuple $t'' \in r$ such that $eval(\varphi, t'') = t'(a)$ and $t''(a') = t'(a')$ for all $a' \in A$. Now, let t''' be the restriction of t'' to P; i.e., $t''' = t''[P]$, which, by Definition 11, also means that $\text{dom}(t''') = P$ and $t''' \in \text{Project}^P(r)$. Then, since $(\text{attrs}(\varphi) \cap A) \subseteq P$ and $t'''(a') = t''(a')$ for all $a' \in P$, it holds that $eval(\varphi, t''') = eval(\varphi, t'')$. As a consequence, there exists a tuple $t^{(4)} \in \text{Extend}^a_\varphi(\text{Project}^P(r))$ such that i) $\text{dom}(t^{(4)}) = \text{dom}(t''') \cup \{a\}$, ii) $t^{(4)}(a) = t'(a)$, and iii) $t^{(4)}(a') = t'''(a')$ for all $a' \in P$. Since $\text{dom}(t''') \cup \{a\} = P \cup \{a\} = \text{dom}(t)$ and $t'(a) = t(a)$ and $t'''(a') = t''(a') = t'(a') = t(a')$ for all $a' \in P$, we can conclude that $t^{(4)} = t$ and, thus, $t \in \text{Extend}^a_\varphi(\text{Project}^P(r))$; i.e., $t \in I_2$.

Now we show that $I_1 \supseteq I_2$, for which we consider a tuple $t \in I_2$ and show that $t \in I_1$. Given that $t \in I_2$, by Definitions 10 and 11, it holds that i) $\mathrm{dom}(t) = \{a\} \cup P$ and ii) there exists a mapping tuple $t' \in \mathrm{Project}^P(r)$ such that $t(a) = eval(\varphi, t')$ and $t(a') = t'(a')$ for all $a' \in P$. Since $t' \in \mathrm{Project}^P(r)$, it holds that i) $\mathrm{dom}(t') = P$ and ii) there exists a mapping tuple $t'' \in r$ such that $\mathrm{dom}(t'') = A$ and $t''(a') = t'(a')$ for all $a' \in P$. Now, let t''' be the mapping tuple for which it holds that i) $\mathrm{dom}(t''') = A \cup \{a\}$, ii) $t'''(a) = eval(\varphi, t'')$, and iii) $t'''(a') = t''(a')$ for all $a' \in A$. Hence, by Definition 10, $t''' \in \mathrm{Extend}^a_\varphi(r)$. Next, let $t^{(4)}$ be the restriction of t''' to $P \cup \{a\}$; i.e., $t^{(4)} = t'''[P \cup \{a\}]$, which, by Definition 11, means that $t^{(4)} \in \mathrm{Project}^{P \cup \{a\}}(\mathrm{Extend}^a_\varphi(r))$. Putting everything together, we have that i) $\mathrm{dom}(t^{(4)}) = P \cup \{a\} = \mathrm{dom}(t)$, ii) $t^{(4)}(a) = t'''(a) = eval(\varphi, t'') = eval(\varphi, t') = t(a)$, and iii) $t^{(4)}(a') = t'''(a') = t''(a') = t(a)$ for all $a' \in P$. As a consequence, $t^{(4)} = t$ and, thus, $t \in \mathrm{Project}^{P \cup \{a\}}(\mathrm{Extend}^a_\varphi(r))$; i.e., $t \in I_1$.

References

1. Arenas-Guerrero, J., Chaves-Fraga, D., Toledo, J., Pérez, M.S., Corcho, O.: Morph-KGC: scalable knowledge graph materialization with mapping partitions. Semantic Web, pp. 1–20 (August 2022). https://doi.org/10.3233/sw-223135
2. Bischof, S., Decker, S., Krennwallner, T., Lopes, N., Polleres, A.: Mapping between RDF and XML with XSPARQL. J. Data Semant. **1**(3), 147–185 (2012)
3. Daga, E., Asprino, L., Mulholland, P., Gangemi, A.: Facade-X: an opinionated approach to SPARQL anything. In: Further with Knowledge Graphs – Proceedings of the 17 th International Conference on Semantic Systems, 6–9 September 2021, Amsterdam, The Netherlands. Studies on the Semantic Web, vol. 53, pp. 58–73. IOS Press (2021). https://doi.org/10.3233/SSW210035
4. Dimou, A., Vander Sande, M., Colpaert, P., Verborgh, R., Mannens, E., Van de Walle, R.: RML: a generic language for integrated RDF mappings of heterogeneous data. In: Proceedings of the 7th Workshop on Linked Data on the Web. CEUR Workshop Proceedings, vol. 1184. CEUR (April 2014). http://ceur-ws.org/Vol-1184/ldow2014_paper_01.pdf
5. Dimou, A., Vander Sande, M., De Meester, B., Heyvaert, P., Delva, T.: RDF Mapping Language (RML) (June 2024). Version 1.1.2; online at https://rml.io/specs/rml/v/1.1.2/
6. Freund, M., Schmid, S., Dorsch, R., Harth, A.: FlexRML: a flexible and memory efficient knowledge graph materializer. In: The Semantic Web, pp. 40–56. Springer Nature Switzerland, Cham (2024)
7. Heyvaert, P., Dimou, A., Chaves-Fraga, D.: RML Implementation Report (February 2022). https://rml.io/implementation-report/
8. Heyvaert, P., Dimou, A., De Meester, B.: RML Test Cases (March 2019). https://rml.io/test-cases/
9. Iglesias, E., Jozashoori, S., Vidal, M.E.: Scaling up knowledge graph creation to large and heterogeneous data sources. J. Web Semant. **75** (2023). https://doi.org/10.1016/j.websem.2022.100755
10. Iglesias-Molina, A., et al.: The RML ontology: a community-driven modular redesign after a decade of experience in mapping heterogeneous data to RDF. In: Proceedings of the 22nd International Semantic Web Conference (ISWC). LNCS,

vol. 14266, pp. 152–175. Springer, Cham (2023). https://doi.org/10.1007/978-3-031-47243-5_9

11. Lefrançois, M., Zimmermann, A., Bakerally, N.: A SPARQL extension for generating RDF from heterogeneous formats. In: The Semantic Web 14th International Conference, ESWC 2017, Portorož, Slovenia, May 28 – 1 June 2017, Proceedings. pp. 35–50. Springer International Publishing, Portoroz, Slovenia (May 2017). https://doi.org/10.1007/978-3-319-58068-5_3
12. Lopes, N., Bischof, S., Decker, S., Polleres, A.: On the semantics of heterogeneous querying of relational, XML and RDF data with XSPARQL. In: Proceedings of the 15th Portuguese Conference on Artificial Intelligence (EPIA 2011), Lisbon, Portugal, pp. 10–13. Citeseer (2011)
13. Min Oo, S., Hartig, O.: An Algebraic Foundation for Knowledge Graph Construction (Extended Version) (2025). CoRR **arXiv/2503.10385**, online at https://arxiv.org/abs/2503.10385
14. Oo, S.M., De Meester, B., Taelman, R., Colpaert, P.: Towards algebraic mapping operators for knowledge graph construction, p. 5 (2023)
15. Rodríguez-Muro, M., Rezk, M.: Efficient SPARQL-to-SQL with R2RML mappings. J. Web Semant. **33**, 141–169 (2015). https://doi.org/10.1016/j.websem.2015.03.001, ontology-based Data Access
16. Rojas, J.A., et al.: Leveraging semantic technologies for digital interoperability in the European railway domain. In: International Semantic Web Conference, pp. 648–664. Springer (2021)
17. Sakor, A., et al.: Knowledge4COVID-19: a semantic-based approach for constructing a COVID-19 related knowledge graph from various sources and analyzing treatments' toxicities. Journal of Web Semantics **75**, 100760 (2023). https://doi.org/10.1016/j.websem.2022.100760
18. Scrocca, M., Comerio, M., Carenini, A., Celino, I.: Turning transport data to comply with EU standards while enabling a multimodal transport knowledge graph. Semantic Web - ISWC **2020**, 411–429 (2020). https://doi.org/10.1007/978-3-030-62466-8_26
19. Sequeda, J.F.: On the semantics of R2RML and its relationship with the direct mapping. In: Proceedings of the 12th International Semantic Web Conference (Posters & Demonstrations Track), vol. 1035, pp. 193–196. ISWC-PD '13, CEUR-WS.org, Aachen, DEU (2013)
20. Sitt Min Oo, Haesendonck, G., De Meester, B., Dimou, A.: RMLStreamer-SISO: an RDF stream generator from streaming heterogeneous data. In: Sattler, U., et al. The Semantic Web – ISWC 2022. ISWC 2022. LNCS, vol. 13489, pp. 697–713. Springer, Cham.. Springer, Cham (2022). https://doi.org/10.1007/978-3-031-19433-7_40
21. Sundara, S., Das, S., Cyganiak, R.: R2RML: RDB to RDF Mapping Language. W3C recommendation, W3C (September 2012). https://www.w3.org/TR/2012/REC-r2rml-20120927/
22. Tailhardat, L., Chabot, Y., Troncy, R.: NORIA-O: an ontology for anomaly detection and incident management in ICT systems. In: The Semantic Web, pp. 21–39. Springer Nature Switzerland, Cham (2024)
23. Van Assche, D., Delva, T., Haesendonck, G., Heyvaert, P., De Meester, B., Dimou, A.: Declarative RDF graph generation from heterogeneous (semi-)structured data: a systematic literature review. J. Web Semant. (2022). https://doi.org/10.1016/j.websem.2022.100753

Inductive Higher Order Embeddings

Giuseppe Pirrò(✉)

Department of Mathematics and Computer Science, University of Calabria, Rende, CS, Italy
giuseppe.pirro@unical.it

Abstract. Nested knowledge graphs (NKGs) encode facts in which subjects or objects can be triples, introducing a hierarchy of relationships beyond traditional entity-to-entity links. This nested structure creates challenges for established knowledge graph embedding (KGE) models, as it requires reasoning not just over entities and their relations, but also over triples as first-class citizens and higher-order (meta) relations connecting them. In addition, real-world scenarios require *inductive* reasoning: at test time, new triples, new entities, and even new predicates may appear unseen during training. Traditional closed-world assumptions of many KGE models fail to accommodate such evolving and dynamic graphs. We introduce INKE (Inductive Nested Knowledge Embeddings), a novel framework that integrates three interconnected graph neural networks (GNNs) to address these challenges. INKE operates on multiple layers of abstraction: (i) A *KG-GNN* produces embeddings for entities and relations, capturing local semantic structures. (ii) A *Line-GNN* refines these embeddings at the triple level, using a line graph representation to model dependencies among triples. (iii) A *Relation-GNN* operates on a relation graph to handle predicate-level embeddings, enabling inductive generalization to unseen predicates and meta-relations. INKE achieves robust and flexible inductive reasoning capabilities in nested knowledge graphs.

Keywords: Knowledge Graphs · Inductive Reasoning · Nested Triples · Graph Neural Networks · Predicate Embeddings

1 Introduction

Knowledge graphs (KGs) have emerged as powerful tools for representing structured information about entities and their relationships. A KG typically encodes facts as triples (h, r, t) where h and t are entities and r is a relation. Over the past decade, *knowledge graph embedding* (KGE) has become a widely adopted technique to support reasoning tasks such as link prediction, question answering, and entity resolution [3,17]. By embedding entities and relations into low-dimensional continuous vector spaces, KGE models enable efficient reasoning about unseen triples, predicting missing links, and evaluating the plausibility of new facts. However, real-world knowledge is not always limited to simple entity-to-entity relations. In many domains—such as event knowledge, scientific literature, temporal facts, and policy documents—*nested structures* arise naturally. A nested triple might look like: $((h_1, r_1, t_1),\ r^n,\ (h_2, r_2, t_2))$ where the subject and/or object of a triple is itself another

E. Curry et al. (Eds.): ESWC 2025, LNCS 15718, pp. 23–40, 2025.
https://doi.org/10.1007/978-3-031-94575-5_2

triple. Such a nested knowledge graph (NKG) represents complex statements like: $((Alice, worksAt, CompanyX), while, (Bob, leads, TeamA))$.

This example encodes a fact stating that "Alice worked at CompanyX while Bob managed TeamA". Here, the relationship $while$ connects entire triples, not just atomic entities. Recent efforts have focused on both the standardization (e.g., RDF-star [10]) and the construction of NKGs [1]. Nested structures add complexity, making tasks like link prediction and reasoning more challenging. Traditional KGE methods often assume a fixed set of entities and relations and focus on direct entity-to-entity patterns. Handling NKGs requires extending beyond this to reason over triples as first-class elements and meta-relations that relate these triples. Some existing approaches (e.g., NestE [18], BiVe [4]) have started to address the problem by learning embeddings using variations of semantic approaches. However, such techniques do not adequately leverage both semantic and structural information. In addition, they do not adequately address the inductive setting where new elements can appear at test time.

Running Example and Architectural Overview. The INKE framework leverages multiple interconnected components to address inductive and nested reasoning challenges in dynamic knowledge graphs. Figure 1 shows an example of a nested knowledge graph along with its management in the INKE framework.

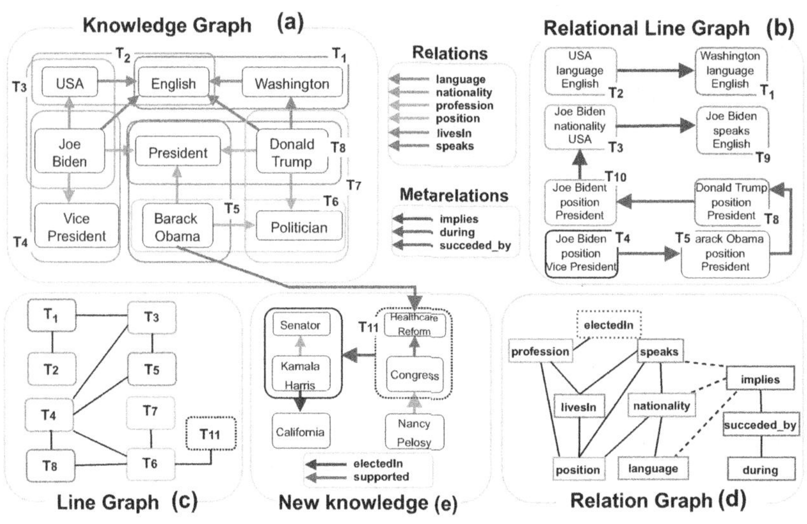

Fig. 1. Overview of Nested Knowledge Graph Data Management in INKE.

Figure 1(a) is the original KG consisting of entities (e.g., Joe Biden, USA, Barack Obama) and their relations (e.g., nationality, profession, position). Relations are color-coded for clarity, linking entities through labeled edges; Fig. 1(b) captures relations between triples. Edges capture meta-relations (e.g., implies, during, succeeded_by), illustrating higher-order semantic relationships between triples; Fig. 1(c) is the structural graph representation of the KG, where triples (e.g., T_1, T_2, T_3) are treated as nodes.

Connections between nodes indicate shared entities in the original KG. Figure 1(d) is a graph showing the interactions between relations and meta-relations based on triple co-occurrence. Based on this data management perspective, INKE's architecture integrates three main modules: *KG-GNN* (Sect. 4.1): Processes the original KG to generate entity and relation embeddings. *Line-GNN* (Sect. 4.2): Refines triple embeddings through the line graph, which connects triples sharing entities. *Relation-GNN* (Sect. 4.3): Propagates predicate embeddings using the relation graph, enabling generalization to unseen relations and meta-relations. These modules interact seamlessly, with attention mechanisms ensuring the model focuses on the most relevant neighbors at each level. This integration allows INKE to dynamically adapt to evolving knowledge graphs, handling new entities, unseen predicates, and nested relationships. Figure 1(e) shows an example of knowledge expansion, where new triples (e.g., Kamala Harris, profession, Senator), new relations, and meta-relations between triples are added.

In contrast, traditional methods like NestE [18] and BiVe [4] fail to integrate such new nodes without retraining. INKE, using its inductive reasoning capability, addresses this challenge through the KG-GNN. It dynamically initializes embeddings for unseen entities and refines them by propagating information from related nodes, such as other politicians or locations, ensuring that inductive reasoning seamlessly incorporates new data. Moreover, the new triples are also added to the Line Graph representation with edges toward other triples sharing some entities (e.g., the new T_{11} becomes a neighbor of T_6 since sharing the entity Barack Obama). Furthermore, as KGs evolve, unseen predicates may also emerge within complex, nested relationships. For instance, *electedIn* does not exist during training. Unlike static approaches, INKE solves this problem through the Relation-GNN, which propagates embeddings within the Relation Graph (Fig. 1(b)), connecting predicates based on co-occurrence semantics or meta-relations. Using attention mechanisms, INKE relates new relations like *electedIn* to known relations such as *profession*, refining their embeddings dynamically.

1.1 Contributions and Outline

We contribute INKE, an Inductive Nested Knowledge Embedding framework that:

1. Operates on three layers of representation via three interconnected GNNs: (i) a KG-GNN over entities and relations, (ii) a Line-GNN over triples, and (iii) a Relation-GNN over predicates and meta-relations.
2. Enables inductive reasoning by dynamically incorporating new entities, triples, and meta-relations at test time, leveraging attention mechanisms and shared substructures to produce embeddings for unseen elements.
3. Incorporates nested triples seamlessly by "flattening" them into embeddings of triple-nodes, allowing higher-order facts to be processed as single nodes in the Line-GNN and Relation-GNN graphs.

INKE effectively bridges local semantics, global structure, and higher-order relationships, making it well-suited for reasoning over complex, nested knowledge graphs. This integrated approach addresses gaps in existing KGE methods, facilitating robust and scalable reasoning in complex, evolving NKGs. The rest of the paper is organized as follows: we review related work (Sect. 2). Then, we provide background on nested KGs and the problem setting (Sect. 3). Next, we describe the INKE framework and its layers in detail (Sect. 4). We discuss experiments (Sect. 5) and then conclude (Sect. 6).

2 Related Work

Knowledge Graph Embedding (KGE) methods are typically divided into *semantic models* and *structural models* (Table 1). Semantic models embed triples by representing the head, relation, and tail independently in a vector space and use scoring functions to assess triple validity. Classical models such as TransE [2] conceptualize relations as translations between entity embeddings, while ConvE [6] uses convolutional layers to model more complex interactions. RotatE [15] captures symmetry and antisymmetry by modeling relations as rotations in a complex vector space. Despite their efficiency, semantic models are inherently limited to capturing direct triple-level interactions, neglecting the broader structural patterns of knowledge graphs. Recent advancements like BiVe [4] and NestE [18] extend semantic modeling to *nested* KGs, enabling the representation of relations between entire triples and accommodating higher-order reasoning.

Table 1. Comparison of Knowledge Graph Embedding Methods.

Method	Nested Triples	Inductive Reasoning	Graph Structure	Model Type
TransE [2]	No	No	No	Semantic
ConvE [6]	No	No	No	Semantic
RotatE [15]	No	No	No	Semantic
BiVe [4]	Yes	No	Partial	Semantic
NestE [18]	Yes	No	Partial	Semantic
RGCN [14]	No	No	Yes	Structural
MINERVA [5]	No	No	Yes	Structural
CompGCN [27]	No	No	Yes	Structural
GraIL [22]	No	Yes	Yes	Structural
SNRI [23]	No	Yes	Yes	Structural
InGram [21]	No	Yes	Partial	Inductive
HINGE [26]	Partial	No	Partial	Hybrid
NBFNet [24]	Partial	No	Yes	Hybrid
INKE	**Yes**	**Yes**	**Yes**	**Inductive/Hybrid**

Structural models, in contrast, leverage the graph's connectivity to propagate information across entities and their neighbors using *Graph Neural Networks (GNNs)*. For example, RGCN [14] extends standard GNNs to handle the relational heterogeneity of knowledge graphs, while MINERVA [5] uses reinforcement learning to learn reasoning paths. CompGCN [27] improves relational reasoning by integrating relations into the message-passing framework. Despite their structural focus, these models fail to handle nested triples and inductive generalization to unseen entities, relations, or entire triples.

Inductive Knowledge Graph Embedding (KGE) methods have recently gained traction due to their ability to address dynamic and open-world scenarios. InGram [21] introduces an inductive framework that constructs a relation graph to propagate embeddings, effectively generalizing to unseen entities by leveraging relational co-occurrence. Similarly, GraIL [22] and SNRI [23] use local subgraph extraction to facilitate inductive reasoning. These approaches demonstrate success in learning embeddings for unseen nodes or triples, but they often lack explicit mechanisms for handling nested relationships or higher-order meta-relations. Recent hybrid methods attempt to combine the strengths of semantic and structural approaches. NBFNet [24] integrates path-based reasoning with GNN-based message propagation, while HINGE [26] incorporates hyper-relational embeddings to model multi-arity relations. However, these models are not inherently designed to address both inductive reasoning and nested knowledge representation simultaneously. In contrast, traditional methods like NestE [18] and BiVe [4] fail to integrate such new nodes without retraining. INKE, using its inductive reasoning capability addresses this challenge through the KG-GNN. It dynamically initializes embeddings for unseen entities and refines them by propagating information from related nodes, such as other politicians or locations, ensuring that inductive reasoning seamlessly incorporates new data. Moreover, the new triples are also added to the Line Graph representation with edges toward other triples sharing some entities (e.g., the new T_{11} becomes a neighbor of T_6 since it shares the entity Barack Obama). Furthermore, as KGs evolve, unseen predicates may also emerge within complex, nested relationships. For instance, *electedIn* does not exist during training. Unlike static approaches, INKE solves this problem through the Relation-GNN, which propagates embeddings within the Relation Graph (Fig. 1(b)), connecting predicates based on co-occurrence semantics or meta-relations.

3 Problem Setting

A Knowledge Graph (KG) is a labeled, directed multigraph $\mathcal{G} = (\mathcal{E}, \mathcal{R}, \mathcal{T})$ where $\mathcal{E}$ is a set of entities, $\mathcal{R}$ is a set of relations, and $\mathcal{T}$ is a set of triples $\{(h, r, t)\}$ with $h, t \in \mathcal{E}$ and $r \in \mathcal{R}$. Real-world knowledge graphs are often richer than simple entity-relation-entity triples. In many scenarios, a fact can involve entire triples as subjects or objects, creating *nested knowledge graphs (NKGs)*.

Definition 1 (Nested Knowledge Graph). $\mathcal{G}_n = (\mathcal{E}, \mathcal{R}, \mathcal{M}, \mathcal{T}_s, \mathcal{T}_n)$ *is a nested knowledge graph, which extends a KG with a set of meta-relations* $\mathcal{M}$ *and a set of nested triples* $\mathcal{T}_n$. *A nested triple* $\langle h^n, r^n, t^n \rangle$ *satisfies the following conditions: (i)* $r^n \in \mathcal{M}$, *(ii) both* h^n *and* t^n *are either standard triples from* $\mathcal{T}_s$ *or entities from* $\mathcal{E}$, *(iii) nested triples are not allowed as arguments; that is, neither* h^n *nor* t^n *can be a nested triple.*

Moreover, real-world KGs evolve over time: new entities, new triples, and even previously unseen predicates emerge. Hence, we assume an *inductive* setting where new elements (entities, triples, predicates) may appear at test time. We consider several tasks to show INKE's (inductive) reasoning capabilities in NKGs:

1. **Triple Prediction:** Given a query $\langle T_i, r^n, ?T\rangle$, where T_i is a known triple and r^n is a (meta-)relation, predict the missing triple or entity $?T$. Crucially, $?T$ may contain unseen entities or represent a nested triple, requiring the model to adapt inductively.
2. **Conditional Link Prediction:** Consider a query $\langle T_i, r^n, (h_j, r_j, ?t)\rangle$. Here, we know a nested structure $\langle T_i, r^n, \cdot\rangle$ and part of a triple $\langle h_j, r_j, ?t\rangle$. The task is to predict the missing tail entity $?t$. Both the entity $?t$ and the predicate r_j can be new, testing the model's ability to integrate unseen elements.
3. **Predicate Prediction:** Given $\langle T_i, ?, T_j\rangle$, where T_i and T_j are known triples or entities, predict the missing relation $?r^n$. The unknown predicate may be a new or unseen meta-relation, testing the model's capacity to infer plausible predicates.
4. **Meta-Relation Prediction:** For higher-order queries, e.g., $\langle\langle T_i, r_i, T_j\rangle, r_{\text{meta}}, ?T_k\rangle$, predict the missing triple $?T_k$ given a meta-relation r_{meta}. This can be seen as a special case of triple prediction.

4 The INKE Architecture

INKE (Inductive Nested Knowledge Embeddings) achieves integrates three interconnected GNN layers (see Fig. 2): (1) KG-GNN (Layer 1): Operates directly on the original knowledge graph, where nodes represent entities and edges represent relations. This layer refines the initial embeddings of entities and relations, capturing local semantic information (Sect. 4.1); (2) Line-GNN (Layer 2): Constructs a *line graph* in which each node corresponds to a triple from the original KG. Edges between these nodes exist if the corresponding triples share at least one entity. The Line-GNN refines triple-level embeddings, representing each triple as a single node. By modeling triples as fundamental units, we can reason over complex nested structures where triples may be subjects or objects themselves (Sect. 4.2); (3) Relation-GNN (Layer 3): Builds a *relation graph*, where nodes represent predicates (including meta-relations) and edges capture co-occurrences or joint participation in higher-order structures. This layer refines predicate-level embeddings, enabling the model to handle unseen predicates inductively and capture higher-order relational patterns (Sect. 4.3).

These three layers progressively refine embeddings: the KG-GNN focuses on entity and relation embeddings at the local level; the Line-GNN aggregates these refined embeddings to produce triple-level embeddings; and the Relation-GNN integrates the resulting information at the predicate level. Through this layered architecture, INKE captures both local semantics and global structural patterns, enabling it to handle complex nested queries and inductively incorporate new elements.

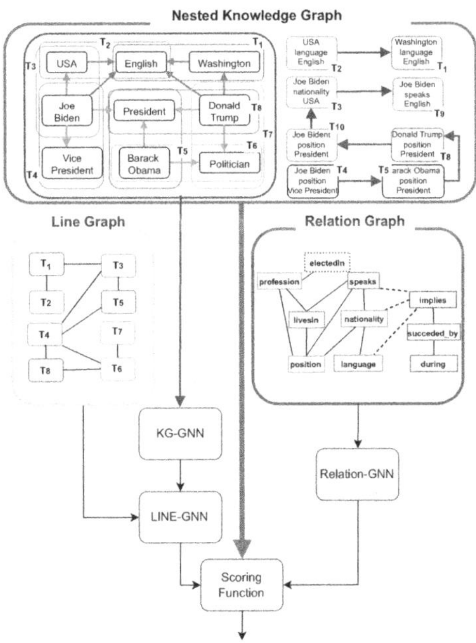

Fig. 2. INKE's architecture.

4.1 Layer 1: KG-GNN (Entity-Level Reasoning)

Let $\mathcal{G} = (\mathcal{E}, \mathcal{R}, \mathcal{T})$ be a KG with entities $\mathcal{E}$, relations $\mathcal{R}$, and triples $\mathcal{T}$. Each triple (h, r, t) initially has embeddings $\mathbf{h}_h, \mathbf{r}_r, \mathbf{h}_t \in \mathbb{R}^d$. The KG-GNN refines entity embeddings using a Graph Attention Network (GAT) [28] and updates relation embeddings:

$$\mathbf{h}_i^{(l+1)} = \text{ReLU}\left(\sum_{j \in \mathcal{N}(i)} \alpha_{ij}^{(l)} W^{(l)} \mathbf{h}_j^{(l)}\right); \mathbf{r}_j^{(l+1)} = \text{ReLU}\left(W_r \left[\mathbf{h}_h^{(l+1)} \parallel \mathbf{h}_t^{(l+1)}\right]\right),$$

where $\alpha_{ij}^{(l)}$ are learned attention coefficients, $W^{(l)}$ and W_r are learned weight matrices, and $\parallel$ denotes concatenation.

4.2 Layer 2: Line-GNN (Triple-Level Reasoning)

The Line-GNN treats each triple as a node in a *line graph*. Two triple-nodes are connected if their corresponding triples share an entity. Initial triple embeddings are formed by concatenating the refined entity and relation embeddings from Layer 1:

$$\mathbf{z}_{t_i}^{(0)} = [\mathbf{h}_s; \mathbf{r}_p; \mathbf{h}_o] \in \mathbb{R}^{3d}.$$

Using a Relational GAT (RGAT), we propagate information among triple-nodes:

$$\mathbf{z}_{t_i}^{(l+1)} = \text{ReLU}\left(\sum_{j \in \mathcal{N}(i)} \beta_{ij}^{(l)} \cdot \text{MLP}_{\text{rel}}\big(\mathbf{z}_{t_j}^{(l)}, r_{ij}\big)\right),$$

where $\beta_{ij}^{(l)}$ are attention weights for triples and r_{ij} encodes relational context (e.g., shared entities). After k layers, we obtain refined triple embeddings $\mathbf{z}_{t_i}^{(k)}$.

4.3 Layer 3: Relation-GNN (Relation-Level Reasoning)

At the predicate level, we construct a *relation graph* $\mathcal{G}_{\text{pred}} = (\mathcal{V}_{\text{pred}}, \mathcal{E}_{\text{pred}})$ where nodes $\mathcal{V}_{\text{pred}}$ represent predicates (from both $\mathcal{R}$ and the set of meta-relations $\mathcal{M}$). The edges $\mathcal{E}_{\text{pred}}$ are defined as follows: (i) for standard triples: two predicates $r_1, r_2 \in \mathcal{R}$ are connected if they share a common entity, i.e., there exist triples (e_i, r_1, e_j) and (e_j, r_2, e_k) or (e_i, r_1, e_j) and (e_k, r_2, e_i) in $\mathcal{T}$; (ii) for nested triples: a predicate $r \in \mathcal{R}$ and a meta-relation $m \in \mathcal{M}$ are connected if r appears within a triple connected by m. Initial predicate embeddings $\mathbf{p}_r \in \mathbb{R}^d$ are similarly initialized (or taken from Layer 1). Using another GAT-based approach:

$$\mathbf{p}_r^{(l+1)} = \text{ReLU}\left(\sum_{u \in \mathcal{N}(r)} \gamma_{ru}^{(l)} W_{\text{pred}}^{(l)} \mathbf{p}_u^{(l)}\right),$$

we refine predicate embeddings, capturing higher-order patterns. Here, $\gamma_{ru}^{(l)}$ are attention weights for predicates and $W_{\text{pred}}^{(l)}$ are learned weight matrices.

4.4 Scoring Mechanism

When encountering a nested triple (T_i, p^n, O), where O (or similarly the subject) may itself be a triple, we *flatten* it by recursively substituting each nested component with the appropriate embedding:

- If the component is an *entity*, we use its refined embedding from the KG-GNN: $\mathbf{h}_O^{(L_1)}$.
- If the component is another *triple*, we use the triple's refined embedding from the Line-GNN: $\mathbf{z}_{T_O}^{(L_2)}$.
- If the component is a *predicate* (including a meta-relation), we use its refined embedding from the Relation-GNN: $\mathbf{p}_{r^n}^{(L_3)}$.

By iteratively applying this substitution, we ensure that every nested component is represented by the correct embedding from the relevant layer. Once the nested triple is fully flattened, we apply a suitable scoring function (e.g., DistMult or a quaternion-based operator) to the resulting embeddings to complete reasoning tasks such as link prediction or meta-relation inference. Generally, let the embeddings from KG-GNN, LINE-GNN, and Relation-GNN be denoted as $\mathbf{h}_{\text{KG}}$, $\mathbf{h}_{\text{LINE}}$, and $\mathbf{h}_{\text{REL}}$, respectively. The scoring function $f(\cdot)$ can be represented as:

$$f(T_i, p^n, T_j) = g\big(\phi(\mathbf{h}_{\text{KG}}, \mathbf{h}_{\text{LINE}}, \mathbf{h}_{\text{REL}})\big),$$

where: (i) $\phi(\cdot)$: Combines the embeddings (e.g., concatenation, weighted sum, or attention mechanism); (ii) $g(\cdot)$: A function that maps the combined embedding to a scalar score (e.g., DistMult). This process preserves higher-order semantics while keeping the computational model manageable and enables INKE to handle complex, inductive scenarios effectively.

Inductive Generalization. INKE supports inductive scenarios seamlessly. For a new triple (h', r', t') at test time, if h' or t' are unseen, INKE initializes their embeddings randomly (or use a heuristic like averaging embeddings of similar nodes). The GNN layers propagate information from known neighbors, enabling the model to assign meaningful embeddings inductively. For a new predicate r_{new}, INKE initializes $\mathbf{p}_{r_{\text{new}}}$ and propagates through the Relation-GNN. Connections to known predicates guide the embedding refinement, allowing INKE to handle new predicates without retraining. This dynamic adaptation ensures model flexibility and effectiveness as the graph evolves.

4.5 Training and Loss Functions

Previous KGE methods typically optimize a single loss function at the triple level. However, INKE operates across three levels (entity/relation, triple, and relation), each capturing different aspects of the graph structure. To ensure consistent and robust embeddings, INKE leverages a combined loss that integrates supervision signals from all three layers. In particular, at the KG level, we use a standard margin-based ranking loss. For each positive triple $(h, r, t) \in T^+$, we generate negative examples $(h', r', t') \in T^-$ by corrupting h or t. We then define:

$$\mathcal{L}_{KG} = \sum_{(h,r,t)\in T^+} \sum_{(h',r',t')\in T^-} \max(0, \gamma - f_{KG}(h, r, t) + f_{KG}(h', r', t')),$$

where f_{KG} is a scoring function using $\mathbf{h}_h^{(L_1)}, \mathbf{r}_r^{(L_1)}, \mathbf{h}_t^{(L_1)}$ from the KG-GNN. At the triple (line graph) level, we want to ensure that triples sharing entities are consistently embedded, and unrelated triples are kept apart. Similar to the KG-level, we define a margin-based loss:

$$\mathcal{L}_{Line} = \sum_{(t_i,t_j)\in E^+_{\text{line}}} \sum_{(t_i,t_k)\in E^-_{\text{line}}} \max(0, \gamma - f_{Line}(t_i, t_j) + f_{Line}(t_i, t_k)),$$

where E^+_{line} are edges in the line graph (representing connected triples) and E^-_{line} are non-edges sampled as negative pairs. The scoring function $f_{Line}(t_i, t_j)$ compares embeddings $\mathbf{z}_{t_i}^{(L_2)}$ and $\mathbf{z}_{t_j}^{(L_2)}$, encouraging connected triples to be closer in embedding space than random, disconnected pairs. At the relation level, we similarly encourage relations that co-occur to have compatible embeddings. Let E^+_{pred} be pairs of predicates that are connected in the relation graph, and E^-_{pred} be negative pairs. We define:

$$\mathcal{L}_{Pred} = \sum_{(r,u)\in E^+_{\text{pred}}} \sum_{(r,v)\in E^-_{\text{pred}}} \max(0, \gamma - f_{Pred}(r, u) + f_{Pred}(r, v)),$$

where $f_{Pred}(r, u)$ measures similarity (e.g., via dot product) between $\mathbf{p}_r^{(L_3)}$ and $\mathbf{p}_u^{(L_3)}$.

Combined Loss Function. To align the learning at all three levels and ensure globally consistent embeddings, we combine the three layer-specific losses into a single objective:

$$\mathcal{L} = \lambda_{KG}\mathcal{L}_{KG} + \lambda_{Line}\mathcal{L}_{Line} + \lambda_{Pred}\mathcal{L}_{Pred}, \tag{1}$$

where λ_{KG}, λ_{Line}, and λ_{Pred} are hyperparameters controlling the relative importance of each component.

4.6 Training and Inferene Regime

We train INKE end-to-end. First, INKE initializes all embeddings (entities, relations, triples, predicates) using, for example, Glorot initialization. Then, INKE samples mini-batches of (nested)triples from T^+ and generate negative samples T^-. For line-level and predicate-level losses, it also samples positive and negative edges from the line and predicate graphs. In the forward pass, INKE propagates embeddings through KG-GNN, Line-GNN, and Relation-GNN and compute $\mathcal{L}_{KG}$, $\mathcal{L}_{Line}$, and $\mathcal{L}_{Pred}$. In the backward pass, INKE computes gradients and update model parameters via gradient-based optimization. It also applies dropout and early stopping. This procedure ensures that INKE learns flexible, generalizable embeddings suitable for reasoning over evolving and nested knowledge graphs. At inference, INKE applies its learned parameters to new queries. Consider a query that involves known and/or unseen entities, triples, or relations. If new entities, triples, or predicates appear, initialize their embeddings (randomly or using average embeddings from neighbors). Then, INKE runs a forward pass through KG-GNN, Line-GNN, and Relation-GNN to obtain final embeddings. For nested triples in the query, INKE recursively flattens them using the triple embeddings from the Line-GNN. Finally, INKE uses the final embeddings and the scoring function f to evaluate candidate answers (e.g., potential missing entities/predicates) and select the top-ranked.

5 Experimental Evaluation

The primary goal was to evaluate INKE's performance on the novel *nested* knowledge graph tasks of triple classification and conditional link prediction with particular emphasis on the inductive setting. In what follows, we describe the experimental setting (Sect. 5.1), then move to the discussion of the evaluation results in different settings: triple prediction (Sect. 5.2), conditional link prediction (Sect. 5.3), and inductive tasks (Sect. 5.4).

5.1 Datasets

We used three state-of-the-art nested knowledge graphs: FBH, FBHE, and DBHE, which contain nested facts and have been constructed and manually curated [4]. FBH and FBHE are derived from FB15K-237 (Freebase), while DBHE is based on DB15K (DBpedia). FBH consists only of nested facts inferred from triple facts, such as *implies*, whereas FBHE and DBHE also include externally sourced knowledge such as *transfers-to*. Table 2 summarizes the datasets used.

Table 2. NKGs Statistics. $\mathcal{G}_n = (\mathcal{E}, \mathcal{R}, \mathcal{R}_{\text{m}}, \mathcal{T}_s, \mathcal{T}_n)$. $|\mathcal{T}'|$ represents the number of atomic triples involved in the nested triples.

Dataset	$\lvert\mathcal{E}\rvert$	$\lvert\mathcal{R}\rvert$	$\lvert\mathcal{T}_s\rvert$	$\lvert\mathcal{R}_{\text{m}}\rvert$	$\lvert\mathcal{T}_n\rvert$	$\lvert\mathcal{T}'\rvert$
FBH	14,541	237	310,117	6	27,062	33,157
FBHE	14,541	237	310,117	10	34,941	33,719
DBHE	12,440	87	68,296	8	6,717	8,206

Baselines. We consider BiVe [4] as well as NestE [18] as our primary baselines since they are specifically designed for NKGs and have shown significant improvements over methods that were not designed to deal with nested triples. For each of these methods, we report the variants with the best performance [18]. We also consider rule-based approaches, including Neural-LP [19], DRUM [13], and AnyBURL [12], which indirectly account for relations between facts using first-order logic-like expressions. Additionally, we include QuatE [20] and BiQUE [7] as state-of-the-art triple-based methods, although they do not directly handle nested facts.

INKE Variants. To process NKGs with INKE, we first precompute the line graphs for the datasets, treating base triples as nodes. Similarly, the relation graph is precomputed for predicate-level reasoning. We implement three distinct variants of INKE:

(1) INKE: Utilizes all three layers of the architecture, ensuring comprehensive reasoning across triple, meta-relational, and predicate levels.
(2) INKE-N: Skips the Line-GNN layer (*Layer 2*) and computes triple embeddings directly from *Layer 1*, where the head, relation, and tail embeddings are concatenated.
(3) INKE-B: Bypasses *Layer 1* and begins with structural refinement at *Layer 2* (Line-GNN), initializing triple embeddings randomly.

These variants are designed not only to handle different aspects of the reasoning pipeline but also to enable *a detailed ablation analysis* of INKE's components. The hyperparameters were tuned via grid search and validation. A learning rate of 0.001 provided stable convergence without overfitting. As for the structural approach (*Layer 2*), ReLU gave the best generalization, while a 0.3 dropout prevented overfitting. Using 2 layers was optimal, as more layers added complexity without significant performance gains. For both base embeddings (*Layer 1*) and final embeddings (*Layer 3*), we adopted the DistMul [25] scoring function. Embedding dimensionality was set to $d = 200$. The contributions to the loss λ_{KG}, λ_{Line}, and λ_{Pred} (Eq. 1) have been set to 0.33.

Metrics. We employed Filtered Mean Rank (MR), Mean Reciprocal Rank (MRR), and Hits@10. The mean performance is reported over 10 random seeds for each method, with the relatively small standard deviations omitted.

Implementation. INKE has been implemented based on OpenKE [9], Apple MLX[1] and PyTorch[2]; the hardware configuration was a Mac Studio with 192 GB of shared

[1] https://github.com/ml-explore/mlx.

[2] https://pytorch.org/.

memory. For the competitors, we reused the best-performing parameters without performing additional searches[3]. The code and datasets are available for download[4].

5.2 Results on Triple Prediction

Most of the KGE baseline methods available are unable to handle triple prediction and conditional link prediction due to their limitations in working with NKGs. Nevertheless, following the approach of BiVe and NestE, we selected some significant baselines and adapted them to work with NKGs. To do so, we separate nested triples from standard triples, thereby constructing *a novel KG where subjects/objects are triples* with meta-relations between triples. Notably, NestE, BiVe, and INKE are the only methods available, to the best of our knowledge, capable of addressing both standard and nested triples *simultaneously*. In the following experiments, we considered the same train/test/validation splits as NestE [18], the state-of-the-art approach.

Table 3. Triple Prediction Results. The best scores are boldfaced. * denotes results taken from [18].

	FBH			FBHE			DBHE		
	MR (↓)	MRR (↑)	Hit@10 (↑)	MR (↓)	MRR (↑)	Hit@10 (↑)	MR (↓)	MRR (↑)	Hit@10 (↑)
QuatE*	145603.8	0.103	0.114	94684.4	0.101	0.209	26485.0	0.157	0.179
BiQUE*	81687.5	0.104	0.115	61015.2	0.135	0.205	19079.4	0.163	0.185
Neural-LP*	115016.6	0.070	0.073	90000.4	0.238	0.274	21130.5	0.170	0.209
DRUM*	115016.6	0.069	0.073	90000.3	0.261	0.274	21130.5	0.166	0.209
AnyBURL*	108079.6	0.096	0.108	83136.8	0.191	0.252	20530.8	0.177	0.214
BiVE-B*	8.63	0.833	0.924	9.53	0.705	0.860	4.66	0.718	0.945
NestE-SB*	**3.34**	**0.922**	**0.982**	**3.05**	**0.851**	**0.962**	**2.07**	**0.862**	**0.984**
INKE-B	4.71	0.708	0.746	4.24	0.646	0.737	2.01	0.668	0.746
INKE-N	3.01	0.704	0.753	2.13	0.656	0.769	2.04	0.656	0.748
INKE	**2.29**	**0.957**	**0.989**	**2.15**	**0.898**	**0.977**	**1.48**	**0.898**	**0.987**

The evaluation results, as presented in Table 3, highlight the effectiveness of INKE in enhancing traditional semantic-based approaches with structural refinement for triple prediction in knowledge graphs. By leveraging its *three-layered architecture*, which integrates *semantic interpretation* (Layer 1) and *structural refinement* (Layer 2) via a line graph-based mechanism, INKE achieves both precision and contextual depth in its predictions. This architecture enables the model to effectively capture broad relational patterns and intricate higher-order dependencies. These results also demonstrate its superiority over state-of-the-art methods and advanced models such as NestE-SB.

Looking at INKE's variants, we note the impact of each architectural choice on the performance. INKE, which fully utilizes all three layers (semantic embedding in

[3] As reported at https://github.com/xiongbo010/NestE.

[4] https://github.com/giuseppepirro/INKE/.

Layer 1, structural refinement via the line graph in Layer 2, and integrated reasoning in Layer 3), achieves superior metrics across all evaluation criteria. The second variant, INKE-B, which bypasses the structural refinement provided by the line graph in Layer 2 and instead directly relies on concatenating the embeddings from Layer 1, shows a noticeable drop in performance. This highlights the critical role of the structural component in refining relational dependencies. INKE-N, which omits Layer 1 entirely and initializes embeddings randomly while relying solely on structural refinement, performs worse than INKE but better than INKE-B in some cases. This suggests that structural refinement alone cannot fully compensate for the absence of semantic embedding, yet it contributes to some level of predictive capability. Overall, these findings reinforce the necessity of the full three-layer architecture for achieving the most accurate and contextually aware triple embeddings, with each layer adding distinct and complementary value to the model.

5.3 Results on Conditional Link Prediction

We evaluated INKE on the task of conditional link prediction (CLP) and compared its performance against state-of-the-art models. Unlike standard link prediction, CLP in *nested* knowledge graphs requires predicting missing links while simultaneously considering the interdependencies and contextual triples surrounding the query. This added complexity demands a model capable of capturing higher-order relationships and meta-relational contexts effectively. We considered the same train/test/validation splits as NestE [18]. The results, summarized in Table 4, demonstrate that INKE consistently outperforms competing models across most benchmarks with the exception of Hit@10 in the FBH dataset, where it performs competitively.

Table 4. Performance on Conditional Link Prediction. The best scores are boldfaced. * denotes results taken from [18].

	FBH			FBHE			DBHE		
	MR (↓)	MRR (↑)	Hit@10 (↑)	MR (↓)	MRR (↑)	Hit@10 (↑)	MR (↓)	MRR (↑)	Hit@10 (↑)
QuatE*	163.7	0.346	0.494	1546.4	0.124	0.189	551.6	0.208	0.309
BiQUE*	111.0	0.423	0.641	90.1	0.387	0.617	29.5	0.378	0.677
Neural-LP*	185.9	0.433	0.648	146.2	0.466	0.716	32.2	0.517	0.756
DRUM*	262.7	0.394	0.555	207.6	0.413	0.620	49.0	0.470	0.732
AnyBURL*	228.5	0.380	0.563	166.0	0.418	0.607	81.7	0.403	0.594
BiVE-Q*	4.33	0.826	0.948	6.56	0.761	0.886	2.69	0.852	0.971
NestE-SB*	1.52	0.934	**0.991**	2.61	0.867	0.951	1.72	0.919	0.990
INKE-B	4.78	0.702	0.875	2.66	0.781	0.856	2.01	0.782	0.839
INKE-N	3.03	0.787	0.978	2.49	0.861	0.861	1.98	0.889	0.917
INKE	**1.36**	**0.956**	0.983	**2.03**	**0.897**	**0.981**	**1.35**	**0.947**	**0.995**

The results presented in Table 4 reconfirm the advantages of INKE and its design choices, as previously discussed. To avoid redundancy, it is worth reflecting on broader implications and comparisons. While INKE achieves state-of-the-art performance across all datasets, particularly excelling in FBHE and DBHE, the consistent outperformance of INKE over its own variants (INKE-B and INKE-N) highlights the importance of integrating all architectural components. Additionally, the strong performance of INKE compared to NestE-SB on DBHE (with a lower MR and a marginally higher MRR) signals its robustness in handling datasets with higher relational complexity. This advantage could stem from the model's ability to refine predictions across multiple layers while balancing semantic and structural dependencies. By contrast, older models like QuatE and BiQUE struggle significantly, as their relatively high MR values demonstrate their inability to handle the intricate relational and nested structures present in FBHE and DBHE. This not only highlights the need for advanced frameworks like INKE but also emphasizes the limitations of earlier designs when applied to modern, nested knowledge graph tasks. Overall, while the table reiterates previous findings, it also underlines INKE's ability to generalize effectively across diverse datasets and outperform both traditional models and competitive baselines.

5.4 Inductive Capabilities

To evaluate INKE's performance in inductive reasoning, we generate datasets that systematically include new entities, relations, and meta-relations in test scenarios. While it is possible to generate numerous dataset variations by varying the proportions of unseen triples, entities, and relations, we focus on a carefully selected subset of datasets to ensure a meaningful and interpretable evaluation. Our selection prioritizes the most challenging cases of inductive reasoning: *Entity Generalization*, where a proportion of test triples involve unseen entities; *Relation Generalization*, where test triples include unseen relations; and *Combined Generalization*, where test triples feature both unseen entities and relations. We denote datasets using the format `DatasetName-Percentage-Type`, where `DatasetName` specifies the dataset (e.g., FBH, FBHE, DBHE), `Percentage` indicates the proportion of test triples involving the specified challenge, and `Type` defines the generalization type (`E` for Entity, `R` for Relation, and `C` for Combined). For example, `FBH-25E` represents a dataset where 25% of test triples involve unseen entities, while `FBH-50R` involves 50% unseen relations, and `FBH-25C` tests combined generalization with 25% unseen entities and relations.

We evaluated the approaches in the conditional link prediction task with the additional challenge that both predicates and entities can be unseen during training. To assess the effectiveness of INKE, we include both inductive and non-inductive approaches in our experiments. Competitors like Neural-LP [19], CompGCN [27], RED-GNN [31], and InGRAM [21] are inherently inductive models that reason over unseen entities and relations by leveraging their graph structures and subgraph patterns. These models serve as baselines to evaluate how well INKE generalizes in inductive settings. Conversely, methods like BiVe [4] and NestE [18] are designed specifically for reasoning over nested knowledge graphs but lack inherent inductive capabilities. These models operate under a transductive assumption, where all entities and relations involved in the nested triples must be observed during training. To evaluate BiVe and NestE in inductive scenarios, we adopt a preprocessing strategy where unseen entities

Table 5. Performance comparison of models under different inductive settings. Metrics include Mean Rank (MR), Mean Reciprocal Rank (MRR), and Hit@10 for datasets with varying proportions of unseen relations and entities.

	FBH-25E			FBH-25R			FBH-25C		
Method	**MR**	**MRR**	**Hit@10**	**MR**	**MRR**	**Hit@10**	**MR**	**MRR**	**Hit@10**
CompGCN	457.7	0.025	0.215	567.6	0.012	0.159	889.4	0.024	0.145
NeuralLP	567.3	0.107	0.223	595.4	0.118	0.194	687.5	0.123	0.287
RED-GNN	68.4	0.174	0.284	103.4	0.145	0.278	474.4	0.124	0.267
InGRAM	29.4	0.445	0.308	12.5	0.354	0.406	67.1	0.312	0.387
BiVe	32.3	0.281	0.362	29.4	0.431	0.523	34.4	0.398	0.478
NestE	9.3	0.562	0.634	8.7	0.382	0.468	22.3	0.478	0.532
INKE	**5.4**	**0.781**	**0.862**	**6.1**	**0.795**	**0.783**	**7.8**	**0.594**	**0.695**

	FBHE-25E			FBHE-25R			FBHE-25C		
Method	**MR**	**MRR**	**Hit@10**	**MR**	**MRR**	**Hit@10**	**MR**	**MRR**	**Hit@10**
CompGCN	626.6	0.063	0.195	536.6	0.014	0.076	852.3	0.074	0.023
NeuralLP	375.4	0.185	0.256	482.4	0.125	0.201	578.6	0.156	0.215
RED-GNN	81.3	0.113	0.145	169.5	0.156	0.234	264.4	0.144	0.293
InGRAM	59.6	0.231	0.346	15.6	0.381	0.342	12.1	0.313	0.403
BiVe	39.2	0.352	0.436	25.4	0.297	0.423	29.4	0.314	0.401
NestE	14.4	0.562	0.665	15.1	0.384	0.462	28.1	0.313	0.428
INKE	**4.15**	**0.761**	**0.864**	**3.14**	**0.821**	**0.875**	**5.26**	**0.614**	**0.711**

	DBHE-25E			DBHE-25R			DBHE-25C		
Method	**MR**	**MRR**	**Hit@10**	**MR**	**MRR**	**Hit@10**	**MR**	**MRR**	**Hit@10**
CompGCN	312.4	0.101	0.251	638.6	0.023	0.150	562.4	0.81	0.113
NeuralLP	471.3	0.171	0.236	421.4	0.111	0.193	314.4	0.185	0.216
RED-GNN	89.5	0.204	0.156	169.3	0.219	0.211	134.1	0.154	0.246
InGRAM	45.1	0.241	0.394	33.1	0.301	0.345	39.1	0.334	0.321
BiVe	39.1	0.251	0.442	25.1	0.362	0.453	34.1	0.394	0.447
NestE	19.2	0.358	0.451	15.2	0.382	0.431	17.1	0.434	0.415
INKE	**4.56**	**0.862**	**0.842**	**3.97**	**0.881**	**0.907**	**6.8**	**0.734**	**0.801**

and relations in the test set are treated as isolated *placeholders* during training. Hence, these models can handle the unseen elements during prediction.

Table 5 shows the results of the evaluation. INKE demonstrates remarkable performance, achieving the highest scores while maintaining the lowest MR values across all datasets and settings. For instance, on FBH-25E, INKE achieves an MRR of 0.781

and Hits@10 of 0.862, far surpassing other models such as NestE (MRR/Hits@10: 0.562/0.634) and BiVe (0.281/0.362). On FBHE-25R, INKE records an MRR of 0.821 and Hits@10 of 0.875, significantly outperforming NestE (0.384/0.462) and BiVe (0.297/0.423). These results underscore INKE's ability to generalize effectively to unseen entities and relations, a critical requirement for real-world inductive reasoning tasks. The combined setting (25C), where both unseen entities and relations must be handled, proves to be the most challenging scenario. This is reflected in the generally lower performance of all models compared to the 25E and 25R settings. However, INKE continues to perform better than competitors. The combined setting's difficulty arises from the requirement to learn relationships among entirely unseen components while integrating contextual information.

Older models such as RED-GNN and NeuralLP struggle to handle the inductive settings, often showing poor MRR and Hits@10 values. Among the baselines, InGRAM is particularly notable due to its inductive capabilities, specifically its ability to generalize to unseen predicates. This model achieves competitive results across several settings. These values highlight InGRAM's effectiveness in handling unseen predicates, making it a strong contender in inductive relation reasoning tasks. However, InGRAM's limitations become evident in more complex scenarios, particularly those involving nested triples. The model's design is not inherently suited to reason over hierarchical structures or nested relationships, which are key in datasets like FBHE and DBHE. Overall, INKE's consistent performance across all datasets and inductive scenarios highlights its robustness and practical applicability.

6 Conclusions and Future Work

We presented INKE, a novel framework for inductive nested knowledge embeddings that integrates three levels of graph neural network (GNN) reasoning: entity-level (KG-GNN), triple-level (Line-GNN), and predicate-level (Relation-GNN). INKE effectively addresses the challenges of nested and evolving knowledge graphs by dynamically incorporating new entities, triples, and predicates at inference time. Nested structures are handled by flattening them into triple-level nodes, while unseen relational patterns are refined through predicate embeddings. INKE achieves its effectiveness by unifying multiple levels of reasoning into a cohesive architecture. The KG-GNN captures fine-grained entity and relation semantics, enabling an understanding of local interactions. The Line-GNN extends this reasoning to triple-level dependencies, facilitating relational inference across triples in the graph. Complementing these, the Relation-GNN refines predicate embeddings, crucial for handling unseen or emerging meta(relations). Attention mechanisms ensure focus on the most relevant graph neighbors, enabling robust and inductive reasoning even in dynamic and evolving settings.

Future work will explore other evaluation scenarios (Sect. 3) like meta-relation prediction, more sophisticated initialization schemes for unseen elements, as well as extensions to temporal or probabilistic reasoning to model evolving knowledge more effectively. We also aim to address under-represented entities [29] and develop explainability tools to enhance interpretability for downstream tasks [30]. Finally, applying INKE to domain-specific knowledge graphs in healthcare, finance, or scientific literature remains an exciting avenue for further research.

References

1. Arenas-Guerrero, J., Iglesias-Molina, A., Chaves-Fraga, D., Garijo, D., Corcho, O., Dimou, A.: Declarative generation of RDF-star graphs from heterogeneous data. Semant. Web (Preprint) 1–19 (2024)
2. Bordes, A., Usunier, N., García-Durán, A., Weston, J., Yakhnenko, O.: Translating embeddings for modeling multi-relational data. In: Burges, C.J.C., Bottou, L., Ghahramani, Z., Weinberger, K.Q. (eds.) Advances in Neural Information Processing Systems 26: Proceedings of the 27th Annual Conference on Neural Information Processing Systems, Lake Tahoe, Nevada, USA, pp. 2787–2795 (2013)
3. Cai, H., Zheng, V.W., Chang, K.: A comprehensive survey of graph embedding: problems, techniques, and applications. Trans. Knowl. Data Eng. **30**(9), 1616–1637 (2018)
4. Chung, C., Whang, J.J.: Learning representations of bi-level knowledge graphs for reasoning beyond link prediction. In: Proceedings of the AAAI Conference on Artificial Intelligence, pp. 4208–4216 (2023)
5. Das, R., et al.: Go for a walk and arrive at the answer: reasoning over paths in knowledge bases using reinforcement learning. In: Proceedings of the 6th International Conference on Learning Representations (ICLR) (2018)
6. Dettmers, T., Minervini, P., Stenetorp, P., Riedel, S.: Convolutional 2D knowledge graph embeddings. In: Proceedings of the AAAI Conference, pp. 1811–1818 (2018)
7. Guo, J., Kok, S.: BiQUE: biquaternionic embeddings of knowledge graphs. In: Moens, M. F., Huang, X., Specia, L., Yih, S. W. (eds.) Proceedings of the 2021 Conference on Empirical Methods in Natural Language Processing (2021)
8. Hamilton, W. L., Ying, Z., Leskovec, J.: Inductive representation learning on large graphs. In: Advances in Neural Information Processing Systems, vol. 30, pp. 1024–1034 (2017)
9. Han, X., Cao, S., Xin, L., Lin, Y., Liu, Z., Sun, M., Li, J.: OpenKE: an open toolkit for knowledge embedding. In: Proceedings of the EMNLP (2018)
10. Hartig, O.: RDF* and SPARQL*: an alternative approach to annotate statements in RDF. In: ISWC (Posters, Demos & Industry Tracks), pp. 1–4 (2017)
11. Kazemi, S. M., Poole, D.: Simple embedding for link prediction in knowledge graphs. In: Advances in Neural Information Processing Systems, vol. 31 (2018)
12. Meilicke, C., Chekol, M.W., Betz, P., Fink, M., Stuckenschmidt, H.: Anytime bottom-up rule learning for large-scale knowledge graph completion. VLDB J. **33**(1), 131–161 (2024)
13. Sadeghian, A., Armandpour, M., Ding, P., Wang, D. Z.: DRUM: end-to-end differentiable rule mining on knowledge graphs. In: Advances in Neural Information Processing Systems, vol. 32 (2019)
14. Schlichtkrull, M., Kipf, T.N., Bloem, P., van den Berg, R., Titov, I., Welling, M.: Modeling relational data with graph convolutional networks. In: Gangemi, A., et al. (eds.) ESWC 2018. LNCS, vol. 10843, pp. 593–607. Springer, Cham (2018). https://doi.org/10.1007/978-3-319-93417-4_38
15. Sun, Z., Deng, Z. H., Nie, J. Y., Tang, J.: RotatE: knowledge graph embedding by relational rotation in complex space. In: Proceedings of the International Conference on Learning Representations (2019)
16. Teru, K., Denis, M.A., Hamilton, W.L.: Inductive relation prediction by subgraph reasoning. In: Proceedings of the International Conference on Machine Learning, pp. 9448–9457. PMLR (2020)
17. Wang, Q., Mao, Z., Wang, B., Guo, L.: Knowledge graph embedding: a survey of approaches and applications. Trans. Knowl. Data Eng. **29**(12), 2724–2743 (2017)
18. Xiong, B., Nayyeri, M., Luo, L., Wang, Z., Pan, S., Staab, S.: NestE: modeling nested relational structures for knowledge graph reasoning. In: Proceedings of the AAAI Conference on Artificial Intelligence, vol. 38, pp. 9205–9213 (2024)

19. Yang, F., Yang, Z., Cohen, W. W.: Differentiable learning of logical rules for knowledge base reasoning. In: Advances in Neural Information Processing Systems, vol. 30 (2017)
20. Zhang, S., Tay, Y., Yao, L., Liu, Q.: Quaternion knowledge graph embedding. arXiv preprint arXiv:1904.10281 (2019)
21. Lee, J., Chung, C., Whang, J.J.. InGram: inductive knowledge graph embedding via relation graphs. In: International Conference on Machine Learning, pp. 18796–18809. PMLR (2023)
22. Teru, K., Denis, E., and Hamilton, W.: Inductive relation prediction by subgraph reasoning. In: International Conference on Machine Learning, pp. 9448–9457. PMLR (2020)
23. Xu, X., Zhang, P., He, Y., Chao, C., Yan, C.: Subgraph neighboring relations infomax for inductive link prediction on knowledge graphs. In: De Raedt, L. (ed.) Proceedings of the Thirty-First International Joint Conference on Artificial Intelligence, IJCAI-22, pp. 2341–2347. International Joint Conferences on Artificial Intelligence Organization (2022). https://doi.org/10.24963/ijcai.2022/325
24. Zhu, Z., Zhang, Z., Xhonneux, L.-P.A.C., Tang, J.: Neural Bellman-Ford networks: a general graph neural network framework for link prediction. In: Advances in Neural Information Processing Systems, vol. 34, pp. 29476–29490 (2021). https://proceedings.neurips.cc/paper/2021/hash/f6a673f09493afcd8b129a0bcf1cd5bc-Abstract.html
25. Yang, B., Yih, W.-T., He, X., Gao, J., Deng, L.: Embedding entities and relations for learning and inference in knowledge bases. In: Proceedings of the 3rd International Conference on Learning Representations (ICLR), San Diego, CA, USA (2015). http://arxiv.org/abs/1412.6575
26. Rosso, P., Yang, D., Cudré-Mauroux, P.: Beyond triplets: hyper-relational knowledge graph embedding for link prediction. In: WWW 2020: The Web Conference 2020, pp. 1885–1896, Taipei, Taiwan, April 20-24, 2020. ACM/IW3C2 (2020). https://doi.org/10.1145/3366423.3380257
27. Vashishth, S., Sanyal, S., Nitin, V., Talukdar, P.P.: Composition-based multi-relational graph convolutional networks. In: 8th International Conference on Learning Representations, ICLR 2020, Addis Ababa, Ethiopia. OpenReview.net (2020)
28. Veličković, P., Cucurull, G., Casanova, A., Romero, A., Liò, P., Bengio, Y.: Graph attention networks. In: International Conference on Learning Representations (ICLR) (2018). https://arxiv.org/abs/1710.10903
29. Zheng, Z., et al.: Low-dimensional hyperbolic knowledge graph embedding for better extrapolation to under-represented data. In: The Semantic Web - 21st International Conference, ESWC 2024, Hersonissos, Crete, Greece, 26–30 May 2024, Proceedings, Part I. Lecture Notes in Computer Science, vol. 14664, pp. 100–120. Springer (2024). https://doi.org/10.1007/978-3-031-60626-7_6
30. Barile, R., d'Amato, C., Fanizzi, N.: Explanation of link predictions on knowledge graphs via levelwise filtering and graph summarization. In: The Semantic Web - 21st International Conference, ESWC 2024, Hersonissos, Crete, Greece, 26–30 May 2024, Proceedings, Part I. Lecture Notes in Computer Science, vol. 14664, pp. 180–198. Springer (2024). https://doi.org/10.1007/978-3-031-60626-7_10
31. Zhang, Y., Yao, Q.: Knowledge graph reasoning with relational digraph. In: Proceedings of the ACM Web Conference 2022, pp. 912–924 (2022)

ReaLitE: Enrichment of Relation Embeddings in Knowledge Graphs Using Numeric Literals

Antonis Klironomos[1,2(✉)], Baifan Zhou[3,4], Zhuoxun Zheng[1,3], Gad-Elrab Mohamed[1], Heiko Paulheim[2], and Evgeny Kharlamov[1,3]

[1] Bosch Center for Artificial Intelligence, Renningen, Germany
{antonis.klironomos,mohamed.gad-elrab,evgeny.kharlamov}@de.bosch.com
[2] University of Mannheim, Mannheim, Germany
heiko.paulheim@uni-mannheim.de
[3] University of Oslo, Oslo, Norway
baifanz@ifi.uio.no
[4] Oslo Metropolitan University, Oslo, Norway

Abstract. Most knowledge graph embedding (KGE) methods tailored for link prediction focus on the entities and relations in the graph, giving little attention to other literal values, which might encode important information. Therefore, some literal-aware KGE models attempt to either integrate numerical values into the embeddings of the entities or convert these numerics into entities during preprocessing, leading to information loss. Other methods concerned with creating relation-specific numerical features assume completeness of numerical data, which does not apply to real-world graphs. In this work, we propose `ReaLitE`, a novel relation-centric KGE model that dynamically aggregates and merges entities' numerical attributes with the embeddings of the connecting relations. `ReaLitE` is designed to complement existing conventional KGE methods while supporting multiple variations for numerical aggregations, including a learnable method. We comprehensively evaluated the proposed relation-centric embedding using several benchmarks for link prediction and node classification tasks. The results showed the superiority of `ReaLitE` (Pronounced as "reality", code: https://github.com/boschresearch/ReaLitE) over the state of the art in both tasks.

1 Introduction

Motivation. Knowledge graphs (KGs) represent information as interconnected entities and their relationships, typically structured in triples (*head*, *relation*, *tail*). Moreover, KGs such as Wikidata [30] or YAGO [20] often incorporate various attributes with numerical values. Knowledge graphs remain incomplete, lacking numerous links between entities. Consequently, various *knowledge graph embedding* (KGE) techniques, *e.g.* TransE [7] have been introduced to predict potential connections between entities. These embedding methods seek to represent the input KG by mapping entities and relations onto a lower-dimensional

E. Curry et al. (Eds.): ESWC 2025, LNCS 15718, pp. 41–58, 2025.
https://doi.org/10.1007/978-3-031-94575-5_3

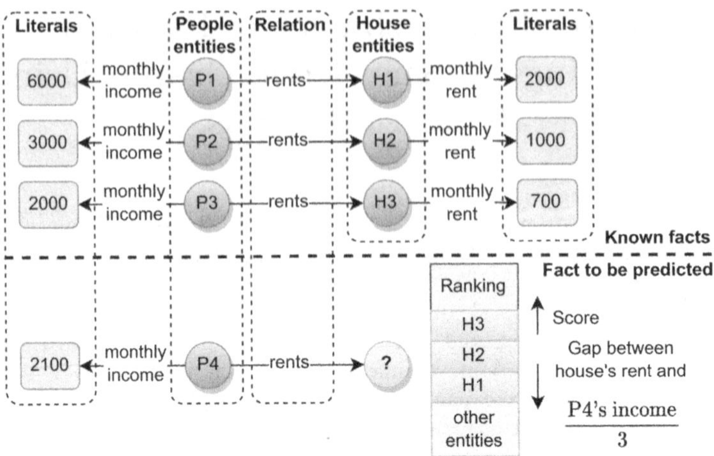

Fig. 1. An example showing the potential benefit of the mathematical relationship between people's income and houses' rent during link prediction.

vector space. This process preserves existing triples in the KG while facilitating the prediction of new links. Subsequently, representations learned through KGEs have found application in diverse tasks, including node classification [21].

Each relation in a KG establishes a connection between pairs of entities. Traditional KG embedding models, such as TransE [7] and ComplEx [29], neglect the literal attributes of entities. Recently, however, the field of multimodal KG completion has gained significant attention [17,26]. Also, the importance of relation representation is stressed by [17,19]. At the same time, various types of relations may exhibit distinct patterns in how numerical attributes are linked to the head and tail of the relation. For instance, a pattern may arise where *people* (head) with a greater *monthly income* (literal) tend to *rent* (relation) *houses* (tail) with higher *monthly rent* (literal) (Fig. 1).

By encoding the relationship between literals of head and tail entities, we can more accurately predict missing links for a head entity by intuitively ranking potential tail entities. The provided example shows that $\textit{monthly income} \approx 3 * \textit{monthly rent}$ for the known triples. Thus, we would incorporate the monthly rent in the ranking of potential tail entities, using the difference to the head entity's $\textit{monthly income}/3$.

State-of-the-Art and Limitations. Existing KGE models that handle numerical literal values employ different strategies [14]. The LiteralE model [15]) incorporates literals into entity embeddings, while the KGA model [31] converts them into entities during KG preprocessing. To our knowledge, KBLRN [13] is the only method attempting to use numeric literals as relation-specific features to enhance link prediction. However, these methods overlook the potential patterns (exemplified in Fig. 1) between different entity attributes for a given relation. KBLRN also faces challenges in incorporating absent literals, affecting the inclusion of numeric information. We argue that there exists an opportunity for improvement by integrating head and tail numeric information in a learnable manner, enabling the model to capture underlying numeric relationships. Since the relation acts as the connection point between head and tail entities, we propose using the relation embedding as a vessel for this combination.

In the realm of node classification, there is limited assessment regarding the performance of triple-scoring KGE methods (e.g., TransE). Additionally, numeric literals are frequently either excluded or handled as conventional nodes [9,23,25]. A recent study proposed a literal preprocessing method based on the current state of the art and evaluated it in node classification [22].

Approach. We introduce `ReaLitE`, a relation-centric method to enhance vanilla KGE models by infusing numeric information from both head and tail entities into the embeddings of connecting relations. The values of each numeric attribute are aggregated separately for the relation's head and tail entities, generating two distinct numeric vectors. The aggregated numeric information is then combined with the relation embedding in a machine-learnable strategy. Such infusion enriches the relation embedding vectors with information about possible correlations between the numerical attributes of its head and tail entities.

Contributions. Our contributions are summarized below:

- We propose `ReaLitE`, an approach that can be combined with any vanilla KGE method with relation embeddings. In addition, we demonstrate the integration of our method into existing KGE frameworks, highlighting its versatility and ease of adoption.
- We experiment with different methods of aggregating numeric literals, including an automated method to learn a combination of multiple aggregation types.
- We evaluate `ReaLitE` extensively and compare it with state-of-the-art in two tasks: link prediction and node classification. For the former, we evaluate on the standard setting of link prediction, along with a more granular relation-focused evaluation. The results show that our approach is comparable or superior compared to the state-of-the-art methods, particularly on the numeric literals with higher correlation and on long-tail relations.

2 Related Work

Literal-Aware Link Prediction. One of the early studies that focused on including numeric literals during link prediction introduced KBLRN: a model that consists of Latent, Relational, and Numerical Features [13]. The inclusion of their model's numeric feature depends on the amount of missing numeric values in the dataset, which is a limitation.

LiteralE was one of the first methods to add literal information into vanilla KGE models by combining literals with the entity embeddings [15]. It does not provide a way to enhance the relation embeddings with numeric information directly. This paper fills that gap and shows the benefits of fusing relation embeddings with numeric literals on link prediction and node classification tasks.

To the best of our knowledge, the latest method for handling numeric literals to perform link prediction is KGA (Knowledge Graph Augmentation) [31]. KGA preprocesses the dataset to transform literals into entities. While that study evaluates multiple models on link prediction, there is a lack of more granular evaluation and evaluation on downstream tasks. We first show the superiority of `ReaLitE` in the typical evaluation setting for link prediction. Then, we investigate the models' performance across different types of relations. We observe significant performance gains favoring `ReaLitE` for long-tail relations and relations with correlated head and tail attributes.

Node Classification. Existing node classification models are mainly message passing (e.g. Graph Convolutional Networks (R-GCNs) [9,25,32]) or feature extraction methods (e.g. RDF2Vec [23]). Triple-scoring embedding methods (e.g., TransE) are rarely evaluated.

A recent study has highlighted that the link prediction performance of a KGE model may not necessarily translate to effectiveness in downstream tasks [35]. To assess KG embeddings, that study uses KG-based recommendation and question-answering downstream tasks. Our work focuses on node classification.

Another work that applies KGE models to downstream tasks uses product knowledge graphs (PKGs) in e-commerce, focusing on essential product relations for applications like marketing and recommendation [33]. Even though that paper evaluates triple-scoring KGE methods such as TransE on node classification, the dataset used (PKG) is inherently not suitable for these methods, as the authors explain in their paper. So, the performance of such methods in that dataset was expected and proven to be low.

Prior research with a more general scope evaluates traditional KGE methods on node classification using a particular KG [1]. In our study, we use multiple datasets that also include literals.

As far as we know, in the downstream task of node classification, the latest study using triple-scoring KGE models with multimodal information in multi-relational KGs is [22]. The authors state that the study's handling of numerical data is partially based on [31] which is the current state of the art in literal-aware KGE methods. So, we choose to extend their study and show that `ReaLitE` outperforms their method in node classification.

3 Problem Description

The formal representation of a KG with numeric literals can be expressed as a set of triples. Each triple consists of a head term (h), a relation or attribute (r or a), and a tail term or literal value (t or v). Officially, a KG can be defined as $G = \{(h, r, t) \mid h, t \in E, r \in R\} \cup \{(e, a, v) \mid e \in E, a \in A, v \in \mathbb{R}\}$. Here, E represents the set of entities, R is the set of relations, and A is the set of data attributes.

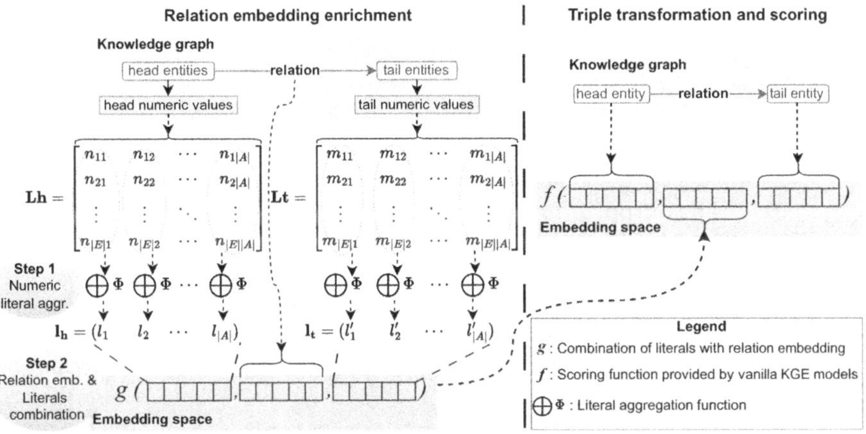

Fig. 2. ReaLitE's pipeline for relation embedding enrichment (left) given a triple to be scored (right). ReaLitE aggregates the head and tail numeric literals (Step 1) and combines the aggregated literals with the relation embedding (Step 2).

Link prediction is a task that predicts the plausibility of a triple (h, r, t) given a graph G. In practice, this task is modeled as a ranking problem where for an input pair of (h, r) the desired output is all entities $e_i \in E$ of G ranked based on the probability P of (h, r, e_i) being a true triple. The ranks are computed using a scoring function $f(h, r, t)$ that produces scores proportional to P.

Another prominent downstream task is *node classification*, which is concerned with predicting the category of a given entity. Formally, given a graph G, and a set of entity labels Y, this task aims to find a function $f(e_i) \rightarrow Y_i$, for each entity $e_i \in E$. A common way of addressing this task is to represent each entity by a pre-trained KGE $emb(e_i)$, and train a classifier $emb(e_i) \rightarrow Y_i$.

4 Method: ReaLitE

Our approach, ReaLitE, involves a two-step procedure: (1) aggregating numeric literals on a per-relation basis and (2) integrating numeric literals into the relation embedding of vanilla KGE methods (Fig. 2). These steps are independent of the scoring function, allowing our approach to complement existing KGE models as shown in the experiments.

Let $\mathbf{L} \in \mathbb{R}^{|E| \times |A|}$ be a matrix of literals, where $|E|$ is the number of entities and $|A|$ is the number of numeric attributes. Each entry $\mathbf{L}_{i,k}$ contains the literal value of the k-th numeric attribute for the i-th entity if a triple with the i-th entity and the k-th numeric attribute exists in the KG, and zero otherwise. L is normalized using min-max normalization, as per the relevant literature [15].

For each relation, two matrices $\mathbf{Lh}, \mathbf{Lt} \in \mathbb{R}^{|E| \times |A|}$ are created, containing the rows of $\mathbf{L}$ corresponding to the relation's head and tail entities, respectively. The remaining rows are zeroed out in $\mathbf{Lh}, \mathbf{Lt}$. Then, an aggregation method is applied column-wise to $\mathbf{Lh}, \mathbf{Lt}$ to produce the vectors $\mathbf{l_h}, \mathbf{l_t} \in \mathbb{R}^{|A|}$. Depending on the configuration, this step uses either an aggregation type (*e.g.*, *mean*) or a learnable linear combination of 11 different aggregation types.

The final part of `ReaLitE`'s architecture involves the function $g : \mathbb{R}^{|A|} \times \mathbb{R}^{D_r} \times \mathbb{R}^{|A|} \rightarrow \mathbb{R}^{D_r}$, where D_r is the dimension of the relation embedding. The function g takes as input: (a) the relation's embedding r and (b) the two literal vectors $\mathbf{l_h}, \mathbf{l_t} \in \mathbb{R}^{|A|}$. The output of g is a vector of the same dimension as the relation embedding. The function g is thoroughly described later in this section.

The resulting vector from g is a literal-enhanced embedding vector, capable of substituting the original embedding vector of the relation within the scoring function. So, every relation embedding r_i is replaced with $\mathbf{r_i}^{lit} = g(\mathbf{l_h}, \mathbf{r_i}, \mathbf{l_t})$ in the scoring functions of respective model The remaining elements of the score functions remain unchanged.

Below we elaborate on the functionality of `ReaLitE` by consecutively explaining both steps of the approach.

4.1 Aggregation of Numeric Literals

To reduce the space complexity and number of trainable parameters of `ReaLitE`, we choose to convert the matrices $\mathbf{Lh}, \mathbf{Lt}$ to vectors. In the process, we consider it crucial to keep the number of columns intact, as it corresponds to the number of numeric attributes ($|A|$). This allows us to incorporate all the available numeric attributes and let the model decide which attributes are more useful than others during training. So, we choose to aggregate the values of each numeric attribute (i.e., each column of $\mathbf{Lh}$ and $\mathbf{Lt}$) to create vectors $\mathbf{l_h}, \mathbf{l_t}$ ('Step 1' of Fig. 2).

$$\mathbf{l_h} = \left(\bigoplus_{i=1}^{|E|} \Phi(\mathbf{Lh}_{i,1}) \cdots \bigoplus_{i=1}^{|E|} \Phi(\mathbf{Lh}_{i,|A|}) \right) \tag{1}$$

$$\mathbf{l_t} = \left(\bigoplus_{i=1}^{|E|} \Phi(\mathbf{Lt}_{i,1}) \cdots \bigoplus_{i=1}^{|E|} \Phi(\mathbf{Lt}_{i,|A|}) \right) \tag{2}$$

We treat the aggregation method denoted by $\bigoplus \Phi$ as a tunable hyperparameter for `ReaLitE`. That is because depending on the skewness of the value distribution for each numeric attribute, the aggregation type that leads to better link prediction performance might vary. The choices for $\bigoplus \Phi$ are commonly used measures of central tendency $\{mean, median, mode\}$ and other descriptive statistics $\{min, max, sum, count\}$, and measures of dispersion $\{variance, std, IQR, range\}$. Additionally, we provide the option to use a learnable linear combination $\mathbf{y}$ of all previous aggregation types (Eq. 3).

$$\mathbf{y} = \sigma(\mathbf{U} \cdot \mathbf{W}_a + \mathbf{b}_1) \tag{3}$$

where, $\forall \phi \in$\{*mean*, *median*, *mode*, *min*, *max*, *sum*, *count*, *variance*, *std*, *IQR*, *range*\}, $\exists$ row $\mathbf{u}$ of the matrix $\mathbf{U}^\intercal$ where:

$$\mathbf{u} = \left(\bigoplus_{i=1}^{|E|} \phi(\mathbf{L}_{i,1}) \cdots \bigoplus_{i=1}^{|E|} \phi(\mathbf{L}_{i,|A|}) \right) \tag{4}$$

The input of linear function $\mathbf{y} \in \mathbb{R}^{|A| \times 1}$ is $\mathbf{U} \in \mathbb{R}^{|A| \times 11}$. The aggregation function is denoted by $\bigoplus \phi$. The learnable parameters are $\mathbf{W}_a \in \mathbb{R}^{11 \times 1}$ and $\mathbf{b} \in \mathbb{R}$, while σ is the sigmoid function. We apply the function in Eq. 3 separately on $\mathbf{Lh}$ and $\mathbf{Lt}$ by substituting $\mathbf{L}$ for each of them in Eq. 4.

4.2 Fusing Literals with Relation Embeddings

The aggregated head and tail numeric literals (vectors $\mathbf{l_h}$ [Eq. 1] and $\mathbf{l_t}$ [Eq. 2]) are combined with the relation embedding via the learnable function g ('Step 2' of Fig. 2). We provide two versions of this function: Linear (Eq. 5) and Gated (Eq. 6). While the former uses a linear transformation, the latter employs a gated mechanism as done in [15], because the authors claim this mechanism lets g use or ignore the numeric information as needed.

$$g_{lin} = \mathbf{W}_r^T [\mathbf{l_h}, \mathbf{r}, \mathbf{l_t}] + \mathbf{b}_2 \tag{5}$$

$$g_{gated} = \mathbf{z} \odot \mathbf{h} + (1 - \mathbf{z}) \odot \mathbf{r} \tag{6}$$

$$\text{where: } \mathbf{z} = \sigma(\mathbf{W}_{zlh}^T \mathbf{l_h} + \mathbf{W}_{zr}^T \mathbf{r} + \mathbf{W}_{zlt}^T \mathbf{l_t} + \mathbf{b}_3)$$
$$\mathbf{h} = h'(\mathbf{W}_r^T [\mathbf{l_h}, r, \mathbf{l_t}])$$

Except for the relation embedding $\mathbf{r} \in \mathbb{R}^{D_r}$, the trainable parameters are $\mathbf{W}_r \in \mathbb{R}^{|A|+D_r+|A| \times D_r}$, $\mathbf{W}_{zr} \in \mathbb{R}^{D_r \times D_r}$, $\mathbf{W}_{zlh}, \mathbf{W}_{zlt} \in \mathbb{R}^{|A| \times D_r}$ and $\mathbf{b}_2, \mathbf{b}_3 \in \mathbb{R}^{D_r}$. With $\odot$ denoting element-wise multiplication, σ as the sigmoid function, and h' as an element-wise nonlinear function (*e.g.*, tanh).

4.3 Model Complexity

Computational Complexity. `ReaLitE` is based on a pipeline that processes the dataset's numeric literals. In particular, the dataset's triples (let N_t be their number) are traversed once (complexity: $O(N_t)$) to fill the matrices $\mathbf{Lh}$ and $\mathbf{Lt}$. Then, the values of each attribute are aggregated for every relation (complexity: $O(|E| \cdot |A| \cdot |R|)$). So, the overall computational complexity of the preprocessing step is $O(N_t + |E| \cdot |A| \cdot |R|)$. During training, the computational overhead introduced by the function g is one matrix multiplication and one vector addition in the case of g_{lin}, and three matrix multiplications and one vector addition in the case of g_{gated}. There is an additional overhead of one matrix multiplication and one vector addition if the learnable function $\mathbf{y}$ is used to aggregate the literals.

Trainable Parameters. `ReaLitE` adds some trainable parameters and thus more complexity to the vanilla KGE method (Table 1). The exact increase of parameters is determined by the function g and the literal aggregation function $\mathbf{\Phi}$. As for the former choice, the additional number of parameters corresponds to the dimensions of the weight matrices $\mathbf{W}$ involved in either Eq. 5 or Eq. 6. Regarding the aggregation method selection, the only case of increased complexity is using the learnable function $\mathbf{y}$. However, since the number of aggregation

Table 1. Number of trainable parameters for LiteralE and for variations of `ReaLitE`. B represents the number of trainable parameters of base models (*e.g.*, TransE). The parentheses in (+12) indicate that this number is added only if the learnable function $\mathbf{y}$ is used to aggregate literals.

Model	Parameters
LiteralE	$B + 2D_e^2 + 2\|A\|D_e + D_e$
`ReaLitE`-g_{lin}	$B + D_r^2 + 2\|A\|D_r + D_r(+12)$
`ReaLitE`-g_{gated}	$B + 2D_r^2 + 4\|A\|D_r + D_r(+12)$

types is fixed at 11 the overhead is a fixed number of parameters determined by the dimensions of $\mathbf{W}_a$ and $\mathbf{b}_1$. The added parameters are compared to LiteralE, the latest literal-aware KGE method that alters the learned embeddings of vanilla KGE methods. The additional complexity over LiteralE is constant, thus negligible.

Space Complexity. During literal preprocessing, for each relation in the dataset, two matrices ($\mathbf{Lh}, \mathbf{Lt}$) are created which leads to storing $2 \cdot |E| \cdot |A| \cdot |R|$ numeric values (complexity: $O(|E| \cdot |A| \cdot |R|)$). These matrices are reduced to vectors $\mathbf{l}_h, \mathbf{l}_t$ by aggregating the values across one dimension, during the literal preprocessing phase. This means that after preprocessing and throughout model training and testing, the number of stored values is $2 \cdot |A| \cdot |R|$.

5 Experimental Setup

This section outlines the experimental setup used to evaluate the proposed `ReaLitE` model. We first describe the datasets used for link prediction and node classification, followed by the models employed for each task. Finally, we present the training and evaluation procedures for both tasks.

5.1 Datasets

Link Prediction Datasets. FB15k-237 [28] is a subset of the Freebase KG [6], focusing on diverse domains such as movies, actors, awards, sports, and sports teams. Originally derived from the FB15k benchmark, FB15k-237 underwent modifications by eliminating inverse relations to enhance the challenge of link prediction. This adjustment aimed to eliminate easily obtainable triples by reversing training triples. The dataset initially lacked numeric literals. It was enriched by [15] using a SPARQL endpoint for Freebase. The data split used in this paper aligns with the latest literal-aware KGE [15,31]. **YAGO15k** [12], stands as a subset derived from YAGO [20], which functions as a general-domain knowledge graph. Initially extended with numeric literals by [18], YAGO15k holds a notable advantage in terms of valid numeric triples compared to FB15k-237. We split the dataset in the same way as [31]. The datasets' metadata is shown in Table 2.

Table 2. Datasets used for link prediction

	Entities	Relations	Triples	Attributes	Literals
FB15k-237	14,541	237	310,116	116	29,220
YAGO15k	15,136	32	138,056	7	23,520

Table 3. Datasets used for node classification

	dmgfull	dmg777k	mdgenre
Entities	593,291	288,379	1,001,791
Relations	62	60	154
Triples	1,850,451	777,124	1,252,247
Numbers	88,168	10,706	14,352
Dates	–	–	113,463
Classes	14	5	12
Nodes	842,550	341,270	349,344

Node Classification Datasets. For this downstream task, to provide a direct comparison with recent relevant literature [22] we use some multimodal datasets of the kgbench collection [4]. Specifically, we use dmgfull (monuments in the Netherlands), dmg777k (a subset of dmgfull), and mdgenre (movie data extracted from Wikidata). Their statistics are shown in Table 3. Although these datasets originally contained data of multiple modalities (e.g., images), we only kept the numbers and dates (converted to the decimal format YYYY.MMDD).

5.2 Models

Link Prediction Models. Similar to literal-aware KGE model [31], we combine `ReaLitE` with five models: TransE [7], DistMult [34], ComplEx [29], RotatE [27], and TuckER [3]. Note that ConvE [10] requires the creation of inverse triples. Since we chose a training strategy that does not require the creation of inverse triples, we do not provide results for ConvE. We conduct experiments to determine the best configuration (including the literal aggregation strategy) for each dataset and KGE model. We report the best results in terms of MRR. We compare our `ReaLitE` to basic KGE models as well as the state-of-the-art models that incorporate numerical values, i.e. KBLN [13], LiteralE [15], and KGA [31].

Node Classification Models. We combine `ReaLitE` with TransE and DistMult to compare our results with those reported in [22]. Additionally, we utilize the literal preprocessing methods presented in the referenced paper to train and evaluate TransE and DistMult. Specifically, we use two preprocessing strategies: KL-REL+LOF, which removes outliers and bins values based on relation similarity, and DATBIN, which converts dates to timestamps for binning [22]. For simplicity, we refer to this combination of preprocessing strategies as MKGA, which is the acronym of that preprocessing framework. As for the classifiers, we use a Support Vector Machine (SVM) [8] and k-Nearest Neighbors (KNN) [11].

5.3 Training

Link Prediction Training. Since `ReaLitE` focuses on enriching the relations' embeddings, we use a training method that preserves the relation types in the training dataset without adding synthetic inverse relations, unlike Local Closed World Assumption (LCWA) [2]. At the same time, due to the small size of the datasets, it is reasonable to perturb the head or tail part of the training triple using all the entities of the dataset. Another requirement is explicitly training the model to predict a triple's head and tail. So, we choose to train `ReaLitE` by scoring heads and tails at once for every true triple, based on the concept initially proposed in [16] as 'instantaneous multi-class log-loss' (implemented in PyKEEN as 'symmetric LCWA'). The chosen loss function is Cross Entropy loss, which generally yields better results than other loss functions [24].

For a training triple (i, j, k), a normalized ground truth tensor y, and a score tensor s, the loss ℓ_{ijk} is given by Eq. 7. The loss consists of two components: the loss for predictions after perturbing the tail entity k (Eq. 8) and the loss for predictions after perturbing the head entity i (Eq. 9).

$$\ell_{ijk} = \ell_{ijk}^{(1)} + \ell_{ijk}^{(3)} \tag{7}$$

$$\ell_{ijk}^{(1)} = -\sum_{k'} y_{ijk'} \log\left(\frac{\exp(s_{ijk'})}{\sum_{k'} \exp(s_{ijk'})}\right) \tag{8}$$

$$\ell_{ijk}^{(3)} = -\sum_{i'} y_{i'jk} \log\left(\frac{\exp(s_{i'jk})}{\sum_{i'} \exp(s_{i'jk})}\right) \tag{9}$$

Node Classification Training. The KGE model trained on link prediction provides embeddings for the KG's entities. We treat these embeddings as features for the entities used to train the classification models. To train the SVM classifier, we use the hinge loss, which is designed to maximize the margin between different classes. The kNN classifier is a non-parametric algorithm that does not involve explicit training with a loss function.

5.4 Evaluation Settings

Overall Link Prediction. For each test triple, the trained model calculates a score after the head entity is replaced with all the dataset entities. This process is repeated for the tail entity. Then, traditional rank-based metrics (i.e., MRR, Hits@1, Hits@10) are calculated based on the position of the test triple in the sorted list of scored triples.

Relation-Focused Link Prediction. Here the test set is split into groups based on their relation type(s), and then the MRR is calculated. The splitting is done in two ways: (1) based on the frequency of the relations in the training set (Table 5), to see the performance on long-tail relations and (2) based on the correlation of numerical attributes between the head and tail entities for each relation in the training set (Table 6), to see if such correlation makes a difference. For (1), we use 2817 (2.55% of training triples) as a frequency threshold for the division of relations. For (2), we calculate the pairwise Pearson's correlation coefficient (denoted by '*corr. coef.*') between head and tail attributes and set a threshold of 0.2 for the division of relations.

Node Classification. All the test entities are labeled using the trained classification model, and then the numbers of true positives, false positives, and false negatives are computed. Afterward, these numbers are used to calculate the micro-F1 score (i.e., the harmonic mean of precision and recall), which is commonly used as a performance measure in classification tasks. For the micro-F1 score, all test samples are treated equally regardless of their class.

6 Results

This section presents the experimental results of combining `ReaLitE` with various vanilla KGE models. We first evaluate the performance of `ReaLitE` on the link prediction task, comparing it against baseline models and state-of-the-art approaches. Subsequently, we investigate the effectiveness of `ReaLitE` on the node classification task, analyzing its impact on downstream applications.

6.1 Link Prediction Results

Overall Link Prediction. Under the standard setting, we observe (Table 4) that `ReaLitE` outperforms the baselines on most backbone KGEs. For FB15k-237, the maximum percentage increase in MRR when using `ReaLitE` over using the vanilla KGE models is 17% (for ComplEx), and the minimum is 6% (for RotatE). For YAGO15k, the respective increases are 10% (for ComplEx) and 7% (for the rest of the base models). Furthermore, the best results for YAGO15k regarding MRR and Hits@1 are obtained with TransE+`ReaLitE` . Since YAGO15k has higher quality numeric literals than FB15k237, we consider the results on YAGO15k to be more indicative. Enhanced by `ReaLitE`, the performance of almost all vanilla KGE models is improved (except for TuckER on FB15k-237). In contrast, LiteralE yields lower metric values than vanilla TransE and TuckER for both datasets; and ComplEx for one dataset. The performance gain by `ReaLitE` is the largest for ComplEx, in which each part (*i.e.*, imaginary and real) of the relation embedding is separately enhanced with numeric literals. This characteristic of ComplEx means that embedding-enriching such as `ReaLitE` and LiteralE are applied twice for this model. Combined with the above observation, this shows the advantage of using the relation-centric `ReaLitE` versus other methods such as the entity-centric LiteralE.

Table 4. Link prediction results. In this context, ReaLitE uses g_{lin} as it yielded better results than g_{gated}. The baseline results are sourced from the current state-of-the-art paper, with the best results per metric and base KGE model in bold.

Model	FB15k-237			YAGO15k		
	MRR	H@1	H@10	MRR	H@1	H@10
TuckER	0.354	0.263	0.536	0.433	0.360	0.571
+LiteralE	0.353	0.262	0.536	0.421	0.348	0.564
+KBLN	0.345	0.253	0.530	0.420	0.349	0.556
+KGA	**0.357**	**0.265**	**0.540**	0.454	0.380	0.592
+ReaLitE	0.347	0.254	0.533	**0.463**	**0.387**	**0.608**
TransE	0.315	0.217	0.508	0.459	0.376	0.615
+LiteralE	0.315	0.218	0.504	0.458	0.376	0.612
+KBLN	0.308	0.210	0.496	0.466	0.382	0.621
+KGA	0.321	0.223	0.516	0.470	0.387	**0.623**
+ReaLitE	**0.335**	**0.242**	**0.520**	**0.491**	**0.422**	0.620
DistMult	0.295	0.212	0.463	0.457	0.389	0.585
+LiteralE	0.309	0.223	0.481	0.462	0.396	0.587
+KBLN	0.302	0.220	0.470	0.449	0.377	0.581
+KGA	0.322	0.233	0.502	0.472	0.402	0.606
+ReaLitE	**0.340**	**0.248**	**0.525**	**0.489**	**0.419**	**0.622**
ComplEx	0.288	0.205	0.455	0.441	0.370	0.572
+LiteralE	0.295	0.212	0.462	0.443	0.375	0.570
+KBLN	0.293	0.213	0.451	0.451	0.380	0.583
+KGA	0.305	0.219	0.478	0.453	0.380	0.591
+ReaLitE	**0.338**	**0.244**	**0.528**	**0.487**	**0.417**	**0.621**
RotatE	0.324	0.232	0.506	0.451	0.370	0.605
+LiteralE	0.329	0.237	0.512	0.475	0.400	0.619
+KBLN	0.314	0.222	0.500	0.469	0.393	0.613
+KGA	0.335	0.242	0.521	0.473	0.392	**0.626**
+ReaLitE	**0.344**	**0.249**	**0.535**	**0.483**	**0.409**	0.624

Relation-Focused Link Prediction. To examine the performance of literal-aware models for the various relation types, we perform link prediction evaluation after filtering the test set of YAGO15k depending on specific relation groups.

In Table 5, we divide the test triples based on the frequency of their relations in the training set. The frequency is the number of training triples that contain the relation. We observe that ReaLitE brings performance increase compared to the baselines, especially significant in group "Long-tail" (except TuckER). This indicates that ReaLitE is better at modeling the long-tail relations, which constitute the majority in KGs. We estimate the exception of TuckER+KGA is because the synthetic relations and entities created by KGA lead to a KG structure that benefits TuckER more than other KGE models, as discussed in [5].

Table 5. MRR for the filtered test set of YAGO15k based on relation types grouped by their frequency. "Long-tail" refers to the test triples that contain relations present in $\leq 2.55\%$ of the training triples. "Frequent" refers to the remaining test triples. "All triples" refers to all test triples. The best results per base model per triple group are in bold. The percentages indicate the MRR increase for the top literal-extended version of a base model versus the second-best per triple group.

Model		Frequent	Long-tail	All triples
TuckER	+LiteralE	0.464	0.282	0.441
	+KGA	0.480	**0.288** (+5%)	0.455
	+ReaLitE	**0.491** (+2%)	0.274	**0.463** (+2%)
TransE	+LiteralE	0.489	0.233	0.456
	+KGA	0.502	0.237	0.469
	+ReaLitE	**0.525** (+5%)	**0.262** (+11%)	**0.491** (+5%)
DistMult	+LiteralE	0.492	0.250	0.461
	+KGA	0.501	0.258	0.470
	+ReaLitE	**0.519** (+4%)	**0.284** (+10%)	**0.489** (+4%)
ComplEx	+LiteralE	0.470	0.253	0.442
	+KGA	0.479	0.256	0.451
	+ReaLitE	**0.514** (+7%)	**0.300** (+17%)	**0.487** (+8%)
RotatE	+KGA	0.494	0.298	0.469
	+ReaLitE	**0.510** (+3%)	**0.302** (+1%)	**0.483** (+3%)

In Table 6, we divide the test triples into groups based on the correlation between the head and tail attributes of their relation in the training set. We observe that ReaLitE brings performance gain for triples with less correlated literals ("Less corr. lit."), this indicates ReaLitE is better for capturing nuanced correlations. In addition, for the triples with more correlated literals ("Corr. lit."), ReaLitE can significantly boost the performance for some methods (especially TransE enjoys an increase of 47% and ComplEx 28%). This means ReaLitE with the mechanism of infusing literals in relations can be very beneficial for some KGE methods. While for TuckER and RotatE, we can see that KGA leads to higher MRR than ReaLitE in group "Corr. lit.". We attribute these models' better performance to the additional relations and entities provided by KGA, which might not suit more complex scenarios as discussed in [5].

Analysis of Literal Aggregation Methods. For the link prediction training, during ReaLitE's hyperparameter optimization (HPO), the choices for literal aggregation were $\{mean, median, mode, min, \mathbf{y}\}$. We did not include the rest of the provided aggregation types to limit the HPO search space. Besides, the provided learnable function $\mathbf{y}$ combines all supported aggregation types. Based on Table 7, the choice of aggregation function that leads to optimal results generally depends on both the dataset and the model. For TransE, RotatE, and TuckER, the aggregation function is the same in both YAGO15k and FB15k-

Table 6. MRR for the filtered test set of YAGO15k based on relation types grouped by literal correlation. "Corr. lit." refers to the test triples where the head and tail attributes are more correlated (the relation has at least one pair of head and tail attributes with $|corr.\ coef.| \geq 0.2$ in the training set). "Less corr. lit." refers to the remaining test triples where the head and tail attributes are less correlated. "All triples" refers to all test triples. The best results per base model per triple group are in bold. The percentages indicate the MRR increase for the top literal-extended version of a base model versus the second-best per triple group.

Model		Less corr. lit.	Corr. lit.	All triples
TuckER	+LiteralE	0.465	0.272	0.441
	+KGA	0.477	+10% **0.298**	0.455
	+ReaLitE	**0.488** (+2%)	0.289	**0.463** (+2%)
TransE	+LiteralE	0.500	0.146	0.456
	+KGA	0.510	0.176	0.469
	+ReaLitE	**0.524** (+3%)	**0.259** (+47%)	**0.491** (+5%)
DistMult	+LiteralE	0.494	0.227	0.461
	+KGA	0.501	0.245	0.470
	+ReaLitE	**0.517** (+3%)	**0.287** (+17%)	**0.489** (+4%)
ComplEx	+LiteralE	0.477	0.193	0.442
	+KGA	0.479	0.245	0.451
	+ReaLitE	**0.511** (+7%)	**0.314** (+28%)	**0.487** (+8%)
RotatE	+KGA	0.488	+12% **0.333**	0.469
	+ReaLitE	**0.509** (+4%)	0.296	**0.483** (+3%)

237. In addition, *mode* is found in 50% of the configurations that yielded the best results per dataset per model, while *mean* and *min* in 20%. The least used aggregation function was the learnable function **y**, which is present in 10% of the configurations. *median* was the only function among the HPO choices that was not selected as part of any configuration.

Table 7. Aggregation function used in the best hyperparameter configurations of ReaLitE. Function **y** is a learnable combination of all supported aggregation types (Eq. 3).

	TransE	DistMult	ComplEx	RotatE	TuckER
FB15k-237	*mode*	**y**	*mode*	*mode*	*mean*
YAGO15k	*mode*	*min*	*min*	*mode*	*mean*

6.2 Node Classification Results

In Table 8, we present an evaluation of KGE models on the node classification datasets. It is apparent that for every dataset, the best result is obtained using

Table 8. Node classification results – Micro-F1 score. In this context, ReaLitE uses g_{gated} as it yielded better results than g_{lin}. The best results per dataset are underlined. The best results per base KGE model per classifier are in bold.

Model	dmg777k		dmgfull		mdgenre	
	KNN	SVM	KNN	SVM	KNN	SVM
TransE	0.506	0.528	**0.649**	**0.662**	**0.634**	0.646
+MKGA	0.439	0.472	0.583	0.573	0.607	0.606
+ReaLitE	**0.609**	**0.638**	0.597	0.605	0.632	**0.662**
DistMult	0.548	0.542	0.619	0.658	0.605	0.622
+MKGA	0.472	0.495	0.605	0.647	0.630	0.640
+ReaLitE	**0.582**	<u>**0.642**</u>	**0.639**	<u>**0.682**</u>	**0.659**	<u>**0.674**</u>

ReaLitE. In particular, for dmg777k, there is a vast increase in F1 score when using ReaLitE compared to the vanilla KGE models. On average, for all combinations of the KGE model and classification model on dmg777k, this increase is 16%. We can observe that when using TransE with KNN or SVM on dmgfull and with KNN on mdgenre, ReaLitE seems to perform worse than the vanilla TransE. However, even in those cases, ReaLitE outperforms MKGA, the literal-aware baseline. In addition, across all used datasets and KGE base models, ReaLitE yields better results when paired with the SVM classifier than when paired with KNN.

7 Conclusion

This paper introduces ReaLitE, a novel approach to enhance Knowledge Graph Embedding (KGE) with literal information. ReaLitE enhances traditional KGEs by incorporating information from the relevant literals of the head and tail entities into the relation embedding. We present variants of ReaLitE with different methods for aggregating numeric literals, including an automated learning approach to combine different aggregations. The experiments demonstrate the superiority of ReaLitE over state-of-the-art methods for literal-aware embeddings, achieving comparable or improved performance in downstream tasks. Notably, detailed analyses reveal that ReaLitE excels in scenarios involving numeric literals with higher correlation and long-tail relations, showcasing its versatility and efficacy in capturing nuanced relationships within knowledge graphs.

Acknowledgement. The work was partially supported by EU projects Graph Massivizer (GA 101093202), Dome 4.0 (GA 953163), SMARTY (GA 101140087), and enRichMyData (GA 101070284).

References

1. Abboud, R., Ceylan, İ.İ.: Node classification meets link prediction on knowledge graphs (2021). https://doi.org/10.48550/arXiv.2106.07297
2. Ali, M., et al.: Bringing light into the dark: a large-scale evaluation of knowledge graph embedding models under a unified framework. IEEE Trans. Pattern Anal. Mach. Intell. **44**(12), 8825–8845 (2022). https://doi.org/10.1109/TPAMI.2021.3124805
3. Balazevic, I., Allen, C., Hospedales, T.: TuckER: tensor factorization for knowledge graph completion. In: Inui, K., Jiang, J., Ng, V., Wan, X. (eds.) Proceedings of the 2019 Conference on Empirical Methods in Natural Language Processing and the 9th International Joint Conference on Natural Language Processing (EMNLP-IJCNLP), pp. 5185–5194. Association for Computational Linguistics, Hong Kong (2019). https://doi.org/10.18653/v1/D19-1522
4. Bloem, P., Wilcke, X., van Berkel, L., de Boer, V.: Kgbench: a collection of knowledge graph datasets for evaluating relational and multimodal machine learning. In: Eighteenth Extended Semantic Web Conference - Resources Track (2020)
5. Blum, M., Ell, B., Ill, H., Cimiano, P.: Numerical literals in link prediction: a critical examination of models and datasets. In: Proceedings of the 23rd International Semantic Web Conference (2024)
6. Bollacker, K., Evans, C., Paritosh, P., Sturge, T., Taylor, J.: Freebase: a collaboratively created graph database for structuring human knowledge. In: Proceedings of the 2008 ACM SIGMOD International Conference on Management of Data, pp. 1247–1250. ACM, Vancouver (2008). https://doi.org/10.1145/1376616.1376746
7. Bordes, A., Usunier, N., Garcia-Duran, A., Weston, J., Yakhnenko, O.: Translating embeddings for modeling multi-relational data. In: Burges, C., Bottou, L., Welling, M., Ghahramani, Z., Weinberger, K. (eds.) Advances in Neural Information Processing Systems, vol. 26. Curran Associates, Inc. (2013)
8. Boser, B.E., Guyon, I.M., Vapnik, V.N.: A training algorithm for optimal margin classifiers. In: Proceedings of the Fifth Annual Workshop on Computational Learning Theory, COLT 1992, pp. 144–152. Association for Computing Machinery, New York (1992). https://doi.org/10.1145/130385.130401
9. Busbridge, D., Sherburn, D., Cavallo, P., Hammerla, N.Y.: Relational graph attention networks (2019). https://doi.org/10.48550/arXiv.1904.05811
10. Dettmers, T., Minervini, P., Stenetorp, P., Riedel, S.: Convolutional 2D knowledge graph embeddings. In: Proceedings of the Thirty-Second AAAI Conference on Artificial Intelligence and Thirtieth Innovative Applications of Artificial Intelligence Conference and Eighth AAAI Symposium on Educational Advances in Artificial Intelligence, AAAI 2018/IAAI 2018/EAAI 2018, pp. 1811–1818. AAAI Press, New Orleans (2018)
11. Fix, E., Hodges, J.L.: Discriminatory analysis - nonparametric discrimination: consistency properties. Technical report, USAF School of Aviation Medicine, Randolph Field, Texas (1951)
12. García-Durán, A., Dumančić, S., Niepert, M.: Learning sequence encoders for temporal knowledge graph completion. In: Riloff, E., Chiang, D., Hockenmaier, J., Tsujii, J. (eds.) Proceedings of the 2018 Conference on Empirical Methods in Natural Language Processing, pp. 4816–4821. Association for Computational Linguistics, Brussels (2018). https://doi.org/10.18653/v1/D18-1516
13. Garcia-Duran, A., Niepert, M.: KBLRN: end-to-end learning of knowledge base representations with latent, relational, and numerical features (2018). https://doi.org/10.48550/arXiv.1709.04676

14. Gesese, G.A., Biswas, R., Alam, M., Sack, H.: A survey on knowledge graph embeddings with literals: which model links better literally? (2020). https://doi.org/10.48550/arXiv.1910.12507
15. Kristiadi, A., Khan, M.A., Lukovnikov, D., Lehmann, J., Fischer, A.: Incorporating literals into knowledge graph embeddings. In: Ghidini, C., et al. (eds.) ISWC 2019. LNCS, vol. 11778, pp. 347–363. Springer, Cham (2019). https://doi.org/10.1007/978-3-030-30793-6_20
16. Lacroix, T., Usunier, N., Obozinski, G.: Canonical tensor decomposition for knowledge base completion. In: Proceedings of the 35th International Conference on Machine Learning, pp. 2863–2872. PMLR (2018)
17. Liu, J., et al.: Beyond entities: a large-scale multi-modal knowledge graph with triplet fact grounding. In: Proceedings of the AAAI Conference on Artificial Intelligence, vol. 38, pp. 18653–18661 (2024). https://doi.org/10.1609/aaai.v38i17.29828
18. Liu, Y., Li, H., Garcia-Duran, A., Niepert, M., Onoro-Rubio, D., Rosenblum, D.S.: MMKG: multi-modal knowledge graphs. In: Hitzler, P., et al. (eds.) ESWC 2019. LNCS, vol. 11503, pp. 459–474. Springer, Cham (2019). https://doi.org/10.1007/978-3-030-21348-0_30
19. Long, X., Zhuang, L., Li, A., Wei, J., Li, H., Wang, S.: KGDM: a diffusion model to capture multiple relation semantics for knowledge graph embedding. In: Proceedings of the AAAI Conference on Artificial Intelligence, vol. 38, pp. 8850–8858 (2024). https://doi.org/10.1609/aaai.v38i8.28732
20. Mahdisoltani, F., Biega, J., Suchanek, F.M.: YAGO3: a knowledge base from multilingual Wikipedias. In: CIDR (2013)
21. Portisch, J., Heist, N., Paulheim, H.: Knowledge graph embedding for data mining vs. knowledge graph embedding for link prediction – two sides of the same coin? Semant. Web **13**(3), 399–422 (2022). https://doi.org/10.3233/SW-212892
22. Preisner, P., Paulheim, H.: Universal preprocessing operators for embedding knowledge graphs with literals (2023). https://doi.org/10.48550/arXiv.2309.03023
23. Ristoski, P., Paulheim, H.: RDF2Vec: RDF graph embeddings for data mining. In: Groth, P., et al. (eds.) ISWC 2016. LNCS, vol. 9981, pp. 498–514. Springer, Cham (2016). https://doi.org/10.1007/978-3-319-46523-4_30
24. Ruffinelli, D., Broscheit, S., Gemulla, R.: You CAN teach an old dog new tricks! On training knowledge graph embeddings. In: International Conference on Learning Representations (2019)
25. Schlichtkrull, M., Kipf, T.N., Bloem, P., van den Berg, R., Titov, I., Welling, M.: Modeling relational data with graph convolutional networks. In: Gangemi, A., et al. (eds.) ESWC 2018. LNCS, vol. 10843, pp. 593–607. Springer, Cham (2018). https://doi.org/10.1007/978-3-319-93417-4_38
26. Shang, B., Zhao, Y., Liu, J., Wang, D.: LAFA: multimodal knowledge graph completion with link aware fusion and aggregation. In: Proceedings of the AAAI Conference on Artificial Intelligence, vol. 38, pp. 8957–8965 (2024). https://doi.org/10.1609/aaai.v38i8.28744
27. Sun, Z., Deng, Z.H., Nie, J.Y., Tang, J.: RotatE: knowledge graph embedding by relational rotation in complex space (2019). https://doi.org/10.48550/arXiv.1902.10197
28. Toutanova, K., Chen, D.: Observed versus latent features for knowledge base and text inference. In: Allauzen, A., Grefenstette, E., Hermann, K.M., Larochelle, H., Yih, S.W.T. (eds.) Proceedings of the 3rd Workshop on Continuous Vector Space Models and Their Compositionality, pp. 57–66. Association for Computational Linguistics, Beijing (2015). https://doi.org/10.18653/v1/W15-4007

29. Trouillon, T., Welbl, J., Riedel, S., Gaussier, E., Bouchard, G.: Complex embeddings for simple link prediction. In: Proceedings of The 33rd International Conference on Machine Learning, pp. 2071–2080. PMLR (2016)
30. Vrandečić, D., Krötzsch, M.: Wikidata: a free collaborative knowledgebase. Commun. ACM **57**(10), 78–85 (2014). https://doi.org/10.1145/2629489
31. Wang, J., Ilievski, F., Szekely, P., Yao, K.T.: Augmenting knowledge graphs for better link prediction (2022). https://doi.org/10.48550/arXiv.2203.13965
32. Wilcke, W.X., Bloem, P., de Boer, V., van t Veer, R.H., van Harmelen, F.A.H.: End-to-end entity classification on multimodal knowledge graphs (2020). https://doi.org/10.48550/arXiv.2003.12383
33. Xu, D., Ruan, C., Korpeoglu, E., Kumar, S., Achan, K.: Product knowledge graph embedding for E-commerce. In: Proceedings of the 13th International Conference on Web Search and Data Mining, WSDM 2020, pp. 672–680. Association for Computing Machinery, New York (2020). https://doi.org/10.1145/3336191.3371778
34. Yang, B., Yih, W.t., He, X., Gao, J., Deng, L.: Embedding entities and relations for learning and inference in knowledge bases. CoRR (2014)
35. Zhang, Y., Li, B., Gao, H., Ji, Y., Yang, H., Wang, M.: Fine-grained evaluation of knowledge graph embedding models in downstream tasks. In: Wang, X., Zhang, R., Lee, Y.-K., Sun, L., Moon, Y.-S. (eds.) APWeb-WAIM 2020. LNCS, vol. 12317, pp. 242–256. Springer, Cham (2020). https://doi.org/10.1007/978-3-030-60259-8_19

Evaluating Approximate Nearest Neighbour Search Systems on Knowledge Graph Embeddings

Gaurav Pandit[1], Michael Röder[2(✉)], and Axel-Cyrille Ngonga Ngomo[2]

[1] Paderborn University, Paderborn, Germany
[2] Data Science Group (DICE), Heinz Nixdorf Institute, Paderborn University, Paderborn, Germany
{michael.roeder,axel.ngonga}@uni-paderborn.de

Abstract. Knowledge Graph Embeddings are often used to bridge symbolic representations of Knowledge Graphs and the sub-symbolic representations that modern Machine Learning algorithms operate on. While embedding models provide users with a mapping from symbolic entities and relations to their sub-symbolic representations, a mapping in the opposite direction typically relies on a nearest neighbor search. Due to the computational complexity of this task, a plethora of approaches for approximate nearest neighbor search have been developed. The majority of these approaches outperform the default brute-force-based approach while providing a high recall. However, previous evaluations of these approaches focused on image data and word embeddings but did not consider Knowledge Graph Embeddings. We close this gap by carrying out a detailed comparison of 22 Approximate Nearest Neighbor Search systems on 16 datasets. In contrast to the state of the art, we fine-tune each approach in each experiment by using Bayesian optimization to ensure the fairness of our experiments. Our results suggest that the overall performance of approaches with respect to runtime and recall is contingent upon the similarity measure used to compare embeddings. Our source code, datasets and results are available at https://github.com/MichaelRoeder/ann-benchmarks/tree/main.

Keywords: Knowledge Graph Embeddings · k-Nearest Neighbor Search · Approximate Nearest Neighbor Search

1 Introduction

Knoweldge Graph Embedding (KGEs) have a wide area of applications [8] ranging from link prediction [11] to Knowledge-Graph-based question answering [5,6] and recommendation systems [16,37]. In several of the aforementioned applications, one or several entities found in the embedded Knoweldge Graph (KG) have to be identified based on the similarity (or distance) of their embedding vector to a given query vector. A brute force comparison of the query vector to

E. Curry et al. (Eds.): ESWC 2025, LNCS 15718, pp. 59–76, 2025.
https://doi.org/10.1007/978-3-031-94575-5_4

all known embedding vectors is a simple, straight-forward solution. However, it comes with a linear time complexity in the number of entities in the input KG and in the number of dimensions of the embeddings. Consequently, brute-force approaches yield a high runtime if

1. there are many vectors in the embedding model and
2. the embedding vectors have a high dimensionality.

This is known as the k nearest neighbors (k-NN) search problem. In recent years, a plethora of algorithms has been proposed that try to approximate the solution for this search problem [2,26,34]. These algorithms are called approximate nearest neighbors search (ANNS) algorithms. The main idea behind ANNS approaches is to accept a small error in the result in exchange for a runtime or memory consumption reduction when compared to exact k-NN search algorithms [26]. Muja et al. [28] report that the choice of the optimal ANNS algorithm highly depends on factors including the data dimensionality and the size and structure of the dataset. Due to their wide applicability, ANNS algorithms have been compared on many different datasets including audio, image, video, text and word embedding datasets in the literature [2,26]. However, to the best of our knowledge, their performance on KGEs has never been studied so far and it is unclear whether previous results hold in this different application area.[1]

Our work tackles this research gap via two main contributions.

1. We provide the first comprehensive evaluation of ANNS algorithms on KGEs. Relying on an ANN benchmarking framework proposed by Aumüller et al. [2], we compare 22 ANNS systems on 16 datasets. These datasets are created using 2 KGs, 2 distance metrics and 4 KGEs models.
2. For the first time, we address the challenge of finding suitable parameters for ANNS systems by using Bayesian optimization. Through this superior approach for parameter tuning, we achieve a fairer comparison of the systems we evaluate than previous grid-search-based works [2].

In the following, we describe related work with a focus on the systems that we evaluate. Section 3 gives a formal problem definition of the problem we tackle and describes the extensions of the ANN benchmarking framework necessary for our experiments. Section 4 comprises a description of our experiments, including the used data, they key performance indicators and the gathered results. We discuss these results in Sect. 5 before we conclude the article in Sect. 6.

2 Related Work

A large number of exact and approximate k-NN search algorithms have been proposed [2,26,34]. In recent years, several articles have been published that

[1] KGEs tend to contain millions of entities but with less dimensions than the image embeddings used in the related work. At the same time, Alshahrani et al. [1] showed that different embedding algorithms use the embedding space in very different ways.

compare these algorithms with respect to their effectiveness and efficiency on different types of data. For example, Wen et al. [26] compare 8 algorithms on 20 datasets with a focus on the Euclidean distance. Their datasets comprise mainly image data but also include audio, video, text and synthetically generated data. Wang et al. [34] compare 15 algorithms on 8 datasets with a focus on graph-based ANNS algorithms. Their evaluation also relies on audio, image, text and video data. Aumüller et al. [2] compare 14 algorithms on 7 datasets and include the angular and Hamming distance. None of these works targets the effectiveness and efficiency of ANNS systems on KGEs.

Aumüller et al. [2] also propose an open-source ANN benchmarking framework that they used for their evaluations.[2] We use this framework as basis for our experiments since it provides the general workflow for a repeatable evaluation and comes with a large number of integrated ANNS systems. Consequently—in contrast to Wen et al. [26]—, we do not alter the algorithm implementations to reduce the impact of optimizations the developers may have integrated into their solutions. Instead, we evaluate ANNS systems that are ready for use. We also include several systems that implement the same base algorithm to see whether the implementation of an algorithm has an impact on our evaluation results.[3]

While there are many ANNS systems available, we ensure the practical feasibility of our study by focusing our evaluation on systems that

1. can handle floating point numbers,
2. support at least the one of the angular and Euclidean distance metrics,
3. have been integrated into the ANN benchmarking framework,
4. come with a working Docker image, and
5. are suggested for benchmarking by the authors of the ANN benchmarking framework.[4]

Table 1 lists the 22 chosen systems, which are briefly described in the following subsections.[5] Note that the list does neither contain hash-based nor neural algorithms. Systems of the former type rely on a hash algorithm like locality sensitive hashing [20,29] to create a list of possible neighbors while the latter are a set of approaches that utilize neural networks to improve existing search algorithms further [25,33], e.g., by learning a hash function [7]. We have to exclude these algorithms since none of them fulfills the criteria above at the time of our evaluation.

[2] The framework is available at https://github.com/erikbern/ann-benchmarks/.

[3] Note that some of the systems offer additional features, e.g., the support of a variety of distance metrics, support for a distributed environment, or the usage of the computational power of GPUs [13]. We won't take these features into account for our experiments as they go beyond the focus of this paper.

[4] Some algorithms, like Annoy (https://github.com/spotify/annoy) have been marked as "disabled" since they have a successor algorithm (in this case, Voyager) that typically performs better.

[5] We also include the pyglass library into our experiments (https://github.com/zilliztech/pyglass). However, it does not achieve a Recall above 0 for any configuration that we try. Hence, we won't discuss it within this article.

Table 1. The ANN systems that we evaluate, their main algorithm, and links to their project page.

Type	System	Algorithm	Link
Graph-based	Elasticsearch	HNSW [27]	↗
	FaissHNSW	HNSW [27]	↗
	Hnswlib	HNSW [27]	↗
	LuceneKNN	HNSW [27]	↗
	N2	HNSW [27]	↗
	NGT-PANNG	PANNG [22]	↗
	NGT-QG	QG [23]	↗
	NN-Descent	NN-Descent [13]	↗
	OpenSearch k-NN	HNSW [27]	↗
	pg-embedding	HNSW [27]	↗
	PyNNDescent	NN-Descent [13]	↗
	QSGNGT	QG [23]	↗
	Vald	QG [23]	↗
	Vespa	HNSW [27]	↗
	Voyager	HNSW [27]	↗
	Weaviate	HNSW [27]	↗
Tree-based	FLANN [28]	kd-trees/k-means tree	↗
	MRPT	MRPT [19]	↗
	Scann	SOAR [32] + VQ [18]	↗
IVF-based	FaissIVF	IVF	↗
	FaissIVFPQ	IVF + PQ	↗
	TinyKNN	IVF + PQ	↗

2.1 Graph-Based Approaches

Graph-based ANNS approaches create a k-NN neighborhood graph, in which data points of the original dataset are the nodes that are connected to their k neighboring vertices with an edge [34]. However, creating an exact k-NN graph is expensive [21]. The NN-Descent algorithm [13] can be used to generate an approximate nearest neighbor graph (ANNG) and is based on the idea that two neighboring vertices are very likely to have the same neighbors. Given an initial ANNG, the algorithm traverses the neighbors of nodes and checks whether their neighbors are also within the k nearest neighbors of the current node. Based on this concept, the algorithm creates an updated version of the given graph. PyNNDescent and NN-Descent are implementations of the NN-Descent algorithm in Python an C++, respectively. Both implementations extend the original algorithm with an initialization based on random projection trees [9] to generate the initial version. Several algorithms of the NGT library are built upon a similar approach to create ANNG [21]. NGT-PANNG [22] is an implementation that includes the pruning of excessive edges in the ANNG. NGT-QG and QSGNGT are further extensions based on quantization graphs (QG), i.e., they make use of product quantization to store approximations of the given vectors that are shorter and, hence, faster to process [23]. Vald is a variant thereof with a focus on distributed setups. One approach to search in an ANNG with a given query vector is to start at one or several seed nodes and to compare the distances of this node and its neighbors to the query node. The closest node is chosen and the search continues until no closer neighbor can be found. Another approach is to further explore nodes within a search space that is narrowed down during the search process [22].

The Hierarchical Navigable Small World (HNSW) algorithm [27] separates the neighborhood graph into several hierarchically connected graphs. The highest layer only contains a small number of randomly chosen points with larger distances while the lower layers contain neighborhoods with small distances. A search in the graph is performed in the same way as before but includes steps from the top layers to the bottom layers. In each layer, the number of neighbors per vertex is limited to ensure a search in logarithmic time. The HNSW algorithm has been widely adopted and in this article, we look at Elasticsearch, FaissHNSW, Hnswlib, LuceneKNN, N2 [27], OpenSearch k-NN, pg-embedding, Vespa, Voyager, and Weaviate, that all rely on this algorithm.[6]

2.2 Tree-Based Approaches

Tree-based approaches separate the high-dimensional space into smaller sub spaces. The separated parts are then organized in a tree structure. MRPT [19] uses multiple random projection trees to separate the search space. The search results from the different trees are combined using voting. FLANN [28] is a system that is based on two different algorithms and decides which to use based on the given data. It can use the multiple randomized kd-trees [3,15] or the priority search k-means tree [17]. The latter is chosen in cases in which the vectors have many dimensions since the multiple randomized kd-trees are known to have a decreasing performance with high dimensional data [28]. Scann is based on SOAR [32], an approach that uses spill trees with multiple representations of the search space to reduce the probability that a nearest neighbor is missed due to the separation of the search space. However, in contrast to previous works, SOAR does not create the trees independently but uses an orthogonality-amplified residual loss. The goal is to optimize each tree to cover cases where other trees perform poorly. In addition, Scann uses a quantization method called vector quantization [18].

---

[6] According to the documentation, OpenSearch k-NN also provides a FaissIVF-based implementation for an ANN search index (https://opensearch.org/docs/2.6/search-plugins/knn/approximate-knn/). However, only the HNSW implementation seems to be integrated into the ANN Benchmarking framework.

2.3 Inverted-File-Based Approaches

Inverted-file-based approaches cluster the set of given vectors into inverted lists. When a search is performed on an inverted file (IVF) index, only vectors from a small number of lists are compared to the given query vector [24]. Faiss-IVF implements this algorithm while FaissIVFPQ and TinyKNN combine the IVF approach with product quantization (PQ).

3 Benchmarking Framework

We define the k nearest neighbor problem as follows: Given a set of vectors V in a space X and a distance function $\delta : X \times X \rightarrow \mathbb{R}^+$, find the set π that contains the k closest vectors to a given query vector u according to δ [2]. Formally, we define π as a k-subset of V such that $\forall v \in \pi\ \forall v' \in V \setminus \pi : \delta(u, v) \leq \delta(u, v')$. An algorithm that provides an approximation of π is called approximate nearest neighbor search (ANNS) algorithm. Within this article we focus on the angular and Euclidean distance functions for δ. The space X is defined by the used Knowledge Graph embedding model.

A dataset for the evaluation of an ANNS system comprises 1. a set of vectors that are to be indexed, 2. a set of query vectors, and 3. the ground truth which of the indexed vectors are the k-NNs of the single query vectors. Aumüller et al. [2] provide an ANN benchmarking framework that already provides the main functionalities that are necessary to carry out our experiments. Given a set of high-dimensional vectors, the framework generates a dataset by splitting the vectors into indexed and query vectors and generates the ground truth based on a brute force search. This search is also used as a baseline for comparison.

We propose two extensions of this framework. First, we use Bayesian optimization to choose parameters of the benchmarked systems. Second, we generate synthetic queries based on the indexed data instead of splitting the data into a train and a test split to account for the comparatively small size of the KGE models we use in our experiments.

3.1 Bayesian Optimization

Nearly all of the ANN systems integrated into the benchmarking framework have one or several parameters that can be used to adapt the system to the given data. The ANN benchmarking framework supports the configuration of a grid search to identify system configurations with different performance characteristics. However, grid search is known to be a disadvantageous option when optimizing parameters in comparison to a random search [4] or hyper parameter optimization [31]. We integrate the latter into the framework using the implementation of [30]. We use the upper confidence bound as acquisition function for the optimizer [35] and use the Recall value achieved by a system's configuration as quality metric for the optimization.

3.2 Query Generation

By default, the ANN benchmarking framework randomly selects 10,000 vectors from the given set of vectors and uses them as query vectors while the remaining vectors will be indexed by the benchmarked system. However, this would remove more than 68% of the vectors in our evaluation, thus leading to a small number of indexed vectors and correspondingly unrealistic evaluation results. At the same time, using several thousand queries supports the reliability of the evaluation results. To circumvent the split of the embedding model vectors, we implement the generation of synthetic queries based on the given embedding vectors. For each vector dimension, we calculate the mean and standard deviation of the pairwise distances of the given vectors. Then, we generate a synthetic query vector by randomly sampling a vector from the set of given vectors. After that, for each vector dimension we sample a value from a Gaussian distribution that is defined by the previously determined mean and standard deviation for this dimension. Then, we randomly choose to add or subtract the sampled value from the sampled vector to calculate the value for the new query vector.

4 Evaluation

4.1 Datasets

We evaluate the 22 systems from Table 1 on 16 datasets. These 16 datasets are created using the angular and the Euclidean distance measures on 8 KGE models. These models are created for the FB15k-237 and Yago3-10 datasets. For each of these two Knowledge Graphs, we reuse 4 Knowledge Graph embedding models that have been pre-calculated using the embedding algorithms DistMult [36], QuatE [38], OMult [10] and Keci [11]. We choose these embedding algorithms because they operate with different vector spaces. The first three algorithms are based on $\mathbb{R}$, $\mathbb{H}$, and $\mathbb{O}$, respectively. In contrast, Keci determines the appropriate Clifford algebra depending on the KG [11]. All embedding models use vectors comprising 128 floats. Each model for FB15k-237 and Yago3-10 contains one vector for each of the Knowledge Graphs' 14,541 and 123,182 entities, respectively. We generate the datasets using these entity vectors as described in Sect. 3. So in addition to the entity vectors, we generate 10,000 synthetic query vectors and create the ground truth.

4.2 Key Performance Indicators

The goal of our analysis is to evaluate the effectiveness and efficiency of the benchmarked systems. The effectiveness is measured as Recall, i.e., we divide the number of correctly retrieved neighbors by $k = 10$.[7] With respect to the system's efficiency, we put our focus on the number of queries that a system is

[7] Note that related work may call the measure Precision, Recall@k or Recall@10. The ANN benchmarking framework also reports the approximative recall. These results can be found in our result files but we will not discuss them in this article.

able to queries per second (QpS). However, we also report the size of the created index and the time the system needs to create it.

As suggested by the related work, we focus on those configurations that lead to the best performances of the benchmarked systems [2,14]. Let S be the set of the benchmarked systems listed in Table 1. Let Θ_s be the set of all possible configurations of the system $s \in S$. Let d be a dataset and M_d be the set of all results that we gathered for all systems on this dataset. A single result m_i in this set is represented as the tuple $m_i = (s_i, \theta_i, r_i, q_i, t_i, b_i)$, where s_i is a system, $\theta_i \in \Theta_s$ is the system's configuration that leads to the measured results r_i, q_i, t_i, and b_i that represent the Recall, QpS, the time needed to create the index and the index size, respectively. We define a comparison operator that identifies configurations that show a lower performance than other configurations with respect to either effectiveness or efficiency and are not worse with respect to the other. Let m_1 and m_2 be two results in M_d and let r_1, r_2, q_1, and q_2 be their recall and QpS scores. We define the comparator $\underset{r,q}{>}$ based on Recall and QpS as follows:

$$m_1 \underset{r,q}{>} m_2 \iff (r_1 > r_2 \wedge q_1 \geq q_2) \vee (r_1 \geq r_2 \wedge q_1 > q_2)\,. \tag{1}$$

Based on this operator, we define a subset of measurements $M_{d,r,q}$ that represent the best performance of the single systems with respect to Recall and QpS as follows:

$$M_{d,r,q} = \left\{ m_i \middle| m_i \in M_d \wedge \nexists m_j \in M_d : m_j \neq m_i \wedge s_j = s_i \wedge m_j \underset{r,q}{>} m_i \right\}, \tag{2}$$

where s_i and s_j are the two systems for which the results with m_i and m_j have been measured, respectively. This subset $M_{d,r,q}$ contains only the best configurations with respect to Recall and QpS that have been identified during our evaluation for all systems for a particular dataset. The curve that they create can be understood as the Pareto front of the systems on the dataset [14].

4.3 Setup

We evaluate the 22 systems listed in Table 1. All experiments are carried out with $k = 10$ on a single machine.[8] For all systems, for which the ANN benchmarking framework would run a grid search, we configure the Bayesian optimizer to search for the configuration with the best Recall in the parameter range that would have been used by the grid search. To keep the comparison of systems fair, we configured the benchmarking framework to use a maximum of 35 runs for each system regardless of the number of parameters the systems may have. We run the queries in batch mode, i.e., systems can answer queries in parallel. This has the following two benefits. First, it is closer to real applications in which for example multiple users interact with a KG-based system in parallel. Second, it enables us to run more experiments and, hence, test more configurations.

[8] A VM with 4 64-bit CPUs and 32GB RAM. Note that the VM does not have a GPU that could be used by the ANNS systems.

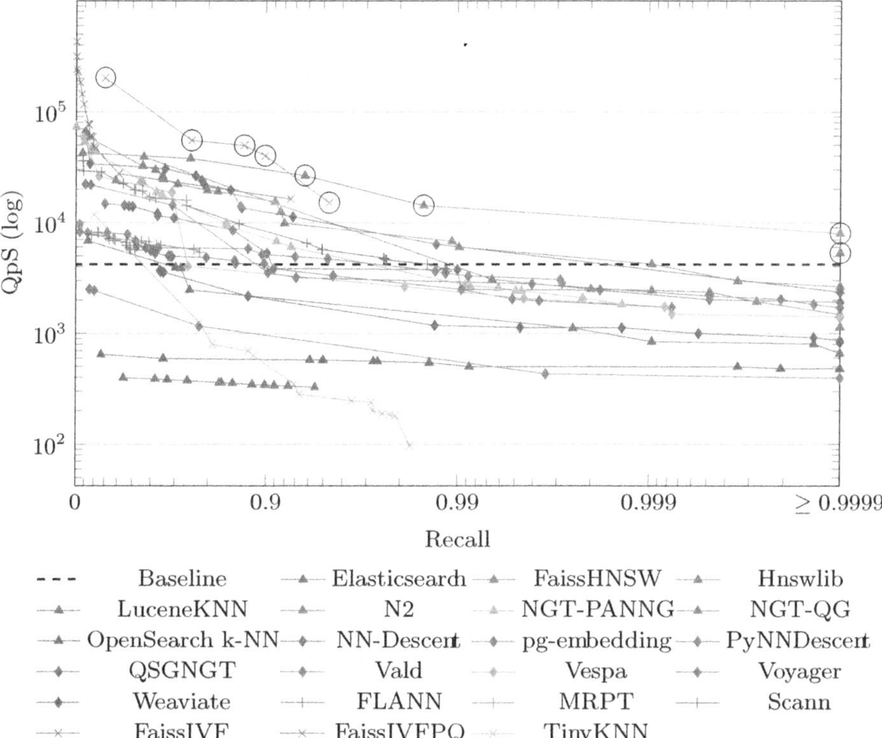

Fig. 1. The best configurations for all tested systems and the performance of the baseline on the FB15k-237 DistMult embeddings with angular distance. The best configurations for this dataset have been encircled (◯). Note that the Recall axis is not linear.

4.4 Results

Our experiment results comprise more than 750 data points per dataset.[9] In previous works, the results have been presented in Recall-QpS plots. However, these plots tend to be crowded when comparing all 22 evaluated systems as it can be seen in Fig. 1. Instead, we depict only the Pareto front of those systems that have at least one configuration that is part of the Pareto front for a particular dataset in Figs. 2 and 3. FaissIVFPQ occurs in every plot. It typically achieves very high QpS scores while the Recall is quite high as well. However, only for some datasets, the Bayesian optimization was able to find configurations of FaissIVFPQ that let to a Recall very close to 1.0. On angular-distance-based datasets, implementations like NGT-QG, QSGNGT or Hnswlib dominate this region of the plots. Many configurations of FaissIVF are not competitive when

[9] All our results are available at https://github.com/MichaelRoeder/ann-benchmarks/tree/main.

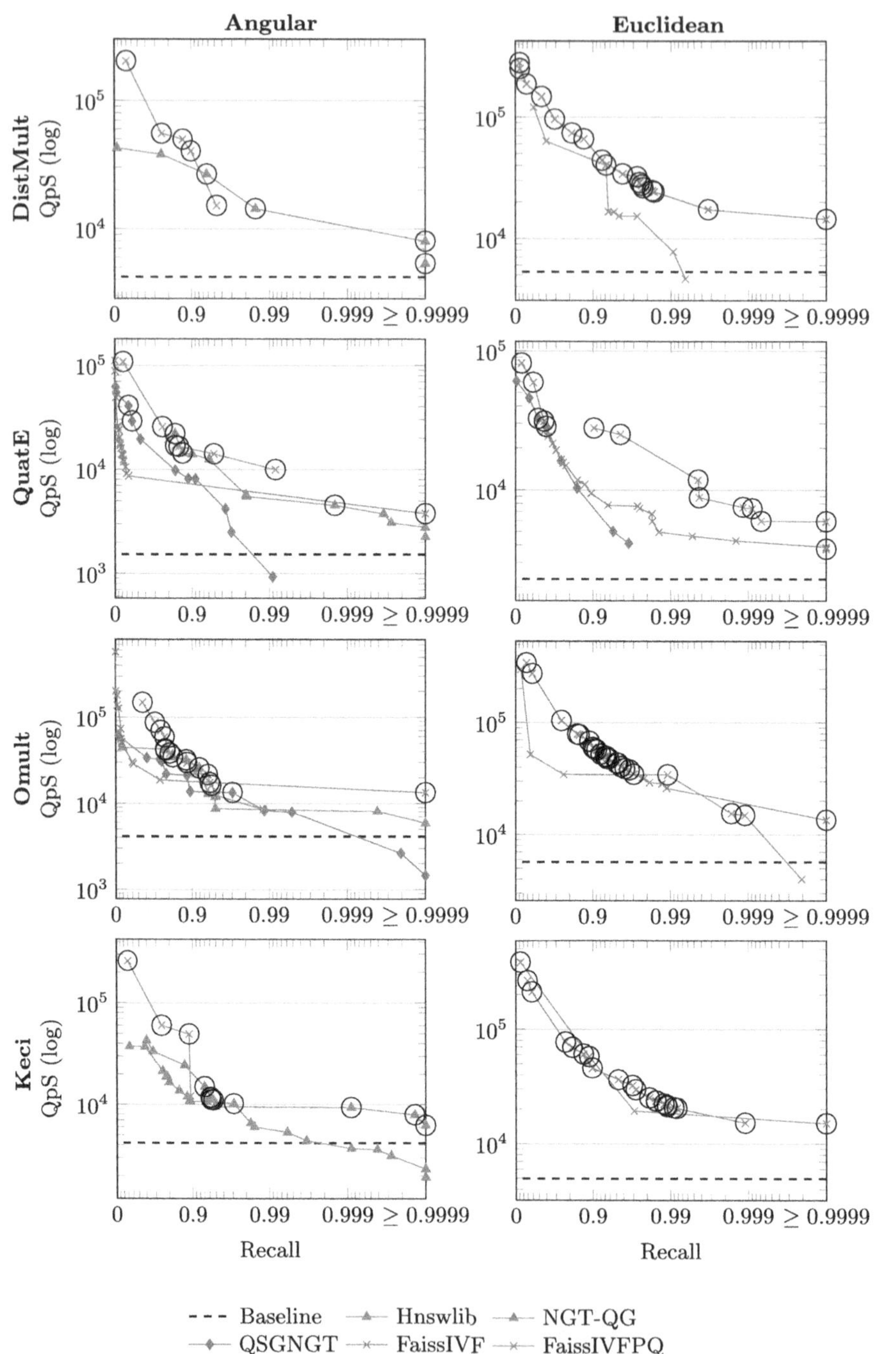

Fig. 2. The Recall-QpS curve of the best-performing systems on the FB15k-237 datasets. Angular and Euclidean datasets can be found in the left and right columns, respectively. The best configurations for the datasets have been encircled (◯).

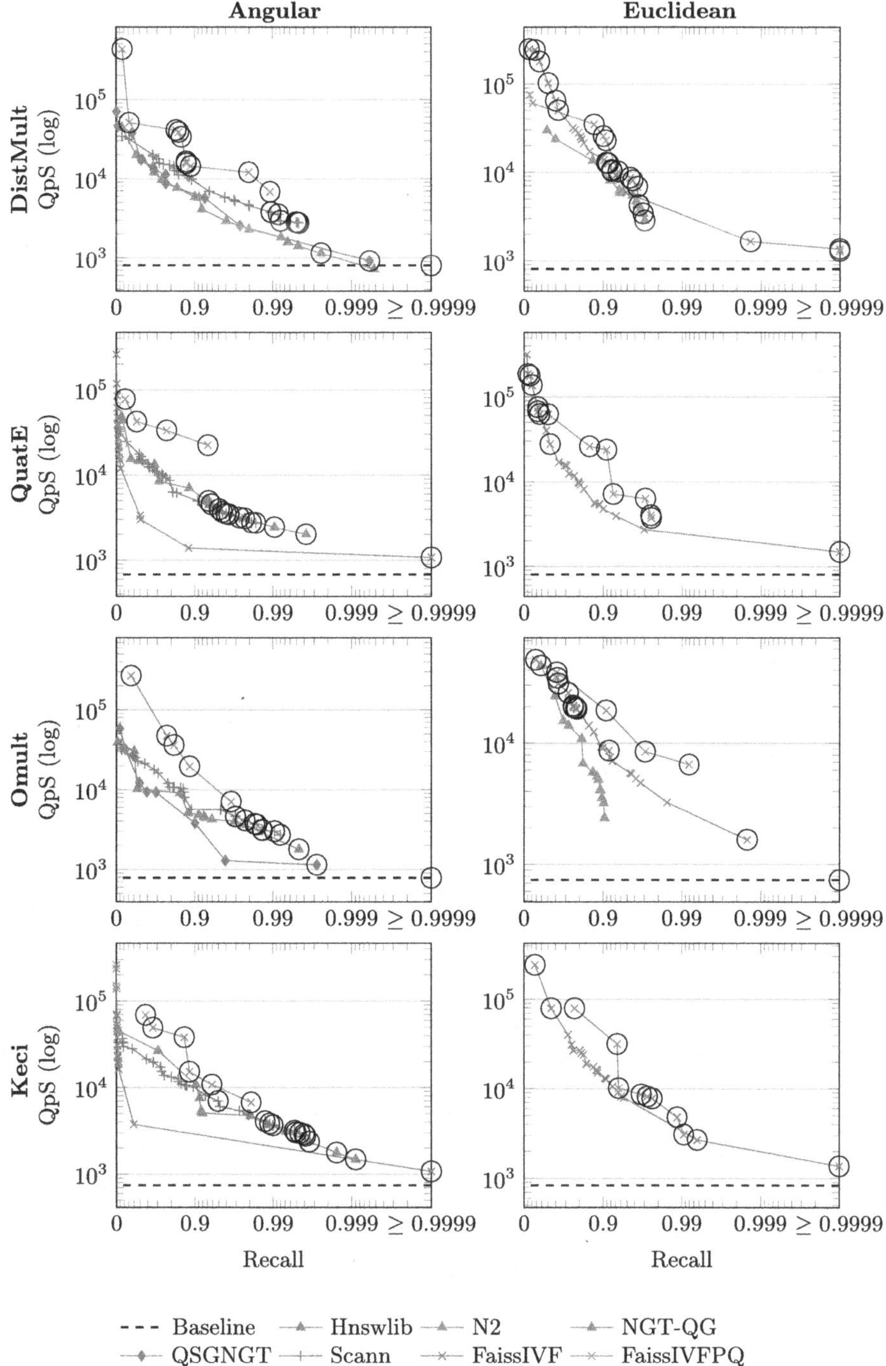

Fig. 3. The Recall-QpS curve of the best-performing systems on the Yago3-10 datasets. Angular and Euclidean datasets can be found in the left and right columns, respectively. The best configurations for the datasets have been encircled (◯).

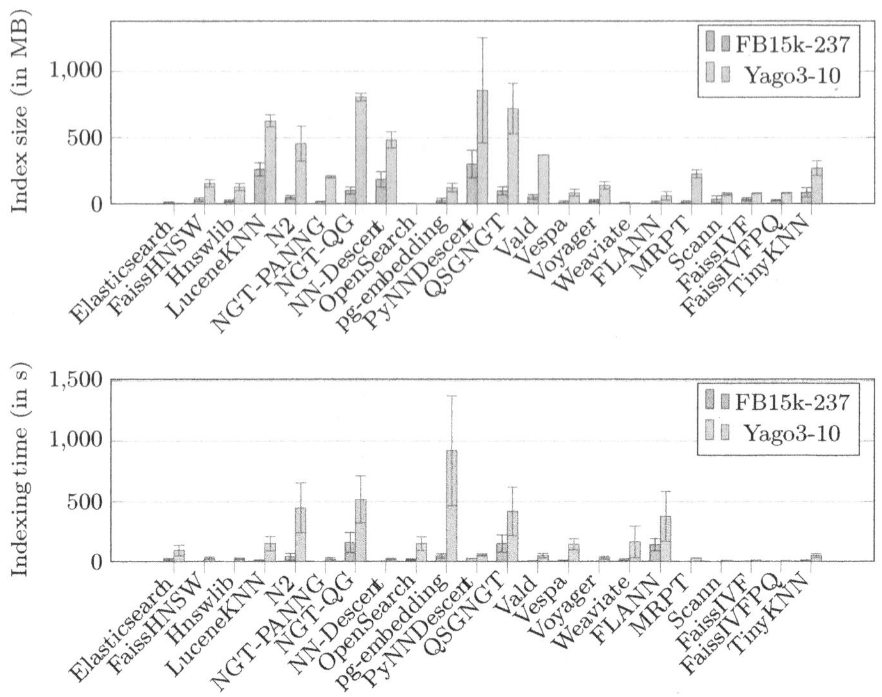

Fig. 4. The average size of the generated indexes on the two knowledge graphs (top) and the average time it took to generate the index (bottom).

used with the angular distance. However, in half of the plots for the angular-distance-based datasets, FaissIVF occurs since it is able to achieve a Recall of 1.0, which rarely is achieved by other approaches. The plots for the Euclidean-distance-based experiments are dominated by FaissIVF and FaissIVFPQ with a similar pattern as before. In many cases, FaissIVFPQ achieves a high QpS rate and a good Recall, but FaissIVF achieves a better Recall.

Figure 4 summarizes the results with respect to the generation of the search indexes. Especially for the bigger KG Yago3-10, the systems show quite different numbers. While the majority of systems stays close or even below the 60MB of the Parquet file, in which the Knowledge Graph embeddings are provided, LuceneKNN, N2, NGT-QG, NN-Descent, PyNN-Descent, and QSGNGT create indexes with a size close to or beyond the 500 MB mark. With respect to the time that is needed to generate the index, pg-embedding is the slowest system with an average runtime of 921 s for Yago3-10 datasets, followed by NGT-QG, N2 and QSGNGT.

Table 2. Pairwise comparisons of systems: the number of datasets on which the system in the row outperformed the system in the column. The row #Worse and the column #Better contain the sum in how many comparisons a system has been worse or better than any other system, respectively.

	BL	Elasticsearch	FaissHNSW	Hnswlib	LuceneKNN	N2	NGT-PANNG	NGT-QG	NN-Descent	OpenSearch-kNN	pg-embedding	PyNNDescent	QSGNGT	Vald	Vespa	Voyager	Weaviate	FLANN	MRPT	Scann	FaissIVF	FaissIVFPQ	TinyKNN	#Better
BL		16	0	0	3	0	0	0	0	8	6	0	0	1	0	0	6	0	0	0	0	0	4	44
Elasticsearch	0		0	0	0	0	0	0	0	0	0	0	0	0	0	0	0	0	0	0	0	0	0	0
FaissHNSW	0	16		0	5	4	0	0	0	7	3	0	0	0	0	0	7	1	3	0	1	0	8	55
Hnswlib	4	16	6		14	5	2	0	4	9	5	6	0	11	9	7	11	7	6	4	1	0	11	138
LuceneKNN	0	16	0	0		0	0	0	0	6	3	0	0	0	0	0	0	0	0	0	0	0	2	27
N2	0	8	3	0	8		0	0	0	8	0	0	1	4	6	5	6	0	0	0	1	0	5	55
NGT-PANNG	0	16	0	0	4	3		0	0	6	1	0	0	0	0	0	2	4	3	0	1	0	13	53
NGT-QG	0	16	4	0	13	6	3		2	13	5	5	0	8	9	10	12	5	7	2	1	0	14	135
NN-Descent	0	16	0	0	12	1	0	0		15	8	2	0	0	0	0	9	5	6	0	1	0	16	91
OpenSearch k-NN	0	16	0	0	0	0	0	0	0		2	0	0	0	0	0	0	0	0	0	0	0	0	18
pg-embedding	0	6	0	0	0	0	0	0	0	0		0	0	0	0	0	0	0	0	0	0	0	0	6
PyNNDescent	0	16	0	0	7	1	0	0	0	9	0		0	0	0	0	5	2	1	0	1	0	14	56
QSGNGT	0	16	1	0	11	5	1	0	0	9	4	3		5	3	3	9	5	6	1	1	0	13	96
Vald	0	16	0	0	1	0	0	0	0	4	0	0	0		0	0	2	1	1	0	0	0	10	35
Vespa	0	16	0	0	14	0	0	0	0	9	6	0	0	0		0	10	2	2	0	1	0	11	71
Voyager	0	16	0	0	13	1	0	0	0	10	8	0	0	0	0		10	3	2	0	0	0	11	74
Weaviate	0	16	0	0	0	0	0	0	0	8	3	0	0	0	0	0		0	0	0	0	0	1	28
FLANN	3	12	0	0	5	0	0	0	0	6	3	0	0	2	0	0	4		0	0	0	0	9	44
MRPT	0	13	1	0	6	0	0	0	0	7	1	0	0	4	0	0	6	0		0	0	0	10	48
Scann	0	15	1	0	1	3	2	0	0	1	0	0	0	6	1	1	0	5	6		1	0	8	51
FaissIVF	10	12	7	7	11	4	8	8	9	12	4	8	6	9	8	8	10	8	9	7		0	11	**176**
FaissIVFPQ	1	15	8	5	8	14	6	7	2	8	0	4	6	9	7	8	8	5	8	5	2		14	150
TinyKNN	0	10	0	0	0	0	0	0	0	1	0	0	0	0	0	0	0	0	0	0	0	0		11
#Worse	18	315	31	12	136	47	22	15	17	156	62	28	13	59	43	42	117	53	60	19	12	**0**	185	

5 Discussion

For a detailed discussion, we analyse the results further. To this end, we remove the configurations from the systems' Pareto curves that have a Recall below 0.1, since although many of these results have a good efficiency (i.e., their QpS is high), they lack effectiveness. This for example removes many of the points in the upper left corner of the plots in Figs. 2 and 3. With these updated curves, we

Table 3. Average ranks according to the pairwise comparisons across the datasets. Best average ranks are marked bold. Systems that are significantly outperformed by the best system are underlined.

System	Distance		KG		Embedding				Overall
	Angular	*Euclidean*	*FB15k-237*	*Yago3-10*	*DistMult*	*QuatE*	*Omult*	*Keci*	
BL	11.19	11.31	10.38	12.13	10.88	12.88	10.00	11.25	11.25
Elasticsearch	23.00	23.00	23.00	23.00	23.00	23.00	23.00	23.00	23.00
FaissHNSW	8.31	15.94	11.56	12.69	12.00	14.13	11.00	11.38	12.13
Hnswlib	**2.25**	7.31	4.44	5.13	5.75	4.38	4.75	4.25	4.78
LuceneKNN	17.13	20.50	18.44	19.19	18.38	18.50	19.13	19.25	18.81
N2	17.88	4.06	11.31	10.63	11.63	11.25	10.25	10.75	10.97
NGT-PANNG	9.81	11.25	9.06	12.00	9.38	8.63	13.38	10.75	10.53
NGT-QG	3.44	5.81	5.81	**3.44**	**3.25**	5.25	6.38	**3.63**	4.63
NN-Descent	4.75	7.81	7.25	5.31	4.75	7.50	6.38	6.50	6.28
OpenSearch k-NN	17.88	20.88	21.13	17.63	18.75	20.25	19.50	19.00	19.38
pg-embedding	18.81	14.31	17.38	15.75	16.88	16.13	17.13	16.13	16.56
PyNNDescent	11.69	8.75	11.06	9.38	10.88	10.25	10.63	9.13	10.22
QSGNGT	5.13	7.69	6.50	6.31	5.13	6.75	6.50	7.25	6.41
Vald	14.63	15.19	16.06	13.75	14.38	14.25	16.13	14.88	14.91
Vespa	6.81	15.69	9.94	12.56	12.50	8.88	11.38	12.25	11.25
Voyager	7.56	14.63	10.31	11.88	10.63	10.88	11.38	11.50	11.09
Weaviate	16.31	19.50	17.56	18.25	18.50	17.00	17.25	18.88	17.91
FLANN	15.69	7.94	8.94	14.69	14.13	12.13	10.75	10.25	11.81
MRPT	17.88	6.56	12.25	12.19	11.88	14.25	10.38	12.38	12.22
Scann	7.94	14.50	13.69	8.75	12.00	11.38	9.38	12.13	11.22
FaissIVF	9.25	**1.25**	**4.19**	6.31	6.50	**2.13**	6.63	5.75	5.25
FaissIVFPQ	6.94	1.75	5.00	3.69	4.13	4.75	**4.00**	4.50	**4.34**
TinyKNN	21.75	20.38	20.75	21.38	20.75	21.50	20.75	21.25	21.06

compare the systems pair-wise on each dataset checking whether a system is able to outperform the other. We define that a system outperforms another system on a particular dataset, if there is no configuration of the second system on the common Pareto front of both systems according to the comparison operator defined in Sect. 4.2. Table 2 shows a summary of these pairwise comparisons over all datasets. For each dataset, we assign ranks to the single systems according to the wins and losses that they achieve in the comparison. Table 3 shows the average ranks of the systems aggregated according to the distance measure, the underlying Knowledge Graph and the KGE used for the different datasets. A Friedman Test acknowledged for all columns in Table 3 that there are significant differences in the systems ranks [12].[10] The post-hoc Nimney Test [12] showed that in each column, several systems are significantly outperformed by the best system in that column.

These results allow several conclusions. The results from the first two columns suggest that in many cases, Hnswlib and FaissIVF are good choices when the angular and Euclidean distances are used, respectively. Similarly, several systems should not be used for the one or the other distance. For example, N2 showed a comparably good performance on two Yago3-10 datasets when used with the Euclidean distance, but was often outperformed on datasets with an angular distance. Similarly, Vespa, Voyager, and Scann are significantly worse on Euclidean datasets while they are not significantly outperformed on angular datasets. Comparing the two distance columns with the other columns shows that the large differences of some of the average ranks that are observed in the first two columns do not repeat. This suggests that the distance measure could be the most important feature when deciding which system to take. However, especially with larger knowledge graphs, the growth of the index size and indexing time depicted in Fig. 4 could become important for several algorithms that achieved good ranks, like N2, NGT-QG or QSGNGT. The four columns of the embeddings only show smaller differences between each other, suggesting that there is only a minor influence of the embedding algorithm on the systems' performance.

Another interesting insight comes from the comparison of the systems that rely on the same ANNS algorithm (see Table 1). For example, Hnswlib is the only HNSW-based implementation in our evaluation that is not significantly outperformed on either angular or Euclidean datasets. Other implementations are either outperformed in one, or like Elasticsearch, LucenKNN, OpenSearch k-NN, pg-embedding, and Weaviate, in both categories. Overall, these last 5 systems together with Vald and TinyKNN perform significantly worse in our evaluation than FaissIVFPQ. The latter achieves the best average rank over all 16 datasets. Hence, we can conclude that FaissIVFPQ can serve as an interesting alternative for Hnswlib and FaissIVF.

6 Conclusion

In this work, we carried out a detailed comparison of 22 Approximate Nearest Neighbor Search systems on 16 KGE-based datasets—a type of data that has been overlooked so far in evaluations of these systems. In contrast to the state of the art, we fine-tuned each approach in each experiment by using Bayesian optimization to ensure the fairness of our experiments. The evaluated systems

[10] We use $\alpha = 0.05$ for all significance tests.

showed significant performance differences in these experiments. Our results suggest that the overall performance of approaches with respect to runtime and recall is contingent upon the similarity measure used to compare embeddings.

Our future work will focus on the best performing systems in this evaluation. While our 16 datasets are based on FB15k-237 and Yago3-10—two Knowledge Graphs that are widely adopted in the research on KGE algorithms—we plan to include larger knowledge graphs to further look into the influence of the Knowledge Graph size on the system's performance.

Acknowledgments. This work has been supported by the Ministry of Culture and Science of North Rhine-Westphalia (MKW NRW) within the project SAIL (NW21-059D) and the Lamarr Fellow Network funded project WHALE (LFN 1-04).

References

1. Alshahrani, M., Thafar, M.A., Essack, M.: Application and evaluation of knowledge graph embeddings in biomedical data. PeerJ. Comput. Sci. (2021). https://doi.org/10.7717/peerj-cs.341
2. Aumüller, M., Bernhardsson, E., Faithfull, A.: ANN-benchmarks: a benchmarking tool for approximate nearest neighbor algorithms. Inf. Syst. **87**, 101374 (2020)
3. Bentley, J.L.: Multidimensional binary search trees used for associative searching. Commun. ACM **18**(9), 509–517 (1975). https://doi.org/10.1145/361002.361007
4. Bergstra, J., Bengio, Y.: Random search for hyper-parameter optimization. J. Mach. Learn. Res. **13**(null), 281–305 (2012)
5. Bordes, A., Chopra, S., Weston, J.: Question answering with subgraph embeddings. In: Moschitti, A., Pang, B., Daelemans, W. (eds.) Proceedings of the 2014 Conference on Empirical Methods in Natural Language Processing (EMNLP), pp. 615–620. Association for Computational Linguistics, Doha (2014). https://doi.org/10.3115/v1/D14-1067, https://aclanthology.org/D14-1067
6. Bordes, A., Weston, J., Usunier, N.: Open question answering with weakly supervised embedding models. In: Calders, T., Esposito, F., Hüllermeier, E., Meo, R. (eds.) Machine Learning and Knowledge Discovery in Databases, pp. 165–180. Springer, Heidelberg (2014)
7. Cao, Z., Long, M., Wang, J., Yu, P.S.: HashNet: deep learning to hash by continuation. In: Proceedings of the IEEE International Conference on Computer Vision (ICCV) (2017)
8. Dai, Y., Wang, S., Xiong, N.N., Guo, W.: A survey on knowledge graph embedding: approaches, applications and benchmarks. Electronics **9**(5) (2020). https://doi.org/10.3390/electronics9050750, https://www.mdpi.com/2079-9292/9/5/750
9. Dasgupta, S., Freund, Y.: Random projection trees and low dimensional manifolds. In: Proceedings of the Fortieth Annual ACM Symposium on Theory of Computing, STOC 2008, pp. 537–546. Association for Computing Machinery, New York (2008). https://doi.org/10.1145/1374376.1374452
10. Demir, C., Moussallem, D., Heindorf, S., Ngonga Ngomo, A.C.: Convolutional hypercomplex embeddings for link prediction. In: Balasubramanian, V.N., Tsang, I. (eds.) Proceedings of The 13th Asian Conference on Machine Learning. Proceedings of Machine Learning Research, vol. 157, pp. 656–671. PMLR (2021). https://proceedings.mlr.press/v157/demir21a.html

11. Demir, C., Ngonga Ngomo, A.C.: Clifford embeddings–a generalized approach for embedding in normed algebras. In: Joint European Conference on Machine Learning and Knowledge Discovery in Databases, pp. 567–582. Springer (2023)
12. Demšar, J.: Statistical comparisons of classifiers over multiple data sets. J. Mach. Learn. Res. **7**, 1–30 (2006)
13. Dong, W., Moses, C., Li, K.: Efficient k-nearest neighbor graph construction for generic similarity measures. In: Proceedings of the 20th International Conference on World Wide Web, WWW 2011, pp. 577–586. Association for Computing Machinery, New York (2011). https://doi.org/10.1145/1963405.1963487
14. Douze, M., et al.: The Faiss library (2024)
15. Friedman, J.H., Bentley, J.L., Finkel, R.A.: An algorithm for finding best matches in logarithmic expected time. ACM Trans. Math. Softw. **3**(3), 209–226 (1977). https://doi.org/10.1145/355744.355745
16. Fu, C., Zhou, M., Xuan, Q., Hu, H.X.: Expert recommendation in OSS projects based on knowledge embedding. In: 2017 International Workshop on Complex Systems and Networks (IWCSN), pp. 149–155 (2017). https://doi.org/10.1109/IWCSN.2017.8276520
17. Fukunaga, K., Narendra, P.: A branch and bound algorithm for computing k-nearest neighbors. IEEE Trans. Comput. **C-24**(7), 750–753 (1975). https://doi.org/10.1109/T-C.1975.224297
18. Guo, R., et al.: Accelerating large-scale inference with anisotropic vector quantization. In: III, H.D., Singh, A. (eds.) Proceedings of the 37th International Conference on Machine Learning. Proceedings of Machine Learning Research, vol. 119, pp. 3887–3896. PMLR (2020). https://proceedings.mlr.press/v119/guo20h.html
19. Hyvönen, V., et al.: Fast nearest neighbor search through sparse random projections and voting. In: 2016 IEEE International Conference on Big Data (Big Data), pp. 881–888. IEEE (2016)
20. Indyk, P., Motwani, R.: Approximate nearest neighbors: towards removing the curse of dimensionality. In: Proceedings of the Thirtieth Annual ACM Symposium on Theory of Computing, STOC 1998, pp. 604–613. Association for Computing Machinery, New York (1998). https://doi.org/10.1145/276698.276876
21. Iwasaki, M.: Proximity search in metric spaces using approximate k nearest neighbor graph. IPSJ Trans. Database **3**(1), 18–28 (2010)
22. Iwasaki, M.: Pruned bi-directed k-nearest neighbor graph for proximity search. In: Amsaleg, L., Houle, M.E., Schubert, E. (eds.) Similarity Search and Applications, pp. 20–33. Springer, Cham (2016)
23. Iwasaki, M.: Fusion of graph-based indexing and product quantization for ANN search. Medium (online) (2023). https://medium.com/@masajiro.iwasaki/fusion-of-graph-based-indexing-and-product-quantization-for-ann-search-7d1f0336d0d0. Accessed 5 Dec 2024
24. Johnson, J., Douze, M., Jégou, H.: Billion-scale similarity search with GPUs. IEEE Trans. Big Data **7**(3), 535–547 (2021). https://doi.org/10.1109/TBDATA.2019.2921572
25. Li, C., Zhang, M., Andersen, D.G., He, Y.: Improving approximate nearest neighbor search through learned adaptive early termination. In: Proceedings of the 2020 ACM SIGMOD International Conference on Management of Data, SIGMOD 2020, pp. 2539–2554. Association for Computing Machinery, New York (2020). https://doi.org/10.1145/3318464.3380600
26. Li, W., et al.: Approximate nearest neighbor search on high dimensional data - experiments, analyses, and improvement. IEEE Trans. Knowl. Data Eng. **32**(8), 1475–1488 (2020). https://doi.org/10.1109/TKDE.2019.2909204

27. Malkov, Y.A., Yashunin, D.A.: Efficient and robust approximate nearest neighbor search using hierarchical navigable small world graphs. IEEE Trans. Pattern Anal. Mach. Intell. **42**(4), 824–836 (2020). https://doi.org/10.1109/TPAMI.2018.2889473
28. Muja, M., Lowe, D.G.: Scalable Nearest neighbor algorithms for high dimensional data. IEEE Trans. Pattern Anal. Mach. Intell. **36** (2014)
29. Naidan, B., Boytsov, L., Nyberg, E.: Permutation search methods are efficient, yet faster search is possible. CoRR abs/1506.03163 (2015). http://arxiv.org/abs/1506.03163
30. Nogueira, F.: Bayesian optimization: open source constrained global optimization tool for Python (2014). https://github.com/bayesian-optimization/BayesianOptimization. Accessed 17 Dec 2024
31. Shahriari, B., Swersky, K., Wang, Z., Adams, R.P., de Freitas, N.: Taking the human out of the loop: a review of Bayesian optimization. Proc. IEEE **104**(1), 148–175 (2016). https://doi.org/10.1109/JPROC.2015.2494218
32. Sun, P., Simcha, D., Dopson, D., Guo, R., Kumar, S.: SOAR: improved indexing for approximate nearest neighbor search. In: Oh, A., Naumann, T., Globerson, A., Saenko, K., Hardt, M., Levine, S. (eds.) Advances in Neural Information Processing Systems, vol. 36, pp. 3189–3204. Curran Associates, Inc. (2023). https://proceedings.neurips.cc/paper_files/paper/2023/file/0973524e02a712af33325d0688ae6f49-Paper-Conference.pdf
33. Wang, G., Ke, X., Hu, J., Li, Q., Shao, M., Fan, J.: Learning to prune: general and efficient approximate nearest neighbor search with direction navigating graph. In: Proceedings of the 2022 5th International Conference on Data Storage and Data Engineering, DSDE 2022, pp. 50–58. Association for Computing Machinery, New York (2022). https://doi.org/10.1145/3528114.3528123
34. Wang, M., Xu, X., Yue, Q., Wang, Y.: A comprehensive survey and experimental comparison of graph-based approximate nearest neighbor search. Proc. VLDB Endow. **14**(11), 1964–1978 (2021). https://doi.org/10.14778/3476249.3476255
35. Wu, J., Chen, X.Y., Zhang, H., Xiong, L.D., Lei, H., Deng, S.H.: Hyperparameter optimization for machine learning models based on Bayesian optimizationb. J. Electron. Sci. Technol. **17**(1), 26–40 (2019). https://doi.org/10.11989/JEST.1674-862X.80904120, https://www.sciencedirect.com/science/article/pii/S1674862X19300047
36. Yang, B., Yih, S.W.T., He, X., Gao, J., Deng, L.: Embedding entities and relations for learning and inference in knowledge bases. In: Proceedings of the International Conference on Learning Representations (ICLR) 2015 (2015). https://www.microsoft.com/en-us/research/publication/embedding-entities-and-relations-for-learning-and-inference-in-knowledge-bases/
37. Zhang, F., Yuan, N.J., Lian, D., Xie, X., Ma, W.Y.: Collaborative knowledge base embedding for recommender systems. In: Proceedings of the 22nd ACM SIGKDD International Conference on Knowledge Discovery and Data Mining, KDD 2016, pp. 353–362. Association for Computing Machinery, New York (2016https://doi.org/10.1145/2939672.2939673
38. Zhang, S., Tay, Y., Yao, L., Liu, Q.: Quaternion knowledge graph embeddings. In: Wallach, H., Larochelle, H., Beygelzimer, A., d'Alché-Buc, F., Fox, E., Garnett, R. (eds.) Advances in Neural Information Processing Systems, vol. 32. Curran Associates, Inc. (2019). https://proceedings.neurips.cc/paper_files/paper/2019/file/d961e9f236177d65d21100592edb0769-Paper.pdf

On Evaluation Metrics for Complex Matching Based on Reference Alignments

Guilherme Santos Sousa[1(✉)], Rinaldo Lima[2], and Cassia Trojahn[1,3]

[1] IRIT, Toulouse, France
guilherme.santos-sousa@irit.fr
[2] Universidade Federal Rural de Recife, Recife, Brazil
rjl4@cin.ufpe.br
[3] Univ. Grenoble Alpes, Inria, CNRS, Grenoble INP, Saint-Martin-d'Hères, France
cassia.trojahn-dos-santos@univ-grenoble-alpes.fr

Abstract. Existing metrics for evaluating complex ontology matching systems often fail to adequately capture the intricacies of (m:n) correspondences. This limitation results in partial or biased alignment quality assessments. This paper introduces a novel metric specifically tailored for complex ontology matching, extending traditional evaluation frameworks by incorporating subgraph similarity measures to ensure structural consistency with reference alignments. It utilizes a tree similarity-based approach, ensuring robustness against common issues such as order variance and detecting incorrect correspondences while adhering to key evaluation properties like completeness and correctness. Empirical experiments conducted on the OAEI complex track datasets demonstrate the superior adaptability of the metric in distinguishing correct structural correspondences compared to conventional and instance-based evaluation methods.

Keywords: Complex ontology matching · evaluation · tree similarity-based approach

1 Introduction

Ontology matching (and more broadly, knowledge graph matching) aims to enable interoperability between knowledge expressed in different schemes. While the field has reached some maturity, most of the matching approaches still focus on generating simple (1:1) correspondences (i.e., those linking one single entity of a source ontology to one single entity of a target ontology, as $Authors \equiv Writer$). However, this type of correspondence is not expressive enough to fully cover different heterogeneities (lexical, semantic, conceptual, granularity). More expressiveness is hence fundamental and complex correspondences (i.e., those involving logical constructors or transformation functions, as e.g., $Accepted\_Paper \equiv Paper \sqcap \exists$ hasDecision.Acceptance). The need for more expressiveness has been recognized across various fields, such as cultural heritage [5], agronomic [10], or still biomedical [4,8].

E. Curry et al. (Eds.): ESWC 2025, LNCS 15718, pp. 77–93, 2025.
https://doi.org/10.1007/978-3-031-94575-5_5

A still open issue, however, is the evaluation of complex correspondences. Existing evaluation metrics do not fully account for structural aspects, only partly exploit reference alignment, or are not always feasible. Although well-known benchmarks contain reference alignments (the result of an expert-driven curation process), no metric comprehensively compares complex correspondences from the reference with those generated by matchers. Current methods rely on manual comparison – as seen in the OAEI Taxon and Complex Conference datasets – or use partial metrics like Entity or Relationship Identification[1], as in the GeoLink benchmark. Concerning the metrics that rely on the use of common instances (A-Box data) [13], such an approach is not always feasible, as ontologies may not be exhaustively populated or may lack an A-Box altogether.

This paper introduces a novel metric specifically designed to evaluate complex ontology alignments based on a reference alignment (gold standard) while capturing their underlying semantics. Building on the evaluation framework proposed in [2], this metric addresses the shortcomings of existing approaches by fully considering the structural intricacies of (m:n) correspondences. By focusing on structural alignment, it offers a more comprehensive and accurate comparison, filling this gap in current methods. The metric assigns a higher score to the correspondence that is structurally closer to the reference, ensuring an accurate evaluation.

The structure of this paper is organized as follows. Section 2 introduces the theoretical foundations of the proposed metric, describing its desired properties, method, and algorithms. Section 3 evaluates the proposed approach through experiments, comparing its performance with state-of-the-art metrics. Section 4 provides a review of related work and existing evaluation metrics for complex ontology matching, highlighting their limitations. Finally, Sect. 5 summarizes the contributions of this study and outlines potential directions for future research.

2 Proposal

The work here is based on the work from [2]. Originally, precision and recall do not account for the proximity of alignments to the expected result, as they only compare exact (1:1) correspondences. This can lead to different alignments receiving the same score despite the varying quality. In that work, it is proposed to use a relaxation of the metrics by first generalizing those metrics by comparing instead the similarity between sets of correspondences and using correspondence proximity metrics, such as how distant entities are from each other, to create a fuzzy metric proximity metric. For that proposal, it is required that the same entity not appear twice in the alignment. For the evaluation of complex alignments, this restriction is incompatible, since the complex correspondences are composed of subgraphs instead of a single entity, where the same entity can appear multiple times in different subgraphs. To solve those constraints, in this work is proposed to use the tree edit distance to compute the correspondence proximity and use an assignment algorithm to enable the same subgraphs to

[1] https://oaei.ontologymatching.org/2020/results/complex/geolink/index.html.

appear multiple times in the alignment. The next sections describe the theoretical foundation of the proposed metric along with its algorithmic implementation.

2.1 Foundations

Complex ontology matching involves identifying and formalizing expressive correspondences between entities in different ontologies. An alignment is defined as a set $A = \{(e_1, e_2) \mid e_1 \in O_1, e_2 \in O_2\}$, where e_1 and e_2 are entity subgraphs from the source ontology O_1 and the target ontology O_2, respectively. These pairs, (e_1, e_2), represent correspondences between semantically related elements in the two ontologies. Alignments are classified as **simple** when $|e_1| = |e_2| = 1$ (1:1) and **complex** when $\max(|e_1|, |e_2|) > 1$ (m:n).

To evaluate the quality of an alignment, a similarity function $f(A_1, A_2) \rightarrow [0, 1]$ is employed. This function measures the degree of similarity between a proposed alignment A_1 and a reference alignment A_{ref}, which acts as the gold standard. The function compares all the entity pairs $(e_1, e_2) \in A_1$ against those in A_{ref}, quantifying how well the proposed alignment captures the intended correspondences. The similarity score produced by f provides an objective metric for evaluating the alignment's adherence to the reference alignment, enabling the measurement and ranking of the matcher's performance.

2.2 Desired Properties of the Proposed Metric

Given the nature of complex alignments, a metric designed to evaluate them must comply with specific properties to ensure comprehensive assessment. For instance, a matcher that produces multiple correspondences, including both correct and incorrect ones, should be assigned a lower performance score compared to a matcher that outputs fewer but entirely correct correspondences. Based on [2], this work introduces a set of properties specifically tailored for complex matching. Unlike the original framework, which assumes that an entity can appear in only one correspondence, our work allows for the same entity to appear multiple times across different correspondences. This distinction acknowledges the fact that various logical combinations, represented as subgraphs that either include the same entity multiple times or involve distinct subgraphs containing the same entity, can convey different semantic meanings.

Considering these considerations, the desired properties of a metric for complex matching, designed to compare two sets of alignments, are as follows:

Definition 1 (Identity). $f(A_1, A_2) = 1$ *when the proposed alignment* A_1 *is identical to the reference alignment* A_2*. This ensures that the similarity function assigns the highest score when the alignments perfectly match.*

Definition 2 (Order Invariance). *Since alignments are sets of entity correspondences, the order in which they are listed should not affect the similarity score. Therefore,* $f(A_1, A_2) = f(A_n, A_2)$*, where* A_n *is a list containing the same correspondences as* A_1 *but in a different order. This reflects that the metric compares the sets of correspondences, not their sequence. In this work, the alignments are considered to be presented as unordered lists.*

Definition 3 (Error Penalization). *The similarity score should decrease if a wrong correspondence is added to the proposed alignment. Specifically, it is expected that* $f(A_n, A_2) < f(A_1, A_2)$, *where* A_n *is* A_1 *with an additional incorrect correspondence or a redundant copy of an already correct correspondence.*

Definition 4 (Incompletness Penalization). *The similarity score should also decrease if correct correspondence is removed from the proposed alignment. Specifically,* $f(A_n, A_2) < f(A_1, A_2)$, *where* A_n *is* A_1 *without a correct correspondence. This property ensures that missing correspondences result in a lower score.*

Definition 5 (Sensitivity to Entity Modification). *Modifying an entity pair in the alignment by adding, deleting, or replacing one of its elements should also decrease the similarity score. Considering,* $f(A_n, A_2) < f(A_1, A_2)$, *where* A_n *is* A_1 *with one entity modified such that the modified pair* (e_n, e_m) *is different from the corresponding pair* (e_1, e_2) *in* A_2. *This property ensures that the metric is sensitive to changes in the specific entities or relationships in the alignment.*

2.3 Evaluation Algorithm

An evaluation algorithm is proposed to address the desired properties of the metric in the context of complex ontology alignment. This proposed algorithm receives as input an alignment file in the EDOAL format [3]. This format is commonly used to store complex alignment results between ontologies as it describes relational entities that can combine multiple entities. It is a subset of RDF/XML and its basic structure is a tree of correspondences containing cells that describe the correspondence between `edoal:entity1` tag from the source and `edoal:entity2` tag from the target with a specific confidence and relation. This algorithm does not consider the type of relation or the degree of confidence by the matcher. One example of a correspondence expressed in EDOAL is given below:

```
<Alignment>
    <map>
      <Cell rdf:about="Reviewer_merge">
        <entity1>
          <edoal:Class>
            <edoal:or rdf:parseType="Collection">
              <edoal:Class rdf:about="&cmt;Reviewer" />
              <edoal:Class rdf:about="&cmt;ExternalReviewer" />
            </edoal:or>
          </edoal:Class>
        </entity1>
        <entity2>
          <edoal:Class rdf:about="&conference;Reviewer" />
        </entity2>
```

```
        <measure rdf:datatype="&xsd;float">1.0</measure>
        <relation>Equivalence</relation>
      </Cell>
    </map>
    ...
</Alignment>
```

The choice of a tree-based algorithm is driven by the fact that EDOAL is inherently tree-structured. Although a tree is a substructure of a graph, and both graph and tree similarity algorithms could theoretically yield similar results, the tree-based approach is more natural and efficient here. While many tree and class-based similarity algorithms exist, their use in automatically evaluating complex correspondences remains unexplored.

The algorithm used to compute the TED in this work is described in [15]. In this case, the children tree is sorted as proposed in [16] since the order of the children in set operation nodes, like an intersection (owl:intersectionOf), must not impact the results. The costs used for the TED algorithm are: insertion and deletion costs are 1 and the update cost is 2. Those costs ensure that the similarity computed between trees composed of single nodes but with different entities is 0.

The proposed evaluation is computed by the Algorithm 1. The first step is loading the correspondences from the matcher output and the reference alignment in lines 2 to 3 of the Algorithm. Then the correspondences are classified into simple and complex based on the number of entities in each subtree in lines 5 to 6. Then the empty score matrix is initialized in line 7.

For each pair in the cartesian product of all correspondences between the evaluated alignment and the reference alignments, a similarity score is computed and filled into the score matrix in lines 8–14. The similarity between all correspondence pairs between A_1 and A_2 is computed as:

$$SL_{i,j} = \frac{\text{tree\_sim}(Q_{1_i}, R_{1_j}) + \text{tree\_sim}(Q_{2_i}, R_{2_j})}{2}, \quad \forall i \in \{1, \ldots, |A_1|\}, j \in \{1, \ldots, |A_2|\} \quad (1)$$

That computation results in a matrix SL where the lines are the correspondence pairs $(Q_1, Q_2)_i$ from the source and the columns are the correspondence pairs (R_1, R_2) from the reference alignment. Now applying the assignment algorithm described in Algorithm 2 in line 15, a set of corresponding maps between A_1 and A_2 is retrieved that maximizes the sum of the corresponding pairs' similarities. In lines 17 and 18, the resulting assignment is classified as simple and complex into four assignments $T_{simple}, T_{complex}, Q_{simple}, Q_{complex}$. Two of them are simple assignments where the source T_{simple} and target Q_{simple} contain simple correspondences, and the other two $T_{complex}$ and $Q_{complex}$ contain the complex assignments. Those different assignments are classified to perform different evaluations based on the number of simple or complex correspondences.

Algorithm 1. Evaluate EDOAL (`evaluate_edoal`)

1: **function** EVALUATEEDOAL($p_1, p_2, w = 0.5,$ `sim_func` = `tree_sim`)
2: $maps_s \leftarrow$ `LoadAlignment`(p_1)
3: $maps_t \leftarrow$ `LoadAlignment`(p_2)
4: Divide alignments into simple and complex correspondences for $maps_s$, $maps_t$:
5: $S_{source}, C_{source} \leftarrow$ splitSimpleComplex$(maps_s)$
6: $S_{target}, C_{target} \leftarrow$ splitSimpleComplex$(maps_t)$
7: `scores` $\leftarrow$ Empty list for similarity scores
8: **for** s_1, s_2 in $maps_s$ **do**
9: `ms` $\leftarrow$ Empty list
10: **for** t_1, t_2 in $maps_t$ **do**
11: `ms.append`((`sim_func`(s_1, t_1) + `sim_func`(s_2, t_2)) / 2)
12: **end for**
13: `scores.append`(ms)
14: **end for**
15: `assigns` $\leftarrow$ `MaximizeAssign`(scores)
16: Separate assignments into simple and complex:
17: $T_{simple}, T_{complex} \leftarrow$ getSourceAssignments(assigns)
18: $Q_{simple}, Q_{complex} \leftarrow$ getTargetAssignments(assigns)
19: `recall` $\leftarrow recall_{avg}$
20: `precision` $\leftarrow prec_{avg}$
21: `fmeasure` $\leftarrow 2 \cdot$ `recall` $\cdot$ `precision`/(`recall` + `precision`)
22: **return** `soft_precision`, `soft_recall`, `soft_fmeasure`
23: **end function**

After that, it is possible to compute average precision and average recall from the resulting set in lines 19 to 21.

To evaluate the performance of the matcher considering only the simple correspondences or the complex ones, a weight w is introduced. This weight ranges from 0 to 1, and when it is 0, only the simple correspondences are considered in the results, and when it's 1, only the complex correspondences are considered. The default value of 0.5 is used to evaluate both correspondence types. To perform this evaluation, assuming that $A_{source} = S_{source} \cup C_{source}$ is the set of correspondences in source and S_{source} is the set of simple correspondences in source and C_{source} is the set of complex correspondences in source. For the target, the set $A_{target} = S_{target} \cup C_{target}$ is assigned respectively. The weighted average precision is computed as:

$$prec_{avg} = \frac{(1-w) \cdot \sum_{s \in T_{simple}} s + w \cdot \sum_{c \in T_{complex}} c}{(1-w) \cdot |S_{source}| + w \cdot |C_{source}|} \tag{2}$$

And the average recall:

$$recall_{avg} = \frac{(1-w) \cdot \sum_{s \in Q_{simple}} s + w \cdot \sum_{c \in Q_{complex}} c}{(1-w) \cdot |S_{target}| + w \cdot |C_{target}|} \tag{3}$$

Also, an aggregated metric such as f-measure can be computed with the averaged precision and recall:

Algorithm 2. Maximum Assignment (max_assign)

```
function MaxAssign(m)
    preferences ← sorted_dict(m) ▷ Sort preferences for each pair
    unassigned ← List of unassigned pairs
    assigned ← {}
    while unassigned is not empty do
        pair ← unassigned.pop(), pair_pref ← preferences[pair]
        if len(pair_pref) = 0 then
            continue
        end if
        next_pref ← pair_pref.pop(0)                    ▷ Highest remaining preference
        if next_pref[0] is in assigned then
            if next_pref[1] > assigned[next_pref[0]][1] then    ▷ Better match
                unassigned.append(assigned[next_pref[0]][0])    ▷ Free current
                assigned[next_pref[0]] ← (pair, next_pref[1])
            else
                unassigned.append(pair)                         ▷ Remain free
            end if
        else
            assigned[next_pref[0]] ← (pair, next_pref[1])       ▷ Assign the pair
        end if
    end while
    return assigned
end function
```

$$f1 = \frac{2 \cdot prec_{avg} \cdot recall_{avg}}{prec_{avg} + recall_{avg}}, (prec_{avg} + recall_{avg}) \neq 0 \tag{4}$$

The proposed evaluation method in this paper is implemented in Python and is available at GitLab[2].

2.4 Properties Verification

In this section, the arguments for the properties of the proposed evaluation method are stated.

Lemma 1 (Identity). *Given alignments A_1 and A_2, if $A_1 = A_2$, then the similarity function $f(A_1, A_2) = 1$.*

Proof. Assume that $A_1 = A_2$. From the definition of the similarity metric, $f(A_1, A_2)$ is computed using the matrix sl of pairwise similarities. Since $A_1 = A_2$, all entity pairs (m_1, m_2) compared in sl will yield $tree\_sim(m_1, m_2) = 1$ when $m_1 = m_2$ and lower similarity for all other pairs. The assignment algorithm $maximize\_assign$ will therefore select matches that maximize the total similarity, which will select the equal pairs. Since all selected pairs have similarity 1, precision and recall are 1, resulting in $f(A_1, A_2) = 1$.

[2] https://gitlab.irit.fr/melodi/ontology-matching/complex/complex-reference-evaluation.

Lemma 2 (Order Invariance). *The similarity function* $f(A_1, A_2)$ *is invariant to the order of entity pairs in* A_1. *Specifically, if* A_n *is a reordering of* A_1, *then* $f(A_1, A_2) = f(A_n, A_2)$.

Proof. Assume A_n is a reordering of A_1. Since all candidate pairs are iterated from the similarity matrix sl, even if a candidate assignment starts at a lower similarity, it will always be replaced by a higher similarity pair, leading to the same sum of similarities. If two distinct pairs, containing different entities but having the same similarity, are assigned in a different order, the total sum remains unchanged. Consequently, the precision and recall values stay consistent, regardless of the initial order.

Lemma 3 (Error Penalization). *If an incorrect correspondence* $(e_1, e_2) \notin A_2$ *is added to* A_1 *to form* A_n, *then* $f(A_n, A_2) < f(A_1, A_2)$.

Proof. Assume A_n is a copy of A_1 with an additional incorrect correspondence. The similarity matrix sl will stay the same as $f(A_1, A_1)$. However, the precision value in this case will be lower, leading to a lower f-measure and then $f(A_n, A_1) < f(A_1, A_1)$.

Lemma 4. (Incompleteness Penalization). *If a correct correspondence is removed from* A_1 *to form* A_n, *then* $f(A_n, A_2) < f(A_1, A_2)$.

Proof. Assume A_n is A_1 with one correct correspondence removed. This reduces the number of high-similarity values in the matrix sl, resulting in a lower optimal assignment score from $maximize\_assign$. Consequently, recall and f-measure decrease.

Lemma 5 (Sensitivity to Entity Modification). *If an entity pair in* A_1 *is modified to form* A_n, *then* $f(A_n, A_2) < f(A_1, A_2)$.

Proof. Assume A_n is A_1 with one entity pair (e_1, e_2) replaced by (e_n, e_2) such that $sim(e_n, e_2) < sim(e_1, e_2)$. The modified pair will have a lower similarity score, leading to reduced values in sl. This decreases the total sum of similarity scores from $maximize\_assign$, lowering precision, recall, and f-measure.

2.5 Specific Case Exploration

To illustrate the main differences between the proposed metrics and the two common approaches, an example is drawn. As stated above, instance-based and Entity Identification are currently used in the OAEI complex track. However, as instance-based metrics act as a proxy metric by measuring the amount of common instances returned by the correspondences, that metric is not directly comparable to the others. So, to illustrate the difference between Entity Identification and the proposed metric, consider the following example:

```
Matcher 1:
IntersectionOf( InverseOf(isWrittenBy), isAuthorOf ) = writePaper
Matcher 2:
IntersectionOf(isWrittenBy, InverseOf(isAuthorOf) ) = writePaper
Matcher 3:
UnionOf(isWrittenBy, InverseOf(isAuthorOf)) = writePaper
Reference alignment:
IntersectionOf( InverseOf(isWrittenBy), isAuthorOf ) = writePaper
```

In the OAEI results page of the year 2020[3] (the one with the most different matchers participating), it remains unclear whether the Entity Identification task accounts for OWL predicates. In the example, if no OWL predicates are considered, since isWrittenBy and isAuthorOf appear both in the reference on the source side and writePaper on the target, all matchers score 1.0 (the maximum score considering the formula found entities / total entities). Since in the example, Matcher 1 is the exact copy of the reference alignment and the others have modifications, the ranking of the matchers doesn't reflect the desired evaluation. If the predicates are considered, the scores vary: In Matcher 1, all entities are present, so the score is 1.0. In Matcher 2, all entities are present but in a different order, yet the score remains 1.0 because Entity Identification doesn't consider the order. In Matcher 3, the inverse appears in the wrong property and, instead of Intersection, a Union is used, so this matcher scores 0.75.

In contrast, our proposed metric uses tree edit distances to measure the similarity between the generated correspondences and the reference alignment. In Matcher 1, all entities are present and in the same order, so the score is 1.0. In Matcher 2, the cost is 2 over 4 entities, so the score is 0.5. And in Matcher 3, the cost is 3 over 4 entities, reaching a score of 0.25. Thus, the proposed metric more effectively ranks correspondences based on their compliance with the reference alignment.

Another example can be considering the comparison for the entity pair illustrated in Fig. 1. Using the proposed metric, four edits are required to transform the tree structure **A** into that of the tree **B**. With the total size of the compared trees being 12, the similarity score for this entity pair is computed as $1 - \frac{4}{12} = 0.33$.

By contrast, the Entity Identification process used, for example, in the populated conference evaluation in the complex track in OAEI 2020, only considers ontology-related entities, filtering the structural entities. In this case, the similarity of those trees will be 0. This evaluation presents an optimistic view of the task, simplifying the task for matchers by ignoring structural requirements; as long as the correct entities are included in the set, the matcher can achieve a higher score without adhering to the proper logical operators that define part of the tree structure.

[3] https://oaei.ontologymatching.org/2020/results/complex/geolink/index.html.

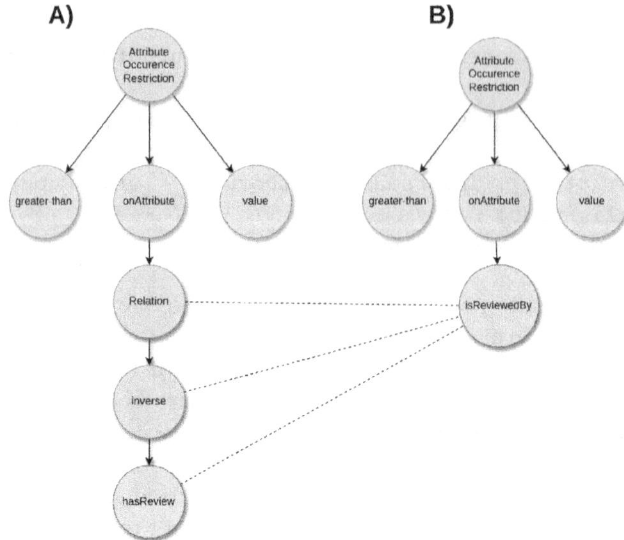

Fig. 1. Example of the similarity comparison applied in the proposed metric. Entities shaded in green are structural nodes while those in blue are ontology-specific nodes. Entities highlighted in green are correspondents, while entities highlighted in red are replaced. (Color figure online)

Based on these observations, it is possible to outline the properties of each metric concerning the proposed definitions for a robust metric for complex matching using reference alignments. These properties are summarized in Table 1. It is possible to see that the proposed metric follows all the proposed definitions, while the relaxed-based metrics fail in property 5 and the Instance-based metrics fail in property 3 and don't apply to property 5.

Table 1. Evaluation of metrics based on adherence to properties defined for proper evaluation of complex alignments. In the table, X marks when the metric doesn't follow the definition and - marks when the definition is not applied.

Metric	Identity	Order Invariance	Error Penalization	Incompleteness Penalization	Sensitivity to Entity Modification
Proposed	✓	✓	✓	✓	✓
Entity Identification	✓	✓	✓	✓	X
Instance-based	✓	✓	X	✓	–

3 Experiments and Discussion

In this section, the results of the most commonly used automatic evaluation metrics are empirically compared. For this experiment, we analyzed the outputs of matchers that participated in the 2020[4] OAEI complex track[5]. In that evaluation, only Taxon was not included since no reference alignment is present for this track. The matchers included in this analysis are AMLC, CANARD, and AROA since those are the unique producing complex alignments in that year.

The evaluation metric proposed in this paper is compared with existing metrics. These include the instance-based evaluator introduced in [9], which is used in the Populated Conference evaluation, and Entity Identification approach used for the Geolink dataset and other datasets within the complex track [11], as well as in related works [1].

To explore the impact of different types of correspondences, the proposed metric was tested with weight values of $w = 0$, $w = 0.5$, and $w = 1.0$, representing simple correspondences, complex correspondences, and a combination of both, respectively. The results for the proposed metric are present in Table 2.

Not all matchers produced alignments for all datasets, and some only generated alignments in the simplified format designed for simple correspondences. Since the proposed metrics are constructed specifically for dealing with EDOAL, certain alignments were unsupported during the loading phase, or the parsed structure resulted in zero scores across all metrics.

When comparing simple and complex evaluation scenarios, it is possible to see that the matchers perform better in aligning simple entities but have lower performance with complex correspondences. In contrast to the proposed metric in this paper, the Entity Identification and recall evaluate the matchers in some datasets with relatively higher performance, as expected. This is because those metrics do not consider structural constraints. For example, in Populated Geolink, CANARD reaches 0.26 f-measure in the balanced proposed metric but gets 0.54 in the relaxed f-measure. But considering only complex alignments is possible to see that in this case, CANARD gets 0.09 f-measure. It gets a higher value in the relaxed metric since that metric ignores structural considerations, leading to higher result values. The same case occurs with AROA, which gets 0.39 in the Proposed metric with $w = 0.5$ and 0.24 considering only complex correspondences, but gets 0.60 in the relaxed f-measure. This highlights the importance of evaluating a system's ability to handle structural complexities (a key capability of modern AI models like LLMs) that comparison similarities fail to capture.

In the instance-based evaluation, which results are in the Populated Conference dataset in Table 2. There were no results for the Geolink, Hydrography, and Populated Enslaved dataset, as it lacks the Competency Question for Alignment (CQA) required for this evaluation. In the populated conference dataset, CANARD outperformed AMLC relative to the instance-based evaluation, but

[4] This year was used in this experiment since in 2024 only one matcher was submitted.

[5] https://oaei.ontologymatching.org/2020/results/complex/index.html.

Table 2. Comparison of metrics across different datasets and matchers. Results of zero in all metrics in some datasets were omitted for brevity. Entity Identification is a subprocess in Relaxed Precision and Recall for the results in this table.

Dataset	Metric	AMLC	CANARD	AROA
Conference	Proposed w = 0.5 (Precision)	0.45	–	–
	Proposed w = 0.5 (Recall)	0.19	–	–
	Proposed w = 0.5 (F1)	0.26	–	–
	Proposed w = 1 (Precision)	0.45	–	–
	Proposed w = 1 (Recall)	0.65	–	–
	Proposed w = 1 (F1)	0.51	–	–
	Precision (Manual)	0.31	–	–
	F-measure (Manual)	0.34	–	–
	Recall (Manual)	0.37	–	–
Populated Conference	Proposed w = 0.5 (Precision)	0.38	0.25	–
	Proposed w = 0.5 (Recall)	0.47	0.52	–
	Proposed w = 0.5 (F1)	0.40	0.33	–
	Proposed w = 0 (Precision)	0.69	0.60	–
	Proposed w = 0 (Recall)	0.46	0.55	–
	Proposed w = 0 (F1)	0.54	0.56	–
	Proposed w = 1 (Precision)	0.21	0.12	–
	Proposed w = 1 (Recall)	0.54	0.43	–
	Proposed w = 1 (F1)	0.27	0.17	–
	Precision (Instance-based)	0.23–0.51	0.25–0.88	–
	Coverage (Instance-based)	0.26–0.31	0.40–0.50	–
Hydrography	Proposed w=0.5 (Precision)	0.02	–	–
	Proposed w = 0.5 (Recall)	0.01	–	–
	Proposed w = 0.5 (F1)	0.01	–	–
	Proposed w = 1 (Precision)	0.05	–	–
	Proposed w = 1 (Recall)	0.06	–	–
	Proposed w = 1 (F1)	0.06	–	–
	Relaxed Precision	0.45	–	–
	Relaxed F-measure	0.10	–	–
	Relaxed Recall	0.05	–	–
Geolink	Relaxed Precision	0.50	–	–
	Relaxed F–measure	0.32	–	–
	Relaxed Recall	0.23	–	–

(*continued*)

Table 2. (*continued*)

Dataset	Metric	AMLC	CANARD	AROA
Populated Geolink	Proposed w=0.5 (Precision)	–	0.45	0.71
	Proposed w = 0.5 (Recall)	–	0.18	0.27
	Proposed w = 0.5 (F1)	–	0.26	0.39
	Proposed w = 0 (Precision)	–	1.00	0.94
	Proposed w = 0 (Recall)	–	0.78	0.82
	Proposed w = 0 (F1)	–	0.88	0.87
	Proposed w = 1 (Precision)	–	0.20	0.57
	Proposed w = 1 (Recall)	–	0.06	0.16
	Proposed w = 1 (F1)	–	0.09	0.24
	Relaxed Precision	0.50	0.89	0.87
	Relaxed F-measure	0.32	0.54	0.60
	Relaxed Recall	0.23	0.39	0.46
Populated Enslaved	Proposed w = 0.5 (Precision)	0.24	0.16	–
	Proposed w = 0.5 (Recall)	0.10	0.10	–
	Proposed w = 0.5 (F1)	0.14	0.12	–
	Proposed w = 0 (Precision)	0.24	0.18	–
	Proposed w = 0 (Recall)	0.47	0.50	–
	Proposed w = 0 (F1)	0.32	0.26	–
	Proposed w = 1 (Precision)	0.23	0.16	–
	Proposed w = 1 (Recall)	0.03	0.03	–
	Proposed w = 1 (F1)	0.06	0.05	–
	Relaxed Precision	0.73	0.42	0.80
	Relaxed F-measure	0.40	0.19	0.51
	Relaxed Recall	0.28	0.13	0.38

in the proposed metric, AMLC has higher results than CANARD, showing that depending on the metric, the ranking of matchers' performance changes. However, this evaluation does not distinguish between the number of simple and complex correspondences returned, making it unclear whether higher results stem from better handling of simple or complex cases. Incorporating this distinction would enhance performance analysis and provide more granular insights into the matchers' capabilities, as was done in the proposed metric. Another characteristic of the instance-based evaluator is that duplicating correspondences and adding them to the alignment does not affect the results. This can lead to issues where proposing multiple small modifications of the same correspondence will not result in reduced precision.

Another observation is the higher resource consumption of instance-based evaluations. These evaluations require CQAs as input for the source and target ontologies, along with CQAs for the related dataset, also with the system outputs. However, precision evaluation does not require CQAs. In contrast, the proposed metric only relies on the output alignment and reference alignment without the need to load the datasets. While obtaining reference alignments can be challenging, the proposed metric is compatible with six out of seven datasets in the 2020 OAEI complex track, except for the Taxon dataset, which lacks a reference alignment. Instance-based metrics, on the other hand, can only evaluate five of the seven datasets and fully analyze only two (populated_conference and Taxon) due to the strict requirements of having both CQAs and instances. For datasets like populated GeoLink, Hydrography, and populated Enslaved, only precision evaluation is feasible. However, this evaluation can complement the proposed one when no reference alignment is present in the dataset.

4 Related Work

In complex ontology matching (the reader can refer to [12] for a survey), manual evaluation remains a technique used to assess matchers' performance due to the intricate nature of the correspondences involved [6,7,14]. Unlike simple matching, where automated metrics like precision and recall can often provide reliable assessments, complex matching requires deeper semantic understanding and contextual interpretation that automated tools have difficulties achieving. Human experts are typically involved in evaluating the alignments' correctness, completeness, and semantic coherence, particularly in scenarios where gold standards are unavailable or incomplete. This manual process is time-consuming and subject to potential biases or inconsistencies.

To enable automatic evaluation of complex ontology matching, one of the commonly used metrics is relaxed precision and recall [2]. These metrics aim to provide flexibility by allowing partial matches between complex correspondences rather than requiring exact equivalence. However, a significant limitation is that, while being a general metric that can be extended, they do not define metrics to measure the similarity between the subgraphs or structures involved in complex correspondences. Consequently, relaxed precision and recall without the proper graph similarity score may fail to capture the full structural and semantic fidelity of complex correspondences, limiting their effectiveness in assessing overall alignment quality.

Other metrics for evaluating complex ontology alignments rely on set-based computations, focusing on the overlap of common entities between the corresponding elements as described in [9]. A common metric applied to the complex matching case based on this principle is the measure of the number of correct entities present in the correspondence, resembling the Jaccard similarity metric. While these metrics are straightforward and computationally efficient, they have limitations in representativity since they don't consider the structural relationships or semantic dependencies between the entities within the subgraphs defined by the correspondences. This omission means that approaches capable of correctly identifying those aggregation relations won't have higher performance than just identifying the composing entities.

Another approach to evaluating complex ontology alignments involves metrics based on Competency Questions for Alignment (CQAs) and instances [13]. CQAs are SPARQL queries written to describe the user needs in terms of alignment, and are often provided as input to the matchers in some datasets. In that proposed evaluator, the quality of the alignment is assessed by comparing the similarity between the CQAs and the rewritten alignments as SPARQL queries. However, this method faces significant challenges, as CQAs are user-defined and require manual creation, making the process time-intensive and dependent on human expertise. Additionally, some metrics evaluate the alignment by measuring the number of common instances retrieved by the rewritten queries. While this can provide valuable insights, it has critical limitations: not all ontologies contain instances, and even when instances exist, not all entities have associated instances. As a result, these metrics fail to comprehensively evaluate all matched concepts, leaving gaps in the assessment of alignment quality, since it may not cover the whole ontology.

5 Conclusion

This paper introduced a novel metric to evaluate complex ontology alignments that consider the structure of matched entities. While some metrics rely on the use of common instances (A-Box data), such an approach is not always feasible, as ontologies may not be exhaustively populated or may lack an A-Box altogether. Our metric provides an alternative in these scenarios. In the OAEI complex track, most datasets include reference alignments, but these are often underexploited—either partially or evaluated manually. Unlike existing approaches, the proposed metric accounts for the structural and semantic relations of (m:n) correspondences by using a tree similarity computation and ensuring adherence to key evaluation properties for alignments written in EDOAL, such as identity, order invariance, and sensitivity to correspondence modifications. The empirical analysis demonstrated the metric's capability to deliver insights into alignment quality compared to traditional and instance-based evaluation methods.

The proposed metric outperforms alternatives in scenarios emphasizing structural evaluation, enabling a more accurate distinction between simple and complex correspondences. It also addresses limitations in other methods, such as their inability to comprehensively evaluate alignments that involve structural components or their reliance on dataset-specific characteristics like instance availability or predefined competency questions. This structural evaluation requires that matchers built for the complex task produce the correct combination structure for the entities. Now with LLMs that combination can be better produced, and having suitable metrics can help the complex evaluation task follow the advancements of techniques in this field. Also, a new evaluation technique can bring new perspectives to the Complex track in OAEI evaluation and help bring more participants.

However, challenges remain. The dependency on high-quality reference alignments can limit applicability in contexts where such alignments are unavailable or prone to errors. Future work will explore approaches to mitigate these dependencies, such as developing probabilistic models to assess alignment quality in the absence of complete gold standards. Furthermore, refinement of the metric to accommodate evolving ontology formats and emerging alignment needs will ensure its continued relevance in advancing ontology matching research.

References

1. Amini, R., Norouzi, S.S., Hitzler, P., Amini, R.: Towards complex ontology alignment using large language models. CoRR abs/2404.10329 (2024). https://doi.org/10.48550/ARXIV.2404.10329
2. Ehrig, M., Euzenat, J.: Relaxed precision and recall for ontology matching. In: Ashpole, B., Ehrig, M., Euzenat, J., Stuckenschmidt, H. (eds.) Integrating Ontologies 2005, Proceedings of the K-CAP 2005 Workshop on Integrating Ontologies, Banff, Canada, October 2, 2005. CEUR Workshop Proceedings, vol. 156. CEUR-WS.org (2005). https://ceur-ws.org/Vol-156/paper5.pdf
3. Euzenat, J., David, J., Atencia, M.: EDOAL: expressive and declarative ontology alignment language (2015). http://alignapi.gforge.inria.fr/edoal.html. Accessed 12 Feb 2024
4. Jouhet, V., Mougin, F., Bréchat, B., Thiessard, F.: Building a model for disease classification integration in oncology, an approach based on the national cancer institute thesaurus. J. Biomed. Semant. **8**(1), 6:1–6:12 (2017). https://doi.org/10.1186/s13326-017-0114-4
5. Nurmikko-Fuller, T., et al.: Building complex research collections in digital libraries: a survey of ontology implications. In: Proceedings of the 15th ACM/IEEE-CE Joint Conference on Digital Libraries, pp. 169–172. ACM (2015). https://doi.org/10.1145/2756406.2756944
6. Atencia, M., Borgida, A., Euzenat, J., Ghidini, C., Serafini, L.: A formal semantics for weighted ontology mappings. In: Cudré-Mauroux, P., et al. (eds.) ISWC 2012. LNCS, vol. 7649, pp. 17–33. Springer, Heidelberg (2012). https://doi.org/10.1007/978-3-642-35176-1_2
7. Ritze, D., Völker, J., Meilicke, C., Sváb-Zamazal, O.: Linguistic analysis for complex ontology matching. In: Shvaiko, P., Euzenat, J., Giunchiglia, F., Stuckenschmidt, H., Mao, M., Cruz, I.F. (eds.) Proceedings of the 5th International Workshop on Ontology Matching (OM-2010), Shanghai, China, 7 November 2010. CEUR Workshop Proceedings, vol. 689. CEUR-WS.org (2010). https://ceur-ws.org/Vol-689/om2010_Tpaper1.pdf
8. Silva, M.C., Faria, D., Pesquita, C.: Complex multi-ontology alignment through geometric operations on language embeddings. In: Endriss, U., et al. (eds.) ECAI 2024 - 27th European Conference on Artificial Intelligence, 19–24 October 2024, Santiago de Compostela, Spain - Including 13th Conference on Prestigious Applications of Intelligent Systems (PAIS 2024). Frontiers in Artificial Intelligence and Applications, vol. 392, pp. 1333–1340. IOS Press (2024). https://doi.org/10.3233/FAIA240632

9. Thiéblin, É.: Automatic generation of complex ontology alignments. (Génération automatique d'alignements complexes d'ontologies). Ph.D. thesis, Paul Sabatier University, Toulouse, France (2019). https://tel.archives-ouvertes.fr/tel-02735724
10. Thiéblin, E., Amarger, F., Hernandez, N., Roussey, C., Trojahn Dos Santos, C.: Cross-querying LOD datasets using complex alignments: an application to agronomic taxa. In: Garoufallou, E., Virkus, S., Siatri, R., Koutsomiha, D. (eds.) MTSR 2017. CCIS, vol. 755, pp. 25–37. Springer, Cham (2017). https://doi.org/10.1007/978-3-319-70863-8_3
11. Thiéblin, É., Cheatham, M., Trojahn, C., Zamazal, O.: A consensual dataset for complex ontology matching evaluation. Knowl. Eng. Rev. **35**, e34 (2020). https://doi.org/10.1017/S0269888920000247
12. Thiéblin, É., Haemmerlé, O., Hernandez, N., Trojahn, C.: Survey on complex ontology matching. Semant. Web **11**(4), 689–727 (2020). https://doi.org/10.3233/SW-190366
13. Thiéblin, É., Haemmerlé, O., Trojahn, C.: Automatic evaluation of complex alignments: An instance-based approach. Semant. Web **12**(5), 767–787 (2021). https://doi.org/10.3233/SW-210437
14. Walshe, B., Brennan, R., O'Sullivan, D.: Bayes-ReCCE: a Bayesian model for detecting restriction class correspondences in linked open data knowledge bases. Int. J. Semant. Web Inf. Syst. **12**(2), 25–52 (2016). https://doi.org/10.4018/IJSWIS.2016040102
15. Zhang, K., Shasha, D.E.: Simple fast algorithms for the editing distance between trees and related problems. SIAM J. Comput. **18**(6), 1245–1262 (1989). https://doi.org/10.1137/0218082
16. Zhang, K., Statman, R., Shasha, D.E.: On the editing distance between unordered labeled trees. Inf. Process. Lett. **42**(3), 133–139 (1992). https://doi.org/10.1016/0020-0190(92)90136-J

Kastor: Fine-Tuned Small Language Models for Shape-Based Active Relation Extraction

Célian Ringwald[1(✉)], Fabien Gandon[1], Catherine Faron[1], Franck Michel[1], and Hanna Abi Akl[1,2]

[1] Université Côte d'Azur, Inria, CNRS, I3S, Nice, France
{celian.ringwald,fabien.gandon,catherine.faron, franck.michel,hanna.abi-akl}@inria.fr
[2] Data ScienceTech Institute, Paris, France

Abstract. RDF pattern-based extraction is a compelling approach for fine-tuning small language models (SLMs) by focusing a relation extraction task on a specified SHACL shape. This technique enables the development of efficient models trained on limited text and RDF data. In this article, we introduce Kastor, a framework that advances this approach to meet the demands for completing and refining knowledge bases in specialized domains. Kastor reformulates the traditional validation task, shifting from single SHACL shape validation to evaluating all possible combinations of properties derived from the shape. By selecting the optimal combination for each training example, the framework significantly enhances model generalization and performance. Additionally, Kastor employs an iterative learning process to refine noisy knowledge bases, enabling the creation of robust models capable of uncovering new, relevant facts.

Keywords: Relation Extraction · Small Language Models · Structured output

1 Introduction

Relation extraction (RE), the task of retrieving relations from unstructured text, was drastically improved by language models and massive corpora aligning texts and facts from Knowledge Bases (KB) – e.g. Wikipedia articles with corresponding Wikidata or DBpedia subgraphs. The use of generative seqToseq models is prevalent today to solve structured output generation, as they are flexible compared to encoder-only models, which require a decoding strategy. In this perspective, fine-tuned encoder-decoder models such as T5 or BART demonstrated good performances in the relation extraction domain [3,26,29], and subsequent works underlined that small models could compete with larger models [16,18,39]. However, these RE models fine-tuned on large-scale datasets do not guarantee good results on rare relations and entities. In addition, the catastrophic forgetting effect [19] makes difficult a second-step adaptation of these models to a more specific domain. On the other hand, distant supervision approaches [2,31] used to align massive text corpora and facts from databases have two significant drawbacks: (1) they may incorporate noise at the model learning stage by coupling a graph describing facts not present in the text and vice versa, and (2) they

E. Curry et al. (Eds.): ESWC 2025, LNCS 15718, pp. 94–115, 2025.
https://doi.org/10.1007/978-3-031-94575-5_6

also affect how the resulting models are evaluated, from a knowledge completion point of view, by labelling as false positives results containing relevant information.

Kastor (Knowledge Active Shape-based extracTOR) answers these issues by proposing the refinement of specialized and frugal small language models (SLMs) focused on specific RDF patterns. The resulting model can produce structured data that could populate a database directly. Building on previous work [27], we propose here to characterize the combination of properties $\mathbb{P}(s^*)$ derived from a given SHACL shape s^*. We reused the shape introduced in [27] targeting the `dbo:Person` class, as it represents $1/6^t h$ of the DBpedia content, by targeting the 7 datatype properties that are most likely to be found in the abstracts. This shows that Kastor already scales. Our framework consists of coupling an initial incomplete and noisy knowledge base $\mathcal{K}$ with language models, and it works in two stages. It first consolidates $\mathcal{K}$ by ensuring strong alignment of the dual base, which will then be sampled to obtain smaller, higher-quality training and evaluation datasets used for training an initial SLM. Kastor's second step integrates a light, active learning process involving a human annotator. This annotator is used to analyze and correct the outputs produced by this first model to produce a gold model, which could extract relevant RDF graphs to complete $\mathcal{K}$. To summarize, this paper addresses the two following research questions:

- **RQ1.** To what extent does a task relying on example-specific achievable patterns improve the performance of relation extraction model?
- **RQ2.** Does an active learning process enhance relation extraction models?

2 Related Work

SLMs are Competitive. The usage of LLMs is called into question today [7,23,36], as they are costly to train, slow in inference and hard to adapt to a specialized domain [15]. Moreover, although LLMs are highlighted for their few-shots abilities, recent work [39] focused on NER challenged this belief. In the context of relation extraction, several works underline the good performances of fine-tuned SLMs over prompted LLMs [5,12,17]. Finally, LLMs only marginally outperform SLMs with the help of great engineering work [4,20,35,40].

Human Feedback is Costly but Necessary. Because of the noise produced by distant supervision methods, the two classical datasets related to the relation extraction task, TACRED and DOCRED, were revised and corrected several times [1,32,37,38]. Wikidata and DBpedia are, by construction, both concerned with coverage and quality issues [8,30], and the datasets recently proposed, such as TREX and REBEL do not spare these issues. For this reason, several works integrate partial human annotation to reduce noise in training datasets [10,24]. The recent concept of LLM-as-judge aims to replace human intervention with LLMs [41]. Nevertheless, this proposal is imperfect, and [21] proposed combining both approaches (CoAnnotation). [25,34] also demonstrated the potential of this synergy.

The Problem of Hallucination. The literature on hallucination shows how language models fail to produce expected values. A first dichotomy is proposed in [13] between the intrinsic hallucination (which contradicts the input) and the extrinsic hallucination (which cannot be verified by the input). [9] goes further in this categorisation by differentiating factuality hallucination, which relates to outputs that could not be supported by the text, and faithfulness hallucinations, which is related to the consistency of the retrieved answer, considering contextual and logical aspects. In the context of the NER task [28] demonstrates that the noise of the training set directly impacts the hallucination rate of fine-tuned SLMs.

Positioning: The shape-based RE extraction framework we propose is a new task that is difficult to compare with the RE state-of-the-art. First, RE models based on the fine-tuning of PLM offer no control over the properties to extract, and they all generate triples following different linearizations [11, 14, 26, 29] that would require additional data transformations that may create additional errors. Second, LLMs, despite their few shot capacities, demonstrate issues when it comes to producing structured output [6, 22].

Contributions: (1) Kastor extends [27] by refining the task definition, and it demonstrates better performance in a wider variety of cases. (2) Kastor integrates this new task definition into a generalized framework, allowing the systematization of shape-based extractors over the chain, from the sample selection to the annotation and later to the PLM finetuning. (3) Kastor proposes a light-active process to build gold datasets and unbiased models, increasing the produced graph's relevance for a KB completion scenario. We also propose a characterization of the errors that (a) allows us to check the generated triples and (b) could be extended in other settings. The produced material is made open and reusable[1]: both the resulting models[2] and the produced datasets[3].

3 The Kastor Framework

3.1 The Original Task: An Extraction Focused on a Maximal Target Shape s^*

We start by formally defining the relation extraction task introduced in [27]. It relies on a training set built from the dual base $\mathcal{K}$ defined from the set $\mathcal{W}$ of Wikipedia abstracts associated with the set $\mathcal{G}$ of DBpedia graphs describing ($desc()$) the same resource e:

$$\mathcal{K} := \{(w, g) \in \mathcal{W} \times \mathcal{G}, \exists e \in IRI \text{ such that } desc_{\mathcal{W}}(e) = w \wedge desc_{\mathcal{G}}(e) = g\} \quad (1)$$

To ensure the quality of the training set, we considered the subset of $\mathcal{K}$ where all the graphs are valid against a SHACL shape s^*. We call this shape maximal, as it matches the largest pattern to be extracted. We note $g \models s^*$ this validation, and $\mathcal{K}_{s^*}$ the corresponding subset:

$$\mathcal{K}_{s^*} := \{(w, g) \in \mathcal{K}, \ g \models s^*\} \quad (2)$$

Finally, to reduce the noise in $\mathcal{K}_{s^*}$, due to mismatch between DBpedia graphs and Wikipedia abstracts, we focused only on the couples (w, g), where the abstract w entails the paired graph g, i.e. the triples of g that can actually be extracted from the paired abstract w. We note it $w \models g$ and we denote the dataset by $\mathcal{K}_{s^*}^{\mathcal{W}\models}$:

[1] https://github.com/datalogism/Kastor.

[2] https://zenodo.org/records/14498940.

[3] https://zenodo.org/records/14382674.

$$\mathcal{K}_{s^*}^{\mathcal{W}\models} := \{(w, g) \in \mathcal{K}_{s^*}\,,\ w \models g\} \tag{3}$$

$\mathcal{K}_{s^*}^{\mathcal{W}\models}$ is used to train a model expected to predict, from an abstract w, a graph $\hat{g}$ valid against s^*. We denote by $\mathcal{M}$ this original model, and baseline:

$$\mathcal{M} : \begin{cases} \mathcal{W} \to \mathcal{G} \\ w \mapsto \hat{g}\,,\ \hat{g} \models s^* \wedge w \models \hat{g} \end{cases} \tag{4}$$

3.2 Example-Specific Patterns and Rule-Based Graph Augmentation

In practice, many abstracts are short and miss some properties defined as mandatory in $s*$. A model not trained to manage such cases could be encouraged to extract relations from an abstract with missing information, leading to hallucinations. So instead of considering a single graph pattern entailed by the SHACL shape s^* for the whole training set, we propose to consider the example-specific "achievable" graph patterns $\pi \in \Pi$ for each pair $(w, g) \in \mathcal{K}$. We denote by $\mathcal{P}(g)$ the set of properties occurring in a graph g:

$$\mathcal{P} : \left| \begin{array}{l} \mathcal{G} \longrightarrow \Pi = 2^I \\ g \longmapsto \pi = \{p_i\,,\ \exists\,(x, p_i, o) \in g\} \end{array} \right. \tag{5}$$

By extension, we denote by $\mathcal{P}(s)$ the set of properties that a shape s constrains a graph with. We denote by $\mathbb{P}(g)$ the powerset of $\mathcal{P}(g)$, respectively by $\mathbb{P}(s)$ the powerset of $\mathcal{P}(s)$, that could be deduced from a given g, respectively with s^*. We call *example-specific patterns* the elements of $\mathbb{P}(g)$ and $\mathbb{P}(s)$.

To compare a graph g with an example-specific pattern π, we note $g \rightarrow \pi$ the fact that all the properties in g are found in π; and we note $g \nrightarrow \pi$ its negation, i.e. the fact that at least one property of π is not found in g. We note $g \leftrightarrow \pi$ the fact that all the properties in g are found in π and vice versa; and we note $g \nleftrightarrow \pi$ its negation. By extension, when it comes to compare g with a shape s, we denote by $g \rightarrow s$ the fact that all the properties in a graph pattern expressed in a shape s are found in g; and by $g \leftrightarrow s$ the fact that all the properties in g are found in the graph pattern expressed in s and vice versa.

We can now relax the constraint in the original task that all the graphs in the training set must be valid against a single common shape s^* (see Eq. 2), in order to define a new training set considering all the example-specific patterns derived from s^*:

$$\mathcal{K}_{\mathbb{P}(s^*)} := \{(w, g) \in \mathcal{K}; g \rightarrow s_i; s_i \in \mathbb{P}(s^*)\} \tag{6}$$

Additionally, we propose to complement $\mathcal{K}_{\mathbb{P}(s^*)}$ by materializing the closure of a set of inference rules $\mathcal{R}$ applied to $\mathcal{G}$. The rationale is as follows: some basic reasoning tasks can easily be deduced from the data by simple inference rules (e.g. deducing a year from a date). Therefore, instead of expecting the language models to learn these basic rules, we apply them in a declarative manner, thus homogenizing the graph and making the learning process easier. The resulting set is denoted $\mathcal{K}_{\mathbb{P}(s^*)}^{\mathcal{R}\models}$:

$$\mathcal{K}_{\mathbb{P}(s^*)}^{\mathcal{R}\models} := \{(w, g')\,,\ (w, g) \in \mathcal{K}_{\mathbb{P}(s^*)}\,,\ \text{where } g' \text{ is the result of applying } \mathcal{R} \text{ to } g\} \tag{7}$$

We ensure the entailment of the graphs by their paired abstracts ($w \models g$) as done in the original design:

$$\mathcal{K}^{\mathcal{WR}\models}_{\mathbb{P}(s^*)} := \{(w,g) \in \mathcal{K}^{\mathcal{R}\models}_{\mathbb{P}(s^*)},\ w \models g\} \tag{8}$$

The resulting set $\mathcal{K}^{\mathcal{WR}\models}_{\mathbb{P}(s^*)}$ is sampled to finetune a new generation of model $\mathcal{M}'$:

$$\mathcal{M}' : \begin{cases} \mathcal{W} \rightarrow \mathcal{G} \\ w \mapsto \hat{g}; \quad \hat{g} \leftrightarrow \mathcal{P}(g) \wedge (w,g) \in \mathcal{K}^{\mathcal{WR}\models}_{\mathbb{P}(s^*)} \end{cases} \tag{9}$$

To train the models $\mathcal{M}$ presented in Eq. 4, and $\mathcal{M}'$ presented Eq. 9, we considered random samples RD which are split into RD_{train}, RD_{eval} and RD_{test}.

We note $\mathcal{M}'_{RD}$, the model $\mathcal{M}'$ trained on dataset RD_{train}, evaluated on RD_{eval} and tested on RD_{test}.

Finally, to characterize the set of distinct example-specific patterns that are actually present in a given dataset D, and compare the variety of patterns in each dataset, we define:

$$\mathbb{P}_D(s) := \{\pi \in \mathbb{P}(s); \exists (w,g) \in D; g \leftrightarrow \pi\} \tag{10}$$

For instance $\mathbb{P}_{RD^1}(s^*)$ will represent the set of example-specific patterns that can be built from s^* and actually found in RD^1.

3.3 Knowledge Distillation: From $\mathcal{K}$ to $\mathcal{K}^{\mathcal{WR}\models}_{\mathbb{P}(s^*)}$

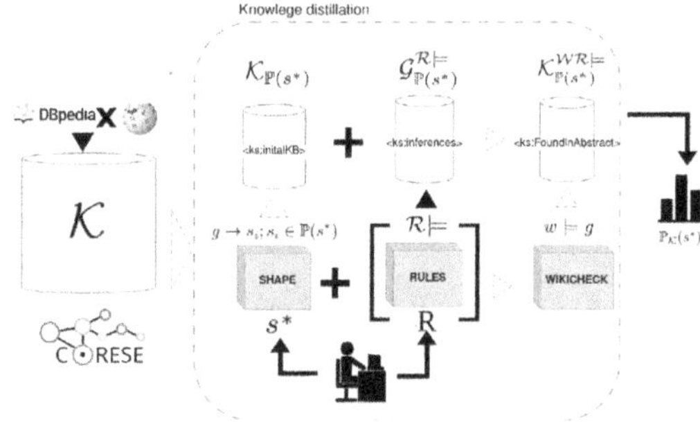

Fig. 1. Knowledge distillation: from the initial base to a refined version based on shape s^*

Figure 1, we consider the **dual base** $\mathcal{K}$ consisting in the 2022.09 DBpedia datadump[4] that gathers 6.109.994 Wikipedia abstracts and their DBpedia graphs. We consider the **shape** s^* described in [27][5] which targets instances of class `dbo:Person` and expresses that they must have one label, one birth year or date, and possibly 4 other optional properties:

[4] https://databus.dbpedia.org/cringwald/collections/kstor.

[5] https://github.com/datalogism/12ShadesOfRDFSyntax#shacl-shape-used.

$\mathcal{P}(s^*) = \{$`rdfs:label`, `dbo:alias`, `dbo:birthName`, `dbo:birthDate`, `dbo:deathDate`, `dbo:birthYear`, `dbo:deathYear`$\}$
and from it we can compute $|\mathcal{P}(s^*)| = 7$ and $|\mathbb{P}(s^*)| = 128$. There are 1.833.493 Person graphs in $\mathcal{G}$ valid against at least one of the 127 non-empty graph patterns in $\mathbb{P}(s^*)$. They are gathered in **the subbase** $\mathcal{K}_{\mathbb{P}(s^*)}$ stored into a named graph `<ks:initalKB>`. We observed in $\mathcal{G}_{\mathbb{P}(s^*)}$ a huge number of birth/death dates compared to the number of birth/death years. To solve this gap, we defined the following set of inference rules (Fig. 1):

$$\mathcal{R}: \begin{cases} \texttt{dbo:deathDate} \models \texttt{dbo:deathYear} \\ \texttt{dbo:birthDate} \models \texttt{dbo:birthYear} \end{cases} \tag{11}$$

These rules, encoded as SPARQL Update queries, are applied to `<ks:initalKB>`, producing an **enriched base** $\mathcal{K}^{\mathcal{R}\models}_{\mathbb{P}(s^*)}$. We note $\mathcal{G}^{\mathcal{R}\models}_{\mathbb{P}(s^*)}$ the 900.000+ *new* triples produced, and we store them in the named graph `<ks:inferences>`.

We filter $\mathcal{K}^{\mathcal{R}\models}_{\mathbb{P}(s^*)}$ with the *wikicheck* module in charge to verify Eq. 8. The **resulting base** $\mathcal{K}^{\mathcal{WR}\models}_{\mathbb{P}(s^*)}$ is stored in the named graph `<ks:foundInAbstract>`. Statistics on the different graphs are given in Table 1. $\mathcal{K}^{\mathcal{WR}\models}_{\mathbb{P}(s^*)}$ filters 40% of the entities described in $\mathcal{K}_{\mathbb{P}(s^*)}$ that have no property value in their graph that can be found in their abstract. This shows the importance of such filtering to avoid training models on facts unsupported by the abstract, which could encourage hallucinations. Note that some properties are often found in the abstracts, such as `dbo:birthYear`, or `dbo:deathDate`, whereas `dbo:alias` is often missing.

Table 1. Number of triples in each consolidated named graph

predicate	$\mathcal{K}_{\mathbb{P}(s^*)}$ <ks:initialKB>	$\mathcal{K}^{\mathcal{R}\models}_{\mathbb{P}(s^*)}$ <ks:initalKB> ∪ <ks:inferences>	$\mathcal{K}^{\mathcal{WR}\models}_{\mathbb{P}(s^*)}$ <ks:foundInAbstract>	Part Found
`dbo:birthName`	120 102	120 102	84 377	70%
`dbo:birthYear`	186 575	**960 977**	**888 745**	92%
`dbo:deathDate`	250 970	250 970	213 424	85%
`dbo:alias`	30 648	30 648	10 571	35%
`dbo:deathYear`	97 817	335 571	315 671	**94%**
`dbo:label`	**987 389**	**987 389**	*596 286*	60%
`dbo:birthDate`	*705 952*	*705 952*	*605 239*	85%
Nb. entities	1 833 493	1 833 493	1 093 886	60%

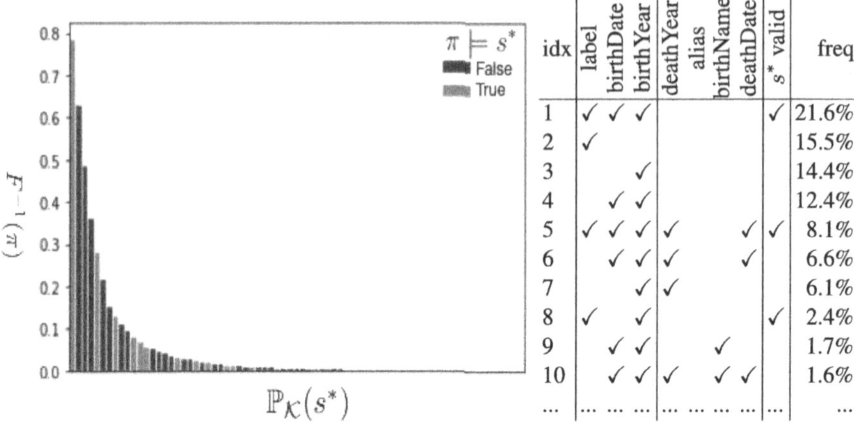

idx	label	birthDate	birthYear	deathYear	alias	birthName	deathDate	s^* valid	freq
1	✓	✓	✓					✓	21.6%
2	✓								15.5%
3			✓						14.4%
4		✓	✓						12.4%
5	✓	✓	✓	✓			✓	✓	8.1%
6		✓	✓	✓			✓		6.6%
7			✓	✓					6.1%
8	✓		✓					✓	2.4%
9		✓	✓			✓			1.7%
10		✓	✓	✓		✓	✓		1.6%
...	...	...	...	...	...	...	...	...	...

Fig. 2. $\mathbb{P}_{\mathcal{K}}(s^*)$ Inverse cumulative frequency distribution

Fig. 3. 10 most frequent $\pi \in \mathbb{P}_{\mathcal{K}}(s^*)$

Reusing the notation from Eq. 10 applied to $\mathcal{K}^{WR\models}_{\mathbb{P}(s^*)}$, we get **an example-specific pattern set** $\mathbb{P}_{\mathcal{K}}(s^*)$ containing $|\mathbb{P}_{\mathcal{K}}(s^*)| = 70$ patterns, which is nearly half of the 127 possible patterns in $\mathbb{P}(s^*)$. The distribution of these patterns, shown in Fig. 2, is typical of a long-tail distribution. All these patterns can be classified as compliant or not with the maximal SHACL shape s^* (cf. colours of the bars on Fig. 2 and column "s^* valid" in Fig. 3). There are only 47 patterns in $\mathbb{P}_{\mathcal{K}}(s^*)$ validating s^* that can be found in the base. The original design of [27], formalised in Eq. 4, was for this reason focused on a subset of $\mathbb{P}_{\mathcal{K}}(s^*)$ (see Eq. 10).

3.4 An Iterative SLM Learning Process with Human in the Loop

Learning from a Small Dataset. In [27] we followed a 5-fold cross-validation training process implying 4 000 training examples, with 250 disjoint examples used for the evaluation and 1 000 test examples. We repeated the experiment to find a better cost-performance balance by testing different fold numbers and sample sizes. We empirically established that a 10-fold cross-validation based on a sample size of 1 000 rotating examples is sufficient to fine-tune an efficient model (cf. Table 2).

Table 2. Impact of the number of folds and sample size on the model performances, sorted by computation time

Nb. folds	Nb ex. train	Nb ex. test	Nb ex. eval	time	$F1^+$
10	900 (90%)	100 (10%)	100 (10%)	**2 h**	0.90
5	800 (80%)	200 (20%)	200 (20%)	3 h	0.89
5	4000 (80%)	1000 (20%)	250 (5%)	5 h	0.91
5	2000 (80%)	500 (20%)	500 (5%)	6 h	0.89
10	2250 (90%)	250 (10%)	250 (10%)	7 h	0.93
10	4500 (90%)	500 (10%)	500 (10%)	13 h	**0.97**
5	4000 (80%)	1000 (20%)	1000 (10%)	14 h	0.91

The Sampler. The sampler is the first piece of the Active Learning process; it selects random and independent subsets from $\mathcal{K}^{\mathcal{WR}\models}_{\mathbb{P}(s^*)}$. We used it to generate two datasets: RD^0 and RD^1 of 1200 examples (including 100 additional examples in case we remove some examples during the annotation phase). The sampler also generated an independent dataset of 600 examples, RD^2, used as a control. We also created the RD^- dataset to serve as a baseline, reproducing the original model (Eq. 4) against which to compare our model (Eq. 9). To obtain it we sampled $\mathcal{K}^{WR\models}_{\mathbb{P}(s^*)}$ with an additional filter selecting only pairs (w, g) where g is valid against s^*.

SLM Training Details. All the trained models follow the design of [27] which is extended: For the pre-trained model, we chose BART-base (140M parameters). We linearized the graphs in the TurtleLight online and factorized syntax, which demonstrated the best performances. The models were finetuned on a single Tesla V100-SXM2-32 GB GPU using the same configuration: an inverse square root scheduler with an initial learning rate of 0.00005, 1000 steps of warmup, and configured with an early stop mode with patience of 5 steps. We also used the same prompt to finetune the models: "`$entity_URI : $Abstract`" where $Abstract is a Wikipedia abstract and $entity_URI the URI of the corresponding entity in DBpedia. We extended the initial code with new metrics definitions, a validation of the TurtleLight syntax via an Extended Backus-Naur Form (EBNF) grammar, the validation of the produced triples against the expected $\mathbb{P}(s^*)$ patterns, the automation of the dataset construction, as well as the new testing process.

Light Active Learning. The process described in Appendix A.3 relates to a single loop annotation, creating a gold dataset and model. It starts by finetuning a first model $\mathcal{M}'_{RD^0}$ on RD^0. This model generates from the abstracts in RD^1 and RD^2 predicted graphs $\hat{g}$, which will be compared to the expected graphs g. We only keep the False Positives triples (FP) and False Negatives (FN) triples of $\hat{g}$ produced by testing $\mathcal{M}'_{RD^0}$ with RD^1 and RD^2. The annotator is a domain expert who has to evaluate the FP/FN triples collected regarding the Wikipedia abstract given in input to the model. A triple is considered erroneous if the datatype property value cannot be found in the text or if it is not strictly equal to the expected value.

Each dataset is then enriched with all the FP+ triples (the correct FP) and corrected by deleting the FN- triples (the erroneous FN) to produce two gold standard datasets: RD^{1+} and RD^{2+}.

Each triple of RD^{1+} and RD^{2+} is evaluated with the NLI and the Triplet Critic models proposed in [11] and [10] and the scores are stored in dedicated named graphs (`<ks:sample1+>` and `<ks:sample2+>`) following the classical RDF design-pattern: `?s ?p _:n. _:n rdf:value ?o; ks:triplet_critic ?tc; ks:xnli ?nli.`

RD^{1+} is finally used to produce a gold model $\mathcal{M}'_{RD^{1+}}$ and RD^{2+} will serve as a control to compare the performances of $\mathcal{M}'_{RD^0}$ and $\mathcal{M}'_{RD^{1+}}$.

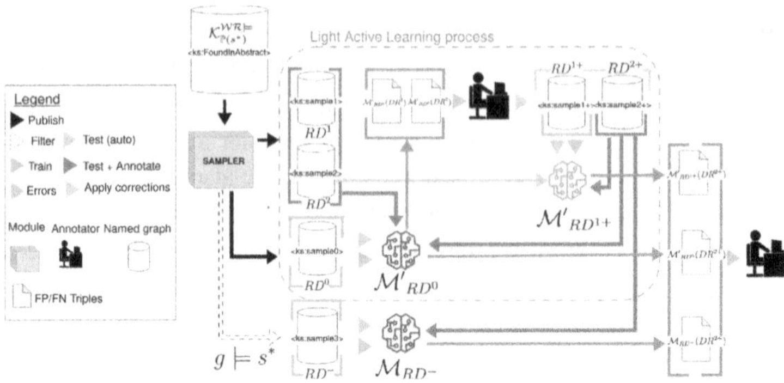

Fig. 4. Overview the experimental set-up including the iterative SLM light active learning process (Color figure online)

Cross-Evaluation of the Models. The complete experimental process followed in this work is presented and summarised in Fig. 4. During the active learning process, we train two models: $\mathcal{M}'_{RD^0}$ and $\mathcal{M}'_{RD^{1+}}$ (cf. orange arrows in Fig. 4). RD^1 and RD^2 are used as test datasets and then corrected after the annotation (cf. dark-blue arrows in Fig. 4). We also tested RD^2 on $\mathcal{M}'_{RD^{1+}}$ to evaluate the gold model in a noisy context. Out of the active process, we train $\mathcal{M}_{RD^-}$. After training, we test all these models with a dedicated subset of D: D_{test} (cf. the light blue arrows in Fig. 5). Finally, we also compare all the produced models with our gold dataset RD^{2+} by annotating all the FP/FN they produced.

4 Experimental Results

4.1 Resulting Datasets

We evaluate a dataset D using several metrics. First, $\overline{|\mathcal{P}(g)|}$ is the average number of properties of the graphs g in a dataset. $|\mathbb{P}_D(s^*)|$ is the number of distinct patterns descending from s^* found in D. We also reused the NLI model[6] [11] and Triplet Critic[7] [10] to estimate if an abstract w contains the information needed to generate the triples of the associated g. These scores, noted $\overline{NLI(w,g)}$ and $\overline{TC(w,g)}$, are defined between 0 and 1. Finally we define r_{s^*} as the rate of graphs following the maximal shape s^*:

$$r_{s^*}(D) = \frac{|D_{s^*}|}{|D|} \tag{12}$$

Table 3 firstly, by considering RD^- (the dataset that only contains graphs validating the maximal shape s^*), we can observe that $|\mathbb{P}_{RD^-}(s^*)| = 16$, which represents only a tiny variety of the patterns realized in the global KB (i.e. $|\mathbb{P}_{\mathcal{K}}(s^*)| = 70$). However, RD^- records a high $\overline{TC(w,g)}$ score compared to other samples. Additionally,

[6] https://huggingface.co/joeddav/xlm-roberta-large-xnli.

[7] https://huggingface.co/Babelscape/mdeberta-v3-base-triplet-critic-xnli.

Table 3. Resulting datasets basic statistics

D	$\|D\|$	$\overline{\|\mathcal{P}(g)\|}$	$\|\mathbb{P}_D(s^*)\|$	$\overline{NLI(w,g)}$	$\overline{TC(w,g)}$	$r_{s^*}(D)$
RD^-	1200	**3.6**	<u>16</u>	0.42	**0.91**	**1.00**
RD^0	1200	<u>2.9</u>	35	<u>0.40</u>	0.55	0.47
RD^1	1200	<u>2.8</u>	**39**	<u>0.40</u>	<u>0.54</u>	0.47
RD^2	600	<u>2.8</u>	30	0.42	0.55	0.49
RD^{1+}	1200	*3.27*	38	*0.59*	*0.75*	0.59
RD^{2+}	599	*3.24*	32	**0.60**	*0.75*	*0.59*

the initial datasets (RD^0, RD^1 and RD^2) all share similar particularities: around 2.9 properties per entity, they realize between 30 and 40 patterns, which is half of the total number of realized patterns of $\mathbb{P}_{\mathcal{K}}(s^*)$. Finally, the corrected datasets (RD^{1+},RD^{2+}) count more properties per entity; compared to their previous versions, they record an increased rate of graphs valid against the maximal shape and higher $\overline{NLI(w,g)}$ and $\overline{TC(w,g)}$ scores.

4.2 Models Performances

We can first underline the frugality of the proposed method in terms of carbon cost and training time (as shown in Table 4). Using *carbontracker*[8], the training cost of these models is competitive: they require less than 10 min of training and the CO_2 footprint associated to their training is limited. To evaluate our models, we propose to define $\mathbb{G}_D$, the graphs expected given a dataset D, and $\widehat{\mathbb{G}_D}$, the set of the predictions obtained from an SLM. These sets allow distinguishing the subset $\widehat{\mathbb{G}}_D^{\text{parsed}}$ of all the predicted graphs that follow the EBNF Turtle Light grammar introduced in Sect. 3.4, and the subset of $\widehat{\mathbb{G}}_D^{\text{URI+}}$ which contains the proper URIs to represent the subjects:

Table 4. Training cost of fine-tuned models

model	$\overline{CO_2(g)}$	$\overline{time}$ (mins)
$\mathcal{M}_{RD^-}$	0,018	7,65
$\mathcal{M}'_{RD^0}$	0,019	8,11
$\mathcal{M}'_{RD^{1+}}$	0,018	7,91

$$\begin{aligned}\widehat{\mathbb{G}}_D^{\text{parsed}} &= \{\hat{g}; (w,g) \in D; \hat{g} \text{ parseable}\}\\ \widehat{\mathbb{G}}_D^{\text{URI+}} &= \{\hat{g}; (w,g) \in D; e \in IRI \text{ the entity described by graph } g;\\ &\quad \hat{e} \in IRI \text{ the entity described by graph } \hat{g}; e = \hat{e}\}\end{aligned} \tag{13}$$

We can now introduce the rate of predicted graphs that were correctly parsed: r_{tll}, as well as r_{URI+}, the rate of correct subject URIs produced:

$$r_{tll} = \frac{|\hat{\mathbb{G}}_D^{parsed}|}{|\mathbb{G}_D|} \qquad r_{URI+} = \frac{|\hat{\mathbb{G}}_D^{URI+}|}{|\hat{\mathbb{G}}_D^{parsed}|} \tag{14}$$

[8] http://carbontracker.info/.

To measure the ability of our models to produce graphs $\hat{g}$ containing exactly the expected property set $\mathcal{P}(g)$, we define the rate of strict property set equivalence:

$$\hat{\mathbb{G}}_D^{\leftrightarrow} = \{\hat{g}; (w,g) \in D; \hat{g} \leftrightarrow \mathcal{P}(g)\} \qquad r_{\hat{\mathbb{G}}_D^{\leftrightarrow}} = \frac{|\hat{\mathbb{G}}_D^{\leftrightarrow}|}{|\hat{\mathbb{G}}_D^{parsed}|} \tag{15}$$

The averaged $\overline{Loss(\hat{\mathbb{G}})}$ based on the cross-entropy helps us evaluate how confident our models are at predicting a given $\hat{g}$ from an abstract w. Moreover, we compute F_1 scores at the macro and micro levels (F_1^+, F_1^-) based on the strict [33] equality of the expected and generated graphs.

Finally, as one of our objectives is to consider and correct the False Positives and False Negatives generated by a model on a given dataset, we also define two final metrics:

$$r_{FP} = \frac{|FP|}{|FP|+|TN|+|TP|+|FN|} \qquad r_{FN} = \frac{|FN|}{|FP|+|TN|+|TP|+|FN|} \tag{16}$$

Table 5. Average performances of fine-tuned models over 10 folds on different test sets, bold values are the best recorded values, italic values represent the second best values, and the underlined values are the worst ones

model	test set	r_{ttl}	r_{URI+}	$\overline{Loss(\hat{\mathbb{G}})}$	F_1^-	F_1^+	r_{FP}	r_{FN}	$r_{\hat{\mathbb{G}}_D^{\leftrightarrow}}$
$\mathcal{M}_{RD^-}$	RD^-_{test}	0.99	1.00	**0.004**	**0.994**	**0.935**	<u>0.0036</u>	<u>0.0021</u>	**0.97**
$\mathcal{M}_{RD^-}$	RD^{2+}	0.99	1.00	<u>0.27</u>	<u>0.887</u>	<u>0.697</u>	**0.0717**	**0.0355**	<u>0.43</u>
$\mathcal{M}'_{RD^0}$	RD^0_{test}	0.99	1.00	0.01	0.978	0.916	0.0155	<u>0.0023</u>	0.92
$\mathcal{M}'_{RD^0}$	RD^2	0.99	1.00	0.07	0.927	0.733	*0.0448*	*0.0162*	0.73
$\mathcal{M}'_{RD^0}$	RD^{2+}	0.99	1.00	0.07	0.94	0.785	0.014	**0.0395**	0.72
$\mathcal{M}'_{RD^{1+}}$	RD^{1+}_{test}	0.99	1.00	**0.004**	**0.991**	0.916	0.0054	0.0031	*0.95*
$\mathcal{M}'_{RD^{1+}}$	RD^2	0.99	1.00	0.12	0.907	0.724	**0.0737**	0.0076	0.62
$\mathcal{M}'_{RD^{1+}}$	RD^{2+}	0.99	1.00	*0.04*	*0.958*	*0.806*	0.026	0.0131	0.80

Table 5 presents the results obtained in the same manner as the other tables in this paper. At first glance, we can notice that all the produced models generate close to syntactically perfect Turtle Light RDF graphs (r_{ttl}) and refer in every case to the correct focused entity URIs (r_{URI+}). The first model $\mathcal{M}_{RD^-}$ records high performances regarding the test set RD^-_{test}: a low $\overline{Loss(\hat{\mathbb{G}})}$, high $F1$ metrics and a significant part of produced graphs follow the expected patterns ($r_{\hat{\mathbb{G}}_D^{\leftrightarrow}}$). Nevertheless, $\mathcal{M}_{RD^-}$ completely failed on RD^{2+}, which reveals an apparent lack of generalisation over graph patterns that do not strictly validate against the maximal shape s^*. The second model, $\mathcal{M}'_{RD^0}$, shows good performances with an average $F1$ macro score of 0.91 and 0.97 at the micro level. Moreover, this model generally reproduces quite well the original patterns

($r_{\hat{\mathbb{G}}_D^{\leftrightarrow}}$) associated with the ground truth examples. Concerning the gold model $\mathcal{M}'_{RD^{1+}}$ it is naturally more adapted to reproduce the triples of $\overline{RD^{2+}}$, as it was trained on a corrected dataset: it records the best $F1$ metrics, low $\overline{Loss(\hat{\mathbb{G}})}$. Finally, we can notice the high rate of generated triples following the initial patterns ($r_{\hat{\mathbb{G}}_D^{\leftrightarrow}}$): twice more than $\mathcal{M}_{RD^-}$ and 10% more than $\mathcal{M}'_{RD^0}$.

5 Error Analysis

Pattern Errors Analysis. We consider the set $\widehat{\mathbb{G}}_D^{\not\leftrightarrow}$ of the predicted $\hat{g}$ whose properties do not correspond to the set of expected properties $\mathcal{P}(g)$, and the corresponding ratios:

$$\widehat{\mathbb{G}}_D^{\not\leftrightarrow} = \{\hat{g}; (w, g) \in D; \hat{g} \not\leftrightarrow \mathcal{P}(g)\} \qquad r_{\widehat{\mathbb{G}}_D^{\not\leftrightarrow}} = \frac{|\widehat{\mathbb{G}}_D^{\not\leftrightarrow}|}{|D|} \qquad r_{\widehat{\mathbb{G}}_D^{\not\leftrightarrow}} = 1 - r_{\widehat{\mathbb{G}}_D^{\leftrightarrow}} \tag{17}$$

We also define in the same way $\widehat{\mathbb{G}}_D^{\rightarrow}$, $r_{\widehat{\mathbb{G}}_D^{\rightarrow}}$, etc. We combine the subsets to measure the pattern extension capacity on a dataset, PEC_D, which is the ratio of predicted graphs that strictly extend the expected pattern:

$$PEC_D = \frac{|\widehat{\mathbb{G}}_D^{\rightarrow}| - |\widehat{\mathbb{G}}_D^{\leftrightarrow}|}{|\widehat{\mathbb{G}}_D^{\not\leftrightarrow}|} \tag{18}$$

Finally we extend the definition of Eq. 10 to consider the patterns found in the graph $\hat{g}$ inferred w.r.t the ground truth (w, g) in a dataset D:

$$\widehat{\mathbb{P}}_D(s) := \{\pi \in \mathbb{P}(s); \exists (w, g) \in D; \hat{g} \leftrightarrow \pi\} \tag{19}$$

The resulting notation allows us to compare the pattern set represented in the expected graph and the predicted one:

$$\begin{aligned} \mathbb{P}_D^{\not\leftrightarrow} &:= \{\pi \in \Pi; \exists (w, g) \in D; \pi := \mathcal{P}(g); \hat{g} \not\leftrightarrow \pi\} \\ \widehat{\mathbb{P}}_D^{\not\leftrightarrow} &:= \{\pi \in \Pi; \exists (w, g) \in D; \pi := \mathcal{P}(\hat{g}); g \not\leftrightarrow \pi\} \end{aligned} \tag{20}$$

For instance, $\mathbb{P}_{RD^1}(s^*)$ represents the set of patterns that can be built from s^* and found in the graph set from RD^1, and $\widehat{\mathbb{P}}_{RD^1}(s^*)$ the set of patterns found in the predictions obtained from RD^1.

Moreover we also extend Eq. 12 to compute the predicted $r_{s^*}(\hat{\mathbb{G}}_D^{\not\leftrightarrow})$ and the expected $r_{s^*}(\mathbb{G}_D^{\not\leftrightarrow})$, that is, the rates of graphs valid against the maximal shape s^*, respectfully for the predictions and the expected values.

Table 6. Focus on the triples generated not following the initial property signature ($\hat{g} \nrightarrow \mathcal{P}(g)$)

$\mathcal{M}$odel (dataset D)	$r_{\hat{\mathbb{G}}_D^{\nleftrightarrow}}$	$\lvert\mathbb{P}_D^{\nleftrightarrow}\rvert$	$\lvert\widehat{\mathbb{P}}_D^{\nleftrightarrow}\rvert$	$r_{s^*}(\mathbb{G}_D^{\nleftrightarrow})$	$r_{s^*}(\hat{\mathbb{G}}_D^{\nleftrightarrow})$	PEC_D
$\mathcal{M}_{RD^-}(RD^{2+})$	**0.57**	**30.5**	10.2	0.29	**1.00**	0.56
$\mathcal{M}'_{RD^0}(RD^2)$	0.27	28.90	19.10	**0.67**	0.53	0.18
$\mathcal{M}'_{RD^0}(RD^{2+})$	0.28	27.70	*23.50*	*0.57*	0.62	0.58
$\mathcal{M}'_{RD^{1+}}(RD^2)$	*0.38*	*29.10*	20.90	0.38	0.59	*0.66*
$\mathcal{M}'_{RD^{1+}}(RD^{2+})$	0.20	28.90	**27.60**	0.46	*0.73*	**0.87**

From Table 6, we can first notice the inability of $\mathcal{M}_{RD^-}$ to reproduce correctly the graphs that were not originally entailing the maximal shape s^* (cf. $r_{s^*}(\mathbb{G}_D^{\nleftrightarrow})$). Moreover the patterns generated by $\mathcal{M}_{RD^-}$ are less varied ($\lvert\widehat{\mathbb{P}}_{RD^-}^{\nleftrightarrow}\rvert$) than the one expected ($\lvert\mathbb{P}_{RD^-}^{\nleftrightarrow}\rvert$) but their are by design all following the maximal shape s^*. Conversely, the generated triples of $\mathcal{M}'_{RD^0}$ ($\lvert\mathbb{P}_{RD^0}^{\nleftrightarrow}\rvert$) are closer to the pattern of the expected graph ($\lvert\widehat{\mathbb{P}}_{RD^0}^{\nleftrightarrow}\rvert$). The correction of RD^2 had a slight impact on the number of triples valid against the shape, but it shows the potential in terms of pattern extension of $\mathcal{M}'_{RD^0}$ by reaching the same levels than PEC_{RD^-}. In addition, our gold model $\mathcal{M}'_{RD^{1+}}$ produces pattern closer to $\mathbb{P}_{RD^{1+}}^{\nleftrightarrow}$ and also tends to produce triples closer to s^*. Finally, this gold model obtains a high PEC_D, promising for knowledge completion.

Table 7. Metrics computed after annotation and averaged over the 10.folds of each model

$\mathcal{M}$odel (dataset D)	FN_-	FN_+	r_{omis}	FP_-	FP_+	r_{disco}
$\mathcal{M}_{RD^-}(RD^{2+})$	2	144	0.005	198	95	0.32
$\mathcal{M}'_{RD^0}(RD^{2+})$	2.9	160	0.006	27	28	0.52
$\mathcal{M}'_{RD^{1+}}(RD^{2+})$	4.5	49.7	**0.01**	33	70	**0.68**

Annotation of the FP/FN Triples. The annotator was asked to verify the false positives (FP) and false negatives (FN) triples produced by each model when tested on RD^{2+}.

Annotating the 10-folds of FN/FP triples took 30 to 40 min per model, depending on the sample size, plus a further 20 to 40 min to classify the mentioned errors. We obtained four sets of triples from our annotation: the erroneous FP- (which could be linked to an error or a hallucination), the discoveries FP+, the correct False negatives FN- and the omissions FN+. We build two metrics from them: the omission rate, relative to the number of the total expected triples ($Nb_{triple\ expected}$) and the rate of discoveries, which consider the number of FP generated that could be considered as relevant, defined as follows:

$$r_{omis} = \frac{|FN_+|}{Nb_{triple\ expected}} \qquad r_{disco} = \frac{|FP_+]}{|FP_+| + |FP_-|}$$

Table 7 shows us the FP produced by $\mathcal{M}_{RD^-}$ are mainly not relevant, by recording at the same time a high omission rate. Inversely, $\mathcal{M}'_{RD^{1+}}$ is very interesting because this one rarely omits facts, and more than half of the FP produced by the models could be considered discoveries. Beyond the analysis of the ratios, if we consider the total number of FP produced, we see that model $\mathcal{M}'_{RD^{1+}}$ produces more FP+ than $\mathcal{M}'_{RD^0}$ however the number of erroneous one ($FP-$) remains more or less the same.

***FP–* classification.** In a second step, the annotator was asked to categorize the FP^- (i.e. the erroneous true positives) depending on the following classification. To illustrate it, an example of an error is given in Appendix A.2, and the resulting distribution of these categories obtained over the testing sets RD^2 and RD^{2+} is drawn in Fig. 5:

- FH - Factual hallucination: the value generated respects the range of the property, e.g. the model output is a year attached to a birthyear property, but this value is not in the abstract and could not be inferred from it.
- AC - Abusive completion: the value generated respects the range of the property, and a part of this sequence is in the text, but the model completes this sequence with plausible tokens that could not be inferred by the text, e.g. the value is a date, but we could only deduce the year from the abstract).
- IAC - Illogical and abusive completion: The generated output is an abusive completion, and the resulting value does not respect the expected range, e.g. a date containing a day superior to 31)
- WV - Wrong value: The generated output is in the text but does not correspond to the targeted predicate, e.g. The value generated is related to the alias when it comes to predicting the label.
- TMI - Typographic minor issue: The generated output is close to the expected one, but it contains a minor typographic difference, e.g. URL encoding errors, uppercase, missing space or special characters.
- SG - Stuttered generation: The output is almost correct, but it contains repeated patterns, e.g. the value returned is a birthname repeated two times.
- ICE - Incomplete Context Error: The output is in the text and corresponds only to a part of the expected value, e.g. the models return a shortened composed birthname.
- LCE - Larger context error: the output is in the text, and the expected value is inside it, but the output contains too much information. e.g. The expected label is an abbreviated birthname, and the prediction is the entire birthname.
- MCE - Mixed context error: the output is false but close to the expected one, and the produced value is composed of a mix of values found in the text that is not the expected one, e.g. The date return is the mix of two dates found in the text.

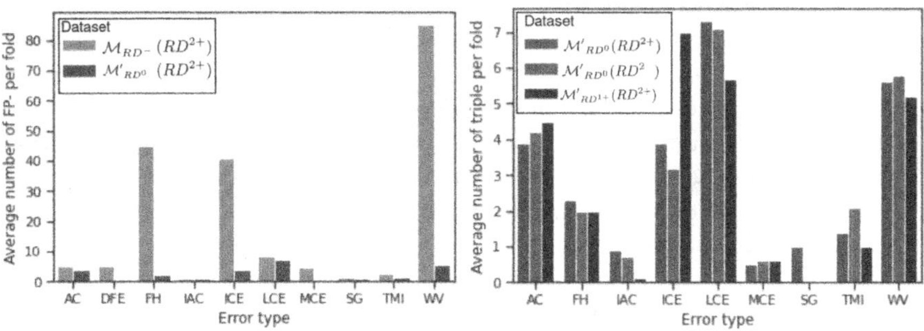

Fig. 5. Errors distribution over the different models and datasets (Color figure online)

First of all, the bar chart on the left shows a huge gap between $\mathcal{M}_{RD^-}$ and $\mathcal{M}'_{RD^0}$. In fact, $\mathcal{M}_{RD^-}$ generates considerably more errors, with a majority of wrong values (WV), a lot of hallucinations (FH) and many incomplete context errors (ICE). The bar chart on the right shows that our new models are more likely to generate abusive completion than hallucination (AC vs. FH). However, the errors remain more or less the same despite our annotation. To address remaining errors, the NLI and the Triplet Critic models [10, 11, 28] could be applied to each triple of $\hat{g}$. But when we applied these models to the manually evaluated FP and FN triples obtained from $\mathcal{M}'_{RD^{1+}}$ on RD^{2+} we showed they do not perform well on that task (Fig. 6). This can be explained by the fact that we are focusing on particular datatype properties. These observations highlight the need to adapt such models to efficiently filter potential hallucinations or recurrent errors highlighted during our analysis. In future work, we shall consider integrating the annotations into the knowledge base, which may help adapt models with contrastive learning approaches after parsing.

6 Discussion

Scalability and Extensibility: The shape complexity that can be handled by our framework is firstly limited by the output size of the SLM, which forces us to focus only on a reduced set of properties. In DBpedia, resources of type `dbo:Person` can described by up to 139 datatype properties, among which exactly 100 are used, but only 25 of them are used in more than 1% of the cases. We chose to focus on the 7 properties that are most likely to be found in the abstracts. We observed that only half of the possible combinations (70/127) exist in our KB after Wikicheck. This remains a manageable number of combinations. Shapes containing a set of less popular properties will lead to an even smaller number of combinations. Additionally, considering approaches based on BART-large, such as REBEL, which can scale to up to 220 types of relations, we might expect to reach similar capabilities with a larger model than our current BART-base model. Considering that the shape targeting `dbo:Person` resources already cover 1/6th of the DBpedia entities, this demonstrates that Kastor can scale to datasets with large numbers of similar instances. In that context our framework is generic and could be adapted to other use cases, relying on a dual base (KG+TXT) and a SHACL shape containing only datatype properties.

A Light Active Learning: We observed that correcting the training set with a single loop and a unique expert annotator is enough to increase the quality of the graphs produced. This annotation iteration of the errors was conducted on 10-folds, thus collecting the errors from 10 different models. This allows us to cover a lot of cases that can occur with datatype properties, although more marginal types of error may still occur. Conversely, further extension to object properties would require extending our typology of errors. We could envisage having several iterations for active learning and using pattern extension capacity or the discovery rate as a stopping criterion. In practice, we would only perform such iterations on the best model. From our point of view, one annotator and one iteration are enough as they increase the F1 performance by 10%. However, this does not reduce the remaining marginal errors, which suggests that iterating would have little effect. Moreover, we protect our setup from noise with different strategies:

the unique expert evaluator, the strict evaluation of the values and the 10-folds annotation. Adding more evaluators would introduce noise and be more costly.

7 Conclusion

We presented Kastor, an open, reusable and extendable framework to perform an RDF-pattern relation extraction task from a noisy and incomplete knowledge base[9]. Our approach firstly demonstrates its frugality: from the model's fine-tuning aspect that requires less than 10 min to a light active learning process, implying an annotator on a small set of FP/FN triples. Concerning our first research question (**RQ1**), we showed that using example-specific achievable patterns improves the performance of the relation extraction model by almost 10% in terms of F_1^+. Moreover, it produces a wider variety of property patterns by avoiding many of the hallucinations plaguing the original design which relied only on graphs valid against a maximal SHACL shape. Additionally, regarding the second research question (**RQ2**), the impact of the active learning process was also demonstrated. It leads to a better model in terms of F_1 scores, better ability to generate diverse patterns, better pattern extension capacities, and a better chance of generating discoveries, i.e. facts that are relevant but not initially present in the KB. Kastor is opening the door to many possible future works: the characterisation of the RDF-pattern distribution gives us the opportunity to better deal with the long-tail thereof. Moreover, the framework's generalizability also allows us to reproduce this current work on any SHACL shape focused on datatype properties and, with further development and system adaptation, on object properties.

Acknowledgments. This work has been supported by the French government, through the 3IA Côte d'Azur Investments (ANR-23-IACL-0001) and by the UCAJEDI (ANR-15-IDEX-01), the OPAL infrastructure and Université Côte d'Azur's Center for High-Performance Computing.

A Appendix

A.1 Ebnf Grammar

```
############## turtle light oneline factorized
root ::= triples+
triples ::= WS? triple WS? "."
triple ::= subj WS? predicateObjectList
predicateObjectList ::= pred objectList ( WS?  ";"  WS? ( pred   WS? objectList)? )*
objectList ::= obj  (  WS? ","  WS? obj  )*
subj ::=  iri
pred ::= iri | "a"
obj ::= iri | string
string ::= WS? "\"" [ \t!#-\[
\bUnALT{}\eUnALT{}
\]-~]* "\"" WS?
iri ::= ":" PN_LOCAL+
WS ::= [ \t\n]
PN_CHARS_BASE ::= [A-Z] | [a-z] | [#x00C0-#x00D6] | [#x00D8-#x00F6] | [#x00F8-#x02FF] |
[#x0370-#x037D] | [#x037F-#x1FFF] | [#x200C-#x200D] | [#x2070-#x218F] | [#x2C00-#x2FEF] |
[#x3001-#xD7FF] | [#xF900-#xFDCF] | [#xFDF0-#xFFFD] | [#x10000-#xEFFFF]
PN_CHARS_U ::= PN_CHARS_BASE | "_"
PN_LOCAL ::= ( PN_CHARS_U | ":" | [0-9] | PLX )
```

---

9 code: https://github.com/datalogism/Kastor
models: https://zenodo.org/records/14498940
dataset https://zenodo.org/records/14382674.

```
( ( PN_CHARS | "." | ":" | PLX )*
( PN_CHARS | ":" | PLX ) ) ?
PLX  ::= PERCENT | PN_LOCAL_ESC
PN_CHARS  ::= PN_CHARS_U | "-" | [0-9] | [#x00B7] | [#x0300-#x036F] | [#x203F-#x2040]
PERCENT  ::= "HEX ::= [0-9] | [A-F] | [a-f]
PN_LOCAL_ESC ::= "\\" ( "_" | "~" | "." | "-" | "!" | "$" | "&" |
"'" | "(" | ")" | "*" | "+" | "," | ";" | "=" | "/" | "?" | "#" | "@" | "
```

A.2 Example of FP Errors by Category

	Context	Predicate	Generated value	Expected value
FH	Marguerite Kathryn Flecknoe is an American voice actress, radio personality, television host and producer.	birthYear	1944	∅
AC	Peter Woon (1931 – May 2014) was a news and current affairs editor at the British Broadcasting Corporation...	deathDate	2014-05-14	
IAC	Frederick Jardine (born 27 September 1941) died 7 october 2019 was a Scottish former professional footballer, ...	deathDate	2019-october 2019	2019-10-07
TMI	Françoise Abanda (born February 5, 1997) is a Canadian professional tennis player. She reached her highest WTA...	label	Fran%C3%A7oise Abanda	Françoise Abanda
SG	Mao Ichimichi (市道 真央, Ichimichi Mao, born February 1, 1992) is a Japanese actress and voice actress. She started her career as a Japanese idol ...	birthName	Mao Ichimichi Mao	Mao Ichimichi
WV	Jeremy Larroux (born 1993), better known as Laylow is a French rapper from Toulouse. In 2018, Laylow released the EPs.RAW and RAW-Z. ..	alias	Jeremy Larroux	Laylow
ICE	Mariano Garchitorena y Chereau (February 12, 1898 - October 1, 1961) was a Filipino politician of Spanish-French descent...	birthName	Mariano Garchitorena	Mariano Garchitorena y Chereau
LCE	Lenilson Batista de Jesús (born May 1, 1981 in Salvador), also known as Lenilson Batista de Souza, Lenilson Batista, or simply Lenílson, is a Brazilian left midfielder.	label	Lenilson Batista de Jesús	Lenflson Batista
MCE	Stephen Edward Smith (September 24, 1927 – August 19, 1990) was the husband of Jean Ann Kennedy...	birthName	Stephen Ann Kennedy	Stephen Edward Smith

A.3 Active Learning Process

Algorithm 1: Light active learning process

Data: $\{RD^0, RD^1, RD^2\} \in \mathcal{K}^{\mathcal{WR}\models}_{\mathbb{P}(s^*)}$
with $RD^0 = RD^0_{test} \sqcup RD^0_{train} \sqcup RD^0_{eval}$
Train $\mathcal{M'}_{RD^0}$ using RD^0_{train};
foreach $D_i \in \{RD^1, RD^2\}$ **do**
    $FP_{D_i} \leftarrow \emptyset$;
    $FN_{D_i} \leftarrow \emptyset$
    foreach $(w, g) \in D_i$ **do**
        $\hat{g} \leftarrow \mathcal{M}_{RD^0}(w)$;
        foreach *triple* $\hat{t} \in \hat{g}$ **do**
            if $\hat{t} \not\subset g$ **then**
                $FP_{D_i} \leftarrow FP_{D_i} \cup \hat{t}$
            end
        end
        foreach *triple* $t \in g$ **do**
            if $t \not\subset \hat{g}$ **then**
                $FN_{D_i} \leftarrow FN_{D_i} \cup t$
            end
        end
        Human annotation of FN_{D_i} and FP_{D_i} as:
        $FN_{D_i} \leftarrow FN^+_{D_i} \cup FN^-_{D_i}$
        $FP_{D_i} \leftarrow FP^+_{D_i} \cup FP^-_{D_i}$
        Gather into κ^+_i only valid triples:
        $D^+_i \leftarrow (D_i \backslash FN^-_{D_i}) \cup FP^+_{D_i}$
    end
    with $RD^{1+} = \kappa^+_1$ and $RD^{2+} = \kappa^+_1$
    and $RD^{1+} = RD^{1+}_{test} \sqcup RD^{1+}_{train} \sqcup RD^{1+}_{eval}$
    Train $\mathcal{M'}_{RD^{1+}}$ using RD^{1+}_{train};
    Evaluate $\mathcal{M'}_{RD^0}$ and $\mathcal{M'}_{RD^{1+}}$ using RD^{2+};
end

A.4 Applied NLI to Annotated Triples

(See Fig. 6)

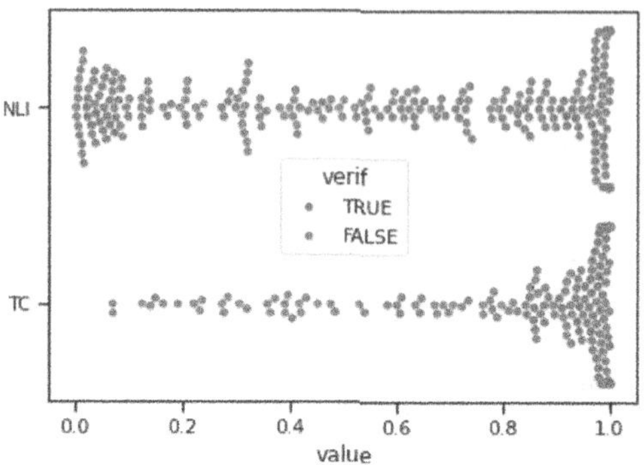

Fig. 6. $\mathcal{M}'_{RD^{1+}}(RD^{2+})$ NLI scores

References

1. Alt, C., Gabryszak, A., Hennig, L.: TACRED revisited: a thorough evaluation of the TACRED relation extraction task. In: Jurafsky, D., Chai, J., Schluter, N., Tetreault, J. (eds.) Proceedings of the 58th Annual Meeting of the Association for Computational Linguistics, pp. 1558–1569. Association for Computational Linguistics, Online (2020). https://doi.org/10.18653/v1/2020.acl-main.142
2. Augenstein, I., Maynard, D., Ciravegna, F.: Relation extraction from the web using distant supervision. In: Janowicz, K., Schlobach, S., Lambrix, P., Hyvönen, E. (eds.) Knowledge Engineering and Knowledge Management, pp. 26–41. Springer, Cham (2014)
3. Dligach, D., Bethard, S., Miller, T., Savova, G.: Exploring text representations for generative temporal relation extraction. In: Naumann, T., Bethard, S., Roberts, K., Rumshisky, A. (eds.) Proceedings of the 4th Clinical Natural Language Processing Workshop, pp. 109–113. Association for Computational Linguistics, Seattle (2022). https://doi.org/10.18653/v1/2022.clinicalnlp-1.12
4. Efeoglu, S., Paschke, A.: Retrieval-augmented generation-based relation extraction. ArXiv abs/2404.13397 (2024). https://api.semanticscholar.org/CorpusID:269292881
5. Gallardo, A.P., Consoli, S., Ceresa, M., Hulsman, R., Bertolini, L.: On constructing biomedical text-to-graph systems with large language models. In: Tiwari, S., et al. (eds.) Joint proceedings of the 3rd International workshop on knowledge graph generation from text (TEXT2KG) and Data Quality meets Machine Learning and Knowledge Graphs (DQM-LKG) Co-located with the Extended Semantic Web Conference (ESWC 2024), Hersonissos, Greece, 26–30 May 2024. CEUR Workshop Proceedings, vol. 3747, p. 12. CEUR-WS.org (2024). https://ceur-ws.org/Vol-3747/text2kg_paper10.pdf
6. Geng, S., Josifoski, M., Peyrard, M., West, R.: Grammar-constrained decoding for structured NLP tasks without finetuning. In: Bouamor, H., Pino, J., Bali, K. (eds.) Proceedings of the 2023 Conference on Empirical Methods in Natural Language Processing, pp. 10932–10952. Association for Computational Linguistics, Singapore (2023). https://doi.org/10.18653/v1/2023.emnlp-main.674, https://aclanthology.org/2023.emnlp-main.674/

7. Grangier, D., Katharopoulos, A., Ablin, P., Hannun, A.: Need a small specialized language model? Plan early! (2024). https://arxiv.org/abs/2402.01093
8. Hofer, M., Obraczka, D., Saeedi, A., Köpcke, H., Rahm, E.: Construction of knowledge graphs: state and challenges (2023). https://arxiv.org/abs/2302.11509
9. Huang, L., et al.: A survey on hallucination in large language models: principles, taxonomy, challenges, and open questions (2023). https://arxiv.org/abs/2311.05232
10. Huguet Cabot, P.L., Tedeschi, S., Ngonga Ngomo, A.C., Navigli, R.: RED$^{\rm fm}$: a filtered and multilingual relation extraction dataset. In: Rogers, A., Boyd-Graber, J., Okazaki, N. (eds.) Proceedings of the 61st Annual Meeting of the Association for Computational Linguistics (Volume 1: Long Papers), pp. 4326–4343. Association for Computational Linguistics, Toronto (2023). https://doi.org/10.18653/v1/2023.acl-long.237
11. Huguet Cabot, P.L., Navigli, R.: REBEL: relation extraction by end-to-end language generation. In: Moens, M.F., Huang, X., Specia, L., Yih, S.W.t. (eds.) Findings of the Association for Computational Linguistics: EMNLP 2021, pp. 2370–2381. Association for Computational Linguistics, Punta Cana (2021). https://doi.org/10.18653/v1/2021.findings-emnlp.204
12. Hussam Ghanem, C.C.: Fine-tuning vs. prompting: evaluating the knowledge graph construction with LLMs (2024). https://ceur-ws.org/Vol-3747/text2kg_paper7.pdf
13. Ji, Z., Lee, N., Frieske, R., Yu, T., Su, D., Xu, Y., Ishii, E., Bang, Y.J., Madotto, A., Fung, P.: Survey of hallucination in natural language generation. ACM Comput. Surv. **55**(12), 1–38 (2023). https://doi.org/10.1145/3571730
14. Josifoski, M., De Cao, N., Peyrard, M., Petroni, F., West, R.: GenIE: generative information extraction. In: Carpuat, M., de Marneffe, M.C., Meza Ruiz, I.V. (eds.) Proceedings of the 2022 Conference of the North American Chapter of the Association for Computational Linguistics: Human Language Technologies, pp. 4626–4643. Association for Computational Linguistics, Seattle (2022). https://doi.org/10.18653/v1/2022.naacl-main.342, https://aclanthology.org/2022.naacl-main.342
15. Kandpal, N., Deng, H., Roberts, A., Wallace, E., Raffel, C.: Large language models struggle to learn long-tail knowledge. In: Proceedings of the 40th International Conference on Machine Learning, ICML 2023. JMLR.org (2023)
16. Lehmann, J., et al.: Large language models for scientific question answering: an extensive analysis of the sciqa benchmark. In: Meroño-Peñuela, A., et al. (eds.) ESWC 2024, Part I. LNCS, vol. 14664, pp. 199–217. Springer, Cham (2024). https://doi.org/10.1007/978-3-031-60626-7_11
17. Lehmann, J., et al.: Large language models for scientific question answering: an extensive analysis of the sciqa benchmark. In: Meroño Peñuela, A., et al. (eds.) The Semantic Web, pp. 199–217. Springer, Cham (2024)
18. Li, B., et al.: Evaluating chatGPT's information extraction capabilities: an assessment of performance, explainability, calibration, and faithfulness. ArXiv abs/2304.11633 (2023). https://api.semanticscholar.org/CorpusID:258297899
19. Li, D., et al.: Overcoming catastrophic forgetting during domain adaptation of seq2seq language generation. In: Carpuat, M., de Marneffe, M.C., Meza Ruiz, I.V. (eds.) Proceedings of the 2022 Conference of the North American Chapter of the Association for Computational Linguistics: Human Language Technologies, pp. 5441–5454. Association for Computational Linguistics, Seattle (2022). https://doi.org/10.18653/v1/2022.naacl-main.398
20. Li, G., et al.: Recall, retrieve and reason: towards better in-context relation extraction. In: Larson, K. (ed.) Proceedings of the Thirty-Third International Joint Conference on Artificial Intelligence, IJCAI-24, pp. 6368–6376. International Joint Conferences on Artificial Intelligence Organization (2024). https://doi.org/10.24963/ijcai.2024/704
21. Li, M., Shi, T., Ziems, C., Kan, M.Y., Chen, N., Liu, Z., Yang, D.: CoAnnotating: uncertainty-guided work allocation between human and large language models for data annotation. In:

Proceedings of the 2023 Conference on Empirical Methods in Natural Language Processing. Association for Computational Linguistics (2023). https://doi.org/10.18653/v1/2023.emnlp-main.92

22. Liu, Y., Li, D., Wang, K., Xiong, Z., Shi, F., Wang, J., Li, B., Hang, B.: Are LLMs good at structured outputs? A benchmark for evaluating structured output capabilities in LLMs. Inf. Process. Manage. **61**(5), 103809 (2024). https://doi.org/10.1016/j.ipm.2024.103809, https://www.sciencedirect.com/science/article/pii/S0306457324001687
23. Lu, Z., et al.: Small language models: survey, measurements, and insights (2024). https://arxiv.org/abs/2409.15790
24. Ma, Y., Wang, A., Okazaki, N.: DREEAM: guiding attention with evidence for improving document-level relation extraction. In: Vlachos, A., Augenstein, I. (eds.) Proceedings of the 17th Conference of the European Chapter of the Association for Computational Linguistics, pp. 1971–1983. Association for Computational Linguistics, Dubrovnik (2023).https://doi.org/10.18653/v1/2023.eacl-main.145
25. van der Meer, M., Falk, N., Murukannaiah, P.K., Liscio, E.: Annotator-centric active learning for subjective NLP tasks (2024). https://arxiv.org/abs/2404.15720
26. Paolini, G., et al.: Structured prediction as translation between augmented natural languages (2021). https://arxiv.org/abs/2101.05779
27. Ringwald, C., Gandon, F., Faron, C., Michel, F., Akl, H.A.: 12 shades of RDF: impact of syntaxes on data extraction with language models. In: Meroño Peñuela, A., et al. (eds.) The Semantic Web: ESWC 2024 Satellite Events, pp. 81–91. Springer, Cham (2025)
28. Rogulsky, S., Popovic, N., Färber, M.: The effects of hallucinations in synthetic training data for relation extraction. arxiv (2024)
29. Rossiello, G., Chowdhury, M.F.M., Mihindukulasooriya, N., Cornec, O., Gliozzo, A.M.: KnowGL: knowledge generation and linking from text (2022). https://arxiv.org/abs/2210.13952
30. Shenoy, K., Ilievski, F., Garijo, D., Schwabe, D., Szekely, P.: A study of the quality of wikidata. J. Web Semant. **72**, 100679 (2022). https://doi.org/10.1016/j.websem.2021.100679
31. Smirnova, A., Cudré-Mauroux, P.: Relation extraction using distant supervision: a survey. ACM Comput. Surv. **51**(5) (2018).https://doi.org/10.1145/3241741
32. Stoica, G., Platanios, E.A., P'oczos, B.: Re-TACRED: addressing shortcomings of the tacred dataset. In: AAAI Conference on Artificial Intelligence (2021). https://api.semanticscholar.org/CorpusID:233296843
33. Taillé, B., Guigue, V., Scoutheeten, G., Gallinari, P.: Let's stop incorrect comparisons in end-to-end relation extraction! In: Webber, B., Cohn, T., He, Y., Liu, Y. (eds.) Proceedings of the 2020 Conference on Empirical Methods in Natural Language Processing (EMNLP), pp. 3689–3701. Association for Computational Linguistics, Online (2020). https://doi.org/10.18653/v1/2020.emnlp-main.301
34. Tsaneva, S., Sabou, M.: Enhancing human-in-the-loop ontology curation results through task design. ACM J. Data Inf. Qual. **16**(1), 4:1–4:25 (2024). https://doi.org/10.1145/3626960
35. Wadhwa, S., Amir, S., Wallace, B.: Revisiting relation extraction in the era of large language models. In: Rogers, A., Boyd-Graber, J., Okazaki, N. (eds.) Proceedings of the 61st Annual Meeting of the Association for Computational Linguistics (Volume 1: Long Papers), pp. 15566–15589. Association for Computational Linguistics, Toronto (2023). https://doi.org/10.18653/v1/2023.acl-long.868
36. Wang, F., et al.: A comprehensive survey of small language models in the era of large language models: techniques, enhancements, applications, collaboration with LLMs, and trustworthiness (2024). https://arxiv.org/abs/2411.03350
37. Yao, Y., et al.: CodRED: a cross-document relation extraction dataset for acquiring knowledge in the wild. In: Moens, M.F., Huang, X., Specia, L., Yih, S.W.t. (eds.) Proceedings of

the 2021 Conference on Empirical Methods in Natural Language Processing, pp. 4452–4472. Association for Computational Linguistics, Online and Punta Cana (2021). https://doi.org/10.18653/v1/2021.emnlp-main.366

38. Yao, Y., et al.: DocRED: a large-scale document-level relation extraction dataset. In: Korhonen, A., Traum, D., Màrquez, L. (eds.) Proceedings of the 57th Annual Meeting of the Association for Computational Linguistics, pp. 764–777. Association for Computational Linguistics, Florence (2019). https://doi.org/10.18653/v1/P19-1074
39. Zaratiana, U., Tomeh, N., Holat, P., Charnois, T.: GLiNER: generalist model for named entity recognition using bidirectional transformer. In: Duh, K., Gomez, H., Bethard, S. (eds.) Proceedings of the 2024 Conference of the North American Chapter of the Association for Computational Linguistics: Human Language Technologies (Volume 1: Long Papers), pp. 5364–5376. Association for Computational Linguistics, Mexico City (2024). https://doi.org/10.18653/v1/2024.naacl-long.300
40. Zhang, B., Reklos, I., Jain, N., Peñuela, A.M., Simperl, E.: Using large language models for knowledge engineering (LLMKE): a case study on wikidata (2023). https://arxiv.org/abs/2309.08491
41. Zheng, L., et al.: Judging LLM-as-a-judge with MT-bench and chatbot arena (2023). https://arxiv.org/abs/2306.05685

Predicting the Road Ahead: A Knowledge Graph Based Foundation Model for Scene Understanding in Autonomous Driving

Hongkuan Zhou[1,2(✉)], Stefan Schimid[1], Yicong Li[3], Lavdim Halilaj[1], Xiangtong Yao[4], and Wei Cao[1,2]

[1] Bosch Corporate Research, Renningen, Germany
{hongkuan.zhou,stefan.schmid5,lavdim.halilaj,wei.cao4}@de.bosch.com
[2] University of Stuttgart, Stuttgart, Germany
[3] University of Montreal, Montreal, Canada
yi.cong.li@umontreal.ca
[4] Technical University of Munich, Munich, Germany
xiangtong.yao@tum.de

Abstract. The autonomous driving field has seen remarkable advancements in various topics, such as object recognition, trajectory prediction, and motion planning. However, current approaches face limitations in effectively comprehending the complex evolutions of driving scenes over time. This paper proposes FM4SU, a novel methodology for training a symbolic foundation model (FM) for scene understanding in autonomous driving. It leverages knowledge graphs (KGs) to capture sensory observation along with domain knowledge such as road topology, traffic rules, or complex interactions between traffic participants. A bird's eye view (BEV) symbolic representation is extracted from the KG for each driving scene, including the spatio-temporal information among the objects across the scenes. The BEV representation is serialized into a sequence of tokens and given to pre-trained language models (PLMs) for learning an inherent understanding of the co-occurrence among driving scene elements and generating predictions on the next scenes. We conducted a number of experiments using the nuScenes dataset and KG in various scenarios. The results demonstrate that fine-tuned models achieve significantly higher accuracy in all tasks. The fine-tuned T5 model achieved a next scene prediction accuracy of 86.7%. This paper concludes that FM4SU offers a promising foundation for developing more comprehensive models for scene understanding in autonomous driving.

Keywords: Foundation Model · Pre-trained Language Model · Scene Understanding · Autonomous Driving

1 Introduction

Multi-modal foundation models are receiving increasing attention in autonomous driving for their ability to perform tasks such as object detection, prediction of future trajectories, and scene understanding [10]. Scene understanding in the

E. Curry et al. (Eds.): ESWC 2025, LNCS 15718, pp. 116–132, 2025.
https://doi.org/10.1007/978-3-031-94575-5_7

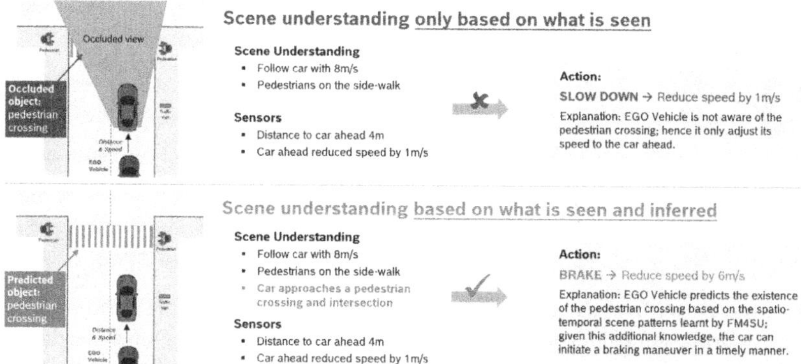

Fig. 1. Motivation – An exemplary driving scene illustrating how FM4SU is able to enrich scene understanding. By learning spatio-temporal patterns within and across diverse driving scenarios, FM4SU can infer missing information (e.g. pedestrian crossing). This enables more informed and consequently safer decision-making for the vehicle.

context of autonomous driving refers to the ability of an autonomous car to perceive its surroundings, interpret the perceived information, and make informed decisions based on that interpretation [11]. Figure 1 exemplifies the potential of employing scene understanding derived from sensory data and coupled with a specifically trained foundation models (FMs) to enhance safety in autonomous driving. The inclusion of additional contextual information, even if it's merely predicted by the FMs, is crucial for an autonomous vehicle to act appropriately.

Learning an exhaustive and easily reusable FMs for scene understanding in autonomous driving poses significant challenges. These complexities stem mainly from the diversity of sensor technologies, such as video and radar, varying ranges and resolutions, as well as sensor deployments (e.g., mounting positions) on vehicles. Consequently, the development and evaluation of a truly large FM from diverse datasets (e.g., nuScenes [2]) becomes a very challenging task. To address these shortcomings, we advocate a novel ontology-based symbolic scene representation as input for learning the foundation model. This supports the integration of heterogeneous autonomous driving datasets into a uniform and compact driving scene representation [12], and thus, facilitates the training of a truly large FM. The advantages of using a symbolic representation are manifold. First, it provides a rich structure and semantics, well suited to incorporate additional domain knowledge via explicit relationships. Second, it allows experts to easily validate the model and use it for the explainability of predictions. Third, it enables further predicting and validation of the results based on the axioms defined in the ontology. Finally, such a high-level scene representation aids in integrating the model-based predictions into diverse downstream tasks [12].

We evaluate our approach on the basis of a real-world autonomous driving dataset, nuScenes [2], and investigate to what extent the trained FM is able to predict masked entities in the scene and a complete next scene. Figure 2 illustrates our learning pipeline. In the first step, we transform the driving scenes

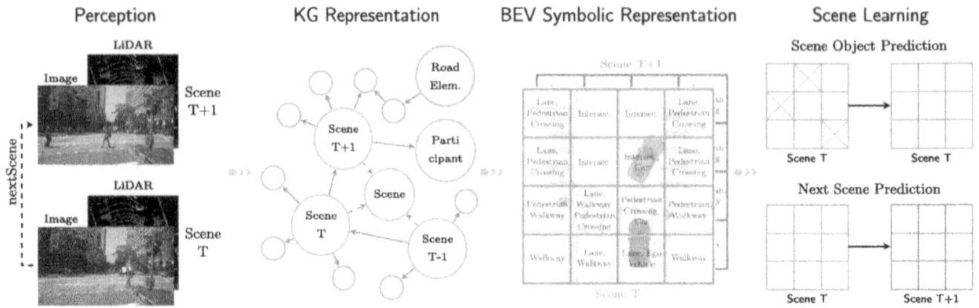

Fig. 2. The Learning Pipeline - comprises four main phases: 1) Perception - captures the information from the environment using various sensors; 2) Knowledge Graph Representation - structures information in the form of entities and relations including rich semantics and domain knowledge; 3) Bird-Eye-View Representation - extracts and transforms information in a matrix representation; and 4) Scene Learning - learns scene evolution based on the co-occurring elements.

(all relevant information extracted by standard object detection and semantic segmentation algorithms) into a knowledge graph (KG). The KG captures the spatio-temporal relations in and across the scenes and the detected entities (both static and dynamic). Based on these semantically enriched scene representations, we extract for each scene a symbolic representation from a Bird's Eye View (BEV) around the Ego Vehicle (EV). These ontology-based scene representations are then used for the training of a large transformer model. Inspired by the recent breakthroughs in large language models (LLMs), we use attention mechanisms to train a large model to understand realistic driving scenes and their spatio-temporal evolution enfolding in the subsequent scene.

We evaluate to what extent large transformer models are able to learn realistic driving scenes and generate predictions on the next scene. Further, we investigate different approaches, including fine-tuning existing large models and different model sizes. The main *contributions* of our work are shown as follows:

- A new methodology for learning a foundation model for scene understanding based on an ontology-based symbolic scene representation. The proposed approach enables easy integration of diverse datasets (different sources, with different sensor modalities, ...) and facilitates the training large FMs.
- We conducted a number of experiments to evaluate to what extent large (language) models are capable of understanding driving scenes without additional visual modalities, such as RGB images or LiDAR, by predicting masked areas in a given scene and/or predicting the next scenes.
- A new dataset (https://github.com/boschresearch/fm4su) for the research community to develop and benchmark new algorithms for learning a foundation model for scene understanding based on the nuScenes dataset.

2 Related Work

The use of LLMs for addressing tasks representable through language has garnered increasing attention in diverse fields, including autonomous driving [3,6,7,29], protein 3D structure representation [14], and robot manipulation [25,34]. In the domain of autonomous driving, LLMs have been employed to tackle challenges such as motion prediction [20,35], traffic visual question answering [22,35], and end-to-end autonomous driving [22,29,33]. Additionally, knowledge graphs (KGs) have been explored as a complementary approach, providing the relationships between entities in traffic scenarios [12,21]. The subsequent subsections provide a detailed discussion of scene understanding, FMs, and KGs in the context of autonomous driving.

2.1 Pre-trained Language Models

PLMs are an early attempt to extract semantic meaning from natural language. ELMo [17] aims to capture context-aware word representations by first pre-training a bidirectional LSTM (biLSTM) network and then fine-tuning the biLSTM network according to specific downstream tasks. Moreover, drawing inspiration from the Transformer architecture [24] and incorporating self-attention mechanisms, BERT [8] takes language model pre-training a step further. It accomplishes this by conducting bidirectional pre-training exercises on extensive unlabeled text corpora. These specially crafted pre-training tasks imbue BERT with contextual understanding, resulting in highly potent word representations.

LLMs become popular as the scaling of PLMs leads to improved performance on downstream tasks. Many researchers study the performance limits of PLMs by scaling the size of models and datasets [13], e.g., the comparatively small 1.5B parameter GPT-2 [18] versus larger 175B GPT-3 [1], 450B Llama [23] and 540B PaLM [5]. Although these models share a similar structure, the enlarged models display different behaviors and show surprising abilities compared to previous works. Considering our available computational resources and the level of open-source accessibility, we opt for the Text-to-Text Transfer Transformer [19] (T5) as a pre-trained language model to build a KG-based FM for scene understanding.

2.2 Scene Understanding in Autonomous Driving

Scene understanding is a critical aspect of autonomous driving, enabling vehicles to perceive and interpret their surroundings for safe and efficient navigation. Recent advancements in autonomous driving models [4,32] leverage multi-modal inputs, including RGB images, LiDAR, and BEV representations. RGB images provide detailed visual context, LiDAR offers precise depth information, and BEV representations integrate these inputs into a structured spatial view, facilitating better scene interpretation. LLMs have emerged as promising tools to address these challenges by incorporating contextual predicting and improving interpretability. For instance, models like PlanAgent [30] employ hierarchical predicting with BEV inputs to generate interpretable motion commands, while DriveMLM [26] integrates multi-view image data and LiDAR point clouds to produce high-level planning decisions. However, due to the inherent complexity of driving scenarios, LLMs sometimes struggle to fully understand intricate environments. Additionally, these approaches face limitations such as slow prediction speeds caused by iterative processes [28] and the lack of diverse, realistic datasets for robust evaluation [9].

2.3 Foundation Models in Autonomous Driving

With the development of LLMs and FMs, more research attention is being paid to integrating the common sense knowledge existing in FMs in the field of autonomous driving. Tian et al. [22] tokenizes multi-view videos, HD-maps, and symbolic representations of objects, enabling better utilization of LLM's generalization capabilities to enhance autonomous vehicle planning in long-tail scenarios. Zhou et al. [35] propose an Embodied Language Model (ELM), a comprehensive framework tailored for agents' understanding of driving scenes with large spatial and temporal spans. DriveGPT4 [29] processes multi-frame video inputs alongside textual queries to predict low-level vehicle control signals and generate textual explanations in response. We observe a growing research trend in tokenizing multi-modal inputs—such as video, LiDAR data, and textual queries—and employing visual question answering in traffic scenes. In such a way, the knowledge embedded in pre-trained FMs is utilized to enhance the generalization capability of autonomous driving pipelines. In this work, we aim to investigate the spatial and temporal scene understanding ability of LLMs in autonomous driving, excluding additional visual modalities such as videos and LiDAR. This approach enables a clearer assessment of the benefits of incorporating LLMs' knowledge in traffic scene analysis.

2.4 Knowledge Graphs in Autonomous Driving

KGs are getting more traction on their potential usage for different tasks in autonomous driving [15,31]. The nuSceneKG [16] is a KG that models all scene participants and road elements explicitly, as well as their semantic and spatial relationships. Based on that, SemanticFormer [21] predicts multimodal trajectories by predicting over a semantic traffic scene graph using a hybrid approach. SocialFormer [27] performs an agent interaction-aware trajectory prediction method that leverages the semantic relationship between the target vehicle and surrounding vehicles by making use of the road topology. In this paper, we build our symbolic BEV representation dataset for scene understanding from the nuSceneKG.

3 Methodology

3.1 Knowledge Graph Representation

We leverage Knowledge Graphs as a medium to capture, organize and interlink the knowledge from the domain. It represents the information in form of triples $G = (E, R, E|V)$ where each element $e \in E$ is an entity, each $v \in V$ is a literal,

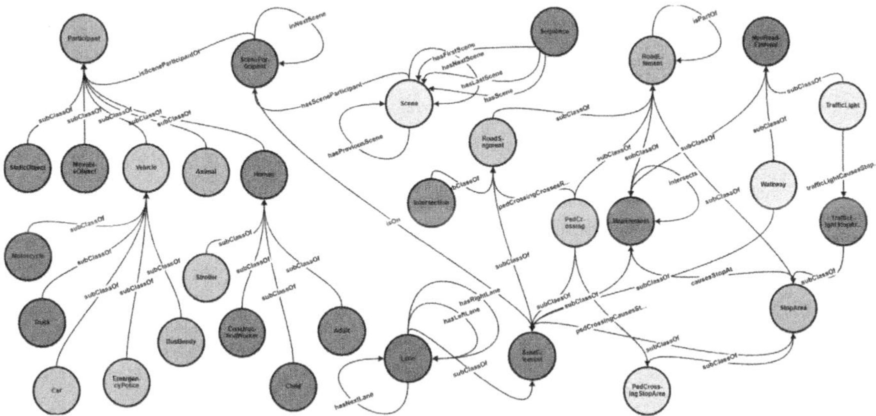

Fig. 3. nuScenes Knowledge Graph - a comprehensive representation of sensory data using ontological concepts with a strong focus on scene understanding. Scene objects O are categorized based on different levels of abstraction and characteristics, such as static and dynamic for fixed and moving objects, respectively.

e.g. number, date or text, and each $r \in R$ is a relation between two entities $<e_1, r, e_2>$ or an entity with a value $<e, r, v>$.

In our approach, we use the nuScenes knowledge graph (nuScenesKG[1]) [16] to represent all information of the nuScenes dataset [2]. The nuScenes dataset is one of the large-scale datasets for autonomous driving, comprising 1,000 driving sequences, each 20 s long, from Boston and Singapore.

The Ontology-(TBox) is structured in two main modules: 1-Agent: encompassing taxonomies of traffic participants: Vehicle (e.g., Car, Bus), Human (e.g., Adult, Child), relations capturing their state and location; and 2-Map: Road topology including its Segments, Lanes, Lane_Snippets and Lane_Slices. The expressivity level is SROIQ(D). Overall, it comprises 42 classes, 10 object properties, and 24 datatype properties. Among them, *Scene* is a key class, denoted as tuple $S_T = (B, C, T, P)$, where B - denotes its state, C - the relations to the predecessor and successor scene, respectively, T - denotes the specific time point at which the scene is captured and P - its participants including their geo-spatial relations. Figure 3 shows a small excerpt of the ontology depicting the core concepts and their relations.

Conversely, the KG-(ABox) persists instances and their respective states over time. Ontological axioms (transitivity, reflexivity and equivalence) enable the reasoning process to make many relations explicit, including the temporal evolution of scenes, participants, and scenery information. We perform additional operations to establish hierarchical relations between Lanes, Lane_Snippets and Lane_Slices. Further, we compute connections between agents (lateral, longitudinal) and their geo-spatial projection to the map topology at specific time points. The KG comprises over 43 million triples in total, capturing the states and interlinks of static and dynamic entities over time.

[1] The ontology and the KG are available at https://zenodo.org/records/10074393.

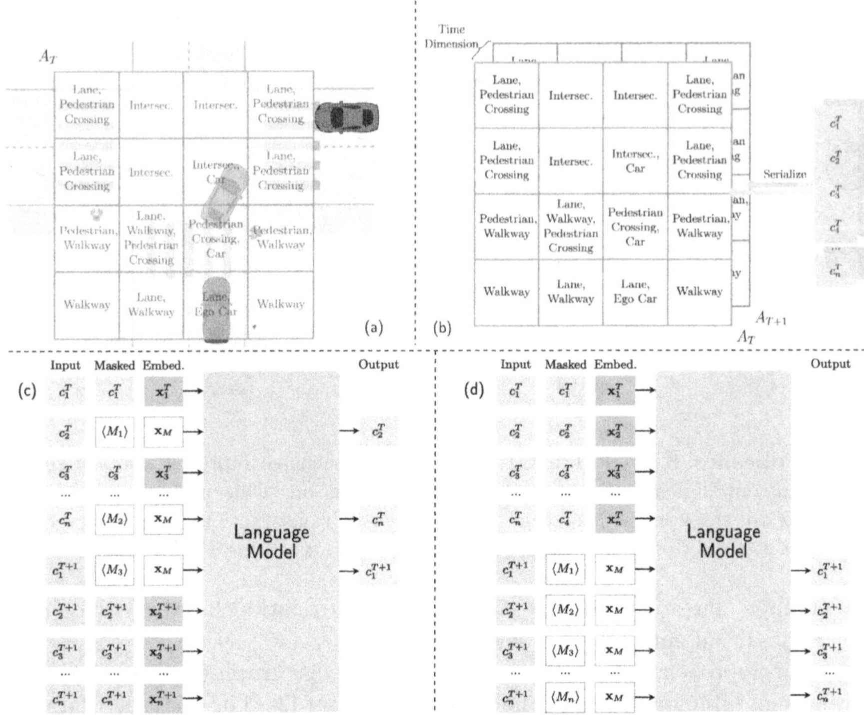

Fig. 4. (a) A matrix of cells with scene objects is extracted from the driving scenes KG. (b) The matrices extracted from the scenes at time steps T and $T+1$ are converted into serialized sequences of tokens. The language model is trained using either (c) scene object prediction or (d) next scene prediction.

3.2 BEV Symbolic Scene Representation

The foundation model training is performed using a BEV-based scene representation extracted from the KG representation. As shown in Fig. 4a, the geo-spatial area around the EV is represented as an area matrix A_T for each driving Scene S_T. The area matrix can be denoted as follows:

$$A_T = \begin{bmatrix} c_{n1}^T & c_{n2}^T & \dots & c_{nm}^T \\ \dots & \dots & \dots & \dots \\ c_{21}^T & c_{22}^T & \dots & c_{2m}^T \\ c_{11}^T & c_{12}^T & \dots & c_{1m}^T \end{bmatrix}, \tag{1}$$

where the dimensions of A_T is $n \times m$ and each matrix cell $c_{ij}^T, \forall i \in \{1, 2, \dots, n\}, \forall j \in \{1, 2, \dots, m\}$ represent a geographical area of height h and width w. The height and width of each cell define the resolution of the matrix, whereas the number of rows and columns define the overall size or area of the scene. The exact area of cell c_{ij}^T is defined by the coordinates $(a_{min}^{ij}, b_{min}^{ij})$ and $(a_{max}^{ij}, b_{max}^{ij})$, which are the positions of diagonally opposite corners of the cell, respectively. As depicted in Fig. 5b, the geographic area of A^T covers the area

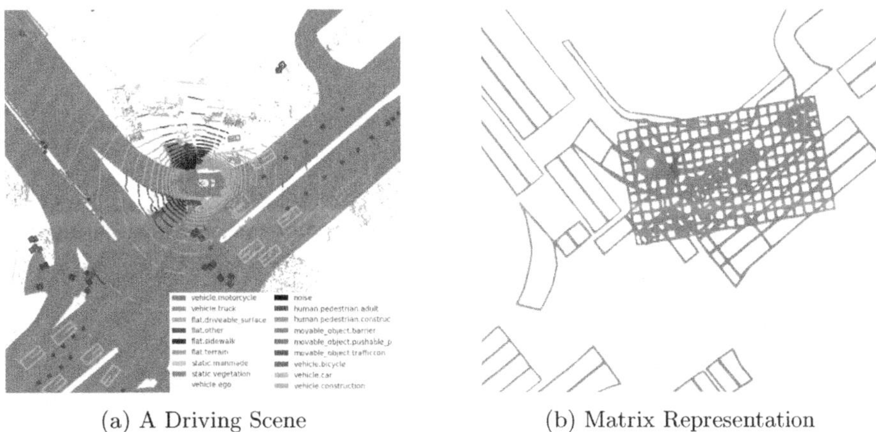

(a) A Driving Scene (b) Matrix Representation

Fig. 5. (a) A nuScenes visualization of a traffic scene using top-down LiDAR data, representing different entities with distinct colors. (b) A BEV representation of map concepts, including lanes, sidewalks, etc., is depicted as polygons. A 20×11 area matrix A to illustrate the scene range in our setup. The pinpoint indicates the EV's location.

ahead, behind, and on both sides of the EV and adopts the orientation of the EV.

Each cell c_{ij}^T comprises a set of scene objects $O = \{o_1, o_2, ..., o_n\}$ that occur within the specific location according to scene representation in the KG. Every object o_k has a geo-location, which is represented by its coordinates (a_k, b_k). An object belongs to a cell c_{ij}^T when the object coordinates are located within the cell area. We can formally define this as follows:

$$o_k \in c_{ij}^T \Leftrightarrow a_{min}^{ij} \leq a_k \leq a_{max}^{ij} \text{ and } b_{min}^{ij} \leq b_k \leq b_{max}^{ij}. \tag{2}$$

As such, the BEV scene representation captures the relative position of the scene objects to the EV and each other in the area matrix. Every scene object o_k has an object type e_k. The object types $E = \{e_1, e_2, ..., e_n\}$ are defined in the ontology and can be referred to by their unique text labels $L = \{l_1, l_2, ..., l_n\}$ respectively. For the extraction of the scene objects from the KG, we utilize geo-SPARQL queries to fetch the respective objects based on their geo-location.

3.3 Serialization

As Fig. 4b shows, the BEV scene representation, where each cell in the matrix c_{ij}^T contains the scene objects $\{o_1, ..., o_n\}$ occurring in the corresponding area, is serialized to a string of tokens.

The serialization method converts two subsequent scenes, S_T and S_{T+1}, into a sequence of tokens. Our approach processes the area matrix A_T in row-major order, starting from c_{11}^T to c_{nm}^T, and then concatenates it with the row-major order of area matrix A_{T+1}, starting from c_{11}^{T+1} to c_{nm}^{T+1}. We use the type label l_k of each object o_k occurring in the cell for the encoding.

At the start of the encoding, <country>, <dist>, and <orientation_diff> tokens are used to indicate that the following token is the country where the EV drives, the traveled distance between scenes S_T and S_{T+1}, measured in meters, as well as the orientation difference of the EV between the scenes S_T and S_{T+1}), measured in degrees. An example is illustrated here,

"<country> US <dist> 4.8 <orientation_diff> 0 <scene_start> lane <col_sep> lane <concept_sep> car <col_sep> pedestrian crossing <col_sep> walkway ... <row_sep> ... <col_sep> intersection <col_sep> turn stop area <col_sep> ...",

where <col_sep> and <row_sep> serve as positional delimiters, separating the columns and rows of the matrix in order to preserve the geo-spatial structure of the scene. <concept_sep> separates different scene objects within a cell.

3.4 Scene Learning

Here, we present two tasks designed to train the FMs for scene comprehension, namely mask prediction and next scene prediction (cf. Fig. 4c).

Scene Object Prediction. The text input formed by a serialized sequence of tokens is randomly masked. An example can be seen as follows,

"<country> US <dist> 4.8 <orientation_diff> 0 <scene_start> lane <col_sep> lane <concept_sep> car <col_sep> <M_1> <col_sep> walkway ... <row_sep> ... <col_sep> <M_2> <col_sep> turn stop area <col_sep> ...",

where <M_1> and <M_2> are the unique sentinel token used to corrupt the text input. The objective is to predict the dropped-out spans, delimited by the sentinel tokens used to replace them in the input plus a final sentinel token <M_3>, as shown below,

"<M_1> pedestrian crossing <M_2> intersection <M_3>".

Next Scene Prediction. Instead of randomly replacing tokens in the sequence with the sentinel token, all tokens representing map concepts and entities in the next scene $T+1$ are masked. Similar to the previously defined mask prediction task, the learning objective is to predict all these map concepts and entities in the next scene $T+1$, delimited by the sentinel tokens.

Learning Loss. We use cross-entropy loss to maximize the probability of correctly predicting ground truth tokens at their respective positions. Specifically, the loss function is defined as follows:

$$L = -\sum_i y_i \log(\hat{y}_i), \tag{3}$$

where y_i is the one-hot encoded ground truth label for i-th token while $\hat{y}_i$ is the predicted probability for i-th token.

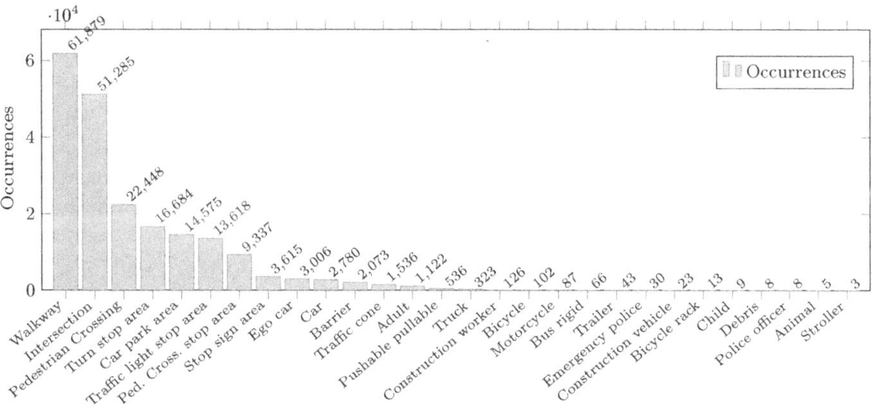

Fig. 6. The occurrence of scene objects in our proposed dataset. The numbers on the bars indicate the occurrence of each corresponding object.

4 Experiments

4.1 Dataset and Data Augmentation

The BEV symbolic scene representations extracted from the nuScenesKG form the backbone of our training dataset. It encompasses approximately 30,000 driving scenes. To prevent data leakage, we divide these driving scenes into training(80%), validation(10%), and testing datasets(10%). The occurrence of scene objects can be seen in Fig. 6. As the figure illustrates, the majority of scene objects are static objects like walkways, intersections, and pedestrian crossings.

A random re-mask strategy is employed for data augmentation, i.e., randomly masking tokens within the training data and reassigning them to different areas (cells) in each scene upon the completion of every epoch. This method ensures that the model is continuously exposed to varying prediction scenarios, thereby enhancing its learning and predictive capabilities for diverse driving scenes.

4.2 Experiment Setting

Language Backbone. We utilize T5, an encoder-decoder language model, to frame language tasks as text-to-text generation problems. T5 is available in various sizes: T5-small, T5-base, T5-large, T5-3B, and T5-11B. Our experiments primarily focus on the T5-base model, with additional experiments conducted using T5-small and T5-large to evaluate the impact of model size on performance. Additionally, we include the randomly initialized T5-base model for comparison.

Scene Range. In our experiments, the range of the BEV symbolic representation extends 30 m in front of the ego vehicle, 10 m behind it, and 12 m to each side. We intentionally give more focus on the space in front of the car as this has a higher relevance for the actions to be taken. In total, each scene we consider has a height of 40 m and a width of 24 m.

Number of Areas in Matrix. The length of the input sequence depends on the resolution of the area matrix (i.e., cell width w and length l, and dimensions $n \times m$ of A): higher resolutions result in more tokens for the model to process. The pre-trained T5 model can handle sequences of up to 4096 tokens without a significant performance drop. In our setup, we use a matrix dimension of 20×11, where each cell represents an area of $2\,\text{m} \times 2\,\text{m} = 4\,\text{m}^2$. Additionally, for ablation studies, we use a lower-dimension matrix of 8×5, where each cell corresponds then to an area of $5\,\text{m} \times 5\,\text{m} = 25\,\text{m}^2$.

Number of Masked Areas. The ratio of mask tokens over all tokens during training is important for the learning efficiency of the model. For the masked prediction task, we randomly mask *three* cells in the current scene and *three* cells in the next scene. For the next scene prediction tasks, there is no masking for the current scene, and the entire next scene is masked. Note that we only mask $14 \times 7 = 98$ central cells. There are two main reasons for that: 1) T5 only supports 100 masking tokens, and increasing the number of predictions would cause a performance drop of T5. 2) It is difficult for the model to predict new incoming objects in the next scene. We replace the margin of the matrix with the empty token <empty> and thus ease the above-mentioned issue.

Training Details. We fine-tune all T5 models using the AdamW optimizer with a learning rate of 0.0001, following a linear schedule without warm up steps. The batch size equals 4. Training for the T5-base model is conducted on a single Nvidia A100 GPU with 80 GB of memory, whereas the T5-large model is trained on a single Nvidia H200 GPU with 141 GB of memory.

4.3 Evaluation Metric

Accuracy, precision, recall, and F1 are computed to evaluate the performance of our proposed model. For the autonomous driving task, it is less risky to predict false positives than to predict false negatives. Predicting a car where none exists may lead to unnecessary avoidance, but failing to predict a car when one is present poses a much higher risk of collision. Hence, recall is a more critical metric for evaluation. All metrics presented in the following sections reflect the performance on the test data.

4.4 Results

Scene Object Prediction. To evaluate the spatial understanding ability of LLMs, we conduct the scene object prediction task. As Table 1 demonstrates, the baseline that fine-tunes the pre-trained LLM achieves good performance on the object prediction task. The overall test accuracy reaches 88.7%. Compared to other LLMs such as LLaMA and ChatGPT, which are widely utilized for scene understanding in autonomous driving, our model demonstrates significantly improved performance. Experiment 1.1 shows that without the implicit

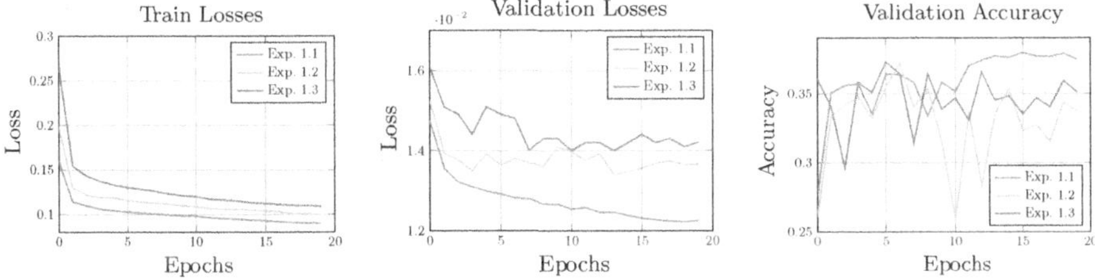

Fig. 7. The losses and accuracy of scene object prediction are reported for Experiment 2, conducted without a dropout strategy, as well as Experiments 3 and 4, which utilized dropout rates of 0.2 and 0.4, respectively.

knowledge from a pre-trained LLM, the prediction performance drops significantly from 88.7% to 37.4%. Figure 7 depicts the training/validation loss and validation accuracy. Experiment 1.2 and 1.3 show that using drop-out to prevent overfitting does not improve the performance either.

Table 1. The accuracy, precision, recall, and F1 score for the scene object prediction task. Llama3.1 and ChatGPT are tested in the zero-shot setting. We use the ChatGPT version '2024-02-15-preview'.

Experiments	Accuracy	Precision	Recall	F1
Exp. 1 - FM4SU(ours)	0.887	0.866	0.744	0.786
Exp. 2 - Llama3.1 (8B)	0.182	0.074	0.069	0.081
Exp. 2.1 - Llama3.1 (70B)	0.224	0.085	0.077	0.068
Exp. 3 - ChatGPT3.5	0.353	0.181	0.123	0.141
Exp. 3.1 - ChatGPT4o	0.412	0.167	0.175	0.161
Exp. 1.1 - w/o pre-training no dropout	0.374	-	-	-
Exp. 1.2 - w/o pre-training dropout rate 0.2	0.327	-	-	-
Exp. 1.3 - w/o pre-training dropout rate 0.4	0.353	-	-	-

Next Scene Prediction. We further conduct next scene prediction experiments to evaluate the spatio-temporal understanding ability of pre-trained LLMs in the driving scenarios. In our setting, the next scene prediction task aims to predict all scene objects in the next scene instead of just a few. Hence, this task is more difficult than the previous tasks. As Table 2 demonstrates, the Exp. 4 baseline, which is further fine-tuned in next scene prediction with the checkpoint in Exp. 1, achieves an overall 86.7% accuracy in the next scene prediction task. Compared to the object prediction in Exp. 1, the precision, recall, and F1 score decrease, while the accuracy slightly drops. One potential reason for having high accuracy while low precision, recall, and F1 is that the model predicts the dominant class more frequently, achieving high accuracy but performing poorly on less frequent or minority classes. To further investigate this, we divide the scene

Table 2. The Accuracy, Training loss, Precision, Recall, and F1 Score for the Task of Next Scene Prediction.

Experiments	Accuracy	Precision	Recall	F1
Exp. 4 - FM4SU (Next Scene Prediction)	0.867	0.618	0.594	0.603
Exp. 4.1 - w/o Scene Object Prediction Training	0.865	0.622	0.598	0.608
Exp. 4.2 - w/o Additional Metadata	0.824	0.590	0.551	0.568

objects into two categories, namely dynamic objects and static objects, and evaluate on these two categories. Dynamic objects refer to moving objects on the road, such as cars, motorcycles, and pedestrians, which are relatively rare in the scene. In contrast, static objects are immovable features like lanes, walkways, and pedestrian crossings, which constitute the majority of the scene. As Fig. 7a and 7b shown, the performance of dynamic object prediction in the next scene prediction drops significantly when compared to the scene object prediction task. These results do not surprise, as spatio-temporal relations of dynamic objects are much harder to learn compared to static objects.

We also investigate if the pre-training of scene object prediction could improve the performance of the next scene prediction. We conduct Exp. 4.1, i.e. fine-tuning on the T5-base checkpoint instead of the checkpoint from Exp. 1. Comparing Exp. 4 and Exp. 4.1 in Table 2, the performance does not vary significantly. However, as shown in Fig. 9, pre-training on scene object prediction achieves much better performance at the beginning, indicating that spatial prior knowledge from scenes leads to in a faster learning process.

Additional metadata also plays an important role in temporal scene understanding performance. Experiment 4.2 is conducted by removing all metadata, including country information, as well as the EV's displacement and orientation shift between consecutive scenes. As Table 2 shows, without these additional metadata, the prediction accuracy, precision, recall, and F1 score drops from 86.7%, 61.8%, 59.4%, 60.3% to 82.4%, 59.0%, 55.1%, 56.8%, respectively. The results confirm our intuition, namely that country information is critical due to differences in driving orientation (right-hand drive vs. left-hand drive), and the ego vehicle's displacement and orientation shift are essential for understanding the relative movement of the surrounding environment.

4.5 Ablation Study

Scene Resolution. We conduct experiments with different scene resolutions for scene object detection tasks. Table 3 demonstrates the baseline with scene resolution 20×11 performs significantly better compared to resolution 8×5. This shows that the higher granularity of information helps the LLM to reach a higher prediction accuracy. Due to their lower resolution, larger cells encompass more scene objects and thus increase the complexity of the prediction task (Fig. 8).

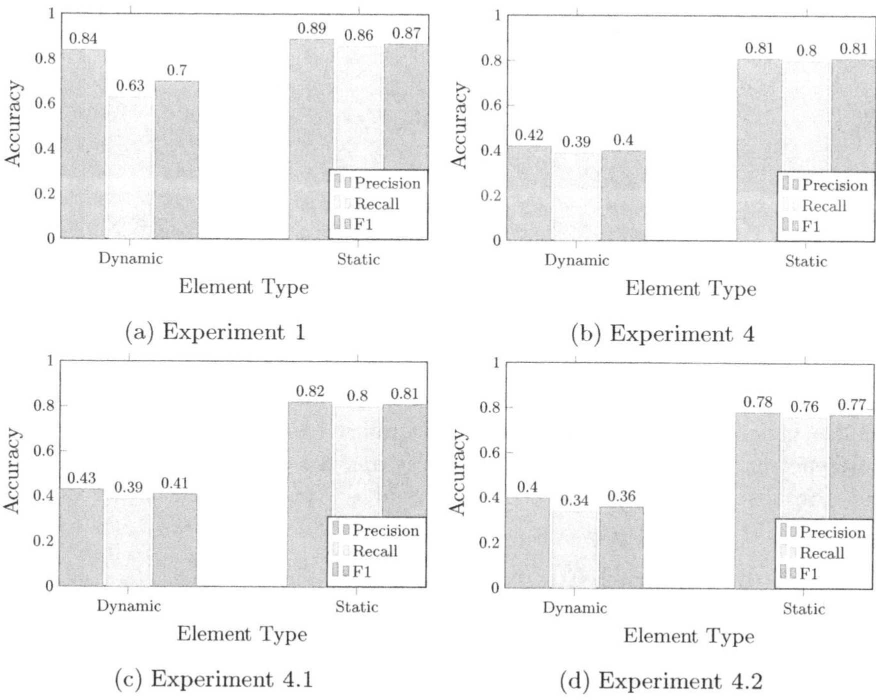

(a) Experiment 1

(b) Experiment 4

(c) Experiment 4.1

(d) Experiment 4.2

Fig. 8. The precision, recall, and F1 score for dynamic/static objects of the scene object prediction (Exp. 1) as well as the next scene prediction task (Exp. 4.1, 4.2, and 4.3). The number of dunamic and static object are 9374 and 371271, respectively.

Table 3. The scene object prediction accuracy with different scene resolution. Experiment 1 and 1.4 are trained with a scene resolution of 20×11 and 8×5, respectively.

Experiments	Accuracy
Exp. 1 - Baseline (Scene Resolution 20×11)	0.887
Exp. 1.4 - Scene Resolution 8×5	0.396

Table 4. The next scene prediction accuracy with different model sizes. Experiment 4.3, 4, 4.4 are trained with T5-small, T5-base, T5-large language backbones, respectively.

Experiments	Accuracy
Exp. 4.3 - with T5-small Backbone	0.770
Exp. 4 - Baseline (with T5-base Backbone)	0.867
Exp. 4.4 - with T5-large Backbone	0.861

Model Size. To examine whether model size influences the performance of the scene prediction task, we conducted experiments using language backbones of

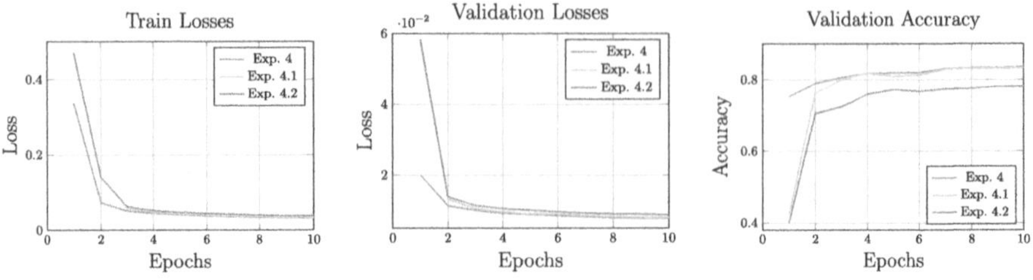

Fig. 9. The training, validation loss and validation accuracy of the next scene prediction task for each Experiment 4, 4.1, 4.2.

varying sizes, specifically T5-small, T5-base, and T5-large. Table 4 shows larger models generally demonstrate better performance. However, given the marginal difference in performance between T5-large and T5-base, we primarily use T5-base for most experiments, as it is more resource-efficient in terms of GPU usage.

5 Conclusion and Future Work

This paper introduced FM4SU, a novel approach for training a FM for scene understanding in autonomous driving. The objective is to overcome the limitations of current methods in comprehending the complex the spatio-temporal evolution of scenery information. The methodology utilizes KGs to capture sensory observations and domain knowledge, as well as a novel BEV symbolic representation of each scene extracted from KG. This BEV scene representation includes the spatio-temporal information among the objects across the scenes. Next, this representation is serialized into a sequence of tokens and given to pre-trained language models for learning an inherent understanding of the co-occurrence among scene elements and generating predictions on the next scenes. FM4SU is evaluated using the T5 model and demonstrates superior performance in both scene object prediction and next scene prediction task, with an accuracy of 88.7% and 86.7%, respectively. Based on our results, we conclude that FM4SU provides a promising framework for developing larger and more exhaustive FMs for scene understanding, and opens avenues for future exploration and downstream tasks in the context of autonomous driving. Further, as a part of our contribution, we have released the BEV scene representation dataset for the nuScenesKG as well as the source code for the extraction and training in order to invite the research community to develop and benchmark new algorithms and FMs for scene understanding based on the nuScenes dataset.

For future work, we strive to integrate into FM4SU other LLMs and perform a wide range of investigations on the performance and scalability aspects. Further, our aim is to assess its performance on other downstream tasks, for example, 3D object detection and trajectory prediction with by incorporating the acquired scene understanding.

References

1. Brown, T.B., et al.: Language models are few-shot learners. CoRR abs/2005.14165 (2020). https://arxiv.org/abs/2005.14165
2. Caesar, H., et al.: nuscenes: a multimodal dataset for autonomous driving. In: 2020 IEEE/CVF Conference on Computer Vision and Pattern Recognition, CVPR 2020, Seattle, WA, USA, 13–19 June 2020, pp. 11618–11628. Computer Vision Foundation/IEEE (2020)
3. Chen, L., et al.: Driving with LLMs: fusing object-level vector modality for explainable autonomous driving. In: IEEE International Conference on Robotics and Automation, ICRA 2024, Yokohama, Japan, 13–17 May 2024, pp. 14093–14100. IEEE (2024). https://doi.org/10.1109/ICRA57147.2024.10611018
4. Chitta, K., Prakash, A., Jaeger, B., Yu, Z., Renz, K., Geiger, A.: Transfuser: imitation with transformer-based sensor fusion for autonomous driving. IEEE Trans. Pattern Anal. Mach. Intell. **45**(11), 12878–12895 (2023). https://doi.org/10.1109/TPAMI.2022.3200245
5. Chowdhery, A., et al.: Palm: scaling language modeling with pathways. J. Mach. Learn. Res. **24**, 240:1–240:113 (2023)
6. Cui, C., et al.: A survey on multimodal large language models for autonomous driving. In: Proceedings of the IEEE/CVF Winter Conference on Applications of Computer Vision (WACV) Workshops, pp. 958–979 (2024)
7. Cui, Y., et al.: Drivellm: charting the path toward full autonomous driving with large language models. IEEE Trans. Intell. Veh. **9**(1), 1450–1464 (2024). https://doi.org/10.1109/TIV.2023.3327715
8. Devlin, J., Chang, M., Lee, K., Toutanova, K.: BERT: pre-training of deep bidirectional transformers for language understanding. In: NAACL-HLT (1), pp. 4171–4186. Association for Computational Linguistics (2019)
9. Ding, X., Han, J., Xu, H., Zhang, W., Li, X.: Hilm-d: towards high-resolution understanding in multimodal large language models for autonomous driving. arXiv preprint arXiv:2309.05186 (2023)
10. Gao, H., Li, Y., Long, K., Yang, M., Shen, Y.: A survey for foundation models in autonomous driving. CoRR abs/2402.01105 (2024). https://doi.org/10.48550/ARXIV.2402.01105
11. Geiger, A., Lenz, P., Stiller, C., Urtasun, R.: Vision meets robotics: the KITTI dataset. Int. J. Robot. Res. **32**(11), 1231–1237 (2013). https://doi.org/10.1177/0278364913491297
12. Halilaj, L., Luettin, J., Monka, S., Schmid, S.: Knowledge graph-based integration of autonomous driving datasets. Int. J. Semant. Comput. **17**(2), 249–271 (2023)
13. Hoffmann, J., et al.: Training compute-optimal large language models. CoRR abs/2203.15556 (2022)
14. Lin, X., et al.: Protokens: a machine-learned language for compact and informative encoding of protein 3D structures. bioRxiv (2023). https://doi.org/10.1101/2023.11.27.568722. https://www.biorxiv.org/content/early/2023/11/28/2023.11.27.568722. https://www.biorxiv.org/content/early/2023/11/28/2023.11.27.568722.full.pdf
15. Luettin, J., Monka, S., Henson, C.A., Halilaj, L.: A survey on knowledge graph-based methods for automated driving. In: KGSWC. Communications in Computer and Information Science, vol. 1686, pp. 16–31. Springer, Cham (2022)
16. Mlodzian, L., et al.: nuscenes knowledge graph - a comprehensive semantic representation of traffic scenes for trajectory prediction. In: Proceedings of the

IEEE/CVF International Conference on Computer Vision (ICCV) Workshops, pp. 42–52 (2023)

17. Peters, M.E., et al.: Deep contextualized word representations. In: Proceedings of the 2018 Conference of the North American Chapter of the Association for Computational Linguistics: Human Language Technologies, Volume 1 (Long Papers), pp. 2227–2237. Association for Computational Linguistics, New Orleans, Louisiana (2018). https://doi.org/10.18653/v1/N18-1202
18. Radford, A., Wu, J., Child, R., Luan, D., Amodei, D., Sutskever, I., et al.: Language models are unsupervised multitask learners. OpenAI Blog **1**(8), 9 (2019)
19. Raffel, C., et al.: Exploring the limits of transfer learning with a unified text-to-text transformer. J. Mach. Learn. Res. **21**, 140:1–140:67 (2020)
20. Seff, A., et al.: Motionlm: multi-agent motion forecasting as language modeling. CoRR abs/2309.16534 (2023)
21. Sun, Z., Wang, Z., Halilaj, L., Luettin, J.: Semanticformer: holistic and semantic traffic scene representation for trajectory prediction using knowledge graphs. IEEE Robot. Autom. Lett. **9**(9), 7381–7388 (2024)
22. Tian, R., et al.: Tokenize the world into object-level knowledge to address long-tail events in autonomous driving. CoRR abs/2407.00959 (2024)
23. Touvron, H., et al.: Llama: open and efficient foundation language models. CoRR abs/2302.13971 (2023). https://doi.org/10.48550/ARXIV.2302.13971
24. Vaswani, A., et al.: Attention is all you need. In: Advances in Neural Information Processing Systems, vol. 30 (2017)
25. Wang, G., et al.: Voyager: an open-ended embodied agent with large language models. Trans. Mach. Learn. Res. **2024** (2024). https://openreview.net/forum?id=ehfRiF0R3a
26. Wang, W., et al.: Drivemlm: aligning multi-modal large language models with behavioral planning states for autonomous driving. arXiv preprint arXiv:2312.09245 (2023)
27. Wang, Z., Sun, Z., Luettin, J., Halilaj, L.: Socialformer: social interaction modeling with edge-enhanced heterogeneous graph transformers for trajectory prediction. CoRR abs/2405.03809 (2024)
28. Wen, L., et al.: Dilu: a knowledge-driven approach to autonomous driving with large language models. arXiv preprint arXiv:2309.16292 (2023)
29. Xu, Z., et al.: Drivegpt 4: interpretable end-to-end autonomous driving via large language model. IEEE Robot. Autom. Lett. **9**(10), 8186–8193 (2024)
30. Zheng, Y., et al.: Planagent: a multi-modal large language agent for closed-loop vehicle motion planning. arXiv preprint arXiv:2406.01587 (2024)
31. Zhou, H., et al.: Visual representation learning guided by multi-modal prior knowledge. CoRR abs/2410.15981 (2024). https://doi.org/10.48550/ARXIV.2410.15981
32. Zhou, H., Sui, A., Cao, W., Shi, L.: What matters to enhance traffic rule compliance of imitation learning for automated driving. arXiv preprint arXiv:2309.07808 (2023)
33. Zhou, H., Sui, A., Shi, L., Li, Y.: Penalty-based imitation learning with cross semantics generation sensor fusion for autonomous driving. In: ITSC, pp. 1876–1883. IEEE (2023)
34. Zhou, H., et al.: Bridging language and action: a survey of language-conditioned robot manipulation. CoRR abs/2312.10807 (2023). https://doi.org/10.48550/arXiv.2312.10807
35. Zhou, Y., et al.: Embodied understanding of driving scenarios. In: ECCV (62). Lecture Notes in Computer Science, vol. 15120, pp. 129–148. Springer, Cham (2024)

ANTS: Abstractive Entity Summarization in Knowledge Graphs

Asep Fajar Firmansyah(✉), Hamada M. Zahera, Mohamed Ahmed Sherif, Diego Moussallem, and Axel-Cyrille Ngonga Ngomo

Data Science Group (DICE), Heinz Nixdorf Institute, Paderborn University, Paderborn, Germany
{asep.fajar.firmansyah,hamada.zahera,mohamed.sherif, diego.moussallem,axel.ngonga}@uni-paderborn.de

Abstract. Entity summarization in knowledge graphs (KGs) aims to generate concise summaries of entities by selecting the most relevant facts (triples) from structured KG data, such as DBpedia or Wikidata. Existing methods focus on selecting triples that are directly *present* in KGs, which inherently contain incomplete data. To address this limitation, we propose ANTS, an abstractive approach that generates entity summaries beyond KG triples. Our approach first identifies relevant entity-related triples using KG embeddings. Furthermore, since KG embeddings struggle with literal values (e.g., numbers, dates), we integrate a large language model (LLM) to augment plausible triples with literal objects. For evaluation, we used ESSUM, a silver-standard dataset combining ESBM-DBpedia and FACES benchmarks. Experimental results demonstrate that ANTS outperforms baseline models, achieving a BLEU score of 5.78, BERTScore F1 of 0.84, and MoverScore of 0.54.

Keywords: Abstractive Entity Summarization · Knowledge Graphs · Large Language Models

1 Introduction

Entity summarization involves generating concise summaries of entities by selecting the most relevant RDF triples in knowledge graphs (KGs) [8]. However, extracting relevant information from KGs is challenging due to their massive data volume [1,19]. As a result, entity summarization has become an essential task for obtaining concise and informative entity [4,8,9]. Furthermore, KGs are often incomplete, which poses additional challenges to generate a comprehensive summaries as well [18]. To address these limitations, our study focuses on *abstractive entity summarization*, i.e., generating summaries that are not only concise and coherent but also includes missing information not explicitly present

A. F. Firmansyah and H. M. Zahera—These authors contributed equally to this work.

E. Curry et al. (Eds.): ESWC 2025, LNCS 15718, pp. 133–151, 2025.
https://doi.org/10.1007/978-3-031-94575-5_8

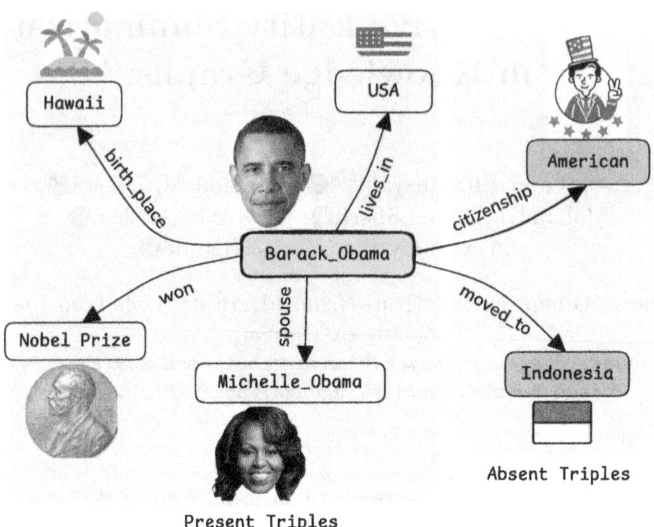

Fig. 1. An incomplete knowledge graph for "Barack Obama". Green nodes indicate present triples, red nodes represent absent triples. (Color figure online)

in KGs. Abstractive entity summarization is particularly beneficial for different applications, such as natural language generation, where the selected RDF triples serve as input to generate a coherent and informative entity descriptions. For instance, it can be employed to generate snippets for search results by calculating triple scores and selecting the most relevant triples [7,12].

Existing methods for entity summarization typically leverage ranking algorithms to select a subset of RDF triples (known as *present triples*) to represent the target entity [8,19,26]. However, these approaches fail to address the inherent incompleteness nature of KGs, where important facts about an entity may be missing, referred as *absent triples*. For instance, in an incomplete KG of `"Barack Obama"` (see Fig. 1), key relationships such as his `"citizenship"` (American) or life events (e.g., `"moved to"` Indonesia) are often missing, resulting in an incomplete summary of the entity `"Barack Obama"`. By integrating *absent triples* into our abstractive summaries, we provide more comprehensive and accurate of entity summarization. In this work, we propose ANTS (Abstractive entity summarization), a novel approach for generating entity summaries in KGs inspired by the foundational principles of abstractive text summarization. Our approach employs knowledge graph embeddings (KGEs) to identify relevant triples related to a target entity. However, existing KGE methods struggle to infer triples with literal objects—such as `⟨Barack Obama, birthDate,"4-Aug-1961"⟩`—due to their limited ability to model numeric or textual values [3,10]. To address this limitation, we employ a large language model (LLM) to augment this missing information, thereby improving the completeness of entity summaries in KGs. Our approach balances completeness via (KGE inference and LLM-generated literals) with precision (through semantic constraints) to ensure logically consistent entity summaries.

Due to the lack of benchmark datasets for abstractive entity summarization, we created two resources: ESSUM and ESSUM-ABSENT. ESSUM is a silver-standard dataset comprising 160 DBpedia entities, with their corresponding silver summaries derived from ESBM [17] and FACES [11], referred to as $\text{ESSUM}_{\text{DBpedia}}$ and $\text{ESSUM}_{\text{FACES}}$, respectively. These summaries are extracted from Wikipedia's entity descriptions, providing a reliable baseline for evaluating abstractive summarization. ESSUM-ABSENT is designed to evaluate a model's ability to infer missing information by randomly removing 20% of triples from the original ESBM and FACES entity descriptions, treating them as *absent triples*. Our experiments demonstrate that ANTS outperforms baseline models, achieving a BLEU score of 5.78, an F1 BERTScore of 0.84 and a MoverScore of 0.54 on $\text{ESSUM}_{\text{DBpedia}}$, as shown in Table 3. Furthermore, integrating LLM with KGE improves summarization performance by up to 1.6%. We summarize the main contributions of our study as follows:

- We propose ANTS, the first abstractive entity summarization approach for KGs that incorporates both present and absent triples.
- Our approach mitigates the limitations of KGEs in modeling literal triples (e.g., dates, numbers) by leveraging LLM, addressing KGs incompleteness.
- We introduce ESSUM and ESSUM-ABSENT, the first silver-standard benchmarks for abstractive entity summarization. ESSUM combines curated summaries from ESBM and FACES, while ESSUM-ABSENT evaluates a model's ability to infer missing triples.
- We release the source code and datasets on the GitHub repository[1] for reproducing our experiments.

2 Related Works

Entity summarization in KGs has been recently leveraged for different downstream tasks such as information retrieval [22] and question answering [14]. Early work mainly focused on extractive methods, which rank and select top-k triples from the input KG as an entity summary. For example, RELIN [5] employs a weighted PageRank model to compute the centrality of each triple in a graph, where nodes are triples and edges are their pairwise similarities. The top-k triples with the highest centrality scores are returned as a final summary. Moreover, FACES [11], augments bag-of-words representation of triples with *WordNet* hypernyms and applies hierarchical clustering to aggregate similar triples. Another line of work, MPSUM [25] that leverages a probabilistic topic model to rank triples based on predicate uniqueness and object significance, then returns the top-k triples as a final output.

In contrast to traditional methods– *relying on hand-crafted features*–, neural approaches leverage deep learning to train end-to-end models that automatically learn to compute the relevance of candidate triples w.r.t. a given entity.

[1] https://github.com/dice-group/ANTS.

For example, ESA [26] uses a bidirectional LSTM network with an attention mechanism to compute the relevance of triples for a given entity. In particular, ESA assigns attention weights to each triple, forming a summary vector that ranks triples by their relevance scores. Moreover, ESA's performance is further enhanced through AutoSUM [27], which incorporates multi-user preferences to refine relevance scoring. DeepLENs [19] is another neural model that includes three components: (1) triple encoding, (2) entity description encoding, and (3) triple scoring, to identify the top-k triples as an entity summary. DeepLENs is trained on a labeled dataset of entity summaries to learn scoring candidate triples based on their interdependencies. Similarly, Gates [8] integrates structural information of KG and text embeddings, encoding triples using ComplEx (for KG embeddings) and GloVe (for textual semantics). It then applies a graph attention network (GAT) to an adjacency matrix representation of the KG, learning importance weights for triple selection.

While these methods advance entity summarization, they remain extractive, limiting their ability to address KG incompleteness to infer *absent triples*. Existing entity summarization methods assume that all relevant facts about an entity are represented in the KG, and then employ ranking algorithms to retrieve a top-k triples as a summary. However, this assumption is often inaccurate due to the incompleteness of KGs, where certain facts may be missing [29]. Consequently, the resulting entity summaries tend to be incomplete. To address this gap, we propose Ants, a novel abstractive approach that combines KGE and LLM. Unlike extractive approaches, Ants generates entity summaries containing both present triples (retrieved via KGEs) and plausible absent triples (inferred by LLMs), thereby generating more comprehensive entity descriptions.

3 Our Approach

3.1 Preliminaries

Let $\mathcal{KG} = (\mathcal{E}, \mathcal{R}, \mathcal{T})$ be a knowledge graph, where $\mathcal{E}$ is a set of all entities, $\mathcal{R}$ is a set of all relations, and $\mathcal{T} \subseteq \mathcal{E} \times \mathcal{R} \times (\mathcal{E} \cup \mathcal{L} \cup \mathcal{C} \cup \mathcal{B})$ represents the set of all triples, with $\mathcal{C}$ (classes), $\mathcal{L}$ (literals), and $\mathcal{B}$ (blank nodes).

Definition 1 (Present Triples): For an entity e and size constraint $k \in \mathbb{N}^+$, *present triples* are a subset $T' \subseteq \mathcal{T}$ of relevant triples that maintains semantic constraints (e.g., domain and range requirements of predicates) to ensure KG consistency [13].

Definition 2 (Absent Triples): *Absent triples* are factually correct assertions relevant to entities in the KG, but missing from $\mathcal{T}$. Absent triples (like the present ones) adhere to the semantic constraints of the KG.

Definition 3 (Abstractive Entity Summarization): This task generates a summary $\mathcal{S}_e$ for an entity $e \in \mathcal{E}$ by selecting the top-k relevant triples from $\mathbf{T} = T' \cup T^*$, where T' (present triples) and T^* (absent triples) are defined as above in Definition 1 and 2. Formally, abstraction summarization can be defined as a function $\mathcal{F} : e \in \mathcal{E} \rightarrow \mathcal{S}_e$, where $\mathcal{S}_e \subseteq \mathbf{T}$, and $|S_e| \leq k \leq |\mathbf{T}|$ where $k \in \mathbb{N}^+$

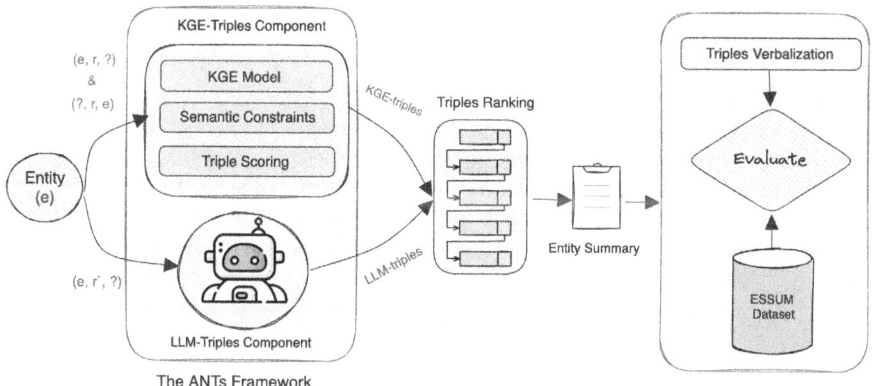

Fig. 2. The ANTS framework, including the evaluation process

3.2 The Architecture

Figure 2 shows the architecture of our approach (ANTS), which takes an entity e as input and return a ranked list of RDF triples $\mathcal{S}_e$ as output. Our approach consists of five components: (i) *KGE-triples*, which employs KGE to infer relevant triples KG for the entity e, including those present in the KG. (ii) *LLM-triples*, which augments the KG with absent triples, particularly those containing literal values, using LLM. (iii) *Triple Ranking* that computes relevance score for each triple based on predicate (i.e., relation) frequency in the KG. (iv) *Entity Summary* is a subset of the top-k triples with the highest relevance scores to form the final summary. (v) *Evaluation* that assesses the quality of entity summaries using automatic evaluation metrics by transforming RDF triples into natural language (see Sect. 4.1 for further detail).

KGE-Triples Component ($\hat{T}_{(kge)}$). This component employs knowledge graph embeddings to predict relevant triples for a given entity e_s. We regard this task as a link prediction, where all relations $r \in \mathcal{R}$ associated with e_s are considered to find the most plausible object entities e'_o, forming triples $\langle e_s, r, e'_o \rangle$. We use the pre-trained CONVE model [6], trained on DBpedia `(version 03-2022)`. As described in [15], this process involves traversing a KG defined by a set of entities $\mathcal{E} = \{e_1, \ldots, e_{N_e}\}$ and a set of relations $\mathcal{R} = \{r_1, \ldots, r_{N_r}\}$. For a given pair (e_s, r), CONVE model computes a scoring function to evaluate the plausibility of candidates triples $\langle e_s, r, e'_o \rangle$, ranking them by relevance to e_s. To handle literal triples, we enhance the CONVE model with LITERALE [15], which replaces entity embeddings e with literal embeddings e^{lit}, as described in [15]. The resulting model, referred to as LITERALE, is represented by the function $f(e_s^{lit}, r, e_o'^{lit})$.

Semantic Constraints. To ensure logical consistency when inferring missing triples, we apply semantic constraints based on domain and range restrictions defined in the KG schema. These constraints leverages ontological knowledge to

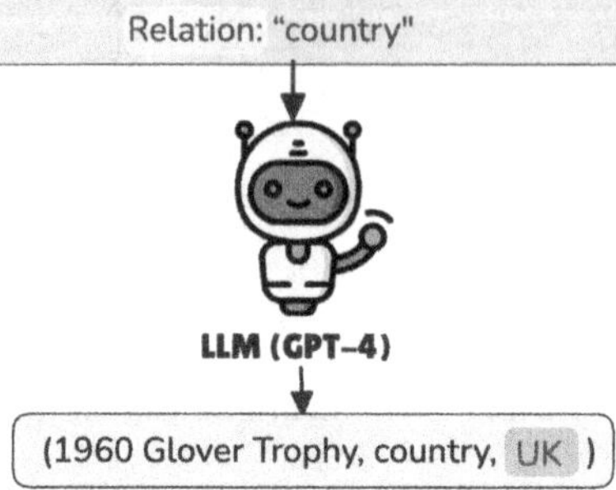

Fig. 3. Prompting GPT-4 to complete a missing triple for the `"1960 Glover Trophy"` entity, where the object `"UK"` is a literal value and is missing in the original KG.

validate the types of entities involved in relationships, as described in [28]. For instance, consider the relationship `born_in`, which connects an entity of type `Person` to an entity of type `Place`. To formalize this, let $\mathcal{E}$ be the set of entities, where each entity e is associated with a class c via the `rdf:type` relationship. The TBox (Terminological Box) defines the KG schema, including the axioms specifying domain and range constraints for properties. A triple $\langle e_s, r, e_o' \rangle$ is considered semantically valid if it satisfies the following conditions:

1. *Domain Constraint*: the subject e_s must belongs to a class c_d, and the TBox must contain the axiom $\langle r, \texttt{rdfs:domain}, c_d \rangle$. Formally,

$$\begin{aligned} \forall r \in \mathcal{R}, \forall c_d \in C, \langle r, \texttt{rdfs:domain}, c_d \rangle \in \text{TBox} \Rightarrow \\ \forall e_s, e_o' \in E, \langle e_s, a, c_d \rangle \Rightarrow \langle e_s, r, e_o' \rangle \text{ is valid} \end{aligned} \tag{1}$$

2. *Range Constraint*: the object e_o' must belong to a class c_r, and the TBox must contain the axiom $\langle r, \text{rdfs:range}, c_r \rangle$. Formally,

$$\begin{aligned} r \in \mathcal{R}, \forall c_r \in C, \langle r, \texttt{rdfs:range}, c_r \rangle \in \text{TBox} \Rightarrow \\ \forall e_s, e_o' \in E, \langle e_o', a, c_r \rangle \Rightarrow \langle e_s, r, e_o' \rangle \text{ is valid} \end{aligned} \tag{2}$$

In essence, a triple $\langle e_s, r, e_o' \rangle$ is valid only if it adheres to both domain and range constraints. The KGE-Triples component returns a list of relevant triples $\hat{T}_{(kge)}$ for the input entity e, ensuring all triples are validated by Eqs. 1 and 2.

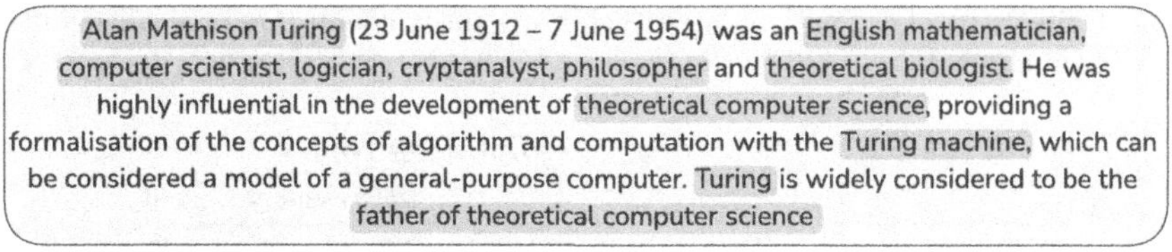

Alan Mathison Turing (23 June 1912 – 7 June 1954) was an English mathematician, computer scientist, logician, cryptanalyst, philosopher and theoretical biologist. He was highly influential in the development of theoretical computer science, providing a formalisation of the concepts of algorithm and computation with the Turing machine, which can be considered a model of a general-purpose computer. Turing is widely considered to be the father of theoretical computer science

Fig. 4. A silver-standard summary example for the entity "`Alan Turing`", with highlighted entities in red boxes. (Color figure online)

LLM-Triples Component ($\hat{T}_{(\text{llm})}$). While KGE models excel in link prediction, they struggle to infer literal triples, where the object entity e'_o is a string value (e.g., dates, names, or numbers) [10]. Recent advancements in LLM have demonstrated their effectiveness in KG completion tasks, including triple classification, relation prediction, and inferring missing triples [29,31]. As shown in Fig. 2, our approach leverages LLMs, such as the Open AI GPT-4, to address the inherent limitation of KGE methods in inferring literal triples. Specifically, we prompt GPT-4 model to complete literal triples. For example, in Fig. 3, we construct a prompt in the format ⟨`subject, relation, ?`⟩, where `?` represent the missing object value. This enables GPT-4 to complete triples for entities such as `1960 Glover Trophy`, leveraging its pre-trained knowledge and retrieval-augmented generation capabilities to infer accurate and contextual information.

Triple Ranking. Our approach combines the outputs of KGE and LLM components to generate abstractive entity summaries. For a given entity e, the summary S_e is defined as $S_e \subseteq \hat{T}_{(kge)} \cup \hat{T}_{(\text{llm})}$. Notably, the triples in S_e may not exist in the original KG, as they are inferred using KGE and LLM components. To rank the triples, we use the frequency of their associated predicates. Specifically, we compute the predicate frequency $f(p)$ for each predicate p in the KG. All triples $t = \langle s, p, o \rangle$ associated to entity e are sorted in descending order of $f(p)$.

Entity Summary. The final summary is obtained from the top-k ranked triples, ensuring that the most relevant information about the target entity are included.

4 Experiment

We conducted our experiments to answer the following research questions:

- **RQ**$_1$: *How does our approach perform in generating abstractive entity compared to state-of-the-art baselines?*
- **RQ**$_2$: *How effectively do large language models improve the inference of literal triples in abstractive summarization?*

Table 1. Statistics of evaluation datasets showing entity counts and triple distributions (present vs. absent).

Dataset		#Entities	#Triples	
			#Present	#Absent
ESSUM	$\text{ESSUM}_{\text{DBpedia}}$	110	3787	-
	$\text{ESSUM}_{\text{FACES}}$	50	2152	-
ESSUM-ABSENT	$\text{ESSUM}_{\text{DBpedia}}$	110	3080	718
	$\text{ESSUM}_{\text{FACES}}$	50	1744	408

4.1 Evaluation Setup

Datasets: Evaluating abstractive entity summarization is challenging due to the lack of benchmarks that consider both present and absent triples. To address this gap, we created two datasets as detailed in Table 1:

- **ESSUM**: A silver-standard dataset combining entities from ESBM-DBPEDIA [17] and FACES [11]. For each entity, we extract sentences with mentioned entities from the first paragraph of its Wikipedia page (e.g., Fig. 4 shows the summary for `Alan Turing`). In our experiment, we created two subsets: (1) $\text{ESSUM}_{\text{DBpedia}}$: 110 entities from ESBM-DBPEDIA, and (2) $\text{ESSUM}_{\text{FACES}}$: 50 entities from FACES
- **ESSUM-ABSENT**: We constructed this dataset by randomly removing 20% of triples from ESBM-DBPEDIA and FACES, both sourced from DBpedia v2015-10. These omitted triples serve as *ground-truth* absent triples to evaluate a model's ability to infer missing facts. To prevent data leakage, we ensured test entities (from DBpedia 2015-10) did not overlap with ConvE or LiteralE's training data (DBpedia 2022). For each entity, 20% of triples were removed before KGE pre-training, ensuring they were never observed during model training. While GPT-4 may have encountered DBpedia during training, we mitigate memorization by validating LLM-generated triples against ESSUM-Absent's ground truth (excluded from the KG) and focusing on inferring missing triples.

Evaluation Protocol: To evaluate the quality of generated summaries, we verbalize RDF triple summaries into natural language using a fine-tuned T5 model[2] and compare them against the ESSUM silver-standard summaries. We adopt the standard evaluation framework from WebNLG[3], employing the following metrics:

- **BLEU** [20]: measures lexical overlap between generated and reference summaries using n-gram precision.

[2] https://github.com/QipengGuo/P2_WebNLG2020.

[3] Evaluation script: https://github.com/WebNLG/GenerationEval.

- **METEOR** [2]: evaluates semantic alignment by incorporating synonymy and stemming.
- **ChrF++** [21]: computes character-level n-gram similarity, capturing morphological and lexical variations.
- **TER** [24]: quantifies the number of edits (insertions, deletions, substitutions) required to match the reference summary.
- **BLEURT** [23]: a learned metric using transformer-based representations to assess semantic similarity.
- **BERTScore** [30]: leverages contextual embeddings from pre-trained BERT model to compare generated and reference summaries.
- **ROUGE-L** [16]: measures overlap of longest common subsequences between summaries, focusing on recall.
- **MoverScore** [32]: combines contextualized embeddings with Earth Mover's Distance to evaluate semantic similarity.

Baselines. We evaluated our approach against four supervised baselines. We exclude the unsupervised methods (RELIN, FACES), since supervised baselines outperform them in entity summarization [8,19]. We summarize each baseline as follows:

- **ESA** [26]: a BiLSTM-based model with attention mechanisms that encodes triples using graph embeddings. ESA is the first deep learning approach for entity summarization by jointly modeling textual and structural KG features.
- **AutoSUM** [27]: a two-module framework with: (1) *Extractor* that encodes entity descriptions and triples via BiLSTM and graph embeddings, and (2) *Simulator* that applies entity-phase attention to significant features and user-phase attention to preference vectors, generating customized extractive summaries.
- **DEEPLENS** [19]: a three-component architecture: (1) Triple encoding with pre-trained word embeddings, (2) Entity description encoding via attention-based bidirectional LSTM, and (3) Triple scoring using multi-perspective matching to select top-k triples.
- **GATES** [8]: combines graph attention networks with ensemble learning. It encodes triples using ComplEx (as KGE model) and GloVe (as text embeddings), applies graph attention to model triple interactions, and uses pointer networks to generate summaries.

Experimental Settings: we setup our experiments in three stages:

- *KGE-Triples Component*: we trained ConvE and LiteralE models on the ESSUM dataset to infer the top-k most relevant triples for each target entity. LiteralE extends ConvE by integrating literal embeddings, enhancing its ability to model numeric and textual attributes.

Table 2. Performance evaluation of our approach and the baselines on the ESSUM dataset ($\mathbf{RQ}_1$).

System	BLEU	METEOR	chrF++	TER	BLEURT
ESSUM$_{\mathbf{DBpedia}}$					
ESA	4.59	0.11	0.24	1.66	0.36
AUTOSUM	5.18	0.11	0.23	2.01	0.35
DEEPLENS	5.25	**0.12**	0.25	1.45	0.36
GATES	5.05	0.11	0.24	1.42	0.36
ANTS (ours)	**5.78**	**0.12**	**0.25**	**1.33**	0.36
ESSUM$_{\mathbf{FACES}}$					
ESA	0.45	0.05	0.11	**0.91**	0.33
AUTOSUM	0.49	0.05	0.12	0.92	0.34
DEEPLENS	0.42	0.05	0.11	0.92	0.32
GATES	0.49	0.05	0.11	0.92	0.32
ANTS (ours)	**0.67**	**0.07**	**0.14**	**0.91**	**0.36**

- *LLM-Triples Component*: We augment literal triples using GPT-4 by prompting it with entity context and relation patterns (e.g., `⟨subject, relation, ?⟩`) to generate plausible missing triples (see Fig. 3). To maintain structural integrity, subject entities are aligned with the KG structure using predefined identifiers (e.g., `Glover_Trophy`), then appended to the KG's prefix (e.g., `http://dbpedia.org/resource/`) to construct the full entity URI. Similarly, we use the KG's property prefix for relations (e.g., `country`) to maintain schema integrity. In contrast, literal objects (e.g., birthdates, names) are inferred as textual or numerical values via LLMs and do not need URIs. While this component enhances the coverage of inferred triples, it introduces potential inconsistencies due to LLM hallucinations. To mitigate this, we incorporate a search-augmented generation feature in the GPT-4 model to retrieve relevant and real-time web data.
- *Triple Ranking and Entity Summary*: Triples from both components were aggregated and ranked by predicate frequency in the KG. The final summary was formed by selecting the top-k triples (k=20), balancing conciseness and coverage based on prior benchmarks [17] .

5 Results and Discussion

5.1 Performance Comparison Against State-of-the-Art Baselines ($\mathbf{RQ}_1$)

Our experiments (see Table 2) show that our approach (ANTS) performs better than existing methods in lexical and semantic metrics. On the general-domain ESSUM$_{\text{DBpedia}}$ dataset, ANTS achieves a BLEU score of 5.78, which is 10.1%

Table 3. Performance evaluation of our approach and the baselines on the ESSUM dataset ($\mathbf{RQ}_1$).

System	BERTScore			ROUGE-L			MoverScore
	Precision	Recall	F1	Precision	Recall	F1	
ESSUM$_{\text{DBpedia}}$							
ESA	0.822	0.834	0.828	0.262	0.273	0.248	0.530
AUTOSUM	0.829	0.837	0.832	0.264	0.276	0.250	0.532
DEEPLENS	0.830	0.837	0.833	0.262	0.270	0.245	0.532
GATES	0.830	0.835	0.832	0.258	0.273	0.245	0.530
ANTS (ours)	**0.848**	**0.837**	**0.842**	**0.272**	**0.294**	**0.263**	**0.540**
ESSUM$_{\text{FACES}}$							
ESA	0.832	0.803	0.817	0.117	0.375	0.168	0.528
AUTOSUM	0.839	0.804	0.821	0.126	0.393	0.180	0.532
DEEPLENS	0.817	0.799	0.808	0.110	0.341	0.159	0.524
GATES	0.831	0.802	0.816	0.120	0.369	0.170	0.528
ANTS (ours)	**0.855**	**0.808**	**0.830**	**0.136**	**0.411**	**0.193**	**0.536**

higher than the best baseline (DeepLENS). It also reduces the translation edit rate (TER) to 1.33, which is 6.8% lower than the baselines. This improvement shows ANTS' ability to infer missing triples while keeping lexical coherence. It matches the baselines in semantic metrics (BLEURT: 0.36) and balances precision (METEOR: 0.12) and fluency (chrF++: 0.25). On the domain-specific ESSUM$_{\text{FACES}}$ dataset, ANTS performs well, achieving a 36.7% higher BLEU score (0.67 vs. 0.49 baselines) and a 40% improvement in METEOR (0.07 vs. 0.05 baselines). The hybrid architecture (KGE+LLM) of ANTS is effective: KGEs keep structural relationships like `locatedIn`, while LLM infer missing literals like `populationTotal`. This dual capability allows ANTS to achieve the lowest TER (0.91) and the highest BLEURT score (0.36). Further analysis using semantic metrics shows ANTS' superiority in contextual alignment. On ESSUM$_{\text{DBpedia}}$, it achieves a BERTScore F1 of 0.842 (1.1% higher than baselines) and a MoverScore of 0.540 (1.5% higher than baselines), indicating stronger semantic matching with reference summaries. The 6.0% improvement in ROUGE-L F1 (0.263 vs. 0.250 baselines) confirms better retention of key concepts like temporal events (e.g., `foundingYear`). For ESSUM$_{\text{FACES}}$, ANTS achieves a ROUGE-L Recall of 0.411$text$–12.2% higher than baselines—while maintaining leading BERTScore F1 (0.830) and MoverScore (0.536).

When evaluated on incomplete KGs (ESSUM-ABSENT) (see Table 5), ANTS recovers 20% of missing triples with a pricision of 0.66. while its precision on present triples (0.98) is slightly lower than that of the baselines (1.0), textscAnts achieves a substantially higher F1-score on absent triples, showing a better balance between novelty and faithfulness. This capability comes from semantic constraints that filter implausible inferences while allowing controlled synthesis of

Table 4. An ablation study for the ANTS components on ESSUM dataset (**RQ$_2$**)

System	Dataset	BLEU	METEOR	chrF++	TER	BLEURT
LLM-Only Models						
ANTS$_{(GPT-3.5)}$	ESSUM$_{DBpedia}$	3.94	0.10	0.21	**1.13**	**0.38**
ANTS$_{(GPT-4)}$	ESSUM$_{DBpedia}$	5.48	0.12	0.25	1.26	0.37
ANTS$_{(GPT-3.5)}$	ESSUM$_{FACES}$	0.36	0.05	0.10	**0.91**	**0.38**
ANTS$_{(GPT-4)}$	ESSUM$_{FACES}$	0.46	0.06	0.12	0.92	0.36
KGE-Only Models WITHOUT SEMANTIC-CONSTRAINTS						
ANTS$_{(ConvE)}$	ESSUM$_{DBpedia}$	1.82	0.09	0.19	1.26	0.32
ANTS$_{(LiteralE)}$	ESSUM$_{DBpedia}$	2.17	0.10	0.20	1.29	0.33
ANTS$_{(ConvE)}$	ESSUM$_{FACES}$	0.13	0.06	0.12	0.92	0.36
ANTS$_{(LiteralE)}$	ESSUM$_{FACES}$	0.18	0.05	0.09	0.93	0.30
KGE-Only Models With Semantic-constraints						
ANTS$_{(ConvE)}$	ESSUM$_{DBpedia}$	2.57	0.11	0.22	1.14	0.36
ANTS$_{(LiteralE)}$	ESSUM$_{DBpedia}$	2.66	0.11	0.21	1.15	0.34
ANTS$_{(ConvE)}$	ESSUM$_{FACES}$	0.15	0.05	0.10	0.92	0.33
ANTS$_{(LiteralE)}$	ESSUM$_{FACES}$	0.16	0.05	0.11	0.92	0.32
KGE Without Semantic-constraints + LLM Models						
ANTS$_{(ConvE+GPT-4)}$	ESSUM$_{DBpedia}$	4.66	0.12	0.24	1.32	0.36
ANTS$_{(LiteralE+GPT-4)}$	ESSUM$_{DBpedia}$	4.51	0.12	0.24	1.37	0.36
ANTS$_{(ConvE+GPT-4)}$	ESSUM$_{FACES}$	0.42	0.06	0.12	0.92	0.36
ANTS$_{(LiteralE+GPT-4)}$	ESSUM$_{FACES}$	0.40	0.06	0.12	0.91	0.36
KGE With Semantic-constraints + LLM Models						
ANTS$_{(ConvE+GPT-4)}$	ESSUM$_{DBpedia}$	5.26	**0.13**	**0.26**	1.24	**0.38**
ANTS$_{(LiteralE+GPT-4)}$	ESSUM$_{DBpedia}$	**5.78**	0.12	0.25	1.33	0.36
ANTS$_{(ConvE+GPT-4)}$	ESSUM$_{FACES}$	0.63	**0.07**	0.13	**0.91**	0.36
ANTS$_{(LiteralE+GPT-4)}$	ESSUM$_{FACES}$	**0.67**	**0.07**	**0.14**	**0.91**	0.36

missing facts. Three key insights in our experiments: (1) The combination of KGEs and LLMs enables Ants to outperform purely extractive methods by 10.1% in BLEU while maintaining semantic fidelity. (2) LLM augmentation is crucial for domain adaptation, yielding 36.7% BLEU gains on specialized datasets. (3) Our constraint-based filtering reduces hallucination risks, enabling reliable generation of missing triples.

5.2 Ablation Study: Impact of LLM (RQ$_2$)

We performed an ablation study to analyze the impact of ANTS' components: KGE models (`ConvE`, `LiteralE`), LLMs (GPT-3.5, GPT-4), and semantic constraints. As shown in Table 4, the first experiment investigates the contribution of KGEs, LLMs, and their combination both with and without semantic

Table 5. Comparing performance of ANTS with baseline models in generating *absent triples* on ESSUM-ABSENT dataset (**RQ**$_2$)

Model	Present Triples			Absent Triples		
	Precision	Recall	F1-Score	Precision	Recall	F1-Score
ESSUM$_{\text{DBpedia}}$						
ESA	0.99	0.31	0.45	0.07	0.01	0.02
AUTOSUM	0.99	0.31	0.45	0.10	0.02	0.03
GATES	**1.0**	**0.32**	**0.46**	0.13	0.02	0.03
ANTS	0.98	0.28	0.42	**0.66**	**0.20**	**0.29**
ESSUM$_{\text{FACES}}$						
ESA	**1.0**	0.24	0.38	0.00	0.00	0.00
AUTOSUM	**1.0**	0.23	0.37	0.00	0.00	0.00
GATES	**1.0**	0.23	0.37	0.02	0.00	0.01
ANTS	**1.0**	**0.29**	**0.44**	**0.56**	**0.12**	**0.18**

Table 6. Performance evaluation in generating absent triple with semantic-constraints setting on ESSUM-ABSENT dataset (**RQ**$_2$)

Model	Present Triples			Absent Triples		
	Precision	Recall	F1-Score	Precision	Recall	F1-Score
ESSUM$_{\text{DBpedia}}$						
ANTS$_{\text{(GPT-4)}}$	0.81	0.10	0.17	0.47	0.12	0.18
ANTS$_{\text{(ConvE)}}$	0.96	0.13	0.22	0.21	0.04	0.07
ANTS$_{\text{(LiteralE)}}$	0.94	0.10	0.18	0.19	0.05	0.07
ANTS$_{\text{(ConvE+GPT-4)}}$	**1.0**	**0.29**	**0.43**	0.57	0.18	0.25
ANTS$_{\text{(LiteralE+GPT-4)}}$	0.98	0.28	0.42	**0.66**	**0.20**	**0.29**
ESSUM$_{\text{FACES}}$						
ANTS$_{\text{(GPT-4)}}$	0.9	0.13	0.22	0.54	0.11	**0.18**
ANTS$_{\text{(ConvE)}}$	**1.0**	0.17	0.29	0.36	0.08	0.13
ANTS$_{\text{(LiteralE)}}$	**1.0**	0.18	0.29	0.00	0.00	0.00
ANTS$_{\text{(ConvE+GPT-4)}}$	**1.0**	0.28	0.43	0.54	**0.12**	**0.18**
ANTS$_{\text{(LiteralE+GPT-4)}}$	**1.0**	**0.29**	**0.44**	**0.56**	**0.12**	**0.18**

constraints. Our approach with the KGE-triple component (ANTS(ConvE) or ANTS(LiteralE)) fails to generate triples with literal information since KGE models are limited in predicting literal values. However, LLMs augment the entity summaries by incorporating these literal triples. Furthermore, semantic constraints ensure that the generated triples conform to the Knowledge Graph's range and domain requirements. These results demonstrate that combining KGE and LLMs yields better performance than using each model individually. Notably, the ANTS$_{(\text{LiteralE+GPT-4})}$ combination achieves the highest BLEU score of 5.78 and 0.67 on the ESSUM$_{\text{DBpedia}}$ and ESSUM$_{\text{FACES}}$ datasets, respectively.

Additionally, Table 6 presents the performance evaluation of different configurations of ANTS models in generating *present* and *absent* triples under constraints settings across two datasets: ESSUM$_{\text{DBpedia}}$ and ESSUM$_{\text{FACES}}$. Generally, the combination of KGE and LLM consistently outperforms other configurations, particularly in generating *absent triples*. For instance, ANTS$_{(\text{LiteralE+GPT-4})}$ demonstrates the best performance in generating *absent triples* on both datasets, achieving F1-score 0.29 and 0.18 and precision values of 0.66 and 0.56 for ESSUM$_{\text{DBpedia}}$ and ESSUM$_{\text{FACES}}$, respectively, indicating that it is the most balanced and effective model. While all models demonstrate high precision in generating present triples, their recall is notably low, resulting in moderate F1 scores. This pattern suggests that while the models are accurate, they often miss relevant triples, especially in the ESSUM$_{\text{FACES}}$ dataset.

6 Conclusion

In this paper, we present our approach (ANTS) for generating abstractive summaries of entities in knowledge graphs. Our approach considers both present and absent triples when generating entity summaries. To do so, we employ two components: *KGE-triples* to infer relevant triples using knowledge graph embeddings and *LLM-triples* that employ LLM to address the limitation of KGE models and infer literal triples. To evaluate our approach, we created ESSUM dataset, the first silver-standard dataset for abstractive entity summarization. Our dataset combines entities from ESBM-DBPEDIA and FACES with human-aligned reference summaries. The experiments demonstrate that our approach outperforms state-of-the-art baselines, achieving a BLEU score of 5.78 (+10.1%) on ESSUM$_{DBpedia}$ and 0.67 (+36.7%) on ESSUM$_{FACES}$. The hybrid architecture of our approach proves particularly effective for handling absent triples, achieving F1-scores 58% higher than individual KGE or LLM components. In our future work, we will explore strategies for dynamic semantic constraints and fine-tune LLMs on specialized KGs to enhance coverage of rare entities. Additionally, we will expand ESSUM to include multilingual and temporal KGs, enabling broader evaluation of abstractive summarization in dynamic contexts.

Acknowledgment. This work was supported by the Deutsche Forschungsgemeinschaft (DFG, German Research Foundation) under "TRR 318/1 2021" - 438445824, the Lamarr Fellow Network programme through project "WHALE" (Grant No. LFN 1-04) funded by the Ministry of Culture and Science of North Rhine-Westphalia (MKW NRW), the European Union's Horizon Europe research and innovation programme under project "ENEXA" (Grant No. 101070305), and the Mora Scholarship from the Ministry of Religious Affairs, Republic of Indonesia.

Table 7. Mean and standard deviation of results over three run experiments on ANTS with LLM-only models.

System	System	BLEU	METEOR	chrF++	TER	BLUERT
ANTS$_{(GPT-3.5)}$	ESSUM$_{DBpedia}$	4.70 ± 0.56	0.11 ± 0.00	0.23 ± 0.01	1.26 ± 0.09	0.37 ± 0.01
ANTS$_{(GPT-4)}$	ESSUM$_{DBpedia}$	$\mathbf{5.58 \pm 0.18}$	$\mathbf{0.12 \pm 0.00}$	$\mathbf{0.25 \pm 0.00}$	$\mathbf{1.24 \pm 0.01}$	0.37 ± 0.00
ANTS$_{(GPT-3.5)}$	ESSUM$_{FACES}$	0.41 ± 0.04	0.06 ± 0.00	0.11 ± 0.01	0.91 ± 0.00	0.37 ± 0.01
ANTS$_{(GPT-4)}$	ESSUM$_{FACES}$	$\mathbf{0.42 \pm 0.04}$	0.06 ± 0.00	$\mathbf{0.12 \pm 0.00}$	0.91 ± 0.01	0.37 ± 0.00

Appendix A

A.1 Error Bars for Experimental Results on ANTS (LLM-Only) Models

This appendix provides the detailed experimental results, including the mean ± standard deviation across three runs for each evaluation metric. The error bars indicate the variability of the results, reflecting the inherent stochasticity of the model. The results include BLEU, METEOR, chrF++, TER, and BLEURT for ANTS models. Table 7 shows that the high variability in GPT-3.5's BLEU scores (±0.56 vs. ±0.18 in GPT-4) suggests that GPT-4 is more stable on ESSUM$_{DBpedia}$. In contrast, GPT-3.5 exhibits greater fluctuations in BLEU and higher TER variability (±0.09 vs. ±0.01), suggesting inconsistent output quality that requires varying levels of modifications. For domain-specific data (ESSUM$_{FACES}$), both models yield consistently low BLEU (up to 0.42) with a small deviation (±0.04). Furthermore, METEOR, chrF++, TER, and BLEURT remain stable across all runs with minimal deviation up to 0.01, indicating that semantic similarity and fluency are not significantly affected by model stochasticity.

A.2 BLEU vs. BLEU-NLTK Differences

Table 8 presents the evaluation results of the ANTS models using BLEU and BLEU-NLTK metrics. The differences observed between BLEU and BLEU-NLTK scores arise from variations in tokenization and smoothing techniques. Specifically, BLEU employs standard tokenization, which is case-insensitive and does not split punctuation, whereas BLEU-NLTK applies stricter tokenization, making it case-sensitive and segmenting contractions and punctuation. For instance, in BLEU, "Alan Turing" is treated as a single token, whereas in BLEU-NLTK, it is split into two separate tokens: "Alan" and "Turing".

A.3 Additional Examples for Completing Triples and Triple Selection

In this section, we present additional examples to demonstrate the effectiveness of the ANTS approach. Traditional entity summarization methods for triple completion rely on selecting a subset of existing triples from a KG. However, these approaches fail to account for missing relationships that could enhance

Table 8. BLEU vs BLUE-NLTK evaluation for ANTS on ESSUM dataset

System	Dataset	**BLEU**	**BLEU$_{\text{NLTK}}$**
ANTS$_{(\text{ConvE+GPT-4})}$	ESSUM$_{\text{DBpedia}}$	5.26	0.05
ANTS$_{(\text{LiteralE+GPT-4})}$	ESSUM$_{\text{DBpedia}}$	**5.78**	**0.06**
ANTS$_{(\text{ConvE+GPT-4})}$	ESSUM$_{\text{FACES}}$	0.63	0.01
ANTS$_{(\text{LiteralE+GPT-4})}$	ESSUM$_{\text{FACES}}$	**0.67**	0.01

the informativeness of the summaries. Our approach addresses this limitation by incorporating absent triples through LLM-based inference. For instance, consider the entity *Barack Obama* in DBpedia. Our KGE-Triple component generates the following triples:

```
dbr:barack_obama dc:subject dbc:activists_from_hawaii
dbr:barack_obama dc:subject dbc:american_people_of_english_descent
dbr:barack_obama dbp:predecessor dbr:george_w._bush
dbr:barack_obama dbp:vicepresident dbr:joe_biden
dbr:barack_obama dbo:almamater dbr:harvard_law_school
...
```

However, our LLM-Triples component infers additional missing triples, such as:

```
dbr:barack_obama dbo:birthPlace Honolulu
dbr:barack_obama dbo:occupation Politician
dbr:barack_obama dbo:termPeriod 2009-2017
dbr:barack_obama dbo:successor Donald Trump
dbr:barack_obama dbo:award Nobel Peace Prize
...
```

After ranking the triples, the final entity summary is as follows:

```
dbr:barack_obama dc:subject dbc:activists_from_hawaii
dbr:barack_obama dc:subject dbc:american_people_of_english_descent
dbr:barack_obama dbo:birthPlace Honolulu
dbr:barack_obama dbo:occupation Politician
dbr:barack_obama dbp:predecessor dbr:george_w._bush
...
```

Finally, the verbalized entity summary is:
Barack Obama was born in Honolulu and was a politician. He was a member of the Democratic Party and served as the President of the United States. He attended Harvard Law School, and his successor was Donald Trump. Barack Obama is married to Michelle Obama, and his parents are Stanley Ann Dunham. He is a follower of Christianity, and his alma mater is Harvard University.

The example above illustrates how ANTS enhances entity summarization by integrating inferred missing triples, thereby improving the completeness and informativeness of the generated summaries.

References

1. Ahmed, A.F., Firmansyah, A.F., Sherif, M.A., Moussallem, D., Ngonga Ngomo, A.C.: Explainable integration of knowledge graphs using large language models. In: Métais, E., Meziane, F., Sugumaran, V., Manning, W., Reiff-Marganiec, S. (eds.) Natural Language Processing and Information Systems, pp. 124–139. Springer, Cham (2023)
2. Banerjee, S., Lavie, A.: Meteor: an automatic metric for MT evaluation with improved correlation with human judgments. In: Proceedings of the ACL Workshop on Intrinsic and Extrinsic Evaluation Measures for MT and/or Summarization, pp. 65–72. ACL (2005)
3. Biswas, R., et al.: Knowledge graph embeddings: open challenges and opportunities. Trans. Graph Data Knowl. **1**(1), 4:1–4:32 (2023). https://doi.org/10.4230/TGDK.1.1.4. https://drops.dagstuhl.de/entities/document/10.4230/TGDK.1.1.4
4. Chen, L., et al.: Entity summarization via exploiting description complementarity and salience. IEEE Trans. Neural Netw. Learn. Syst. (2022)
5. Cheng, G., Tran, T., Qu, Y.: Relin: relatedness and informativeness-based centrality for entity summarization. In: International Semantic Web Conference, pp. 114–129. Springer, Cham (2011)
6. Dettmers, T., Minervini, P., Stenetorp, P., Riedel, S.: Convolutional 2D knowledge graph embeddings. In: AAAI 2018/IAAI 2018/EAAI 2018. AAAI Press (2018)
7. Ermilov, T., Moussallem, D., Usbeck, R., Ngomo, A.C.N.: Genesis: a generic RDF data access interface. In: Proceedings of the International Conference on Web Intelligence, WI 2017, pp. 125–131. Association for Computing Machinery, New York (2017). https://doi.org/10.1145/3106426.3106514
8. Firmansyah, A.F., Moussallem, D., Ngomo, A.C.N.: Gates: using graph attention networks for entity summarization. In: Proceedings of the 11th on Knowledge Capture Conference, pp. 73–80 (2021)
9. Firmansyah, A.F., Moussallem, D., Ngomo, A.: ESLM: improving entity summarization by leveraging language models. In: Meroño Peñuela, A., et al. (eds.) The Semantic Web, pp. 162–179. Springer, Cham (2024)
10. Gesese, G.A., Biswas, R., Alam, M., Sack, H.: A survey on knowledge graph embeddings with literals: which model links better literally? Semant. Web **12**(4), 617–647 (2021)
11. Gunaratna, K., Thirunarayan, K., Sheth, A.: Faces: diversity-aware entity summarization using incremental hierarchical conceptual clustering. In: Proceedings of the Twenty-Ninth AAAI Conference on Artificial Intelligence, pp. 116–122. AAAI Press (2015)
12. Hasibi, F., Balog, K., Bratsberg, S.E.: Dynamic factual summaries for entity cards. In: Proceedings of the 40th International ACM SIGIR Conference on Research and Development in Information Retrieval, SIGIR 2017, pp. 773–782. Association for Computing Machinery, New York (2017). https://doi.org/10.1145/3077136.3080810
13. Hogan, A., et al.: Knowledge graphs. ACM Comput. Surv. **54**(4) (2021). https://doi.org/10.1145/3447772
14. Jalota, R., Vollmers, D., Moussallem, D., Ngomo, A.C.N.: Lauren - knowledge graph summarization for question answering. In: 2021 IEEE 15th International Conference on Semantic Computing (ICSC), pp. 221–226 (2021). https://doi.org/10.1109/ICSC50631.2021.00047

15. Kristiadi, A., Khan, M.A., Lukovnikov, D., Lehmann, J., Fischer, A.: Incorporating literals into knowledge graph embeddings. In: Ghidini, C., et al. (eds.) The Semantic Web – ISWC 2019, pp. 347–363. Springer, Cham (2019)
16. Lin, C.Y.: ROUGE: a package for automatic evaluation of summaries. In: Text Summarization Branches Out, pp. 74–81. Association for Computational Linguistics, Barcelona, Spain (2004). https://aclanthology.org/W04-1013
17. Liu, Q., Cheng, G., Gunaratna, K., Qu, Y.: ESBM: an entity summarization benchmark. In: European Semantic Web Conference, pp. 548–564. Springer, Cham (2020)
18. Liu, Q., Cheng, G., Gunaratna, K., Qu, Y.: Entity summarization: state of the art and future challenges. J. Web Semant. **69**, 100647 (2021). https://doi.org/10.1016/j.websem.2021.100647. https://www.sciencedirect.com/science/article/pii/S1570826821000226
19. Liu, Q., Cheng, G., Qu, Y.: Deeplens: deep learning for entity summarization (2020)
20. Papineni, K., Roukos, S., Ward, T., Zhu, W.J.: BLEU: a method for automatic evaluation of machine translation. In: Proceedings of the 40th Annual Meeting on Association for Computational Linguistics (2002)
21. Popović, M.: chrF++: words helping character n-grams. In: Proceedings of the Second Conference on Machine Translation, pp. 612–618 (2017)
22. Safavi, T., Belth, C., Faber, L., Mottin, D., Müller, E., Koutra, D.: Personalized knowledge graph summarization: from the cloud to your pocket. In: 2019 IEEE International Conference on Data Mining (ICDM), pp. 528–537 (2019). https://doi.org/10.1109/ICDM.2019.00063
23. Sellam, T., Das, D., Parikh, A.P.: Bleurt: learning robust metrics for text generation. In: Proceedings of ACL (2020)
24. Snover, M., Dorr, B., Schwartz, R., Micciulla, L.: A study of translation edit rate with targeted human annotation (2006)
25. Wei, D., Gao, S., Liu, Y., Liu, Z., Huang, L., Hu, S.: Mpsum: predicate-based matching for RDF triples with application to LDA. In: EYRE@CIKM (2018)
26. Wei, D., et al.: ESA: entity summarization with attention (2019)
27. Wei, D., et al.: AutoSUM: automating feature extraction and multi-user preference simulation for entity summarization. In: Lauw, H.W., Wong, R.C.-W., Ntoulas, A., Lim, E.-P., Ng, S.-K., Pan, S.J. (eds.) PAKDD 2020. LNCS (LNAI), vol. 12085, pp. 580–592. Springer, Cham (2020). https://doi.org/10.1007/978-3-030-47436-2_44
28. Weyns, M., Bonte, P., Steenwinckel, B., Turck, F.D., Ongenae, F.: Conditional constraints for knowledge graph embeddings. In: Alam, M., Buscaldi, D., Cochez, M., Osborne, F., Recupero, D.R., Sack, H. (eds.) Proceedings of the Workshop on Deep Learning for Knowledge Graphs (DL4KG2020) co-located with the 17th Extended Semantic Web Conference 2020 (ESWC 2020), Heraklion, Greece, 02 June 2020 - moved online. CEUR Workshop Proceedings, vol. 2635. CEUR-WS.org (2020). https://ceur-ws.org/Vol-2635/paper3.pdf
29. Yao, L., Peng, J., Mao, C., Luo, Y.: Exploring large language models for knowledge graph completion. CoRR abs/2308.13916 (2023). https://doi.org/10.48550/ARXIV.2308.13916
30. Zhang, T., Kishore, V., Wu, F., Weinberger, K.Q., Artzi, Y.: Bertscore: evaluating text generation with BERT. In: 8th International Conference on Learning Representations, ICLR 2020, Addis Ababa, Ethiopia, 26–30 April 2020. OpenReview.net (2020). https://openreview.net/forum?id=SkeHuCVFDr

31. Zhang, Y., Chen, Z., Zhang, W., Chen, H.: Making large language models perform better in knowledge graph completion. CoRR abs/2310.06671 (2023). https://doi.org/10.48550/ARXIV.2310.06671
32. Zhao, W., Peyrard, M., Liu, F., Gao, Y., Meyer, C.M., Eger, S.: MoverScore: text generation evaluating with contextualized embeddings and earth mover distance. In: Inui, K., Jiang, J., Ng, V., Wan, X. (eds.) Proceedings of the 2019 Conference on Empirical Methods in Natural Language Processing and the 9th International Joint Conference on Natural Language Processing (EMNLP-IJCNLP), pp. 563–578. Association for Computational Linguistics, Hong Kong, China (2019). https://doi.org/10.18653/v1/D19-1053. https://aclanthology.org/D19-1053

Delete/Rederive with Marking for Update Streams

Moritz Illich(✉) and Birte Glimm

Ulm University, James-Franck-Ring, Ulm, Germany
{moritz.illich,birte.glimm}@uni-ulm.de

Abstract. Given a set of facts and inference rules, a materialized dataset is obtained by extending the facts with all logical consequences w.r.t. the rules. Classically, sequential updates of a materialized dataset are processed separately without any direct interference. Effectively dealing with streams of updates where changes may occur in large numbers and high frequencies, however, demands a more sophisticated approach that is able to consider several updates at once, such that we may perform computations in parallel or prevent their repetition to accelerate the overall processing. Driven by this goal, we extend the classical Delete/Rederive (DRed) algorithm with *marking*: while processing one update, we directly consider also the next update and mark each fact in the dataset that is added or deleted by the latter. A marked fact that is used in a derivation may then conditionally mark the derived fact to indicate that it will change in the next update too. The proposed approach allows for reducing the overall number of applied rules by directly performing some of the computations that are relevant for the next update before it is actively processed. We show the correctness of the algorithm and provide a prototypical implementation that we evaluate for both synthetic and real data. The evaluation demonstrates a reduction of needed CPU time by about 25% in average compared to classical DRed.

Keywords: materialization · incremental reasoning · update stream

1 Introduction

A common reasoning task for rule-based formalisms, like Datalog [1], is the so-called *materialization*, where a dataset is extended by all its logical consequences based on a set of inference rules. In such a materialized dataset, every implicitly derivable fact is made explicitly accessible, such that queries may be answered without the need for additional reasoning. Optimized Datalog engines also allow for efficiently querying ontologies with large ABoxes (sets of facts) [5], where the description logic ontology axioms are rewritten into Datalog rules. Such an approach is, for example, employed by the highly efficient RDFox reasoning engine [12]. A problem of materialization is, however, that changes in the original dataset have to be reflected in the materialization too. Since re-computing

E. Curry et al. (Eds.): ESWC 2025, LNCS 15718, pp. 152–168, 2025.
https://doi.org/10.1007/978-3-031-94575-5_9

everything from scratch is usually expensive, *incremental materialization maintenance* algorithms have been developed, which try to efficiently adapt a materialized dataset. In detail, these algorithms focus on the actual changes in a materialization by computing new consequences based on added facts, as well as removing every fact that can no longer be derived due to introduced deletions.

Incremental materialization algorithms such as the Delete/Rederive (DRed) and counting approach [8] process one update at a time, i.e., a new update is not taken into account until the current materialization is completely updated. This works well as long as the updates are not very frequent and little interference occurs between them. Consider, however, the scenario of autonomous driving, where data from sensors (GPS, cameras, ...) is to be combined with data from high-definition maps for tasks such as route planning or determining the lane on which the vehicle is located. In this context, Datalog rules provide for a unified conceptual view over different map data formats and enriched with logical conclusions [14]. High-definition maps are, however, so large in size that only map tiles in the surrounding of the vehicle are dynamically loaded. In addition to loading or unloading map tiles, changing data from various sensors or updated GPS positions constantly have to be integrated without delays, such that new updates often arrive while another update is still being processed.

Since directly integrating new updates into the current processing comes with several challenges [10], we propose an approach that still deals with each update one by one, but where we directly take a look at the next available update to identify related operations. This allows for exploiting that in many scenarios changes happen in a continuous way and semantically relate to each other. For instance, while a leaving vehicle on the neighboring lane might no longer block an overtake maneuver, another vehicle might approach to take the vacant spot and, thus, still block the maneuver. The principle idea is to *mark* facts in the dataset that are added or deleted by the next update. These marks are then passed on to facts in further derivations. This way, we can directly determine some deletions and re-additions of implicit facts for the next update in parallel to the current update's processing and, hence, reduce the number of rule applications that would usually have to be performed when the next update is processed. To realize this idea, we extend the classical DRed algorithm with marking and directly take a whole stream of updates as input.

In the remaining sections, we start with defining some preliminaries such as Datalog and materialization maintenance, before continuing with a description of related work, including the classical DRed approach, in Sect. 3. In Sect. 4, we then provide a general explanation of our idea to extend DRed with marking and formally prove the correctness of the presented algorithm. In Sect. 5, we evaluate the approach and compare it to classical DRed for both synthetic and real datasets. We finally conclude in Sect. 6.

2 Basics and Preliminaries

To formally define *Datalog*, we fix countable, disjoint sets of *predicates*, *constants* and *variables*. A *term* is a constant or a variable. An *atom* has the form

$p(t_1, \ldots, t_k)$, where p is a k-ary predicate and each t_i, $1 \leq i \leq k$, is a term. An atom is *ground* if it does not contain variables. A *fact* is a ground atom and a *dataset* is a finite set of facts. A Datalog *rule* r is a logical implication of the form $B_1, \ldots, B_k \rightarrow H$ where $B_1, \ldots, B_k$ are called *body atoms*, and H is a *head atom*. We use $\mathsf{body}(r)$ and $\mathsf{head}(r)$ to denote the set of body atoms and the head atom of r, respectively. A rule is *safe* if variables that appear in the head also appear in a body atom. A *Datalog program* is a finite set of safe rules. Predicates that occur in the head of a rule are called *intensional* (IDB) predicates; all other predicates are *extensional* (EDB). A fact is *implicit* if it has an IDB predicate, and *explicit* if it has an EDB predicate.

A *substitution* σ is a partial mapping from variables to constants. For α a term, an atom, a rule, or a set of rules, $\alpha\sigma$ is the result of replacing each occurrence of a variable x in α on which σ is defined with $\sigma(x)$. If r is a rule and σ is a substitution mapping all variables of r to constants, then $r\sigma$ is an *instance* of r. We say that a set of facts I instantiates a rule r if there exists a substitution σ s.t. $\mathsf{body}(r)\sigma = I$. Given a Datalog program P and a dataset I, we define $\rho(P, I)$ as the set of all rule instances that can be created by instantiating rules in P with subsets of I. A fact is called *derivable* if it appears as head in a rule instance of $\rho(P, I)$. For a program P, we define $P(I) = \bigcup_{r \in \rho(P,I)} \{\mathsf{head}(r)\}$.

2.1 Materialization Maintenance

Let E be a finite dataset (of explicit facts). Then, let $I_0 = E$; for each $i \geq 1$, let $I_i = I_{i-1} \cup P(I_{i-1})$, and let $I_n = I_{n+1}$ for some $n \geq 1$. The set I_n is the *materialization* of P w.r.t. E, denoted as $\mathsf{mat}(P, E)$.

Let E, E^-, and E^+ be finite datasets with $E^- \subseteq E$, $E \cap E^+ = \emptyset$, and $E^+ \cup E^- \neq \emptyset$. The tuple $U = (E^-, E^+)$ is called an *update* for E, where E^- denotes the facts explicitly deleted from E, and E^+ the facts explicitly added to E, respectively. Applying the update U on E leads to the updated dataset $E' = (E \setminus E^-) \cup E^+$. We allow only EDB predicates in updates. This is w.l.o.g. as we can replace a k-ary IDB predicate p that is to be used in an update by a new k-ary predicate p' and add rules of the form $p(t_1, \ldots, t_k) \rightarrow p'(t_1, \ldots, t_k)$ such that p becomes an EDB predicate (see, e.g., [4]).

Materialization maintenance is the task of computing the updated materialization $\mathsf{mat}(P, (E_0 \setminus E^-) \cup E^+)$ for a given materialization $\mathsf{mat}(P, E_0)$ and an update $U = (E^-, E^+)$. Let $\hat{U} = \langle U_1, U_2, \ldots \rangle$, with $|\hat{U}| \geq 1$, denote a *stream of updates* for a dataset E_0, where for each $U_i = (E_i^-, E_i^+) \in \hat{U}$, we have $E_i^- \subseteq E_{i-1}$ and $E_{i-1} \cap E_i^+ = \emptyset$, resulting in $E_i = (E_{i-1} \setminus E_i^-) \cup E_i^+$. For a stream of updates $\hat{U}$, materialization maintenance leads to a stream of materialized datasets $\langle \mathsf{mat}(P, E_1), \mathsf{mat}(P, E_2), \ldots \rangle$ with $E_i = (E_{i-1} \setminus E_i^-) \cup E_i^+$ for each $U_i = (E_i^-, E_i^+) \in \hat{U}$.

3 Related Work

Our work relates most to the classical DRed algorithm, which we introduce in the following, before we discuss further related approaches.

One way to incrementally adapt a materialization is the *Delete/Rederive* (DRed) algorithm [8], which works for both non-recursive and recursive Datalog programs. The algorithm takes as input a materialization $M = \mathsf{mat}(P, E)$ together with one update $U = (E^-, E^+)$ and produces as output the updated materialization $M' = \mathsf{mat}(P, (E \setminus E^-) \cup E^+)$. The processing is separated into an overdeletion, a rederivation, and an insertion phase, where the first two handle the deletions E^-, while the last one takes care of the additions E^+.

During *overdeletion*, all facts from M that are contained in or derived by some fact from E^- are deleted. This procedure guarantees that we remove every fact from M that can no longer be derived in M', but it potentially also (over)deletes facts that have alternative derivations and that should not be affected by the deletion. To fix this, overdeletion is followed by the *rederivation* phase, where we re-add each deleted fact that can still be derived by the remaining facts. Finally, the *insertion* phase adds E^+ and iteratively computes all newly derivable facts.

Aside from classical DRed, which we use as base for our work, and alternative approaches like counting [8], Backward/Forward [13], or variations thereof [9,11], other related works may be found in the context of parallelized materialization maintenance [12], which can also be transferred to DRed [11]. DynamiTE [17] enables parallel materialization for RDF stream reasoning, while IMaRS [3,6] supports incremental reasoning over streams with window-based expiration times. Ren and Pan [15] deal with update streams on $\mathcal{EL}^{++}$ ontologies based on a truth maintenance system. In addition, Terdjimi et al. [16] present a tag-based approach which also prevents repeated rule applications for dynamically changing data by memorizing how a fact was derived and caching deleted facts for quick re-insertions. In all of these works, however, we only process one update at a time, without taking other available updates into account.

4 DRed with Marking for Update Streams

With classical DRed, we can process a whole sequence of updates by applying the algorithm on each update one by one and always using the resulting, adapted materialization of the current update as initial input for the next update. In order to make this process more efficient, we want to identify situations where we can reduce the number of (typically expensive) rule applications. Here we choose a similar approach to the semi-naive evaluation [2] by preventing repetitions. For that, we extend the DRed algorithm in such a way that rule instances which are relevant for the processing of both the current and the next update are not applied twice, thus, reducing the overall number of performed rule applications. The case where a rule instance is considered again in the next update occurs when a currently computed fact addition or deletion is reversed by the next update, as demonstrated in the following example:

Example 1. Assume we have a rule $A(x), B(x) \rightarrow C(x)$ and two consecutive updates $U_1 = (\emptyset, \{B(c)\})$ and $U_2 = (\{A(c)\}, \emptyset)$, which we sequentially apply on a dataset $I = \{A(c)\}$. Then the rule is applied twice: first for U_1 to derive $C(c)$ due to adding $B(c)$, and then again for U_2 to delete $C(c)$ due to deleting $A(c)$.

The general reason why we have to regard the rule twice in the example is that if a fact F changes in the next update, due to being added or deleted, this change also affects facts that can be derived by F, such that those derived facts may be added or (over)deleted as well, and in order to identify those derived facts we have to, again, take a look at the applicable rules that include the changed fact F in their body. Accordingly, if we know which facts will change in the next update, we can also detect which rules will be repeated. Furthermore, if we apply a rule and already know whether and how its body facts will change for the next update, we can also directly determine how the derived fact in the rule's head will change and, thus, avoid the need for a repeated rule application. Therefore, we extend DRed with *marking* in order to both indicate and determine such facts that will change due to the next update.

Example 2 (cont.). Consider again the situation from the previous Example 1, but assume that we already take a look at the update U_2 during the processing of U_1. Thus, we can directly see that the fact $A(c)$ will be deleted by the upcoming U_2 and indicate this by marking $A(c)$. When we now process U_1 and use the facts $A(c)$ and $B(c)$ to derive $C(c)$, we also pass the mark of $A(c)$ on to $C(c)$, since the prospective deletion of $A(c)$ will affect its derived facts too. Once the processing of U_1 is done and we move on to U_2, we directly know that $C(c)$ has to be deleted due to its mark, without the need to consider the rule again.

As demonstrated in the example, what enables us to prevent a repeated rule application in the end is the marking of implicit, derived facts. In DRed, there exist two different kinds of derivations: in the overdeletion phase, we derive facts to be deleted, whereas in the rederivation and insertion phase, we derive facts to be added, respectively. Accordingly, we have to distinguish between two different kinds of markings too:

Definition 1. *Given two sequential updates U_1 and U_2, we say that a fact is* marked positively *if it is affected by a deletion of U_1 but re-added by U_2, whereas a fact is* marked negatively *if it is present in the adapted dataset of U_1 but affected by a deletion of U_2.*

Positive marks for implicit facts are computed during the overdeletion phase and require that the marked fact is derivable for the next update. The latter potentially means that the marked, (over)deleted fact might be already derivable during the rederivation phase of the current update, in which case the fact does not need to be marked anymore as it is directly re-added before we even deal with the next update. If this is not the case, then the fact is not present in the dataset when we start with the processing of the next update, such that we have to re-add it during the next update's insertion phase. Consequently, a derived fact should only be marked if we definitely know that the rule which we use to derive the fact in the current update's overdeletion phase is also applicable in the insertion phase of the next update. In particular, this requires that every fact of the applied rule's body is also present in the dataset during the next update's insertion phase. This holds, when every body fact affected by a deletion is marked (positively), while the remaining unaffected body facts are explicit and not marked (negatively). The reason why the second condition focuses on explicit facts is because we can generally not tell whether an implicit fact does not have to be deleted and, thus, is still present for the next update, until both the overdeletion and rederivation phase are completed.

The computation of *negative marks* happens during the rederivation and insertion phase. Negative marks determine facts that are to be deleted during the next update's overdeletion phase. First, note that rules used for rederivation and insertion only contain body facts that are not affected by a deletion of the current update. Hence, they are potentially still applicable in the next overdeletion phase. Second, we observe that a body fact might be marked negatively from an upcoming deletion in the next update. For rules with at least one such (negatively) marked body fact, we then also (negatively) mark the derived head fact during rederivation or insertion. A demonstration is provided by Example 2.

To summarize, marking a derived, deleted fact allows us to prevent rule applications that would usually be needed for the next update's insertion phase, whereas marking a derived, added fact allows us to prevent rule applications for the next overdeletion phase. What is not influenced by markings, however, is the next update's rederivation phase. The reason for this is that despite the reduction of overdeletion rule applications, we still obtain the same set of (over)deleted facts and, hence, the same number of rederivations afterward.

For the sake of simplicity, we only take a look at one additional update here and exclude any further successors. Otherwise, facts could have various marks, potentially at the same time, that each refer to a different upcoming update. Computing marks for derived facts would, consequently, become more complex and introduce more overhead, especially since it is much harder to tell if a certain rule will even be applicable for an upcoming update that still has some other pending updates of the stream in front of it.

4.1 Algorithm

The approach of extending DRed with marking is formalized in Algorithm 1. We further assume two procedures MARK(S) and UMMARK(S), which add and remove marks from the facts in the given set S, respectively, a function GET-MARKED(S), which returns the marked facts from S, and a function GET-NEXT($\hat{U}$), which returns the next update in the stream $\hat{U}$ if available. Note that we do not need to actively distinguish between marking a fact positively or negatively, since positive marks only refer to deleted facts, while negative marks only refer to added, present facts. Algorithm 1 takes as input a Datalog program P, a (possibly empty) materialized dataset $I = \mathsf{mat}(P, E_0)$, and a stream of updates $\hat{U} = \langle U_1 = (E_1^-, E_1^+), U_2 = (E_2^-, E_2^+), ...\rangle$. The algorithm's output is a stream of materialized datasets $\langle \mathsf{mat}(P, E_1), \mathsf{mat}(P, E_2), ...\rangle$ with $E_i = (E_{i-1} \setminus E_i^-) \cup E_i^+$ for $i \geq 1$, where each dataset refers to an update in $\hat{U}$. Note that we do not deal with specific queries, but consider the general case of computing the materialization for each update in the stream. Further note that the program P stays the same, i.e., we do not deal with changes to the set of rules.

Algorithm 1. DRed with Marking

Require: Datalog program P, materialized dataset I, stream of updates $\hat{U}$
Ensure: stream of materialized datasets

$U_1 = (E_1^-, E_1^+) \leftarrow$ null; $U_2 = (E_2^-, E_2^+) \leftarrow$ null; phase $\leftarrow 1$
repeat
  if $U_1 =$ null **then repeat** $U_1 \leftarrow$ GETNEXT($\hat{U}$) **until** $U_1 \neq$ null
  if $U_2 =$ null **then**
    $U_2 \leftarrow$ GETNEXT($\hat{U}$)
    if $U_2 \neq$ null **then** MARK($((I \cup E_1^+) \cap E_2^-) \cup (E_1^- \cap E_2^+)$)
  if phase $= 1$ **then** ▷ Overdeletion
    if $\exists r \in \rho(P, I) : \mathsf{body}(r) \cap E_1^- \neq \emptyset$ and $\mathsf{head}(r) \notin E_1^-$ **then**
      if $\forall F \in \mathsf{body}(r)$: either ($F \in E_1^-$ and F is marked) or ($F \notin E_1^-$ and F is explicit and not marked) **then** MARK($\{\mathsf{head}(r)\}$)
      $E_1^- \leftarrow E_1^- \cup \{\mathsf{head}(r)\}$
    else phase $\leftarrow$ phase $+ 1$
  if phase $= 2$ **then** ▷ Rederivation
    if $\exists r \in \rho(P, I) : \mathsf{body}(r) \cap E_1^- = \emptyset$ and $\mathsf{head}(r) \in E_1^-$ **then**
      if $\exists F \in \mathsf{body}(r)$: F is marked **then** MARK($\{\mathsf{head}(r)\}$)
      $E_1^- \leftarrow E_1^- \setminus \{\mathsf{head}(r)\}$
    else phase $\leftarrow$ phase $+ 1$
  if phase $= 3$ **then** ▷ Apply deletions and prepare insertions
    if $U_2 \neq$ null **then**
      $E_2^+ \leftarrow E_2^+ \cup$ GETMARKED(E_1^-)
      UNMARK($E_1^- \cup E_2^+$)
    $I \leftarrow (I \setminus E_1^-) \cup E_1^+$
    phase $\leftarrow$ phase $+ 1$
  if phase $= 4$ **then** ▷ Insertion
    if $\exists r \in \rho(P, I) : \mathsf{head}(r) \notin I$ **then**
      if $\exists F \in \mathsf{body}(r)$: F is marked **then** MARK($\{\mathsf{head}(r)\}$)
      $I \leftarrow I \cup \{\mathsf{head}(r)\}$
    else phase $\leftarrow$ phase $+ 1$
  if phase $= 5$ **then** ▷ Prepare processing of next update
    if $U_2 \neq$ null **then**
      $E_2^- \leftarrow E_2^- \cup$ GETMARKED(I)
      UNMARK($I \cup E_2^-$)
    write I to output; $U_1 \leftarrow U_2$; $U_2 \leftarrow$ null; phase $\leftarrow 1$
until $\hat{U}$ ends and $U_1 =$ null

As the classical DRed algorithm and as indicated by the comments in the algorithm, the general procedure consists of the sequential overdeletion and rederivation phase to identify each fact that has to be removed from the dataset, as well as the subsequent insertion phase which adds each newly derivable fact to the dataset. The main differences are that we now (i) perform the computations in a big loop, (ii) consider two updates at the same time, and (iii) deal with fact marking. The loop is needed since we work with a whole stream of updates. In order to have access to both the current and the next update, we use two variables U_1 and U_2 in the algorithm, where U_1 always refers to the current update, while U_2 refers to the next update in the stream. The related sets E_1^-, E_1^+ and E_2^-, E_2^+ are treated as variables too, which particularly means that replacing U_1 or U_2 by a new update will also replace the related sets appropriately.

Due to potential delays between updates in the stream, we might, however, not always have direct access to the next update. Since waiting for the next update would prolong the processing of U_1, the GETNEXT function instantly returns null if no new update is available. To make sure that we fetch the next update once it actually is available (cf. lines 5 and 6), the overdeletion, rederivation, and insertion phases only apply one rule per loop iteration (see lines 8, 13, 24), such that we may quickly integrate the next update in the following iteration. Despite that, the sequential order of the phases is still guaranteed due to the gradually increased phase variable (cf. lines 7 and 11), which is reset to 1 after the processing of the current update ends (see line 32).

The main purpose of U_2 is to indicate what facts change in the next update and, hence, have to be marked. Explicit facts are directly marked as soon as U_2 is instantiated (see line 6). This includes facts that are already in the dataset or explicitly inserted by U_1, but explicitly deleted by U_2 (cf. $(I \cup E_1^+) \cap E_2^-$), as well as facts that are explicitly deleted by U_1, but also immediately re-added by U_2 (cf. $E_1^- \cap E_2^+$). The marking of implicit, derived facts happens in lines 9, 14, and 25 and demands that the requirements discussed earlier and stated in the same lines are satisfied appropriately.

Because the marks tell us which facts change in the next update U_2, we use them to extend the update's sets E_2^+ and E_2^- (see lines 19 and 30) such that those facts may directly be introduced during the next update's processing, which starts after assigning U_2 to U_1 in line 32. Since a rule is only applied for overdeletion and insertion if its head is not yet present in the related, adapted set of facts (see lines 8 and 24), the direct introduction of implicit facts subsequently reduces the number of applied rules.

4.2 Correctness

In the following, "overdeletion" specifically refers to the computations that are performed in Algorithm 1 when phase $= 1$, "rederivation" to those when phase $=$ 2, and "insertion" to those when phase $= 4$. Furthermore, we say that "a rule is applied during overdeletion (and marks a fact)" if the rule satisfies the conditions in line 8 (and 9). The same goes for rederivation based on line 13 (and 14), as well as insertion with line 24 (and 25). Given two sequential updates U_a and U_b, we additionally define that the "processing of U_a" refers to loop iterations in Algorithm 1 where $U_1 = U_a$ and $U_2 = U_b$, while the "processing of U_b" refers to the subsequent, continuing iterations where $U_1 = U_b$.

We prove correctness of Algorithm 1 by showing that, despite the marking extension, we still obtain the same results as classical DRed. For that, we start with proving that the algorithm correctly adapts a materialized dataset if the input stream only consists of a single update:

Lemma 1 (Single update). *Given a Datalog program P, a materialized dataset $I = \mathsf{mat}(P, E_0)$, and a stream of updates $\hat{U} = \langle U \rangle$ with $U = (E^-, E^+)$, Algorithm 1 returns as output a stream $\langle \mathsf{mat}(P, E) \rangle$ with $E = (E_0 \setminus E^-) \cup E^+$.*

Proof. Since we only have a single update, we do not mark any explicit facts in line 6. Furthermore, the conditions in line 9 as well as lines 14 and 25 cannot be satisfied. Hence, implicit facts are not marked either in the same lines. Accordingly, the algorithm behaves like classical DRed:

While $\mathsf{phase} = 1$, we check in line 8 for rule instances where a body fact is in the set E_1^- of deleted facts and then also add the head to E_1^-. Due to the loop, this is repeated until every fact that can be derived based on some deleted fact has been identified, and we get $\mathsf{phase} = 2$. While $\mathsf{phase} = 2$ in the loop, line 13 considers every rule instance that allows a fact from E_1^- to be derived by facts that are not in E_1^-, in which case the fact is removed from E_1^- too. Hence, this fact may also be used to rederive other facts as well. Once we cannot find any more applicable rule instances, phase is set to 3. At this point, E_1^- only contains explicit facts from E^- together with those that require at least one fact from E_1^- for their derivation and, thus, have to be deleted, as done in line 21. In the same line, we also introduce the newly added facts E^+ from our update. In the following, we have $\mathsf{phase} = 4$ and identify every derivable fact that is not yet in the dataset in line 24, before adding this fact in line 26. This step is repeated until we cannot find any further new derivable fact and, hence, obtain the adapted materialized dataset as with classical DRed. □

Knowing that we obtain correct results if no facts are marked, we next show that the marking of facts does actually not have any influence on the computed dataset, so that we still obtain the same results when more than one update is present in the stream:

Lemma 2 (No influence on current dataset). *Given two sequential updates U_a and U_b from $\hat{U}$ in Algorithm 1, the marks that are computed due to U_b during the processing of U_a do not influence the dataset I that is returned at the end of processing U_a.*

Proof. In the algorithm, there are five situations where a marked fact is relevant: in line 9, in lines 14 and 25, in lines 20 and 31, in line 19, and in line 30. The first two cases only decide if another fact has to be marked, but do not add or delete any facts. Similarly, the third case merely removes marks from facts. In the last two cases, we just extend the sets E_2^+ and E_2^- of the next update, which do not have any influence on the current dataset I either. □

Even though the marking of facts does not influence the dataset of the currently processed update, it still changes the additions and deletions that are directly introduced by the next update (see lines 19 and 30). To prove that we obtain a correctly adapted materialized dataset for the next update too, we, therefore, show that the additional additions and deletions based on marked implicit facts are also computed in the classical case where we do not extend the next update.

Lemma 3 (No influence on next insertions). *Let U_a and U_b be two sequential updates from $\hat{U}$ in Algorithm 1 and let r be a rule that is applied in the overdeletion phase and marks a fact during the processing of U_a, but* $\mathsf{head}(r)$ *is not derived during rederivation or insertion of U_a. Then r is applicable in the insertion phase during the subsequent processing of U_b if we do not add* $\mathsf{head}(r)$ *to E_2^+ in line 19.*

Proof. Not adding the fact $\mathsf{head}(r)$ to E_2^+ means that it will also not be in I for U_b's insertion phase. What remains to be shown is that the body facts of r are also present in I for U_b's insertion, s.t. $r \in \rho(P, I)$. Since r marks a fact, it satisfies the conditions from line 9, which means that its body only contains explicit, non-marked facts $S_0^+ \not\subseteq E_1^-$ and marked facts $S_1^- \subseteq E_1^-$. Explicit facts can only be directly deleted and marked by an update, hence, S_0^+ still appears in I for U_b. The same holds for each fact in S_1^- that is explicit, because the mark indicates its re-insertion by U_b. Accordingly, if $\mathsf{body}(r)$ only contains explicit facts, we can still apply r for U_b's insertion. Hence, every marked, implicit fact in E_1^- that was derived by such a rule with only explicit body facts also occurs in I for U_b after this rule has been applied. This, furthermore, means that a rule r where S_1^- contains implicit facts can be applied too. □

Lemma 4 (No influence on next deletions). *Let U_a and U_b be two sequential updates from $\hat{U}$ in Algorithm 1 and let r be a rule that is applied in the rederivation or insertion phase and marks a fact during the processing of U_a. Then r is applicable in the overdeletion phase during the subsequent processing of U_b if we do not add* $\mathsf{head}(r)$ *to E_2^- in line 30.*

Proof. Not adding the fact $\mathsf{head}(r)$ to E_2^- means that it will also not be in E_1^- for U_b's overdeletion phase. Since r marks a fact, it satisfies the conditions from line 14 and 25, which means that its body contains at least one marked fact. If the marked fact is explicit, it occurs in U_b's set E_b^- and E_1^-, respectively, thus allowing r to be applied in U_b's overdeletion based on line 8. Hence, every marked, implicit fact that was derived by r will be added to U_b's set E_1^- too and, thus, enables the application of other rules that contain those implicit facts as well. The same argumentation can be applied for further derived implicit facts. □

Combining all of the above results, we can finally prove correctness:

Theorem 1. *Given a Datalog program P, a materialized dataset $I = \mathsf{mat}(P, E_0)$, and a stream of updates $\hat{U} = \langle U_1, U_2, ...\rangle$, Algorithm 1 returns as output a stream $\langle \mathsf{mat}(P, E_1), \mathsf{mat}(P, E_2), ...\rangle$ with $E_i = (E_{i-1} \setminus E_i^-) \cup E_i^+$ for each $U_i = (E_i^-, E_i^+) \in \hat{U}$.*

Proof. We show this by induction: As base case, we consider the first update $U_1 \in \hat{U}$, for which we obtain $\mathsf{mat}(P, E_1)$ based on Lemma 1 and 2. For the induction step, we consider some update $U_{i+1} \in \hat{U}$ that was extended to an update $U'_{i+1} = (E'^-_{i+1}, E'^+_{i+1})$ due to marking during the processing of $U_i \in \hat{U}$. By induction hypothesis, we get $\mathsf{mat}(P, E'_{i+1})$ as result. Based on Lemma 3 and 4, we know, however, that $\mathsf{mat}(P, E'_{i+1}) = \mathsf{mat}(P, E_{i+1})$. □

$$edge(x,y) \rightarrow path(x,y)$$
$$edge(x,y), path(y,z) \rightarrow path(x,z)$$

(a) Rules for Datalog program P_{trans}

$$edge(x,y) \rightarrow edge1(x,y)$$
$$edge1(x,y) \rightarrow edge2(x,y)$$
$$edge2(x,y) \rightarrow edge3(x,y)$$
$$edge3(x,y) \rightarrow edge4(x,y)$$

(b) Rules for Datalog program P_{seq}

Fig. 1. Rules for the Datalog programs used in synthetic data tests

$$\begin{aligned} nextInWay(x, y_1, z_1), nextInWay(x, y_2, z_2), neq(z_1, z_2) &\rightarrow connection(z_1, z_2)\\ nextInWay(x, y_1, z_1), nextInWay(x_2, x, z_2), neq(z_1, z_2) &\rightarrow connection(z_1, z_2)\\ nextInWay(x_1, y, z_1), nextInWay(y, y_2, z_2), neq(z_1, z_2) &\rightarrow connection(z_1, z_2)\\ nextInWay(x_1, y, z_1), nextInWay(x_2, y, z_2), neq(z_1, z_2) &\rightarrow connection(z_1, z_2)\\ connection(x, y), connection(y, z) &\rightarrow connection(x, z)\\ \text{with } neq(x, y) := x \neq y & \end{aligned}$$

Fig. 2. Rules for the Datalog program used in real data tests

5 Evaluation

We implemented DRed with marking and compared it to classical DRed by measuring the needed CPU time and number of rule applications in each processing phase, based on synthetic and real test cases.

5.1 Test Data

For synthetic tests, we consider a graph represented by a set of directed edges, where updates add and delete edges. The facts/edges in each update are chosen randomly with the restrictions that we only add edges which are not already in the current graph and which refer to nodes, represented as positive integers, within a certain range, and analogously only delete edges that actually occur in the graph. Every update within a stream adds and deletes the same amount of facts, which we call the update's size, except for the first update, which we use to initialize the dataset.

Rules are provided by two separate programs P_{trans} and P_{seq} shown in Fig. 1. P_{trans} computes transitive paths in the graph, while P_{seq} produces simple sequences of rule applications by repeatedly renaming the predicate of a fact. In particular, P_{trans} enables facts to be derived potentially in several ways, causing many rederivations, while P_{seq} does not allow alternative derivations at all, which is the optimal case for DRed due to the lack of any rederivations.

As a real test case, we deal with a task related to autonomous driving, where we perform reasoning on map data that is dynamically loaded based on GPS positions. For that, we utilized a GitHub project[1] that represents data from OpenStreetMap[2] as Datalog facts and generates a stream of updates for a given

[1] https://github.com/mrupp-sudo/gps-osm-project/.
[2] https://www.openstreetmap.org/.

GPS track to indicate how the map data of a specified area around a GPS position changes between the individual positions of the track. The project already contains a GPS track, which describes a route from the main station to the central library in the German city Ulm ($track_0$), and which was created with the online tool Kurviger[3]. We used the same tool to get two further GPS tracks, one describing a route from the same main station to Ulm University ($track_1$) and the other from the main station in Stuttgart to the airport in Stuttgart ($track_2$).

For our test case, we focus on the provided facts $nextInWay(N_1, N_2, W)$, which state that node N_1 is followed by node N_2 on the way W. We defined appropriate rules that, first, detect connections between ways by looking if facts from different ways have a common node and, second, compute transitive relations between connected ways. The corresponding Datalog rules are shown in Fig. 2. The update streams that we generated for each of the three above GPS tracks were also limited to only contain *nextInWay*-facts and, furthermore, restricted to a certain type of way. In detail, we focus on footways for $track_0$, cycleways for $track_1$, and secondary and tertiary highways for $track_2$. For each GPS position, we considered map data within a radius of 50 m.

5.2 Test Execution and Results

In order to conduct the evaluation, we created a Java project, which is also available online.[4] In detail, we used Java to provide the update streams, while both the classical and our extended DRed algorithm were implemented for each tested Datalog program in Constraint Handling Rules [7] based on SWI-Prolog[5]. During the tests, we measured the time spent by the CPU to process the whole stream, along with the number of applied rules in each DRed phase. For our extended DRed approach, we additionally counted the number of positive and negative marks that were computed for both explicit and implicit facts. The following results were obtained on a Windows 11 PC with an AMD Ryzen 7 3700X 3.59 GHz CPU and 16 GB RAM, using SWI-Prolog 9.3.15 with a 4 GB stack.

Synthetic Data Results. The synthetic test results are average values from three runs that used different random seeds and were repeated five times. The streams always contained 50 updates and started with an update that adds 100 facts to the initially empty dataset. For this initial update, we first computed the whole materialization before the remaining updates were processed. We generated eight streams for every test run, each with a different update size ranging from 10 to 80. For the tests with program P_{trans}, edges were generated for a graph with at most 20 nodes, whereas program P_{seq} worked with up to 100 nodes.

[3] https://kurviger.com/en.
[4] https://github.com/M-Illich/dred-mark-eval.
[5] https://www.swi-prolog.org/.

Table 1. Test measurements for the synthetic data with P_{trans}

size	time [in seconds]		overdeletions		rederivations	insertions		positive marks		negative marks	
	old	new	old	new	both	old	new	ex.	im.	ex.	im.
10	18.91	32.87	19,454	14,871	19,308	545	545	13	0	480	4,583
20	52.55	40.60	18,505	8,402	15,926	2,912	2,912	52	0	960	10,103
30	108.40	81.10	18,416	3,493	11,472	7,288	7,288	116	2	1,399	14,923
40	130.52	97.65	18,934	3,193	12,759	6,575	6,575	213	3	1,816	15,741
50	129.84	106.16	19,342	3,336	15,172	4,570	4,569	339	5	2,215	16,006
60	132.04	109.98	19,501	3,251	16,567	3,327	3,327	480	3	2,604	16,250
70	110.56	100.59	19,561	3,559	17,803	2,158	2,158	671	2	2,973	16,002
80	105.81	92.80	19,580	3,778	18,198	1,782	1,782	840	4	3,293	15,802

The results for P_{trans} are shown in Table 1, where "old" refers to classical DRed and "new" to DRed with marking, respectively. Taking a look at the measured CPU times in the "time" columns, we see first that DRed with marking can be slower than the classical algorithm (seen for update size 10). This is due to the fact that, on the one hand, our implementation is merely a prototype without any optimizations and, on the other hand, some updates, especially small ones, might only lead to a small amount of marked derived facts and, hence, a small amount of reduced rule applications, so that the additional overhead of marking is not compensated. This changes in our tests, however, for update sizes larger than 10, where we can reduce the processing time by around 20% on average. The reason for this becomes obvious when looking at the number of rules applied during overdeletion, which is significantly smaller for the marking approach. In particular, we see that the number of prevented rule applications correlates with the number of implicit facts (im. – implicit, ex. – explicit) that were marked negatively, as displayed in the last table column.

Besides, we observe that for the update sizes 30 to 50 the number of prevented overdeletions is larger than the number of performed rederivations. This means that DRed with marking can actually have fewer rule applications than some alternative approach which only avoids rederivations, like counting [8], Backward/Forward [13], or variations thereof [9,11], for instance.

While the rederivation phase is not affected by marking, the insertion phase theoretically could be. In practice, however, this rarely seems to be the case. The main reason here is that we do not just have a smaller number of positively marked explicit facts, but also stricter conditions under which a positive mark is passed on to a derived fact, which results in a very small amount of positively marked, implicit facts. Furthermore, since those positive marks are determined during overdeletion, it is still possible that a positively marked fact is actually derived in the remaining rederivation or insertion phase too, such that the fact is already added before the next update is processed and the direct insertion due to marking does not have an effect anymore. This explains why the number of positively marked, implicit facts does not correlate with the (barely existing) amount of reduced insertions.

Table 2. Test measurements for the synthetic data with P_{seq}

size	time [in seconds]		overdeletions		rederivations	insertions		positive marks		negative marks	
	old	new	old	new	both	old	new	ex.	im.	ex.	im.
10	8.63	8.35	1,960	1,317	0	2,360	2,359	1	1	480	643
20	16.49	14.87	3,920	2,380	0	4,317	4,316	1	1	960	1,540
30	25.38	22.30	5,880	3,351	0	6,265	6,253	6	12	1,440	2,529
40	38.85	33.01	7,827	4,289	0	8,228	8,215	7	13	1,917	3,538
50	60.53	49.66	9,709	5,108	0	10,173	10,148	12	25	2,377	4,601
60	101.57	83.11	11,588	6,149	0	12,129	12,078	20	51	2,844	5,439
70	145.06	117.82	13,457	6,964	0	14,069	14,025	23	44	3,311	6,493
80	215.85	175.97	15,283	7,999	0	16,005	15,942	31	63	3,767	7,284

The results for the second program P_{seq} are shown in Table 2 and lead to similar conclusions. The main difference is, however, that the rules in P_{seq} do not allow alternative derivations for facts. This means, on the one hand, that we do not have any rederivations, while on the other hand, each positively marked, implicit fact does now indeed lead to a reduction of rule applications in the insertion phase. Since each rule in P_{seq} only contains a single body atom, the condition for computing positive marks is also less strict than for P_{trans}, hence, leading to more positively marked, implicit facts. Nevertheless, they can still not compete with the number of negative marks, which, again, result in a large reduction of rule application in the overdeletion phase.

Real Data Results. As real data tests, we processed the three update streams that we generated from our three GPS tracks. The differences to the synthetic data stream are that each stream has a distinct length, while the number of added and deleted facts is generally not the same for an update and also varies between different updates in the stream. The exact numbers for each stream are in Table 3. We performed the same measurements as for the synthetic tests and computed the average values of five repetitions. The results are shown in Table 4, where "old" again refers to classical DRed and "new" to DRed with marking.

Overall, the marking approach enabled a reduction of CPU time by about 27% on average for our real data streams. Nevertheless, the relatively long processing time for the first GPS track still emphasizes the need for further optimizations in our implementation. As in the synthetic tests, rule applications for overdeletion could be reduced significantly due to the large number of negatively marked, implicit facts, while insertions could only be reduced in one case by a small amount due to the rare occurrence of positively marked facts.

Table 3. Number statistics for the real data update streams

	updates	max. added facts	max. deleted facts	avg. added facts	avg. deleted facts
$track_0$	55	172	114	18	17
$track_1$	83	152	167	14	14
$track_2$	114	65	74	9	8

Table 4. Test measurements for the real data

	time [in seconds]		overdeletions		rederivations	insertions		positive marks		negative marks	
	old	new	old	new	both	old	new	ex.	im.	ex.	im.
$track_0$	577.58	440.08	7,799	4,534	3,299	4,638	4,636	9	2	910	3,265
$track_1$	24.10	17.16	4,954	2,402	2,317	2,653	2,653	11	0	1,208	2,552
$track_2$	18.94	13.58	2,910	1,274	1,206	1,704	1,704	0	0	1,017	1,636

6 Conclusion

Initially motivated by the problem where updates to a dataset appear faster than they can be fully processed due to streams with large update sizes and frequencies, like in the context of autonomous driving, we examined a way to improve incremental materialization maintenance for whole streams of updates. The idea here is that we immediately take a look at the next update during the current processing in order to detect and prevent re-occurring rule applications. For that, we extended the classical Delete/Rederive algorithm with marking for facts, such that we can directly consider and compute changes of the next update without introducing additional rule applications.

We provided a prototypical implementation of our approach and evaluated it for both synthetic and real data. The test results showed that we can greatly lower the number of rule applications for overdeletion, sometimes even beyond the number of rederivations, and, thus, reduce the overall processing time. Notable improvements for rederivations or insertions, however, did not occur.

For future work, we want to see how other approaches, like counting [8] or Backward/Forward [13], might benefit from marking, and also investigate more use cases. In addition, the presented algorithm and implementation are still open for optimizations, where a focus on overdeletion and potentially rederivation seems appropriate. Besides, the general goal is to examine further ways where a concurrent processing of updates in a stream is feasible, with a potential immediate and direct integration of new updates for optimal responsiveness.

References

1. Abiteboul, S., Hull, R., Vianu, V.: Foundations of Databases. Addison-Wesley (1995)
2. Bancilhon, F.: Naive evaluation of recursively defined relations. In: On Knowledge Base Management Systems: Integrating Artificial Intelligence and Database Technologies, pp. 165–178. Topics in Information Systems. Springer, Cham (1986). https://doi.org/10.1007/978-1-4612-4980-1_17
3. Barbieri, D.F., Braga, D., Ceri, S., Della Valle, E., Grossniklaus, M.: Incremental reasoning on streams and rich background knowledge. In: Aroyo, L., et al. (eds.) ESWC 2010. LNCS, vol. 6088, pp. 1–15. Springer, Heidelberg (2010). https://doi.org/10.1007/978-3-642-13486-9_1
4. Bry, F., et al.: Foundations of rule-based query answering. In: Antoniou, G., et al. (eds.) Reasoning Web 2007. LNCS, vol. 4636, pp. 1–153. Springer, Heidelberg (2007). https://doi.org/10.1007/978-3-540-74615-7_1
5. Carral, D., González, L., Koopmann, P.: From horn-SRIQ to datalog: a data-independent transformation that preserves assertion entailment. In: Proceedings of the AAAI Conference on Artificial Intelligence, vol. 33, pp. 2736–2743. AAAI Press (2019). https://doi.org/10.1609/aaai.v33i01.33012736
6. Dell'Aglio, D., Valle, E.D.: Incremental reasoning on RDF streams. In: Linked Data Management, pp. 413–435. Chapman and Hall/CRC (2014). http://dellaglio.org/preprints/14b1.pdf
7. Frühwirth, T.: Constraint Handling Rules. Cambridge University Press (2009)
8. Gupta, A., Mumick, I.S., Subrahmanian, V.S.: Maintaining views incrementally. In: Proceedings of the 1993 ACM SIGMOD International Conference on Management of Data, pp. 157–166. ACM Press (1993). https://doi.org/10.1145/170035.170066
9. Hu, P., Motik, B., Horrocks, I.: Optimised maintenance of datalog materialisations. In: Proceedings of the AAAI Conference on Artificial Intelligence, vol. 32, pp. 1871–1879. AAAI Press (2018). https://doi.org/10.1609/aaai.v32i1.11554
10. Illich, M., Glimm, B.: Fully dynamic materialization maintenance. In: Kutz, O., Lutz, C., Ozaki, A. (eds.) Proceedings of the 36th International Workshop on Description Logics (DL 2023). CEUR Workshop Proceedings, vol. 3515. CEUR-WS.org (2023). https://ceur-ws.org/Vol-3515/paper-15.pdf
11. Motik, B., Nenov, Y., Piro, R., Horrocks, I.: Maintenance of datalog materialisations revisited. Artif. Intell. **269**, 76–136 (2019). https://doi.org/10.1016/j.artint.2018.12.004
12. Motik, B., Nenov, Y., Piro, R., Horrocks, I., Olteanu, D.: Parallel materialisation of datalog programs in centralised, main-memory RDF systems. In: Proceedings of the AAAI Conference on Artificial Intelligence, vol. 28, pp. 129–137. AAAI Press (2014). https://doi.org/10.1609/aaai.v28i1.8730
13. Motik, B., Nenov, Y., Piro, R.E.F., Horrocks, I.: Incremental update of datalog materialisation: the backward/forward algorithm. In: Proceedings of the AAAI Conference on Artificial Intelligence, vol. 29, pp. 1560–1568. AAAI Press (2015). https://doi.org/10.1609/aaai.v29i1.9409
14. Qiu, H., Ayara, A., Glimm, B.: Ontology-based processing of dynamic maps in automated driving. In: Proceedings of the 12th International Joint Conference on Knowledge Discovery, Knowledge Engineering and Knowledge Management, IC3K 2020, Volume 2: KEOD, Budapest, Hungary, 2–4 November 2020, pp. 98–107. SCITEPRESS (2020). https://doi.org/10.5220/0010133900980107

15. Ren, Y., Pan, J.Z.: Optimising ontology stream reasoning with truth maintenance system. In: Proceedings of the 20th ACM International Conference on Information and Knowledge Management, pp. 831–836. ACM Press (2011). https://doi.org/10.1145/2063576.2063696
16. Terdjimi, M., Médini, L., Mrissa, M.: Web reasoning using fact tagging. In: Companion Proceedings of the The Web Conference 2018, WWW 2018, pp. 1587–1594. International World Wide Web Conferences Steering Committee, Republic and Canton of Geneva, CHE (2018). https://doi.org/10.1145/3184558.3191615
17. Urbani, J., Margara, A., Jacobs, C., van Harmelen, F., Bal, H.: DynamiTE: parallel materialization of dynamic RDF data. In: Alani, H., et al. (eds.) ISWC 2013. LNCS, vol. 8218, pp. 657–672. Springer, Heidelberg (2013). https://doi.org/10.1007/978-3-642-41335-3_41

Balancing Privacy and Utility: Semantic Anonymization of Time-Aware Knowledge Graphs

Prachi Naik and Vinu E. Venugopal(✉)

International Institute of Information Technology Bangalore, Bengaluru, India
{prachi.naik,vinu.ev}@iiitb.ac.in

Abstract. Time-aware Knowledge Graphs (TKGs) extend traditional Knowledge Graphs (KGs) by incorporating temporal dimensions, enabling improved handling of time-sensitive data in domains such as finance, healthcare, and organizational management. However, temporal data raises privacy concerns due to potential pattern recognition and semantic similarity attacks, which are often overlooked by current anonymization methods. Existing techniques inadequately address semantic context and fail to determine optimal k and l values in k-anonymity and l-diversity frameworks. To address these challenges, we propose a novel approach leveraging sentence embeddings to capture semantic nuances and applying k-anonymity and l-diversity to enhance privacy. We formulate a Multi-Objective Optimization Problem to select optimal k and l values by minimizing quasi-identifier distances, maximizing sensitive attribute diversity, and reducing information loss. Experiments on three real-world datasets demonstrate that our method achieves lower information loss and superior anonymization compared to benchmarks. Further, downstream task evaluations using KG embedding models reveal that anonymized KGs outperform non-anonymized counterparts in Mean Reciprocal Rank (MRR) and Hits@n metrics, although with slightly reduced triple confidence scores. These results highlight the effectiveness of our approach in balancing robust anonymization with practical utility.

Keywords: Knowledge Graphs · Anonymization · Privacy · Sentence Embeddings · Multi-Objective Optimization Problem · Knowledge Graph Embedding Models

1 Introduction

Knowledge Graphs (KGs), such as Microsoft Academic graph [36], ICEWS [7], YAGO [29], Freebase [2], etc., organize information in the form of a directed graph, where nodes represent entities, and edges denote the relationships between these entities. This format allows the visualization of intricate connections within the data. By organizing data in this interconnected manner, KGs support a myriad of applications, including question-answering systems [1], recommender

E. Curry et al. (Eds.): ESWC 2025, LNCS 15718, pp. 169–187, 2025.
https://doi.org/10.1007/978-3-031-94575-5_10

systems [40], semantic search, reasoning, and decision-making processes across diverse domains like cyber-security, education and so on [42]. As the complexity and dynamism of real-world data grow, static KGs often fall short in capturing the evolving nature of knowledge. This limitation has led to the emergence of time-aware knowledge graphs (TKGs) [7,13,22].

TKG builds on the concept of traditional knowledge graphs by incorporating temporal data, since many facts are not static but highly ephemeral. For example, the triple $\langle$Michael Jackson, diedIn, Los Angeles$\rangle$ occurred on June 25, 2009 (2009-06-25); similarly, the triple $\langle$Ronaldo, playsFor, Real Madrid C.F.$\rangle$ is valid only during the period 2002–2007. This not only captures the event and the entities involved but also specifies the exact date when it happened or the duration of that activity. This temporal dimension enables more sophisticated analyses, such as predicting future events, understanding historical trends, and capturing temporal patterns in data. Many works such as TKG completion [15], time-aware link prediction [16], temporally aware KG embeddings generation [4,37], reasoning in TKGs [33] have already been done which makes TKGs a well-explored yet continually evolving research area, offering a rich foundation of real-world applications. However, the inclusion of temporal data introduces new privacy concerns. By leveraging the temporal snapshots contained within these TKGs, attackers can easily deduce an individual's identity and uncover sensitive information. Addressing these privacy issues is crucial for the *responsible* use of TKGs. Ensuring privacy protection not only safeguards individual rights but also aligns with legal and ethical standards [18], fostering trust in data-driven technologies [28].

Current TKG anonymization methods [6,11,12,32] overlook the underlying meaning of facts in knowledge graphs before anonymizing. Additionally, while existing techniques [11] ensure that sensitive attributes in groups of size k meet l-diversity requirements, they fail to account for semantic diversity, leaving them vulnerable to *semantic similarity attacks*. Furthermore, no established method exists to determine the optimal values for k and l.

In this paper, we propose a novel approach of *Semantic Anonymization*, where we use sentence transformers (such as SBERT [27]) to capture meaning of the *quasi-identifiers* (QIDs) and the *sensitive attributes* (SAs). Further, we leverage k-anonymity [30] and l-diversity [21] principles on the embeddings, which ensure that a user (represented as node in KG) cannot be identified with a confidence higher than $\frac{1}{k}$ and users' sensitive values cannot be inferred with a confidence higher than $\frac{1}{l}$ (Refer Sect. 4 for formal definitions). Since models like SBERT capture both syntax and semantics and enable finer distinctions between QIDs, we can preserve meaningful patterns in anonymized data by grouping similar embeddings; that is, clustering together the data points that are semantically close. Moreover, in our observation, use of distance metrics like cosine similarity on embeddings yields more meaningful groupings than traditional methods, which rely on exact matches or simple distance measures. We also select ideal values for k and l for which, we formulate an optimization problem having three objective functions, minimize distances between QIDs in group of k individuals, maximize distances between sensitive attributes in that group thus preventing *semantic similarity attack*, and minimize average information loss.

The rest of this paper is organized as follows. Survey on related work is done in Sect. 2. In Sect. 3, we illustrate our approach with a simple example. Section 4 explains the formal definitions of the terminologies, notations, and information loss metric used in our paper. Our algorithm and selection of optimal k and l values are explained in Sect. 5. Experimental setup and experimental results are presented in Sect. 6. Finally, we conclude the paper and explain future work in Sect. 7.

2 Related Work and Background

Anonymization has been extensively explored in the literature, with key techniques such as k-anonymity, l-diversity, and Differential Privacy aiming to balance privacy protection with data utility. Early work on anonymization focused on relational databases. Sweeney's k-anonymity model [30] ensures that each individual in the database cannot be distinguished from at least $k-1$ others. Building on this, l-diversity [21] enhances protection by ensuring diversity within sensitive attributes (SAs). t-closeness [20] further extends these methods, requiring the distribution of sensitive attributes within a group of k records (with matching quasi-identifiers) to closely resemble the overall dataset, preventing inference attacks. Differential privacy [14], which adds noise to datasets, faces challenges in preserving accuracy, especially for structured data like knowledge graphs (KGs). Additionally, it requires customized algorithms for each analysis type, reducing flexibility. For these reasons, we focus on k-anonymity in this work.

Recent advances in anonymization cover both structured and unstructured data. For text, [25] utilizes knowledge graphs for distant supervision in k-anonym- ization, while [39] explores pseudonymization using pre-trained language models to replace personal identifiers. Another approach, presented in [8], focuses on accuracy-guided k-anonymization to prevent inference attacks on sensitive features in training data. In social networks, anonymization techniques such as clustering algorithms [31] achieve k-anonymity by grouping similar nodes and modifying the graph to protect against degree-based attacks. Additionally, [41] introduces methods to mitigate *neighborhood attacks* by coding vertex neighborhoods and anonymizing similar vertices together.

Anonymizing knowledge graphs has also received attention. In [6], a data-independent approach is presented, applying privacy policies as SPARQL queries and using suppression methods. Further, in [32], the authors group sensitive attributes based on ontologies but lacks sufficient semantic diversity, which can expose privacy. For instance, subdividing "Christianism" into "Catholicism" and "Protestantism" does not provide enough anonymity, as attackers can still deduce the general category. k-AttributeDegree (k-ad), introduced in [10], uses an Attribute and Degree Information Loss Metric (ADM) to cluster individuals and anonymize KGs. However, this method was further extended in [11] to handle temporal knowledge graphs (TKGs). Authors of [12] incorporated l-diversity but ignored semantic factors in anonymization. Their method also discards useful nodes and adds fake ones, leading to increased information loss. Another approach by authors of [35] introduces a parameter to control actual diversity in tree-structured SAs, although this method does not extend to numerical attributes. In contrast, the work in [24] enhances l-diversity with (l, d)-semantic diversity, ensuring a minimum semantic distance between sensitive attributes to strengthen privacy protection.

To date, no existing work incorporates semantic factors for both quasi-identifiers (QIDs) and sensitive attributes (SAs) in TKG anonymization. Additionally, many approaches fail to provide sufficient diversity, leaving them vulnerable to *homogeneity* and *background knowledge* attacks. Current models also overlook semantic similarity among SAs, exposing them to *semantic similarity* attacks. Furthermore, optimal selection of parameters such as k, l, and similarity/dissimilarity values is often not addressed. These gaps highlight the need for improved anonymization models that balance privacy and utility, especially in the context of TKGs.

3 Illustration of the Proposed Approach

Our approach to anonymizing TKGs is demonstrated using the example of a corporate knowledge graph, as depicted in Fig. 1 and 2. The knowledge graphs, G_1 and G_2, in Fig. 1 contain multiple nodes and edges. For clarity, we have used different colors and shapes to represent the nodes. The red ellipses denote individuals, while rectangles represent the attributes (or object property values) of those individuals. Green rectangles indicate non-sensitive attributes, whereas blue rectangles represent sensitive attributes. For instance, the attribute 'hasEth' is treated as sensitive because it corresponds to an individual's ethnicity, a characteristic that could lead to discrimination, privacy violations, or harm if misused. Other examples of sensitive attributes (SAs) include religion, sexual orientation, diseases, and financial information.

In the anonymization process, the first step is to remove direct identifiers (DIDs) corresponding to each individual, as these can uniquely identify a person in the original KGs (refer to Fig. 1). We use suppression or removal techniques [23] for this purpose. For example, employee IDs (EIDs) can be removed since they are not typically used in analytical or decision-making tasks. Another DID in our example is the employee's name, which can either be suppressed or removed; for simplicity, we mask it with '***'. In Fig. 2, the anonymized versions of the original KGs ($\overline{G}_1$ and $\overline{G}_2$) are shown.

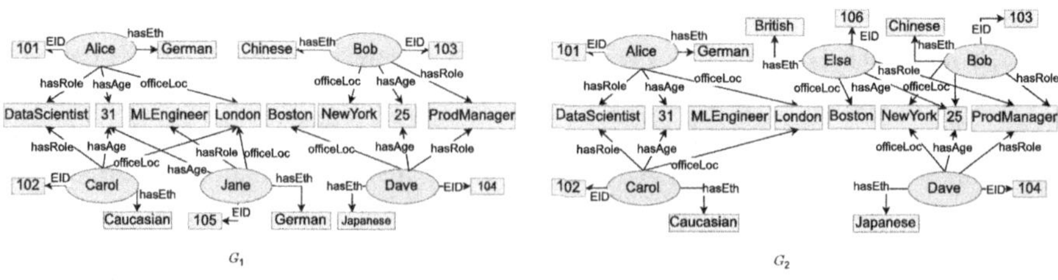

Fig. 1. G_1 and G_2 represent the original corporate KG at time $t = 1$ and $t = 2$ respectively (Color figure online)

The next critical step is identifying quasi-identifiers (QIDs)—such as role, age, and office location (Fig. 1)—which are attributes that, although not unique individually, can be combined with some other data source, using record linking techniques, to re-identify individuals. This constitutes a *linkage attack* [26]. For instance, an attacker who knows Alice is 31 years old, works as a data scientist in London, could use this information to query the KG. Despite the suppression of Alice's name, the combination of these QIDs could still uniquely identify her. Linkage attack is a specific case of re-identification attack, relying on quasi-identifiers to match datasets. *Re-identification* [9] occurs when anonymisation is reversed or de-anonymised, bringing the identifying information to light, by linking datasets, using prior or background knowledge or by comparing longitudinal data to find patterns. In our example, all attributes, except for DIDs and SAs, are treated as QIDs. Alice's set of QIDs is represented as {*DataScientist, 31, London*}.

To counter re-identification and linkage attacks, k-anonymity [30] is applied. This ensures that each record in a dataset is indistinguishable from at least $k-1$ other records by grouping individuals with similar QIDs. Each group contains at least k individuals, making it difficult for an attacker to pinpoint a specific individual within the dataset. Even if they attempt to link external data, they will be unable to ascertain which individual belongs to the dataset, as each QID group represents multiple people. We implement this k-anonymity technique within our *Semantic Anonymization* method, ensuring that individuals cannot be identified with confidence greater than $\frac{1}{k}$. To achieve this, QIDs are combined into comma-separated sentences. For example, *'DataScientist, 31, London'* and *'ProdManager, 25, NewYork'*. These sentences are then grouped based on the semantic similarity of their content. We use SentenceTransformer [27] to compute embeddings of these sentences, which are then clustered using a clustering algorithm [17] to group entities with semantically similar attributes. For example, *'DataScientist, 31, London'* and *'MLEngineer, 31, London'* are clustered together due to the similarity between the job roles *DataScientist* and *MLEngineer*. Prior approaches to KG anonymization [11,12] grouped entities based on identical QID values rather than the semantic meaning of the QIDs, resulting in significant information loss for analytics and decision-making tasks. Similarly, *'ProdManager, 25, Boston'* and *'ProdManager, 25, New York'* are clustered together based on the geographic proximity between Boston and New York. We ensure that each group has a minimum size of k and a maximum size of $2k-1$. As a result, in $\overline{G}_1$ (Fig. 2), *Alice*, *Carol*, and *Jane* (orange ellipses) form a group of size $k=3$. Similarly, *Bob* and *Dave* are grouped together, but to maintain $k=3$, a fake node (dotted blue ellipse) is added, forming a group of blue ellipses in $\overline{G}_1$ (Fig. 2).

In a k-anonymized group, if all individuals share the same sensitive attribute (e.g., ethnicity), an attacker can still infer that attribute, leading to a *Homogeneity Attack*. For example, if everyone in a group speaks fluent German, an attacker could infer that they share the ethnicity "*German*". Similarly, an attacker with prior knowledge (e.g., someone knowing a group of people can speak German) can perform a *Background Knowledge Attack* to infer sensitive attributes. To prevent these attacks, we apply l-diversity [21], ensuring that each group contains at least l unique sensitive attributes. In our example, the group of *Alice*, *Carol*, and *Jane* contains two unique SAs—"*German*" and "*Caucasian*"—which satisfy $l = 2$. Please refer Sect. 5.1 where we explain about creation of *valid groups* that are k-anonymous and l-diverse.

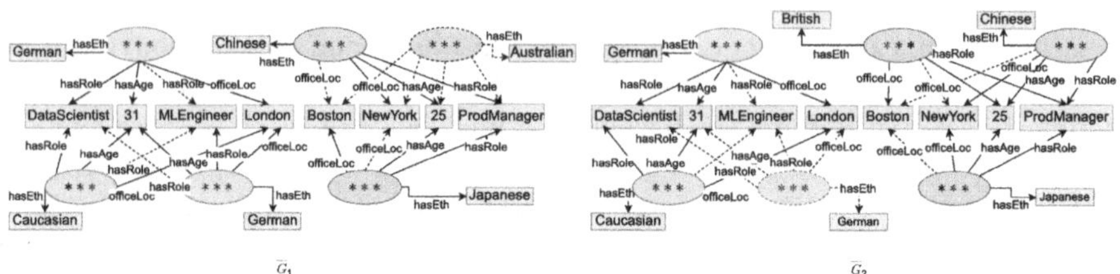

Fig. 2. $\overline{G}_1$ and $\overline{G}_2$ represent anonymized forms of G_1 and G_2 respectively (Color figure online)

In G_1 (Fig. 1), *Bob* and *Dave* have SAs of "*Chinese*" and "*Japanese*", respectively. Since they have similar QIDs, they are grouped together in $\overline{G}_1$ (blue ellipses in Fig. 2). l-diversity ensures that the probability of identifying any individual's SA is less than or equal to $1/l$. However, when SAs are semantically similar, l-diversity may fail to provide sufficient diversity, leading to a *Semantic Similarity Attack*. For example, if the anonymized group satisfies 2-diversity but consists only of ethnicities "*Chinese*" and "*Japanese*", an attacker could infer that an individual in that group is of Asian descent. To address this, our algorithm maximizes the semantic distance between SAs, adding fake nodes where necessary to ensure diversity. In $\overline{G}_1$ (Fig. 2), a fake node with the SA "*Australian*" is added to the group containing *Bob* and *Dave*. Our approach takes care of this also by maximising distances among SAs in a group (refer to Sect. 5.2).

Thus far, the anonymization process has been applied to a single snapshot at time $t = 1$. In dynamic settings, data providers may publish subsequent snapshots, such as at time $t = 2$, reflecting additions, deletions, and updates. These snapshots are referred to as TKGs. In our example, at time $t = 2$, Jane leaves the organization, Elsa joins, and Dave changes his office location from Boston to New York (changes shown in G_2 Fig. 1). We assume that adversaries with background knowledge have access to this series of anonymized snapshots.

For node deletions, prior works [11,12] removed the entire group if a node was deleted to prevent inference attacks. However, this leads to substantial data loss. Our algorithm addresses this by adding a fake node to the group to replace the deleted node, preserving the group's structure and retaining valuable data. In $\overline{G}_2$ (Fig. 2), a fake node replaces Jane, retaining the integrity of the orange ellipses group. In cases of node addition, such as when Elsa joins the organization, prior works introduced $k - 1$ fake nodes to prevent adversaries from identifying the newly added node. However, our algorithm clusters the newly added node with similar individuals, reducing the need for fake nodes. In $\overline{G}_2$ (Fig. 2), Elsa is added to the cluster containing Bob and Dave, replacing the fake node from $\overline{G}_1$. Lastly, for attribute updates, such as Dave's office location change in G_2 (Fig. 1), previous approaches added fake edges to both old and new attributes, generalizing the attributes and leading to information loss. In our method, Dave is re-clustered with individuals sharing similar QIDs, preventing both over-generalization and inference attacks.

To optimize anonymization, we evaluate all combinations of k and l based on three objective functions: semantic similarity between QIDs, semantic dissimilarity between SAs, and information loss.

4 Terminologies, Notations and Information Loss Metric

Table 1. Notations used in the paper

Symbol	Meaning
n	Number of individuals
t	Time instance
$G_t, \overline{G}_t$	A KG and its anonymized version at time t
$g_t, \overline{g}_t$	A sequence of continuous KGs and their anonymized version at times: 1, 2, ..., t
V_t, V_t^U, V_t^A, $\overline{V_t}$, $\overline{V_t}^U$, $\overline{V_t}^A$	A set of nodes, users, and attributes' values in G_t and $\overline{G}_t$
$sen(c)$	Set of sensitive values of a cluster 'c'
n_{fake}	Number of fake nodes/individuals to be added in a cluster
$C_t, \overline{C}_t$	Set of all clusters and set of valid clusters at time instance t
$\|c\|$	Size of cluster
$s_1, s_2, \ldots, s_n$	Comma-separated n sentences
$e_1, e_2, \ldots, e_n$	Embeddings of comma-separated sentences $s_1, s_2, \ldots, s_n$ resp.
E_t	Set of embeddings $e_1, e_2, \ldots, e_n$
S_t	Set of sentences $s_1, s_2, \ldots, s_n$
EM	Embedding Model (Sentence Transformer)
CA	Clustering Algorithm

Table 1 provides the notations, and the following present definitions of the key terminologies used throughout this paper.

Definition 1 Sensitive attributes. *An attribute whose value for any individual must be kept confidential from those who do not have direct access to the original data.*

Definition 2 Quasi-identifier. *A set of non-sensitive attributes* $\{Q_1, \cdots, Q_w\}$ *in a knowledge graph (KG) that can be linked with external data to uniquely identify at least one individual in the general population.*

Let us denote the set of all quasi-identifiers by QI.

Definition 3 ***k*-anonymity.** *A knowledge graph G satisfies k-anonymity if for every individual $t \in T$, there exist $k-1$ other individuals $t_1, t_2, \ldots, t_{k-1} \in T$ such that $t[C] = t_1[C] = t_2[C] = \cdots = t_{k-1}[C]$ for all $C \in QI$.*

The q^*-block represents the set of individuals' data in $\overline{G}$ where the quasi-identifier values generalize to q^*.

Definition 4 ***l*-diversity.** *A q^*-block is l-diverse if it contains at least l "well-represented" values for the sensitive attribute S. A KG is l-diverse if every q^*-block is l-diverse.*

To evaluate the quality of anonymized KGs, we measure the average information loss of users in anonymized KGs. We utilized and modified the evaluation metrics (AL, AIL and AAIL) proposed in [12] to adapt our setting.

Definition 5 Information Loss. *The Average Attribute Information Loss (AAIL) quantifies the information loss in an anonymized KG $\overline{G_t}$ by comparing user attribute values in $\overline{G_t}$ with their original counterparts in G_t. Higher differences indicate greater loss, normalized to [0,1]. AAIL is defined as:*

$$AL(G_t, \overline{G_t}, u) = \frac{1}{|R_t^{UA}|} \sum_{r_a \in R_t^{UA}} \frac{|\overline{I_t^a}(r_a, u) - I_t^a(r_a, u)|}{|dom_t(r_a) - I_t^a(r_a, u)|},$$

$$AAIL(G_t, \overline{G_t}) = \frac{1}{|\overline{V}_t^U|} \sum_{u \in \overline{V}_t^U} AL(G_t, \overline{G_t}, u),$$

where AL (Attribute Loss) measures the loss for user u, $I_t^a(r_a, u)$ and $\overline{I_t^a}(r_a, u)$ are the values of attribute r_a for u in G_t and $\overline{G_t}$, respectively, and $dom_t(r_a)$ is the domain of r_a in G_t. R_t^{UA} is a set of user-to-attribute relationship types in G_t.

As our approach adds fake individuals and removes some users, we use a further metric AIL (Average Information Loss) to measure not only the average information loss of anonymized users but also that of fake/removed ones. AIL measures normalized information loss of attributes and the loss of each fake/removed user is measured as 1. Finally, AIL takes the average of all the measured information loss of all users who are in both G_t and $\overline{G_t}$

$$AIL(G_t, \overline{G_t}) = \frac{1}{|U_t^{\cup}|} \sum_{u \in U_t^{\cup}} \begin{cases} AL & \text{if } u \in U_t^{\cap}, \\ 1 & \text{otherwise}, \end{cases}$$

where $U_t^{\cup} = \overline{V}_t^{\cup} \cup V_t^{\cup}$ and $U_t^{\cap} = \overline{V}_t^{\cup} \cap V_t^{\cup}$.

5 Semantic Anonymization of TKGs

In this section, we explain our approach in stepwise manner. First, we create *valid groups* that satisfy both k-anonymity and l-diversity. To enhance clarity, we provide both an algorithmic explanation (Algorithm 1) and a corresponding flowchart (Fig. 3). In the subsequent stage, we focus on selecting the optimal values of k and l by formulating a multi-objective optimization problem.

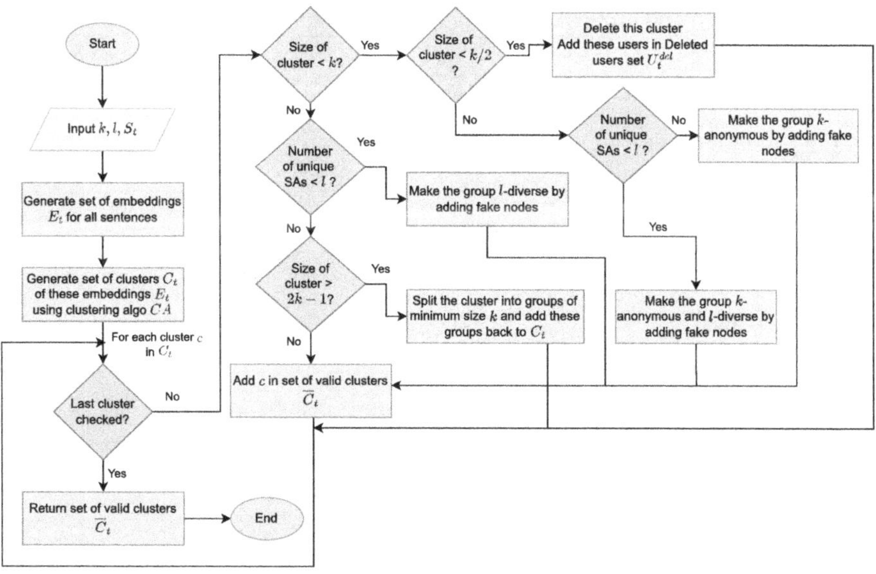

Fig. 3. Flowchart of Valid Groups Formation algorithm (Color figure online)

Creation of Valid Groups. In this subsection, we explain the implementation of forming *valid groups* with the aid of an algorithm and a flowchart (Fig. 3). *Valid groups* refer to groups of individuals with a minimum size of k and a maximum size of $2k-1$, where the individuals within each group possess l unique sensitive attributes. As depicted in the flowchart (Fig. 3), which corresponds to Algorithm 1, the process begins by receiving k, l, and S_t (a set of comma-separated quasi-identifier sentences $s_1, s_2, \ldots, s_n$) as inputs (represented by the yellow parallelogram). Using a sentence transformer, the algorithm generates E_t, a set of embeddings $e_1, e_2, \ldots, e_n$, corresponding to the embeddings of the sentences $s_1, s_2, \ldots, s_n$ (step written in blue rectangle). To form groups of embeddings representing similar quasi-identifier sentences, we make use of clustering algorithm (denoted as CA) [17] such as K-Medoids. This gives clusters of varying sizes, each containing quasi-identifiers that are semantically similar—i.e., individuals within each cluster share semantically similar characteristics. To ensure that these clusters conform to the size constraints of k to $2k-1$ and contain l diverse sensitive attributes, we iterate over all clusters and make necessary adjustments (represented in red diamond). For every cluster $c \in C_t$, we first check if the size of the cluster is less than k, then we check if the number of unique SAs are less than l, and lastly, we check if the size of the cluster is greater than $2k-1$ (represented by three vertical red diamonds). According to whatever condition passed, we perform the next steps of adding necessary fake nodes and making the cluster/group valid (k-anonymous and l-diverse) or deleting nodes. The valid clusters/groups are added into $\overline{C_t}$. The flowchart concludes when a set of *valid clusters*, $\overline{C_t}$, is obtained, satisfying both k-anonymity and l-diversity.

Algorithm 1 Valid groups creation

```
Input: k, l, S_t
Output: \overline{C_t}
1: E_t ← EM(S_t)
2: C_t ← CA(E_t, k)
3: for c ∈ C_t do
4:     if |c| < k then
5:         if |c| < k/2 then:
6:             C_t^t ← C_t^t − c
7:             U_t^{del} ← U_t^{del} ∪ U_t^c
8:         else:
9:             if |sen(c)| < l then:
10:                n_{fake} = max(l − |sen(c)|, k − |c|)
11:                \bar{c} ← makeValidCluster(c, n_{fake})
12:                \overline{C_t} ← \overline{C_t} ∪ \bar{c}
13:            else:
14:                n_{fake} = k − |c|
15:                \bar{c} ← makeValidCluster(c, n_{fake})
16:                \overline{C_t} ← \overline{C_t} ∪ \bar{c}
17:            end if
18:        end if
19:    else if |sen(c)| < l then
20:        n_{fake} = l − |sen(c)|
21:        \bar{c} ← makeValidCluster(c, n_{fake})
22:        \overline{C_t} ← \overline{C_t} ∪ \bar{c}
23:    else if |c| > 2 * k − 1 then
24:        C_t^{split} ← splitGroup(c, k)
25:        C_t ← C_t ∪ C_t^{split}
26:    else
27:        \overline{C_t} ← \overline{C_t} ∪ c
28:    end if
29: end for
30: return \overline{C_t}
```

Initially, the algorithm computes a set of embeddings E_t for all sentences $s_1, s_2, \ldots, s_n$ containing quasi-identifiers (line 1) using EM, a sentence embedding model. Subsequently, clustering algorithm CA is applied to generate clusters of similar sentences (line 2), effectively grouping data based on similar quasi-identifiers. These clusters in set C_t, may not inherently satisfy k-anonymity and l-diversity. The algorithm then iterates over each cluster $c \in C_t$, verifying whether the cluster is valid or not (lines 3–29). If the cluster size $|c|$ is less than k (line 4) and the number of unique sensitive attributes $|sen(c)|$ is less than l, the algorithm adds a maximum of $k - |c|$ and $l - |sen(c)|$ fake entities to the cluster to satisfy both k-anonymity and l-diversity (line 11). If the l-diversity condition is already met, the algorithm only adds $k - |c|$ fake entities to ensure the cluster is valid (line 15). However, if $|c|$ is less than $k/2$, the algorithm removes the cluster and transfers its members to U_t^{del} to prevent introducing more than $k/2$ fake entities (line 6). If the k condition is satisfied but the l condition is not, the algorithm adds $l - |sen(c)|$ fake nodes and makes that cluster valid (line 21). Here, the algorithm ensures that the fake nodes introduced possess sensitive values distinct from those already present, thereby fulfilling the l condition. If the cluster size $|c|$ exceeds $2k - 1$, the algorithm splits the cluster into $|c|/k$ parts and adds them to C_t for re-evaluation. If none of the above conditions satisfies, that means the cluster is already k-anonymous and l-diverse. We directly add c into the set of valid clusters $\overline{C}_t$. Finally, the algorithm returns $\overline{C}_t$.

Selection of Optimal *k* and *l* Values. The second part of our contribution focuses on determining the optimal pair of values for k and l. For this purpose, we formulate a Multi-Objective Optimization Problem (MOOP), that optimizes three functions and gives us the values of k and l that yield the optimal results across all three functions. In this Multi-Objective Optimization Problem (MOOP), one of the objective function is to maximize cosine-similarity [19] among QID sentences within a group so that the properties of individuals in that group are as similar as possible. Second objective function is to minimize cosine-similarity among sensitive attributes embeddings to avoid semantic similarity attack (explained in Sect. 3). It is important to note that the addition of fake nodes and the removal of outlying nodes can lead to significant information loss (refer to Sect. 4 Definition 5). Hence, we also include minimization of average information loss as part of the optimization criteria.

We first calculate values of all three objective functions for various combinations of k and l. Then, we employ the NSGA-2 [5] algorithm to efficiently handle the trade-offs between the values of these conflicting objectives.

$$\text{Maximize } SE(k,l) = \frac{1}{|G_v|} \sum_{g \in G_v} \frac{1}{\binom{|g|}{2}} \sum_{i=0}^{|g|} \sum_{j=i}^{|g|} \cos\_\text{sim}(QEmb(i), QEmb(j))$$

$$\text{Minimize } SA(k,l) = \frac{1}{|G_v|} \sum_{g \in G_v} \frac{1}{\binom{|g|}{2}} \sum_{i=0}^{|g|} \sum_{j=i}^{|g|} \cos\_\text{sim}(SAEmb(i), SAEmb(j))$$

$$\text{Minimize } AIL(G_t, \overline{G_t}) = \frac{1}{|U_t^{\cup}|} \sum_{u \in U_t^{\cup}} \begin{cases} AL & \text{if } u \in U_t^{\cap} \\ 1 & \text{otherwise} \end{cases}$$

Subject to $l \leq k$.

where k: valid variable for k-anonymity, l: valid variable for l-diversity, G_t: original KG, $\overline{G_t}$: anonymized KG using k and l, G_v: set of valid groups created using k and l. Refer to Sect. 4, Definition 5 for Information Loss calculation.

6 Evaluation

This section provides an overview of the experimental setup, datasets, and methodologies used to evaluate the proposed approach. The code for our experiments is available in our GitHub repository[1].

Experimental Setup and Datasets. Since temporal snapshots are not readily available, we generated them using existing temporal datasets. To evaluate the effectiveness of our algorithm, we utilized three real-world datasets containing timestamps for facts: YAGO2 [13], ICEWS14 [7], and Freebase [2]. These datasets include temporal information, where each triple is associated with specific dates. For example, in YAGO2, the facts are of the form: ⟨ Jamie O'Hara, playsFor, Arsenal F.C., occursSince, 1998⟩, ⟨ Jamie O'Hara, playsFor, Arsenal F.C., occursUntil, 2003⟩. Our objective was to create time slices (snapshots) based on these dates using the "Mean" method, as described in [12]. In the "Mean" method [12], snapshots are generated by grouping time instances (e.g., 1991, 1993, etc.) such that the standard deviation of the number of edges in each time-instance graph (time-snapshot) is minimized. The total number of time instances (years) must exceed the desired number of snapshots, ensuring that data from at least one year is included in each snapshot. After generating the snapshots, we transform all attributes of individuals from the triples in each snapshot into comma-separated sentences, treating all attributes as quasi-identifiers. These sentences are labeled as $s_1, s_2, \ldots, s_n$. Each snapshot thus contains a collection of these sentences corresponding to their respective years.

Impact on the Average Information Loss. In this section, we evaluate the impact of our approach on Time-Aware Knowledge Graphs and compare its performance with the (k, l)-Sequential Attribute Degree ((k, l)-sad) method [12]. Figure 4 shows that as k increases, the Average Information Loss (AIL) generally decreases. For $k = 2$, the AIL remains stable around 0.26 across snapshots. As k increases to 4, 6, 8, and 10, the AIL decreases, with $k = 10$ exhibiting the lowest values (below 0.18). This indicates that higher k values improve privacy while reducing information loss, although the effect varies by snapshot. The higher AIL for $k = 2$ arises because clusters smaller than $k/2$ are discarded. Since the number of clusters equals the number of individuals divided by k, smaller k (e.g., $k = 2$) results in more clusters, increasing the likelihood of singletons. These singleton clusters are deleted, raising AIL. Introducing a threshold for the maximum allowable distance between a singleton and existing clusters could reassign such points, reducing deletions and lowering AIL, though this might compromise privacy. Figures 6 and 7 compare AIL values from our method (blue line) with those of (k, l)-sad (orange line) across snapshots. Our method consistently achieves lower AIL values. While (k, l)-sad performs better at $t = 0$, its AIL increases significantly from $t = 1$ onwards due to the deletion of nodes from

[1] https://github.com/naikprachi27/SemanticAnonymization2/.

invalid clusters. With a threshold of 0, these nodes cannot be reassigned, leading to higher AIL. For details on the addition of fake nodes and node removal in both methods, refer to Sect. 3. The reason for sharp drop at $t = 8$ in Fig. 7 is the property of (k, l)-sad to assign re-inserted individuals to their old clusters with same signature and attributes, minimizing discarded and fake entities.

Figure 5 illustrates the AIL values generated by four sentence embedding models: `all-MiniLM-L6-v2`, `paraphrase-MiniLM-L12-v2`, `distilbert-base-nli-stsb-mean-tokens`, and `all-mpnet-base-v2`. Most models exhibit similar trends, with slight variations in AIL depending on the k and l combinations. Increasing l generally stabilizes AIL values across models, with reduced variability at higher l values. Figures 8 and 9 present the results of the Multi-Objective Optimization Problem (MOOP) solved using NSGA-2 (Non-Dominated Sorting Genetic Algorithm II) [5]. After running the Valid Groups Creation Algorithm (Sect. 5.1) across various k and l combinations, we computed QID sentence embeddings (SE), sensitive attribute (SA) similarities, and AIL values (see Sect. 5.2). The results identified optimal k and l values where the objective functions achieved the best trade-offs, balancing privacy and data utility.

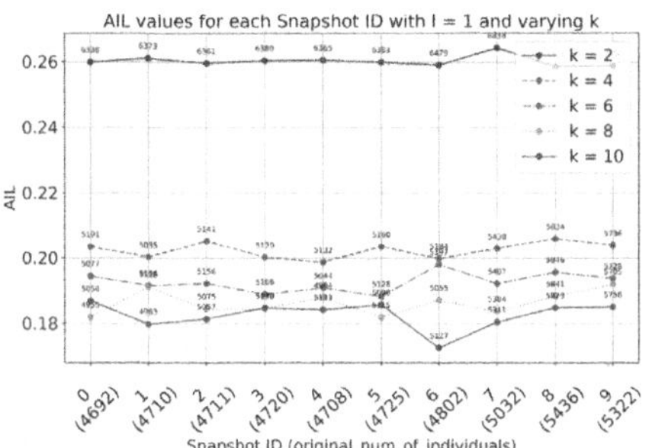

Fig. 4. AIL differences with variable k and $l = 1$ (Color figure online)

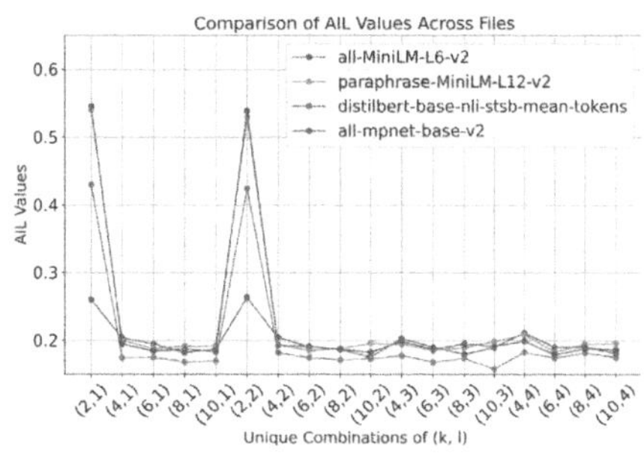

Fig. 5. Comparison between different embedding models (Color figure online)

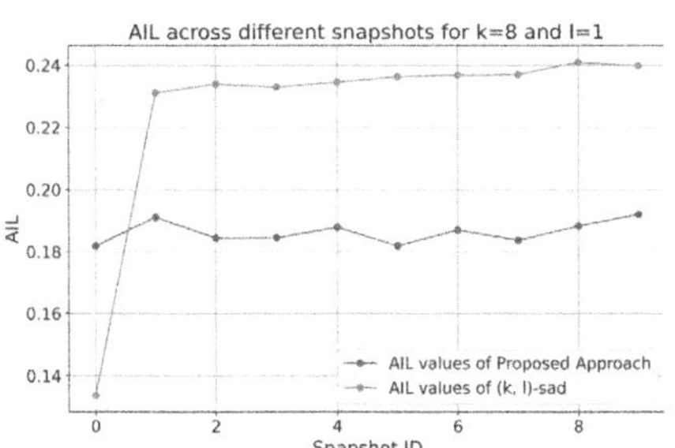

Fig. 6. AIL difference when $k = 8$ and $l = 1$ (YAGO) (Color figure online)

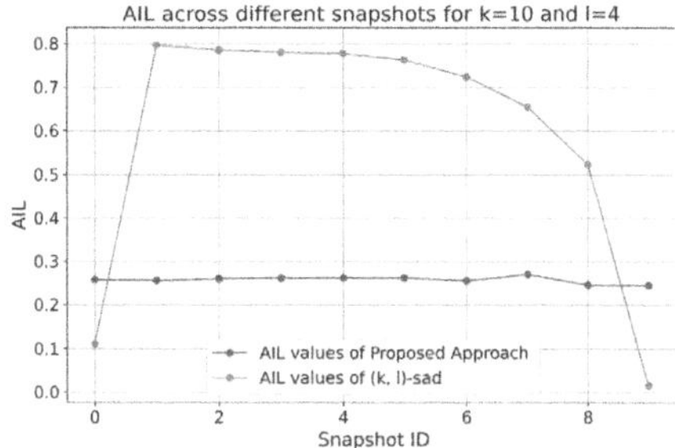

Fig. 7. AIL difference when $k = 10$ and $l = 4$ (Freebase) (Color figure online)

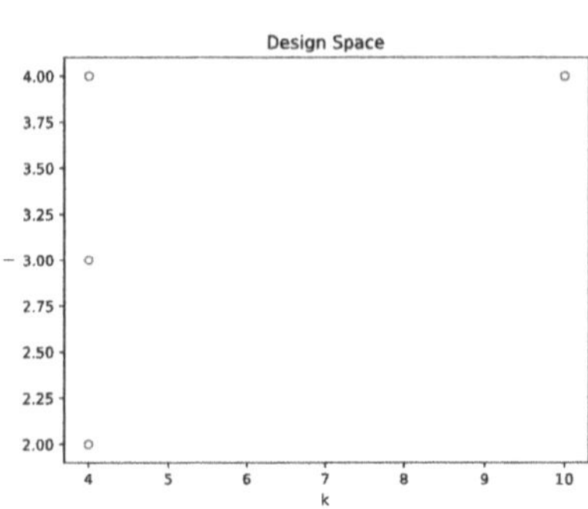

Fig. 8. Design Space (Color figure online)

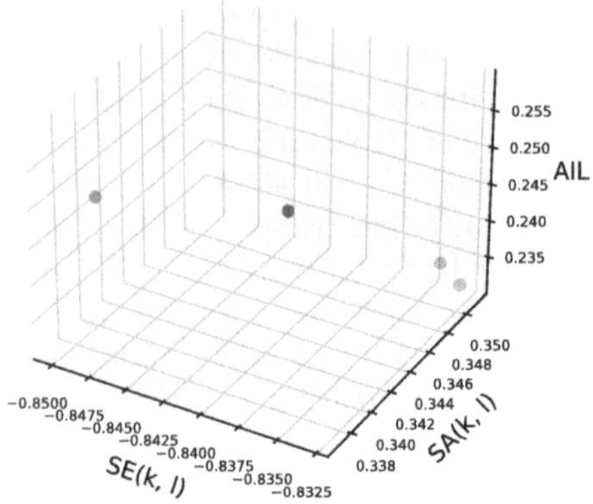

Fig. 9. Objective Space (Color figure online)

Figure 8 (Design Space) maps the k and l combinations explored during optimization. Red circles indicate Pareto-optimal solutions, where no objective can be improved without compromising another. These points represent the most effective trade-offs between conflicting goals, providing the optimal k and l combinations. Figure 9 (Objective Space) visualizes the trade-offs between the three objectives: SE similarities (QID sentence embeddings), SA similarities (sensitive attributes), and AIL. The 3D plot highlights the Pareto-optimal solutions, showing how the objectives are balanced across different k and l combinations.

Impact on the Performance of KG Embedding Models. In this section, we present the impact of our anonymization approach on the performance metrics of embedding models. We trained three Knowledge Graph Embedding models- TransE [3], DistMult [38] and ComplEx [34], on both anonymized and non-anonymized datasets (Table 2), assessing their performance using Mean Reciprocal Rank (MRR), Hits@1, Hits@5, Hits@10 metrics.

Table 2. Comparison of original and anonymised datasets across Train, Test, and Valid data.

Dataset	Train Data						Test Data		Valid Data	
	#Original Triples	Discarded Individuals	Fake Nodes	Total Individuals	Total Triples	Invalid Triples	Individuals	Triples	Individuals	Triples
Anonymised FB13	50,682	5,520	6,518	51,680	392,914	2,860	14,011	18,559	4,309	4,633
Original FB13	50,682	0	0	50,682	138,631	0	14,011	18,559	4,309	4,633
Anonymised YAGO	55,017	4,885	7,691	57,823	545,317	298	2,930	3,066	2,928	3,042
Original YAGO	55,017	0	0	55,017	150,833	0	2,930	3,066	2,928	3,042
Anonymised ICEWS14	5,395	285	761	5,871	204,966	80	1,781	7,056	1,835	6,933
Original ICEWS14	5,395	0	0	5,395	46,275	0	1,781	7,056	1,835	6,933

During the anonymization process, each individual in a group of size k shares the group's attributes. Some individuals may need to be removed, and synthetic individuals may be introduced to maintain the required group characteristics. Table 3 shows low performance values due to a large number of corrupted triples, ensuring true test triples rank highest [3,34]. More corruptions make ranking harder, lowering MRR and Hits@n without implying poor performance. For example, MRR=0.01 (avg. rank = 100) is strong with 1M corruptions but weak with only 100. To maximize corruptions, we replaced both subject and object with all unique test entities, doubling the number of corrupted triples.

Table 3. Performance comparison of models on FB13, YAGO, and ICEWS14 datasets with Normal and Anonymised data.

Model	Dataset	MRR		Hits@1		Hits@5		Hits@10	
		Normal	Anon.	Normal	Anon.	Normal	Anon.	Normal	Anon.
TransE	FB13	0.1541	0.1891	0.1272	0.1535	0.1756	0.2207	0.2076	0.2557
DistMult	FB13	0.0036	0.0582	0.0011	0.0484	0.0043	0.0675	0.0074	0.0757
ComplEx	FB13	0.0052	0.0673	0.0020	0.0599	0.0061	0.0734	0.0100	0.0787
TransE	YAGO	0.0147	0.0841	0.0067	0.0662	0.0225	0.0990	0.0259	0.1115
DistMult	YAGO	0.0071	0.0345	0.0039	0.0243	0.0078	0.0435	0.0124	0.0523
ComplEx	YAGO	0.0092	0.0504	0.0064	0.0450	0.0099	0.0533	0.0134	0.0575
TransE	ICEWS14	0.2496	0.4219	0.0094	0.3344	0.5405	0.5114	0.6592	0.6035
DistMult	ICEWS14	0.3738	0.3520	0.2686	0.2771	0.4892	0.4196	0.5870	0.5157
ComplEx	ICEWS14	0.4593	0.4492	0.3665	0.3730	0.5675	0.5298	0.6385	0.6067

Experiments on three datasets (refer Table 3) showed that anonymized datasets generally outperformed non-anonymized ones in performance metrics, with MRR increasing by 455% and Hits@10 by 288%, suggesting that anonymization preserves data utility. The magnitude of performance improvements varies

Table 4. Mean scores for Normal and Anonymised datasets using TransE, DistMult, and ComplEx models.

Data Type	Dataset	TransE		DistMult		ComplEx	
		Model	Mean Score	Model	Mean Score	Model	Mean Score
Normal	FB13	TransE	−5.0963	DistMult	0.0039	ComplEx	−0.0135
Anonymised	FB13	TransE	−6.8080	DistMult	−0.0855	ComplEx	−0.0756
Normal	YAGO	TransE	−5.0691	DistMult	0.0836	ComplEx	0.0473
Anonymised	YAGO	TransE	−5.2249	DistMult	0.1045	ComplEx	0.0569
Normal	ICEWS14	TransE	−4.4923	DistMult	1.2617	ComplEx	1.0412
Anonymised	ICEWS14	TransE	−5.0696	DistMult	1.1616	ComplEx	0.0000

due to differences in dataset characteristics, the number of corrupted triples, and model-specific properties. However, as shown in Table 4, the anonymized datasets had an average 208% decrease in confidence. This shows that the model's ability to distinguish between similar groups of k individuals with shared attributes decreases, reflected in confidence score, as it loses access to unique relations for each individual. For ranking-related queries, the model performs well due to relative ranking. Both ranking and confidence should be considered to assess the quality of the model. This trade-off highlights the need for an ideal anonymization process that enhances performance while maintaining confidence in the results. The application of a confidence threshold could improve the reliability of the ranking.

7 Conclusion and Future Work

In this work, we have introduced *Semantic Anonymization* of Time-aware Knowledge Graphs in which, we incorporated sentence embeddings into the anonymization process to make use of meaning behind the quasi-identifiers and sensitive attributes and hence enhancing the effectiveness of k-anonymity and l-diversity techniques. We also semantically diversified the sensitive attributes within a group of k to avoid *semantic similarity attacks*. Additionally, we formulated a Multi-Objective Optimization Problem (MOOP) to find optimal pairing of k and l values with three objectives- increasing closeness among groups of k individuals, increasing dissimilarity among sensitive attributes within a group of k individuals and minimising average information loss. Our approach outperformed previous methods, demonstrating a reduction in AIL by 56%. Moreover, experiments showed that KG embedding models trained on anonymized KGs achieved improved performances, preserving privacy while maintaining data utility. We further plan to extend this work by introducing t-closeness [20] in semantic anonymization which will protect from *bias attacks*, which occurs when an attacker exploits the distribution of sensitive attributes within a dataset to infer sensitive information about individuals. Another extension is the support of multiple sensitive attributes, allowing data providers to specify multiple sensitive attributes (for example disease and ethnicity). We can also extend this work by including a threshold variable, which will let the user choose the privacy level of the anonymization.

References

1. Bakhshi, M., Nematbakhsh, M., Mohsenzadeh, M., Rahmani, A.: Sparseqa: sequential word reordering and parsing for answering complex natural language questions over knowledge graphs. Knowl.-Based Syst. **235**, 107626 (2021). https://doi.org/10.1016/j.knosys.2021.107626
2. Bollacker, K., Evans, C., Paritosh, P., Sturge, T., Taylor, J.: Freebase: a collaboratively created graph database for structuring human knowledge. In: SIGMOD 2008, pp. 1247–1250. Association for Computing Machinery, New York (2008). https://doi.org/10.1145/1376616.1376746

3. Bordes, A., Usunier, N., Garcia-Durán, A., Weston, J., Yakhnenko, O.: Translating embeddings for modeling multi-relational data. In: Proceedings of the 26th International Conference on Neural Information Processing Systems, NIPS 2013, vol. 2, pp. 2787–2795. Curran Associates Inc., Red Hook (2013)
4. Dasgupta, S.S., Ray, S.N., Talukdar, P.: HyTE: hyperplane-based temporally aware knowledge graph embedding. In: Proceedings of the 2018 Conference on Empirical Methods in Natural Language Processing, pp. 2001–2011. Association for Computational Linguistics, Brussels, Belgium (2018). https://doi.org/10.18653/v1/D18-1225
5. Deb, K., Pratap, A., Agarwal, S., Meyarivan, T.: A fast and elitist multiobjective genetic algorithm: NSGA-II. IEEE Trans. Evol. Comput. **6**(2), 182–197 (2002). https://doi.org/10.1109/4235.996017
6. Delanaux, R., Bonifati, A., Rousset, M.-C., Thion, R.: Query-based linked data anonymization. In: Vrandečić, D., et al. (eds.) ISWC 2018. LNCS, vol. 11136, pp. 530–546. Springer, Cham (2018). https://doi.org/10.1007/978-3-030-00671-6_31
7. García-Durán, A., Dumancic, S., Niepert, M.: Learning sequence encoders for temporal knowledge graph completion. CoRR abs/1809.03202 (2018)
8. Goldsteen, A., Ezov, G., Shmelkin, R., Moffie, M., Farkash, A.: Anonymizing machine learning models (2020). https://doi.org/10.48550/arXiv.2007.13086
9. Henriksen-Bulmer, J., Jeary, S.: Re-identification attacks—a systematic literature review. Int. J. Inf. Manag. **36**(6, Part B), 1184–1192 (2016). https://doi.org/10.1016/j.ijinfomgt.2016.08.002. https://www.sciencedirect.com/science/article/pii/S0268401215301262
10. Hoang, A.-T., Carminati, B., Ferrari, E.: Cluster-based anonymization of knowledge graphs. In: Conti, M., Zhou, J., Casalicchio, E., Spognardi, A. (eds.) ACNS 2020. LNCS, vol. 12147, pp. 104–123. Springer, Cham (2020). https://doi.org/10.1007/978-3-030-57878-7_6
11. Hoang, A.T., Carminati, B., Ferrari, E.: Privacy-preserving sequential publishing of knowledge graphs, pp. 2021–2026 (2021). https://doi.org/10.1109/ICDE51399.2021.00194
12. Hoang, A.T., Carminati, B., Ferrari, E.: Time-aware anonymization of knowledge graphs. ACM Trans. Priv. Secur. (2022). https://doi.org/10.1145/3563694
13. Hoffart, J., Suchanek, F.M., Berberich, K., Weikum, G.: Yago2: a spatially and temporally enhanced knowledge base from Wikipedia. Artif. Intell. **194**, 28–61 (2013). https://doi.org/10.1016/j.artint.2012.06.001. Artificial Intelligence, Wikipedia and Semi-Structured Resources
14. Ji, S., Mittal, P., Beyah, R.: Graph data anonymization, de-anonymization attacks, and de-anonymizability quantification: a survey. IEEE Commun. Surv. Tutor. **19**(2), 1305–1326 (2017). https://doi.org/10.1109/COMST.2016.2633620
15. Jiang, T., et al.: Towards time-aware knowledge graph completion. In: Proceedings of COLING 2016, the 26th International Conference on Computational Linguistics: Technical Papers, pp. 1715–1724. The COLING 2016 Organizing Committee, Osaka, Japan (2016)
16. Jiang, T., et al.: Encoding temporal information for time-aware link prediction. In: Proceedings of the 2016 Conference on Empirical Methods in Natural Language Processing, pp. 2350–2354. Association for Computational Linguistics, Austin, Texas (2016). https://doi.org/10.18653/v1/D16-1260
17. Jin, X., Han, J.: K-Medoids Clustering, pp. 564–565. Springer, Boston (2010). https://doi.org/10.1007/978-0-387-30164-8_426

18. Kazim, E., Denny, D., Koshiyama, A.: AI auditing and impact assessment: according to the UK information commissioner's office. AI Ethics 1–10 (2021). https://doi.org/10.1007/s43681-021-00039-2
19. Lahitani, A.R., Permanasari, A.E., Setiawan, N.A.: Cosine similarity to determine similarity measure: study case in online essay assessment. In: 2016 4th International Conference on Cyber and IT Service Management, pp. 1–6 (2016). https://doi.org/10.1109/CITSM.2016.7577578
20. Li, N., Li, T., Venkatasubramanian, S.: t-closeness: privacy beyond k-anonymity and l-diversity. In: 2007 IEEE 23rd International Conference on Data Engineering, pp. 106–115 (2007). https://doi.org/10.1109/ICDE.2007.367856
21. Machanavajjhala, A., Gehrke, J., Kifer, D., Venkitasubramaniam, M.: L-diversity: privacy beyond k-anonymity. In: 22nd International Conference on Data Engineering (ICDE 2006), p. 24 (2006). https://doi.org/10.1109/ICDE.2006.1
22. Mahdisoltani, F., Biega, J.A., Suchanek, F.M.: Yago3: a knowledge base from multilingual Wikipedias. In: Conference on Innovative Data Systems Research (2015)
23. Majeed, A., Lee, S.: Anonymization techniques for privacy preserving data publishing: a comprehensive survey. IEEE Access (2020). https://doi.org/10.1109/ACCESS.2020.3045700
24. Oishi, K., Sei, Y., Tahara, Y., Ohsuga, A.: Semantic diversity: privacy considering distance between values of sensitive attribute. Comput. Secur. **94**, 101823 (2020). https://doi.org/10.1016/j.cose.2020.101823
25. Papadopoulou, A., Lison, P., Øvrelid, L., Pilán, I.: Bootstrapping text anonymization models with distant supervision. In: Proceedings of the Thirteenth Language Resources and Evaluation Conference, pp. 4477–4487. European Language Resources Association, Marseille, France (2022)
26. Powar, J., Beresford, A.R.: SoK: managing risks of linkage attacks on data privacy. Proc. Priv. Enhancing Technol. **2023**, 97–116 (2023). https://api.semanticscholar.org/CorpusID:257417122
27. Reimers, N., Gurevych, I.: Sentence-BERT: sentence embeddings using Siamese BERT-networks. In: Proceedings of the 2019 Conference on Empirical Methods in Natural Language Processing. Association for Computational Linguistics (2019)
28. Schwabe, D.: Trust and privacy in knowledge graphs. In: Companion Proceedings of The 2019 World Wide Web Conference, WWW 2019, pp. 722–728. Association for Computing Machinery, New York (2019). https://doi.org/10.1145/3308560.3317705
29. Suchanek, F.M., Kasneci, G., Weikum, G.: Yago: a core of semantic knowledge. In: Proceedings of the 16th International Conference on World Wide Web, WWW 2007, pp. 697–706. Association for Computing Machinery, New York (2007). https://doi.org/10.1145/1242572.1242667
30. Sweeney, L.: k-anonymity: a model for protecting privacy. Int. J. Uncertain. Fuzziness Knowl.-Based Syst. **10**(5), 557–570 (2002). https://doi.org/10.1142/S0218488502001648
31. Thompson, B., Yao, D.: The union-split algorithm and cluster-based anonymization of social networks, pp. 218–227 (2009). https://doi.org/10.1145/1533057.1533088
32. Thouvenot, M., Curé, O., Calvez, P.: Knowledge graph anonymization using semantic anatomization, pp. 4065–4074 (2020). https://doi.org/10.1109/BigData50022.2020.9377824
33. Trivedi, R., Dai, H., Wang, Y., Song, L.: Know-evolve: deep temporal reasoning for dynamic knowledge graphs. In: Precup, D., Teh, Y.W. (eds.) Proceedings of

the 34th International Conference on Machine Learning. Proceedings of Machine Learning Research, vol. 70, pp. 3462–3471. PMLR (2017)
34. Trouillon, T., Welbl, J., Riedel, S., Gaussier, E., Bouchard, G.: Complex embeddings for simple link prediction. In: Proceedings of The 33rd International Conference on Machine Learning. Proceedings of Machine Learning Research, vol. 48, pp. 2071–2080. PMLR, New York (2016)
35. Wang, H., Han, J., Wang, J., Wang, L.: (l, e)-diversity - a privacy preserving model to resist semantic similarity attack. J. Comput. **9**, 59–64 (2014)
36. Wang, K., Shen, Z., Huang, C., Wu, C.H., Dong, Y., Kanakia, A.: Microsoft academic graph: when experts are not enough. Quant. Sci. Stud. **1**(1), 396–413 (2020). https://doi.org/10.1162/qss_a_00021
37. Xu, C., Nayyeri, M., Alkhoury, F., Shariat Yazdi, H., Lehmann, J.: TeRo: a time-aware knowledge graph embedding via temporal rotation. In: Proceedings of the 28th International Conference on Computational Linguistics, pp. 1583–1593. International Committee on Computational Linguistics, Barcelona, Spain (Online) (2020). https://doi.org/10.18653/v1/2020.coling-main.139
38. Yang, B., Tau Yih, W., He, X., Gao, J., Deng, L.: Embedding entities and relations for learning and inference in knowledge bases (2015)
39. Yermilov, O., Raheja, V., Chernodub, A.: Privacy- and utility-preserving NLP with anonymized data: a case study of pseudonymization. In: Proceedings of the 3rd Workshop on Trustworthy Natural Language Processing (TrustNLP 2023), pp. 232–241. ACL, Toronto, Canada (2023). https://doi.org/10.18653/v1/2023.trustnlp-1.20
40. Zhang, L., Kang, Z., Sun, X., Sun, H., Zhang, B., Pu, D.: Kcrec: knowledge-aware representation graph convolutional network for recommendation. Know.-Based Syst. **230**(C) (2021). https://doi.org/10.1016/j.knosys.2021.107399
41. Zhou, B., Pei, J.: Preserving privacy in social networks against neighborhood attacks. In: 2008 IEEE 24th International Conference on Data Engineering, pp. 506–515 (2008). https://doi.org/10.1109/ICDE.2008.4497459
42. Zou, X.: A survey on application of knowledge graph. **1487**(1), 012016 (2020). https://doi.org/10.1088/1742-6596/1487/1/012016

Training-Free Score Calibration for Complex Query Decomposition

Simon Ott[1(✉)], Melisachew Wudage Chekol[2], Christian Meilicke[3], and Heiner Stuckenschmidt[3]

[1] AIT Austrian Institute of Technology, Vienna, Austria
simon.ott@ait.ac.at
[2] Utrecht University, Utrecht, The Netherlands
[3] University of Mannheim, Mannheim, Germany

Abstract. Answering complex queries on incomplete knowledge graphs poses significant challenges, as models must infer their answers despite gaps in the available data. Previous research has addressed this problem by developing end-to-end architectures specifically designed for complex query answering. These models are difficult to interpret and require extensive data and computational resources for training. Alternatively, some approaches have focused on leveraging existing neural link predictors, which have been designed for simple queries, to handle complex queries. This approach reduces the amount of training examples needed and offers more transparent reasoning. However, the output scores of the neural link predictors may require calibration for effective interaction during the reasoning process and a special adaption function has to be learned to achieve this. In this work, (i) we show that depending on the query type, standard normalization methods are equally as effective as learning an adaption function. (ii) Furthermore, we replace the neural link predictor with a rule-based approach that does not require any score calibration. With such an approach we achieve new state-of-the-art results and increase the mean reciprocal ranks from 35.1% to 37.1% averaged across datasets and query types. (iii) We conduct comprehensive empirical analysis to support our claims (The code and data for all our experiments can be accessed here: https://figshare.com/s/4f1fbd5f5d2c4aca7c2e).

1 Introduction and Related Work

Knowledge graphs (KGs) are structured representations of information where data is organized as a network of entities and the relationships between them. KGs offers a flexible and powerful way to model complex real-world knowledge, making KGs valuable across diverse domains. Some KGs might be designed to describe a rather specific domain with a restricted vocabulary and a small set of facts, while other KGs may encompass millions of entities and billions of facts. Examples include general-purpose knowledge bases like DBpedia [3], Freebase [7], Wikidata [38], and YAGO [33], which cover a broad range of domains.

E. Curry et al. (Eds.): ESWC 2025, LNCS 15718, pp. 188–207, 2025.
https://doi.org/10.1007/978-3-031-94575-5_11

In more specialized areas, KGs such as Hetionet [18] and Bio2RDF [13] support life sciences, while WordNet [22] focuses on linguistic relationships.

All of these KGs have in common that they might not be a complete description of the domain that they try to describe [12]. Even if we restrict the domain description to the vocabulary already used in the KG, many triples will still be missing. The task of knowledge graph completion, often interchangeably referred to as link prediction, is a well-studied machine learning task. It is usually framed as the task of predicting missing relationships between the entities (partially) described in the KG. Suppose for example that our KG contains information about movies, actors and other related entities. Typical queries in KG completion might include the following:

Which movies are categorized under the genre *comedy*? (1)

Which movies has *Keanu Reeves* starred in? (2)

Note that the KG can be incomplete. It might already have some movies belonging to the genre *comedy*, while some movies might be missing. The same might be the case for the movies that Keanu Reeves starred in. The answer to these queries is usually an ordered list of candidate entities. A KG completion model orders these candidates with the help of a function, which assigns a real-valued score to each candidate. Aside from the order induced by these scores, most of these models do not associate any semantic meaning with them, which will be important later on.

In recent years, this task has been extended to complex query answering, which involves addressing complex queries that combine various operators, such as chains of link predictions (projections), intersections, and unions. With respect to our previous example, the following illustrates an intersection query:

In which *comedy* movies has *Keanu Reeves* starred in? (3)

The task of complex query answering would be trivial if the knowledge graph would be complete. The queries can just be matched to the knowledge graph to retrieve the query results. With respect to query (3), we would query the knowledge graph for all comedy movies and for all movies in which *Keanu Reeves* starred in. The result set of the complex query would be the intersection of the result sets of the simple queries.

However, as argued above, knowledge graphs frequently exhibit incompleteness due to various factors. Thus, it is not possible to solve the problem by directly combing the answer of two simple queries, which return only results that are already stated in the KG. Instead, there are two possible and clearly different ways to solve the problem.

One possible approach is based on an end-to-end architecture that learns to answer complex queries by training on different complex query structures directly [17,28,29]. Several approaches achieve this by embedding queries into the same vector space as entities, ensuring that entities answering the query are positioned close to the query embedding [10,16,17]. Variants of this method

Table 1. Some candidate predictions by a neural link predictor for two simple queries *genre(?,comedy)* and *acted(?,kreeves)* which are the building blocks of Query (3). Candidates which are already stated in the given knowledge graph are marked with a ✓.

Candidate	genre(?,comedy)	acted(?,kreeves)
Thumbsucker	✓	✓
Borat	✓	0.376
Waynes World	2.767	0.421
Bill & Ted's Adventure	0.922	1.656
Matrix Reloaded	0.283	✓

embed queries as geometric shapes, such as boxes [28] or cones [42]; or as probability distributions [29]. GNN-QE [43] modifies a method for knowledge graph completion [45]. They learn to answer complex queries by using them as examples to train a combination of graph neural networks and fuzzy logic operators. Approaches that follow this paradigm require training a resource-intensive model, which requires extensive computational resources and large amounts of training data. Moreover, it lacks explainability as it does not generate intermediate results in terms of answers to the simple queries, which are the building blocks of the complex query. An alternative approach is called complex query decomposition (CQD) [1,23]. Within such an approach in a first step the complex query is decomposed into several simple queries. These queries can then be answered, in a second step, by any standard knowledge graph completion model. The results of the simple queries need to be aggregated in a third step using fuzzy operators for negation, intersection and union. However, the scores of neural models are not directly trained to interact with each other nor can they be interpreted as probabilities. We can illustrate the resulting problem when aggregating these scores with an example shown in Table 1. Within this table we see hypothetical results for the two queries: (a) movies that belong to the genre comedy and (b) all movies in which Keanu Reeves starred in. The table lists some candidates that are already stated in the KG (marked with a ✓) together with candidates proposed by a neural KG completion method (numerical score). The values in the table illustrate that it is a non-trivial task to compute an aggregated score that allows to entail a ranking for the complex query. How should we, for example, combine the two values in the third or fourth row? And which candidate should be ranked higher? How do we deal with cases, where a candidate is stated for one of the two queries (second and fifth rows)?

In [2], the authors tackle these problems by learning special score *adaption* model, which they call CQD^{A} [2]. This model is applied to the scores in Table 1 in order to normalize them. The normalized scores are then combined with the appropriate fuzzy operator. Contrary to the end-to-end learning approach, the results of complex queries remain explainable as the method can explain the results of a complex query in terms of the results of its building blocks. However, the approach requires still to learn a score normalization model that is specific for each dataset and each neural model that is used within the overall workflow.

As our first main contribution we propose an alternative to both approaches presented above: instead of training an end-to-end model or a model that learns a normalization function, we propose to apply standard normalization functions that can be applied to each neural model without any training. The results obtained by this approach are on par (43.0%) with the top results achieved by CQD^{A} [2] (42.9%) and greatly improve the results of plain CQD (36.4%) when comparing the mean reciprocal rank values averaged over all datasets and over positive queries (queries without negation); for queries containing negation CQD^{A} still seems to have a beneficial effect. This illustrates that the results published in [2] are partially misleading as they did not report the naive baseline that we propose in this paper.

While neural models require score calibration, there are also rule-based methods that generate scores with a probabilistic semantics [20,21]. In a second set of experiments we show that we obtain the best results for complex query answering, when we directly use the scores of rule-based methods without any calibration and increase the mean reciprocal rank values averaged over all datasets and query types including negations from 35.1% to 37.1% (42.9% to 45.5% not considering queries containing negation).

While many researchers conclude their papers with the insight that neural models might not be explainable but achieve better results compared to symbolic or classic machine learning methods, our experiments point to a different direction. Using a simple parameter-free aggregation method with an input generated by a fully explainable symbolic method achieves results that are better than or on par with current state of the art.

2 Preliminaries

Knowledge Graphs. A knowledge graph (KG) is a structured representation of information using entities and relationships. Formally, a KG is a tuple $\mathcal{G} = (\mathcal{E}, \mathcal{R}, \mathcal{T})$ where $\mathcal{E}$ denotes the set of entities/nodes, $\mathcal{R}$ denotes the set of relations, and $\mathcal{T}$ denotes the set of triples, where for every triple $(h, r, t) \in \mathcal{T}$, $h, t \in \mathcal{E}$ and $r \in \mathcal{R}$. Examples of triples are (*Max, livesIn, England*) or (*Dutch, language, Netherlands*). $\mathcal{T}$ can also be represented in First-Order Logic (FOL) as a set of atomic sentences involving binary predicates or facts, formally, $\{r(h,t) \mid (h,r,t) \in \mathcal{T}\}$.

2.1 Knowledge Graph Completion

The task of knowledge graph completion (KGC) involves predicting the tail entity for a given incomplete triple of the form $r(h, ?T)$ or the head entity for a triple of the form $r(?T, t)$. These represent atomic FOL queries, where $r \in \mathcal{R}$ is a relation, $?T$ denotes the unknown entity (or variable) to be predicted, $t \in \mathcal{E}$ is a known tail entity, and $h \in \mathcal{E}$ is a known head entity. This task is analogous to answering atomic FOL queries. For KGC involving atomic queries, specifically $r(h, ?T)$ and $r(?T, t)$, there are broadly two main approaches: neural link predictors and rule-based methods. These methods assign a score for a predicted fact. Since KGC involves predicting missing links ($?T$), it is also referred to as (neural) link prediction.

Neural Link Predictors. A plethora of neural link prediction approaches have appeared over the years which can mainly be divided in approaches utilizing graph neural networks (GNN) [35,44,45]) and knowledge graph embeddings (KGE) [4,8,11,25,34,37,41]). Approaches based on knowledge graph embeddings aim to learn vector representations of entities and relations by optimizing a specific scoring function $\varphi_r(h,t) = s(e_h, e_r, e_t)$ where s is the scoring function and $e_h, e_r, e_t \in \mathbb{R}^d$ embedding vectors in a d-dimensional space. A higher score of a triple indicates a greater likelihood that the model considers it true. For example TransE [8], an early but intuitive approach, defines the scoring function as negative distance between e_t and e_h translated by e_r. This results in learned embeddings that approximately satisfy the relation $e_h + e_r \approx e_t$.

Different training strategies and loss functions have been proposed for training neural link predictors [39]. All training strategies include creating negative samples from positive triples. One such strategy is negative sampling, where for each positive triple $r(h,t) \in \mathcal{T}$ a set of negatives is created by randomly corrupting either the head $\{r(h',t)|h' \in \mathcal{E}\}$ or the tail entity $\{r(h,t')|t' \in \mathcal{E}\}$. Other strategies include 1vsAll, where all entities of the graph are used to corrupt the head or tail position of the positive triple $r(h,t)$ and KvsAll, which is similar to 1vsAll with the difference that triples having the same link prediction query structure $r(h, ?T)$ or $r(?T, t)$ are merged together as a single task. A variety of loss functions can be used to train neural link predictors. Two of the commonly used loss functions are binary cross entropy (BCE) and cross entropy (CE). The targets (or labels) are set to 1 for positive triples and 0 for negative triples. The choice of a loss function can have a big impact on the model performance and depends on the choice of scoring function and the data [31].

Rule-Based Approaches. Rule-based methods for KGC identify explicit patterns and regularities within a knowledge graph. These patterns, expressed as rules, logical formulas, or clauses, can be learned from a KG. Various approaches have been developed to mine such rules, such as AMIE [14,15] or AnyBURL [20,21]. Examples of rules that can be learned using AnyBURL, an approach that has proven to have very high KGC performance [30], are

$$rl_1 : \text{speaks}(?X, ?Y) \leftarrow \text{lives}(?X, ?A), \text{lang}(?A, ?Y)$$
$$rl_2 : \text{speaks}(?X, d) \leftarrow \text{married}(?X, ?A), \text{born}(?A, a)$$

The rule rl_1 represents: 'if $?X$ lives in location $?A$, where language $?Y$ is spoken, then $?X$ also speaks $?Y$', while rule rl_2 represents: 'if $?X$ is married to $?A$, who was born in a (amsterdam), then $?X$ also speaks d (dutch). Here uppercase letters represent variables and lowercase letters represent constants. The left-hand side is called the head of the rule, while the right-hand side is called the

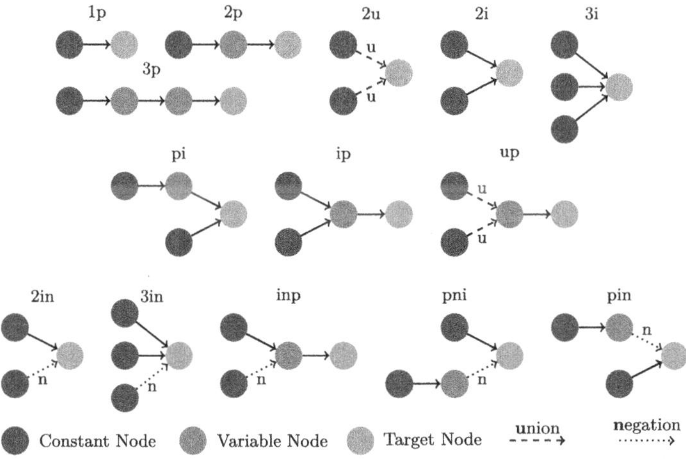

Fig. 1. The 14 different query structures present in the benchmark datasets.

body of the rule that consists of one or multiple body atoms. rl_1 is called a cyclic rule where both variables in the head are present in the body of the rule, while rl_2 is called an acyclic rule having a constant in the head and in the body.

In rule-based KGC, the predictive quality of rules can vary significantly. Some rules are highly specific, generating a limited number of correct triples, while more general rules may produce a larger set of triples with a lower proportion of correct predictions. To assess the performance of these rules, various metrics have been proposed, with rule confidence being one of the most commonly utilized. For a given rule rl, its confidence $\text{conf}(rl)$ is computed as follows:

$$\text{conf}(rl) = \frac{\text{TP}}{\text{TP} + \text{FP} + l_c}$$

TP (True Positives) refers to the number of triples correctly predicted by the rule that already exist in the KG. FP (False Positives) refers to triples predicted by the rule that do not exist in the KG. l_c is a small constant introduced to prevent overfitting, particularly for rules that generate only a few correct triples.

Given an atomic query $r(h, ?T)$, a set of rules can be applied to the knowledge graph to retrieve predictions (answers to $?T$). Since a particular predicted entity can be predicted by multiple rules, the confidences of those rules have to be aggregated. The aggregation of these confidences is an ongoing research [6,26,27]. The most canonical approach for aggregating the predictions of multiple rules is the maximum aggregation, where the confidence of an entity is that of the highest-confident rule that predicted it. As was shown in [21] this aggregation can be improved by recursively reranking all entities in a group of same-confident entities by the confidence of the next highest rule.

Table 2. Fuzzy query score aggregation using the Gödel and product semantics.

Gödel semantics	Product semantics
$\top_G : Q \rightarrow [0,1]$	$\top_p : Q \rightarrow [0,1]$
$\top_G(\neg Q) = 1 - \top_G(Q)$	$\top_p(\neg Q) = 1 - \top_p(Q)$
$\top_G(Q_1 \wedge Q_2) = \min(\top_G(Q_1), \top_G(Q_2))$	$\top_p(Q_1 \wedge Q_2) = \top_p(Q_1) \cdot \top_p(Q_1)$
$\top_G(Q_1 \vee Q_2) = \top_G(\neg(\neg Q_1 \wedge \neg Q_2))$	$\top_p(Q_1 \vee Q_2) = \top_p(\neg(\neg Q_1 \wedge \neg Q_2))$

2.2 Complex Query Answering

Complex queries can be constructed by combining atomic FOL queries using logical connectives such as conjunction ($\wedge$), disjunction ($\vee$), and negation ($\neg$). The syntax for constructing these complex queries is provided below:

$$Q := r(h, ?V) \mid r(?V_1, ?V_2) \mid \neg Q \mid (Q_1 \wedge Q_2) \mid (Q_1 \vee Q_2) \mid \exists ?V.Q$$

where $?V$, $?V_1$ and $?V_2$ denote variables, some of which are existentially quantified and one is projected. We also use the symbol $?T$ to denote the projected (target) variable which is the unknown entity to be predicted. Q, Q_1 and Q_2 are complex queries. Note that in the grammar, we disregard the order of variables and constants, which implies that we also consider the reversed order of variables and constants. In this work, we consider a subset of complex queries that have certain structures. The templates of these queries are given in Fig. 1. As an example, the template '1p' denotes the class of queries of the form: $r(h, ?T)$; and the template 'up' denotes the class of queries of the form: $\exists ?V.(r_1(h_1, ?V) \vee r_2(h_2, ?V)) \wedge r_3(?V, ?T)$, where $r_1, r_2, r_3 \in R$, $h, h_1, h_2 \in \mathcal{E}$, $?V$ is an existentially quantified variable and $?T$ is the target variable.

Complex query decomposition (CQD) is a framework for answering complex queries by reasoning over sets of entities in the embedding space [1,23]. It decomposes a given complex query into multiple atomic queries, which can be answered using neural link predictors, and aggregates their scores using fuzzy logic semantics. As neural link predictors only need to be trained on 1-hop queries, the approach needs less training data than end-to-end architectures.

Fuzzy Query Score Aggregation. Fuzzy logic extends classical Boolean logic by generalizing conjunction ($\wedge$) and disjunction ($\vee$) through the use of binary operations called t-norm and t-conorm. The t-norm (resp. t-conorm) operation generalizes conjunction (resp. disjunction) in fuzzy logic. In this work, we consider two different semantics: Gödel and product, to recursively compute the score of a complex query. For an atomic query Q, its fuzzy truth value (score) lies in the range $[0, 1]$ (line 1 in Table 2). Moreover, to compute the score of a (complex) query, it is first grounded by instantiating the variables with the predicted entities, after which the scores are recursively aggregated as defined in Table 2.

Complex Query Decomposition as an Optimization Problem. Using t-norm for conjunction, t-conorm for disjunction and fuzzy negation operator to aggregate the predictions of atomic queries, answering a complex query can be considered as an optimization problem to find the set of variable instantiations that maximize the overall score of the target entity. CQD introduces two approaches for optimization: continuous and combinatorial. Since the continuous optimization variant is not relevant to this work, we refer to the respective paper [1] for more details. The better performing **combinatorial optimization (CQD-Beam)** aims at identifying the optimal set of variable-to-entity assignments directly using a neural link predictor. This optimal set of variable substitutions is searched using a procedure akin to beam search by traversing the dependency graph of a query and in each step keeping only the top-k entities with the highest score using a neural link predictor. Consider the query $r_1(h, ?V_1) \wedge r_2(?V_1, ?T)$, where h is a constant entity and $?V_1, ?T$ are variables. The score of the query can be computed using: $\top_G(\varphi_{r_1}(h, ?V_1) \wedge \varphi_{r_2}(?V_1, ?T))$. Starting with the constant h the dependency graph of the query is traversed similar to depth-first search, only further considering the top-k entities in each step, that have the highest score according to the neural link predictor. This results in a maximum of k^2 predictions for $?T$.

Score Calibration. Neural link predictors are good at predicting 1-hop queries (1p); however, due to the way they are trained, they can learn to rank entities (which suffices for answering 1-hop queries) rather than learning scores that capture the interactions among different components of the query. To be able to meaningfully aggregate the scores using the t-norm or t-conorm, the scores are assumed to have the same scale. For example, consider the query from the introduction $\text{genre}(\text{comedy}, ?T) \wedge \text{acted}(?T, \text{kreeves})$ and the Gödel t-norm $\top_G(Q_1 \wedge Q_2)$ to aggregate the scores for a conjunction. Both queries independently could get answered correctly by a link predictor with a ranked list of entities. However if it is true that $\forall t \in \mathcal{E} : \varphi_{\text{genre}}(\text{comedy}, t) < \varphi_{\text{genre}}(t, \text{kreeves})$ the conjunction query would be dominated by the predictions of the query $\text{genre}(\text{comedy}, ?T)$ and hence determine the final result. This imbalance of scores needs adjustment to better reflect the semantic importance of both atomic queries. CQD^{A} [2] learns an adaption model for the scores of a neural link predictor $\varphi_r(h, t)$ to recalibrate them to reflect the semantics of the query: $\rho_\theta(\varphi_r(h, t)) = (1 + \alpha) * \varphi_r(h, t) + \beta$, where $\theta = \{\alpha, \beta\}$ are the parameters that are learned. θ can then be conditioned on the embedding representation of the head, relation and tail $\theta = \psi(e_h, e_r, e_t)$, where ψ is a (deep) neural network. Finally they use a training set of complex queries to learn the parameters of the adaption function using the 1vsAll training strategy and cross entropy loss. In their experiments, they found that conditioning θ on the relation only, i.e., $\theta = \psi(e_r)$, yields better results than using any combination of entity and relation representation [2].

3 Training-Free Score Calibration

Standard Normalization Techniques for Neural Link Predictors. In this study we investigate whether a special adaption function is needed to calibrate the scores or if simple normalization techniques suffice. As previously introduced,

CQD^{A} learns a special score adaption model to map the scores of a neural link predictor to the range $[0, 1]$, while the normalization function employed by CQD alone remains unclear. Depending on the loss function and training strategy used, the semantics of the score of a neural link predictor changes. Currently, the state-of-the-art approach to train neural link predictors is using a cross entropy loss [31], a multi-class loss function that is computed by comparing the model distribution (the softmax distribution over scores) and the data distribution (the labels of triples normalized to sum to 1). However, since link prediction is generally a multi-label classification task, softmax normalization might not always be the best choice when the scores are aggregated.

We examine the performance of three well-known methods for scaling the scores of a neural link predictor to the range $[0, 1]$: *sigmoid*, *min-max* and *softmax*. The normalization functions are applied to prediction scores of each atomic query prior to aggregation using fuzzy logic operators. Given an atomic query $r(h, ?V)$ a neural predictor assigns scores for every entity to be a substitution (grounding) of $?V$. The scores can be denoted by $x = [\varphi(h, t') \mid t' \in \mathcal{E}]$. To map this vector of prediction scores $x = (x_1, \ldots, x_{|\mathcal{E}|}) \in \mathbb{R}^{|\mathcal{E}|} \rightarrow (0, 1)^{|\mathcal{E}|}$, the *sigmoid* normalization uses the sigmoid function to rescale the scores and is defined as $\mathrm{sigmoid}(x_i) = 1/(1 + e^{-x_i})$. Min-max normalization is a simple method that does linear interpolation of scores where 0 is the minimum score and 1 is the maximum score, the definition is $\mathrm{minmax}(x_i) = (x_i - \min(x))/(\max(x) - \min(x))$. The softmax normalization function transforms scores into a probability distribution. It is defined as $\mathrm{softmax}(x_i) = e^{x_i}/(\sum_{j=1}^{K} e^{x_j})$.

Rule-Based Approach Does not Require Score Calibration. While continuous optimization cannot be used with symbolic rule-based approaches because these approaches do not learn embeddings of entities and relations, combinatorial optimization can be applied to rule-based approaches similarly as to neural link predictors. However, instead of transforming entities and relations into vector embeddings, each step of the reasoning process remains symbolic and distinct.

The confidence metric of rules reflects the proportion of true positives relative to the total predictions made by the rules. This metric is semantically meaningful, as it measures the likelihood of a prediction being correct and its calculation aligns with the frequentist interpretation of probability. The probabilistic nature of rule confidences should enhance the fuzzy aggregation of predictions as t-norms, Table 2, have a clear relation to probability. The product t-norm represents the joint probability of two independent events Q_1 and Q_2 being true and the product t-conorm represents the probability of either Q_1 or Q_2 being true if Q_1 and Q_2 are non-mutually exclusive and independent (also called Noisy-OR in literature [14,21]). Both assume independency. However as pointed out in [21] rule sets can contain a lot of dependent rules, and the maximum aggregation of rule confidences often outperforms the Noisy-OR aggregation approach. The maximum aggregation is equivalent to the Gödel t-conorm. Since confidences have greater semantic value than scores learned by neural link predictors, combinatorial optimization with a rule-based approach should facili-

Table 3. Numbers of training validation and test queries of different query types. The training set only consists of a subset of query types: p/i designates the types 1p, 2p, 3p, 2i, 3i and n are the query types that include negations 2in, 3in, inp, pin and pni.

Dataset	Training		Validation		Test	
	p/i	n	1p	Others	1p	Others
FB15k	273,710	27,371	59,097	8,000	67,016	8,000
FB15k-237	149,689	14,968	20,101	5,000	22,812	5,000
NELL995	107,982	10,798	16,927	4,000	17,034	4,000

tate better aggregation through fuzzy operators, leading to improved predictive performance.

4 Experimental Study

In this section, we present the design and results of our analysis. We perform experiments using CQD on top of a neural model, where we use standard score normalization instead of training a complex score calibration model. Further, we replace the neural model by a rule-based approach that does not require any score calibration. We compare the results of both methods against the results achieved by other approaches including current state-of-the-art models. To further support our claim that a rule-based approach does not require score normalization, we learn a specific calibration model for the rule-based approach and show that it has no (positive) impact on predictive performance.

4.1 Settings

Datasets. We follow previous research in complex query answering and evaluate our approaches using the benchmark datasets introduced in [29] and were derived from the commonly known knowledge graph completion datasets FB15k [8], FB15k-237 [36] and NELL995 [40]. FB15k and FB15k-237 were both extracted from Freebase [7], while FB15k-237 is a revised version of FB15k and is considered more difficult [36]. NELL995 was created using the Never-Ending Language Learning system [24], which continuously crawls the web extracting semantic triples. As noted in [32], many of the triples in NELL995 are nonsensical, which calls into question its suitability as a knowledge graph completion benchmark. The datasets for complex query answering derived from FB15k, FB15k-237 and NELL995 in [29] each consists of sets of complex queries adhering to 14 different query structures. The different query structures are depicted in Fig. 1. Each dataset and query type consists of a training, validation and testing set of queries and answers. The training set consists of a subset of query types: 1p, 2p, 3p, 2i, 3i, 2in, 3in, inp, pin and pni. Statistics on the different types of query structures is given in Table 3.

Evaluation Protocol. For each dataset and query type we report the mean reciprocal rank (MRR). For its definition, we refer readers to [30]. All results reported are approaches evaluated under filtered setting, where the answer sets of the validation and testing set were filtered from entities that can simply be retrieved by applying the query to the training set. Similar to [23], we report an avg_{all} metric, which is the MRR averaged across all datasets and query types, avg_{p} averaged across all existential positive first order (EPFO) queries (1p, 2p, 3p, 2i, 3i, pi, ip, 2u, up) and avg_{n} averaged across queries containing negation. These average scores are macro averages that assign to each dataset and each query type the same weight.

Fuzzy Operators and Normalization Method. Within our approach there are six possible combinations for a neural link predictor with respect to the fuzzy operator semantic (Gödel vs. product) and normalization method (sigmoid, min-max, softmax). We determine the best combination for each dataset and query type using the validation set. We refer to the CQD approach that uses this standard normalization (SN) setting together with a knowledge graph embedding model (KGE) as $\text{CQD}^{\text{KGE+SN}}$. For the approach where we replace the neural model with a rule-based approach, we do not need a normalization method. Here, we determine only the best t-norm. We refer to this approach as CQD^{Rule}.

Other Approaches/Models. First of all, we compare the results of these simple CQD approaches against other approaches based on CQD: the original CQD [23] and CQD_A [2]. We include also results for end-to-end architecture in our experiments using the following models: GQE [17], Query2Box [28], BetaE [29], ConE [42], FuzzQE [9] and GNN-QE [43]. For details on these approaches, we refer to their original papers.

Configurations for Link Predictors. For the rule-based approach, we learn rules using AnyBURL [21] for 1000 s for each dataset. We learn cyclic rules of a maximum length of 3 body atoms and acyclic rules of length 1. We only keep rules that predict at least 2 correct predictions and have a confidence of at least 0.0001. These settings and learning duration are taken from [21] and have proven to be robust. For applying the rules and retrieving their predictions we use PyClause [5]. As a neural link predictor, we follow previous work on complex query decomposition [1,2,23] and use ComplEx [37] with N3 regularization [19]. For the hyperparameters used for training the model we refer to [2].

Efficiency. Efficiency comparisons regarding runtime (training and inference) and computational resource usage between neural link predictors and AnyBURL can be found in [20]. In general, inference with AnyBURL is approximately 10 to 100 times slower than with neural link predictors; however, it requires significantly less training time and provides the added benefit of explainability compared to KGE models.

Table 4. Averaged MRR over all datasets using the testing sets. avg_{all} is averaged over all query types, avg_p over the 9 EPFO queries not containing negation and avg_n over queries containing negation. Best results are in bold.

	avg_{all}	avg_p	avg_n
GNN-QE	34.5	42.8	19.5
CQD	-	36.4	-
CQD^A	35.1	42.9	21.3
CQD^{KGE+SN}	30.9	43.0	9.2
CQD^{Rule}	**37.1**	**45.5**	**21.9**

4.2 Results

Table 4 shows the results, aggregated over datasets and query types, while Table 5 gives more fine-grained evaluation results and shows the MRR for each dataset and query type. In the following discussion we refer to both tables.

CQD^{KGE+SN}. Standard normalization techniques for neural link predictors achieve the same predictive performance on EPFO queries as CQD^A, which has to train an adaption model. For CQD^{KGE+SN} we measure an avg_p of 43.0%, while CQD^A achieves 42.9%. This means that for these types of queries, training a specific adaption model is unnecessary, however standard normalization is still needed as plain CQD only achieves an avg_p of 36.4%. Compared to the best end-to-end approach, GNN-QE, which achieves 42.8%, CQD^{KGE+SN} demonstrates also equally strong performance. On queries containing negation however, CQD^{KGE+SN} performs worse. It achieves an avg_n of 9.2%, while CQD^A obtains 21.9%. This indicates that, for these types of queries, the adaption model is beneficial for neural link predictors.

CQD^{Rule}. The approach of using the rule-based AnyBURL as predictor for CQD, denoted CQD^{Rule}, improves current state of the art from 35.1% (CQD^A) to 37.1%, when comparing avg_{all}. When comparing avg_p and avg_n, CQD^{Rule} achieves an avg_p of 45.5% and an avg_n of 21.9%. The next best results are an avg_p of 42.9% and avg_n of 21.3% produced by CQD^A (CQD with learned adaption model). This means that without learning a specific adaption function, the rule-based approach is able to achieve better results on both positive and negative query types. The original CQD that also uses no learned adaption function achieves only 36.4% avg_p (43.0% CQD^{KGE+SN}), which indicates that rule confidence is better suited to be aggregated using fuzzy operators. Rule confidence also can be negated better than scores generated by neural link predictors. The original CQD did not publish results for queries containing negation, however CQD^{KGE+SN} (which is CQD with the best standard score normalization function according to the validation set) can also only achieve 9.2% avg_n. Furthermore, CQD^{Rule} also outperforms the best end-to-end architecture GNN-QE on all metrics: avg_{all}, avg_p and avg_n.

Table 5. MRR results for the 14 different query structures of the FB15k, FB15k-237 and NELL dataset. Results with the highest scores are marked in bold, second highest scores are underlined. The $\text{CQD}^{\text{KGE+SN}}$ and CQD^{Rule} are the approaches established in this study. The results for CQD are from [23] and the results for CQDA from [2], All the remaining results were taken from [43].

Model	avg_p	avg_n	1p	2p	3p	2i	3i	pi	ip	2u	up	2in	3in	inp	pin	pni
								FB15K								
GQE	28.0	-	54.6	15.3	10.8	39.7	51.4	27.6	19.1	22.1	11.6	-	-	-	-	-
Q2B	38.0	-	68.0	21.0	14.2	55.1	66.5	39.4	26.1	35.1	16.7	-	-	-	-	-
BetaE	41.6	11.8	65.1	25.7	24.7	55.8	66.5	43.9	28.1	40.1	25.2	14.3	14.7	11.5	6.5	12.4
ConE	49.8	14.8	73.3	33.8	29.2	64.4	73.7	50.9	35.7	55.7	31.4	17.9	18.7	12.5	9.8	15.1
GNN-QE	72.8	38.6	88.5	69.3	58.7	79.7	83.5	69.9	70.4	**74.1**	61.0	44.7	41.7	**42.0**	30.1	34.3
CQD-CO	46.9	-	89.2	25.3	13.4	74.4	78.3	44.1	33.2	41.8	21.9	-	-	-	-	-
CQD-Beam	58.2	-	89.2	54.3	28.6	74.4	78.3	58.2	67.7	42.4	30.9	-	-	-	-	-
CQD^{A}	70.4	**42.8**	89.2	64.5	57.9	76.1	79.4	70.0	70.6	68.4	57.9	**54.7**	**47.1**	37.6	35.3	24.6
$\text{CQD}^{\text{KGE+SN}}$	71.1	15.3	89.2	65.7	57.7	77.1	80.6	72.3	66.2	72.3	59.4	15.7	30.1	12.4	5.8	12.4
CQD^{Rule}	**76.7**	41.2	**89.3**	**74.5**	**71.8**	**80.2**	**84.6**	**77.0**	**75.7**	70.3	**67.3**	44.8	37.6	39.2	**40.9**	**43.5**
								FB15K-237								
GQE	16.3	-	35.0	7.2	5.3	23.3	34.6	16.5	10.7	8.2	5.7	-	-	-	-	-
Q2B	20.1	-	40.6	9.4	6.8	29.5	42.3	21.2	12.6	11.3	7.6	-	-	-	-	-
BetaE	20.9	5.5	39.0	10.9	10.0	28.8	42.5	22.4	12.6	12.4	9.7	5.1	7.9	7.4	3.5	3.4
ConE	23.4	5.9	41.8	12.8	11.0	32.6	47.3	25.5	14.0	14.5	10.8	5.4	8.6	7.8	4.0	3.6
GNN-QE	26.8	10.2	42.8	14.7	11.8	**38.3**	**54.1**	31.1	18.9	16.2	13.4	10.0	**16.8**	9.3	7.2	7.8
CQD-CO	21.8	-	**46.7**	9.5	6.3	31.2	40.6	23.6	16.0	14.5	8.2	-	-	-	-	-
CQD-Beam	22.3	-	**46.7**	11.6	8.0	31.2	40.6	21.2	18.7	14.6	8.4	-	-	-	-	-
CQD^{A}	25.7	10.7	**46.7**	13.6	11.4	34.5	48.3	27.4	20.9	**17.6**	11.4	**13.6**	**16.8**	7.9	8.9	5.8
$\text{CQD}^{\text{KGE+SN}}$	25.5	6.1	**46.7**	13.3	11.2	34.6	48.3	26.5	19.9	**17.6**	11.5	5.8	12.4	5.3	2.3	4.6
CQD^{Rule}	**27.5**	**11.1**	43.7	**15.8**	**15.1**	36.5	50.7	**31.5**	**22.4**	16.9	**15.2**	10.4	16.2	**10.8**	**9.5**	**8.6**
								NELL995								
GQE	18.6	-	32.8	11.9	9.6	27.5	35.2	18.4	14.4	8.5	8.8	-	-	-	-	-
Q2B	22.9	-	42.2	14.0	11.2	33.3	44.5	22.4	16.8	11.3	10.3	-	-	-	-	-
BetaE	24.6	5.9	53.0	13.0	11.4	37.6	47.5	24.1	14.3	12.2	8.5	5.1	7.8	10.0	3.1	3.5
ConE	27.2	6.4	53.1	16.1	13.9	40.0	50.8	26.3	17.5	15.3	11.3	5.7	8.1	10.8	3.5	3.9
GNN-QE	32.3	**13.3**	53.3	18.9	14.9	42.4	52.5	30.8	18.9	15.9	12.6	9.9	14.6	11.4	6.3	6.3
CQD-CO	28.8	-	**60.4**	17.8	12.7	39.3	46.6	30.1	22.0	17.3	13.2	-	-	-	-	-
CQD-Beam	28.6	-	**60.4**	20.6	11.6	39.3	46.6	25.4	23.9	17.5	12.2	-	-	-	-	-
CQD^{A}	32.3	**13.3**	**60.4**	22.9	16.7	43.4	52.6	**32.1**	**26.4**	**20.0**	**17.0**	**15.1**	**18.6**	15.8	**10.7**	6.5
$\text{CQD}^{\text{KGE+SN}}$	32.2	6.2	**60.4**	22.7	18.1	**43.6**	**53.0**	31.3	24.4	19.9	16.8	5.7	10.7	7.7	2.2	4.6
CQD^{Rule}	**32.4**	**13.3**	59.7	**23**	**21.6**	42	52.1	31.9	25.7	18.4	16.9	13.3	17.8	**15.9**	10.5	**8.9**

As shown in Table 5, CQD^{Rule} achieves the biggest improvements on FB15k, with an avg_p of 76.7%, compared to the second-best method, GNN-QE, which achieves 72.8%. Also on FB15k-237 it could achieve significant improvements (+1.8% points (pp) compared to CQD^{A} and +0.7pp compared to GNN-QE), while the performance on NELL995 is en-par to the current state of the art

(+0.1pp compared to CQD^{A} and GNN-QE). CQD^{Rule} seems to perform particularly good on queries 2p and 3p, where it could improve the performance significantly on all three benchmark datasets. Comparing the performance on negative queries, the highest improvements of CQD^{Rule} are on FB15k-237, where the avg_n could be improved by +0.4pp compared to the next best approach CQD^{A}. On NELL995, CQD^{Rule} performs en-par with CQD^{A} and GNN-QE, while on FB15k CQD^{Rule} performs better than GNN-QE (+2.6pp), however can not achieve the performance of CQD^{A} which scores 1.6pp higher.

Best Combination of t-norm and Normalization Method. For $\text{CQD}^{\text{KGE+SN}}$ we evaluated which combination of normalization approach and t-norm has the best performance on the validation set for each dataset and query type. Interestingly, the best configurations vary across query types, but are mostly consistent across datasets. Table 6 in Appendix A shows the validation MRR results for each configuration. For nearly all query types, the product t-norm performs best, except for queries that contain the disjunction operator, where the Gödel t-norm works best. For the query types 2p and 3p, sigmoid and product t-norm seem to work very well. The intersection query types 2i and 3i and query types containing negation are best normalized using the softmax. For the disjunction query type 2u the minmax normalization works best across all datatypes.

For the rule-based approach, where we only search for the best t-norm, on almost every query type and dataset the product t-norm performs better. It is only slightly outperformed by the Gödel t-norm on the FB15k-237 dataset with the 2u query type. Table 7 in Appendix A shows the MRR results for each t-norm using the validation sets.

Rule Confidences are Calibrated by Default. We evaluated if confidences of rules are calibrated by default or if they could benefit from an adaption function akin to CQD^{A}. We applied the same strategy as CQD^{A} and use the training sets of the query types 2i, 3i, 2in and 3in to learn an adaption function $\rho_\theta(c_{rl}) = c_{rl} * (1+\alpha) + \beta$ for the confidence scores. To enable more stable training, we convert the confidence values to logits using $c_{rl} = \ln(\frac{\text{conf}(rl)}{1-\text{conf}(rl)})$.

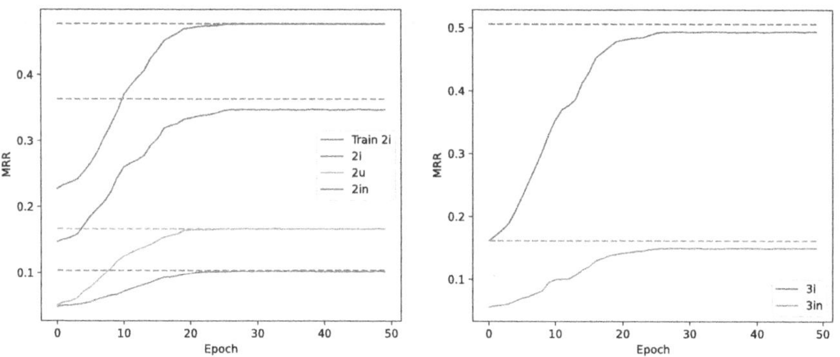

Fig. 2. Traces of the MRR when fitting an adaption function on rule confidences.

For conditioning the parameters $\theta = \{\alpha, \beta\}$, we explored three approaches: (1) as entities and relations in rule-based systems have no embeddings we interpret the relation as a one-hot encoded categorical variable $v_r = [\mathbb{I}(i = r)]_{i=1}^{k}$ and condition the parameters on the relation directly $\theta = \psi(v_r)$, (2) we use the embeddings of entities and relations of the neural link predictor and condition θ on the relation embedding using $\theta = \psi(e_r)$ and (3) on the source entity and relation embedding $\theta = \psi(e_s, e_r)$ with $e_s, e_r \in \mathbb{R}^k$. We perform a hyperparameter search to find the best parameters using the following options: learning rate in $[10^{-4}, 10^{-1}]$ (logarithmic scale), regularization weight in $[10^{-5}, 10^{-5}]$, batch size in $\{64, 128, 256, 512\}$, training strategy in $\{\text{KvsAll}, \text{1vsAll}\}$, optimizer in $\{\text{Adam}, \text{Adagrad}\}$, loss in $\{\text{CE}, \text{BCE}\}$ and number of hidden layers in $\{0, 1\}$. If a hidden layer is used we use a hidden dimensionality of 500 and a ReLU non-linearity, the same as CQD^{A} [2].

Interestingly, no trial run of the hyperparameter search produced better results with the trained adaption function than without it. For the tested query types 2i, 3i, 2in, 3in, and 2u the adaption function was only able to recover the same or worse performance as without the adaption function. Figure 2 shows traces of their MRR values on the test sets when fitting an adaption function on rule confidences over 100 epochs. The dashed line indicates the MRR scores obtained by setting $\alpha = 0$ and $\beta = 0$, which corresponds to the performance without the adaption function. We also included the MRR trace of the query type 2i using the training set, to show that also on the training set itself the adaption function can only recover the same performance.

5 Conclusion

We proposed two simple solutions for the problem of answering complex queries over incomplete KGs. Both are based on the Complex Query Decomposition framework: (i) instead of training a model for normalizing the scores of the neural link predictors, which are applied to the simple queries that are the building blocks of the complex query, we apply standard normalization methods. Our experimental results show that our method is on average slightly better than training a score calibration model, which has so far been the state of the art. However, we detected also that our simple approach does not work well on queries that contain negation. For these queries the predictive performance is clearly better when training an adaption model. Further research is required to better understand the difference between queries with and without negation. (ii) Furthermore, we show that using a rule-based approach instead of a neural link predictor as the atomic query predictor improves the mean reciprocal rank (MRR) averaged across all datasets and query types (including negation) from 35.1% to 37.1%. This rule-based approach requires no score calibration, while having the additional advantage of being fully explainable.

We are aware that our approach is not built on a complex neural architecture nor introduces a novel conceptualization or innovative algorithm. Instead, we analyze the predictive quality of (i) a baseline that has been overlooked in the experimental results of previous papers on complex query answering and (ii) a rule-based approach that inherently does not require score normalization. Our paper closes this gap and demonstrates that the baseline performs on par with most existing methods, while the rule-based approach achieves new state-of-the-art results.

Acknowledgments. This work was funded by the Austrian security research program KIRAS of the Federal Ministry of Finance (BMF) through the DAGMAR project (grant No. 52224305). ChatGPT was used to enhance the readability of some of the text and improve the language of this paper.

A Validation Results

Table 6 shows the mean reciprocal rank values for all six combinations of fuzzy operator semantic (product vs. Gödel) and standard normalization method (sigmoid, minmax and softmax) on the validation sets. The best combination varies across query types, however stay roughly consistent across datasets.

Table 6. MRR results of the grid search on the normalization functions sigmoid (s), minmax (mm) and softmax (sm) and the t-norms product (p) and Gödel (g) using the validation set. The best configurations were taken for CQD$^{\text{KGE+SN}}$.

	1p	2p	3p	2i	3i	pi	ip	2u	up	2in	3in	inp	pin	pni
	FB15k-237													
s-p	44,7	**11,3**	**9,2**	27,4	39,2	21,8	16,0	11,1	8,7	0,0	0,0	0,5	0,1	1,1
s-g	44,7	10,7	8,3	25,0	34,4	19,2	13,1	12,0	**9,3**	0,0	0,0	0,2	0,0	0,6
mm-p	44,7	9,1	5,6	32,4	47,5	**23,5**	16,1	11,9	6,5	2,6	6,1	2,8	1,1	**3,4**
mm-g	44,7	9,2	5,6	29,5	40,5	22,2	14,1	**12,3**	7,0	0,9	0,5	1,3	0,4	3,4
sm-p	44,7	10,0	7,5	**32,4**	**47,5**	23,5	**17,0**	11,2	8,0	**4,0**	**10,5**	**5,5**	**1,9**	3,2
sm-g	44,7	8,7	6,2	15,4	20,1	13,4	9,8	11,1	6,8	0,1	0,1	0,2	0,0	1,9
	FB15k													
s-p	79,1	**35,5**	**31,1**	54,6	66,7	**49,3**	**43,7**	26,8	19,3	0,0	0,0	0,5	0,0	1,2
s-g	79,1	34,7	29,4	53,8	65,0	47,2	40,7	32,8	**28,8**	0,0	0,0	0,2	0,0	0,6
mm-p	79,1	25,5	15,1	56,6	70,7	46,1	40,7	31,9	15,1	5,9	9,0	3,7	2,3	7,8
mm-g	79,1	25,5	14,9	55,8	68,7	45,8	39,2	**33,5**	17,8	1,0	0,5	1,0	0,5	6,2
sm-p	79,1	29,5	21,2	**57,6**	**71,1**	46,8	42,1	31,7	22,1	**9,7**	**19,5**	**9,1**	**4,0**	**7,9**
sm-g	79,1	24,3	15,5	45,7	53,9	36,9	33,1	31,5	17,2	0,3	0,1	0,3	0,0	6,3
	NELL995													
s-p	58,8	**20,8**	**18,8**	35,7	47,6	27,1	23,8	15,9	9,8	0,1	0,1	1,6	0,0	1,2
s-g	58,8	19,5	16,8	34,3	43,8	24,6	19,7	**16,6**	**14,0**	0,0	0,0	0,4	0,0	0,7
mm-p	58,8	17,2	13,0	40,3	55,5	**29,7**	25,5	16,6	9,0	3,8	5,7	4,6	1,4	4,2
mm-g	58,8	17,4	12,9	35,8	46,1	28,4	21,2	16,5	11,2	1,8	0,5	2,7	0,5	**4,5**
sm-p	58,8	18,6	15,9	**41,7**	**56,5**	29,3	**25,8**	16,1	11,8	**5,4**	**10,8**	**7,6**	**2,4**	4,2
sm-g	58,8	16,8	14,9	29,5	35,5	25,6	17,6	15,8	10,3	0,2	0,1	0,1	0,0	2,3

Table 7. MRR results for the different t-norms: product (p) and Gödel (g) using the validation set and the rule-based approach. The best configurations were taken for CQD^{Rule}.

	1p	2p	3p	2i	3i	pi	ip	2u	up	2in	3in	inp	pin	pni
							FB15k-237							
p	**42,0**	**12,7**	**12,5**	**33,0**	**49,0**	**26,8**	**18,7**	11,2	**11,7**	**0,1**	**1,3**	**9,4**	**7,4**	**6,5**
g	42,0	12,3	11,8	30,2	41,2	24,5	15,6	**11,2**	11,4	0,1	1,1	8,2	7,2	6,4
							FB15k							
p	**79,0**	**37,4**	**34,1**	**58,3**	**72,0**	**52,5**	**45,8**	**34,1**	**30,7**	**19,7**	**16,2**	**19,0**	**16,6**	**19,8**
g	79,0	36,8	32,9	57,3	69,7	51,0	43,4	34,0	30,5	19,7	14,4	17.7	16,2	19,1
							NELL995							
p	**57,9**	**21,3**	**21,4**	**40,1**	**54,9**	**30,3**	**26,1**	**15,9**	**13,9**	**14,3**	**18,3**	**17,0**	**12,0**	**11,1**
g	57,9	19,5	18,6	35,2	46,5	25,8	19,5	15,7	13,5	14,3	17,1	15,3	11,8	11,0

Table 7 compares the MRR values of the two fuzzy operator semantics: product and Gödel. The product semantic performs best on every query type and dataset, with the exception of the query type 2u on the dataset FB15k-237, where Gödel semantic performs slightly better.

References

1. Arakelyan, E., Daza, D., Minervini, P., Cochez, M.: Complex query answering with neural link predictors. In: International Conference on Learning Representations (2021). https://openreview.net/forum?id=Mos9F9kDwkz
2. Arakelyan, E., Minervini, P., Daza, D., Cochez, M., Augenstein, I.: Adapting neural link predictors for data-efficient complex query answering. In: Thirty-Seventh Conference on Neural Information Processing Systems (2023). https://openreview.net/forum?id=1G7CBp8o7L
3. Auer, S., Bizer, C., Kobilarov, G., Lehmann, J., Cyganiak, R., Ives, Z.: Dbpedia: a nucleus for a web of open data. In: The Semantic Web, pp. 722–735. Springer (2007)
4. Balazevic, I., Allen, C., Hospedales, T.: TuckER: tensor factorization for knowledge graph completion. In: Proceedings of the 2019 Conference on Empirical Methods in Natural Language Processing and the 9th International Joint Conference on Natural Language Processing (EMNLP-IJCNLP), Hong Kong, China, pp. 5185–5194. Association for Computational Linguistics (2019). https://doi.org/10.18653/v1/D19-1522. https://www.aclweb.org/anthology/D19-1522
5. Betz, P., Galarraga, L., Ott, S., Meilicke, C., Suchanek, F.M., Stuckenschmidt, H.: Pyclause-simple and efficient rule handling for knowledge graphs. In: IJCAI, demo track. Ijcai.org (2024)
6. Betz, P., Meilicke, C., Stuckenschmidt, H.: Supervised knowledge aggregation for knowledge graph completion. In: The Semantic Web: 19th International Conference, ESWC 2022, Hersonissos, Crete, Greece, 29May–2 June 2022, Proceedings, pp. 74–92. Springer (2022)

7. Bollacker, K., Evans, C., Paritosh, P., Sturge, T., Taylor, J.: Freebase: a collaboratively created graph database for structuring human knowledge. In: Proceedings of the 2008 ACM SIGMOD International Conference on Management of Data, pp. 1247–1250. ACM (2008)
8. Bordes, A., Usunier, N., Garcia-Duran, A., Weston, J., Yakhnenko, O.: Translating embeddings for modeling multi-relational data. In: Advances in Neural Information Processing Systems, pp. 2787–2795 (2013)
9. Chen, X., Hu, Z., Sun, Y.: Fuzzy logic based logical query answering on knowledge graphs. In: Proceedings of the AAAI Conference on Artificial Intelligence, vol. 36, pp. 3939–3948 (2022)
10. Das, R., Neelakantan, A., Belanger, D., McCallum, A.: Chains of reasoning over entities, relations, and text using recurrent neural networks. In: Proceedings of the 15th Conference of the European Chapter of the Association for Computational Linguistics: Volume 1, Long Papers. Association for Computational Linguistics (2017)
11. Dettmers, T., Minervini, P., Stenetorp, P., Riedel, S.: Convolutional 2D knowledge graph embeddings. In: Proceedings of the AAAI Conference on Artificial Intelligence, pp. 1811–1818 (2018)
12. Dong, X., et al.: Knowledge vault: a web-scale approach to probabilistic knowledge fusion. In: Proceedings of the 20th ACM SIGKDD International Conference on Knowledge Discovery and Data Mining, pp. 601–610 (2014)
13. Dumontier, M., et al.: Bio2rdf release 3: a larger connected network of linked data for the life sciences. In: Proceedings of the 2014 International Conference on Posters & Demonstrations Track, ISWC-PD 2014, Aachen, DEU, vol. 1272, pp. 401–404. CEUR-WS.org (2014)
14. Galárraga, L., Teflioudi, C., Hose, K., Suchanek, F.M.: Fast rule mining in ontological knowledge bases with AMIE+. VLDB J. **24**(6), 707–730 (2015)
15. Galárraga, L.A., Teflioudi, C., Hose, K., Suchanek, F.: Amie: association rule mining under incomplete evidence in ontological knowledge bases. In: Proceedings of the 22nd International Conference on World Wide Web, pp. 413–422. ACM (2013)
16. Guu, K., Miller, J., Liang, P.: Traversing knowledge graphs in vector space. In: Proceedings of the 2015 Conference on Empirical Methods in Natural Language Processing, pp. 318–327 (2015)
17. Hamilton, W., Bajaj, P., Zitnik, M., Jurafsky, D., Leskovec, J.: Embedding logical queries on knowledge graphs. In: Advances in Neural Information Processing Systems, vol. 31 (2018)
18. Himmelstein, D.S., et al.: Systematic integration of biomedical knowledge prioritizes drugs for repurposing. eLife **6**, e26726 (2017). https://doi.org/10.7554/eLife.26726
19. Lacroix, T., Usunier, N., Obozinski, G.: Canonical tensor decomposition for knowledge base completion. In: International Conference on Machine Learning, pp. 2863–2872. PMLR (2018)
20. Meilicke, C., Chekol, M.W., Betz, P., Fink, M., Stuckeschmidt, H.: Anytime bottom-up rule learning for large-scale knowledge graph completion. VLDB J. **33**(1), 131–161 (2024). https://doi.org/10.1007/s00778-023-00800-5
21. Meilicke, C., Chekol, M.W., Ruffinelli, D., Stuckenschmidt, H.: Anytime bottom-up rule learning for knowledge graph completion. In: Proceedings of the Twenty-Eighth International Joint Conference on Artificial Intelligence, pp. 3137–3143. Ijcai.org (2019)

22. Miller, G.A.: WordNet: a lexical database for English. In: Speech and Natural Language: Proceedings of a Workshop Held at Harriman, New York, 23–26 February 1992 (1992). https://aclanthology.org/H92-1116
23. Minervini, P., Arakelyan, E., Daza, D., Cochez, M.: Complex query answering with neural link predictors (extended abstract). In: Raedt, L.D. (ed.) Proceedings of the Thirty-First International Joint Conference on Artificial Intelligence, IJCAI-22, pp. 5309–5313. International Joint Conferences on Artificial Intelligence Organization (2022). https://doi.org/10.24963/ijcai.2022/741
24. Mitchell, T., et al.: Never-ending learning. Commun. ACM **61**(5), 103–115 (2018)
25. Nickel, M., Tresp, V., Kriegel, H.P.: A three-way model for collective learning on multi-relational data. In: ICML, vol. 11, pp. 809–816 (2011)
26. Ott, S., Betz, P., Stepanova, D., Gad-Elrab, M.H., Meilicke, C., Stuckenschmidt, H.: Rule-based knowledge graph completion with canonical models. In: Proceedings of the 32nd ACM International Conference on Information and Knowledge Management, CIKM 2023, pp. 1971–1981. Association for Computing Machinery, New York (2023). https://doi.org/10.1145/3583780.3615042
27. Ott, S., Meilicke, C., Samwald, M.: Safran: an interpretable, rule-based link prediction method outperforming embedding models. In: 3rd Conference on Automated Knowledge Base Construction (2021). https://openreview.net/forum?id=jCt9S_3w_S9
28. Ren, H., Hu, W., Leskovec, J.: Query2box: reasoning over knowledge graphs in vector space using box embeddings. In: International Conference on Learning Representations (2020). https://openreview.net/forum?id=BJgr4kSFDS
29. Ren, H., Leskovec, J.: Beta embeddings for multi-hop logical reasoning in knowledge graphs. Adv. Neural. Inf. Process. Syst. **33**, 19716–19726 (2020)
30. Rossi, A., Barbosa, D., Firmani, D., Matinata, A., Merialdo, P.: Knowledge graph embedding for link prediction: a comparative analysis. ACM Trans. Knowl. Discov. Data (TKDD) **15**(2), 1–49 (2021)
31. Ruffinelli, D., Broscheit, S., Gemulla, R.: You can teach an old dog new tricks! on training knowledge graph embeddings. In: International Conference on Learning Representations (2020). https://openreview.net/forum?id=BkxSmlBFvr
32. Safavi, T., Koutra, D.: CoDEx: a comprehensive knowledge graph completion benchmark. In: Proceedings of the 2020 Conference on Empirical Methods in Natural Language Processing (EMNLP), pp. 8328–8350. Association for Computational Linguistics, Online (2020). https://doi.org/10.18653/v1/2020.emnlp-main.669. https://www.aclweb.org/anthology/2020.emnlp-main.669
33. Suchanek, F.M., Kasneci, G., Weikum, G.: Yago: a core of semantic knowledge. In: Proceedings of the 16th International Conference on World Wide Web, pp. 697–706. ACM (2007)
34. Sun, Z., Deng, Z.H., Nie, J.Y., Tang, J.: Rotate: knowledge graph embedding by relational rotation in complex space. In: Proceedings of the ICLR 2019 (2019)
35. Teru, K.K., Denis, E., Hamilton, W.L.: Inductive relation prediction by subgraph reasoning. arXiv Learning (2020)
36. Toutanova, K., Chen, D.: Observed versus latent features for knowledge base and text inference. In: Proceedings of the 3rd Workshop on Continuous Vector Space Models and their Compositionality, pp. 57–66. Association for Computational Linguistics (2015)
37. Trouillon, T., Welbl, J., Riedel, S., Gaussier, É., Bouchard, G.: Complex embeddings for simple link prediction. In: International Conference on Machine Learning, pp. 2071–2080 (2016)

38. Vrandecic, D., Krötzsch, M.: Wikidata: a free collaborative knowledgebase. CACM **57**(10), 78–85 (2014)
39. Wang, Q., Mao, Z., Wang, B., Guo, L.: Knowledge graph embedding: a survey of approaches and applications. IEEE Trans. Knowl. Data Eng. **29**(12), 2724–2743 (2017)
40. Xiong, W., Hoang, T., Wang, W.Y.: DeepPath: a reinforcement learning method for knowledge graph reasoning. In: Palmer, M., Hwa, R., Riedel, S. (eds.) Proceedings of the 2017 Conference on Empirical Methods in Natural Language Processing, Copenhagen, Denmark, pp. 564–573. Association for Computational Linguistics (2017). https://doi.org/10.18653/v1/D17-1060. https://aclanthology.org/D17-1060
41. Yang, B., Yih, W.t., He, X., Gao, J., Deng, L.: Embedding entities and relations for learning and inference in knowledge bases. arXiv preprint arXiv:1412.6575 (2014)
42. Zhang, Z., Wang, J., Chen, J., Ji, S., Wu, F.: Cone: cone embeddings for multi-hop reasoning over knowledge graphs. Adv. Neural. Inf. Process. Syst. **34**, 19172–19183 (2021)
43. Zhu, Z., Galkin, M., Zhang, Z., Tang, J.: Neural-symbolic models for logical queries on knowledge graphs. In: International Conference on Machine Learning, pp. 27454–27478. PMLR (2022)
44. Zhu, Z., et al.: A*net: a scalable path-based reasoning approach for knowledge graphs. arXiv preprint arXiv:2206.04798 (2022)
45. Zhu, Z., Zhang, Z., Xhonneux, L.P., Tang, J.: Neural bellman-ford networks: a general graph neural network framework for link prediction. In: Advances in Neural Information Processing Systems, vol. 34 (2021)

Information-Aware Entity Indexing in Knowledge Graphs to Enable Semantic Search

Samuel García[1(✉)] and Carlos Bobed[2]

[1] Technical University of Munich, Heilbronn, Germany
samuel.garcia@tum.de
[2] University of Zaragoza, Zaragoza, Spain
cbobed@unizar.es

Abstract. Knowledge Graphs (KGs) have been broadly adopted as representation models to capture the world's information in a flexible way. This has allowed them to be the backbone of many different information systems, ranging from Web search engines to integration solutions for in-house enterprise systems. Given that they contain knowledge about entities, we need mechanisms to efficiently search for them, which leads to the task of Entity Retrieval. Searching for particular entities in a KG is usually tackled by applying traditional Information Retrieval (IR) techniques: indexing the KG by building Virtual Documents for each entity and applying classic IR models. However, this requires a deep analysis of the ontology of the KG (e.g., its properties and types) and the distribution of the actual data populating it. Manually analyzing these aspects in large KGs to build custom-tailored indexes becomes unfeasible, requiring instead supervised Machine Learning approaches.

In this work, we propose an approach to creating a multi-field and information-aware KG index in a completely unsupervised way. By building on different information measures, our approach clusters the properties of the KG by capturing their relative importance at different levels. This allows us to build a retrieval system for all the entities in the KG, or focus on a subset of them based on their types (i.e., building a vertical search engine). Experimental results on the DBpedia and IMDb KGs show that our retrieval performance rivals that of the state of the art, without requiring either: 1) any manual analysis of the graph regardless of its size, type hierarchy or general complexity of its properties, or 2) the use of any supervised Machine Learning techniques in order to fine-tune fielded retrieval models. Thus, we additionally eliminate the need to create annotated data such as relevance query sets.

Keywords: Entity Retrieval · Semantic Search · Knowledge Graphs

1 Introduction

Knowledge Graphs (KGs) [14] make it possible to capture extensive knowledge and data about a single or several domains in a flexible and integrated way.

E. Curry et al. (Eds.): ESWC 2025, LNCS 15718, pp. 208–226, 2025.
https://doi.org/10.1007/978-3-031-94575-5_12

While they have been around for many years under different names, such as Semantic Networks [28], they have attracted the attention of other research communities thanks to their adoption as the backbone of the main Web search engines (e.g., Google helped coin the current term back in 2012). Moreover, they act as a source of structured and possibly curated data for many Machine Learning models [29]. Given the considerable complexity and size of widely-used KGs such as the DBpedia [16], whose current core release[1] contains more than 900 million triples, we need to be able to explore and search the information contained within them appropriately.

Within the Question Answering field [8], there exist many approaches that aim to, based on a user's natural language input, build a formal query to retrieve information within the KG. However, while performant, these approaches present several issues when it comes to exploring a KG from usability and development perspectives. On the one hand, when retrieving multiple plausible answers to such a formal query, commonly expressed in SPARQL, query relevance and ordering are aspects that are usually overlooked or, at least, difficult to assess. This is due to SPARQL query results consisting of a set of variable mappings (or resources), which are typically not ordered under any particular criteria. A common workaround for this is to rely on static node centrality measures such as PageRank's variant RDFRank [9], which still do not factor in query relevance. On the other hand, we might need to re-train a Natural Language Processing (NLP) pipeline for a new domain if we change the underlying KG, as we will usually need to adapt at least the Named Entity Linking and the Relation Linking modules [5,27][2]). This latter case is usual when dealing with in-house solutions where KGs are used to integrate information from different underlying information systems [23,26].

A different approach to exploring and searching in a KG consists in, given a user query, focusing only on directly retrieving particular entities of the KG[3]. In particular, this paradigm usually adopts techniques from classic Information Retrieval, building an inverted index of the KG and employing retrieval models such as BM25 [25] to efficiently access documents via query terms[4]. In order to obtain such inverted indexes, we can build a Virtual Document (VDoc) [3] for each of its entities. The simplest VDoc would consist of a concatenation of all the textual elements that describe an entity (i.e., its attributes and all other entity names it is related to). By separating this text into different fields, generating a multi-field VDoc, we can better capture the structure of the entities to be

[1] https://www.dbpedia.org/resources/latest-core, last accessed 16th March 2025.

[2] Modules in charge of mapping the user's input to particular entities, either concepts or instances, and properties in a KG.

[3] As defined and pointed out by Balog [2], an entity is "a uniquely identifiable object or thing, characterized by its name(s), type(s), attributes, and relationships to other entities". The retrieval of such entities, a task known as Entity-Oriented Search, is crucial for 40-70% of queries in Web search [2].

[4] Such query terms are usually the ones contained in keyword queries (*Formula 1 driver championship*) or natural language queries (*Formula 1 drivers who won a championship*).

searched and the relevance of each field to the search. However, adopting a fielded representation poses some drawbacks, as it currently requires to: 1) select the properties of the entities that are relevant for the search, 2) set the granularity of the indexing (i.e., we might group several entity properties in the same field for storage and access efficiency purposes), and 3) define the relative weights of each field to assess their contribution to each entity's search relevance. For this reason, existing approaches require domain experts to assess the importance of each property and attribute in the KG, which is costly when dealing with large KGs. Additionally, they need to employ supervised learning techniques such as Learning to Rank [17] methods to be able to tune the parameters of the retrieval model, such as the field weights. This requires a reference relevance assessment in the form of an annotated queryset, usually both difficult and expensive to obtain.

In this work, we propose a new approach to create a multi-field representation of entities contained in a Knowledge Graph. Instead of relying on a manual analysis of the object and data properties contained in the KG (i.e., its relationships and attributes), we define different Importance Metrics built upon structural and information quantity features. These metrics provide the relative importance of each property, and are then used to cluster them into any given number of fields in an unsupervised way. This essentially allows to automatically create effective multi-field VDocs without requiring any prior manual analysis. We empirically show that this approach also eliminates the need to employ supervised learning techniques to adjust the field weights: the assignment of uniformly spaced weight values according to the cluster's centroid values suffices for a performant setup. Last but not least, we show how to use our Importance Metrics to create index slices for entities of a given type (i.e., vertical search engines) within the KG. The main contributions of this work are:

- We define a new set of Importance Metrics, independent of the KG domain, and design a novel strategy for using them to build multi-field VDocs ready to be fed into any indexing and retrieval system.
- We integrate them in a complete search system, further described in this work, which we make publicly available and ready to be used over four different indexing systems: ElasticSearch[5], Galago[6], pyTerrier[7], and Lucene[8].
- We empirically show the benefits of our approach, achieving SoA-comparable results without requiring neither manual nor supervised learning approaches to fine tune the retrieval models.

The rest of the work is as follows: First, in Sect. 2, we present and analyze related works. Then, in Sect. 3, we define the metrics we are going to exploit in our proposal to cluster the properties. Section 4 presents the indexing system, detailing how the metrics are calculated and included in the KG processing

[5] https://www.elastic.co/elasticsearch, last accessed 16th March 2025.

[6] https://lemurproject.org/galago.php, last accessed 16th March 2025.

[7] https://github.com/terrier-org/pyterrier, last accessed 16th March 2025.

[8] https://lucene.apache.org/, last accessed 16th March 2025.

pipeline. Section 5 shows the complete evaluation we have carried out to validate our proposal and its results, and, finally, in Sect. 6, we summarize the conclusions and draw some possible future work.

2 Related Work

Focusing on Question Answering or fact retrieval over KGs, we can find a diverse breadth works focused on keyword interpretation [4] (i.e., building a formal query out of a set of keywords). Here, influenced by keyword search over databases (e.g., BANKS [1]), we can find works such as SPARK [34], Q2Semantic [32] and SemSearchPro [30]. However, these works are detached from an entity-centric interpretation, as their query models assume that every single element in the input must be a matching element in the graph. More recently, we can find the work by Marx et al. [19] (CACAO), which enables the use of keyword factual queries over Knowledge Graphs via a spread activation method, although it inherits the drawbacks of previous approaches for entity exploration and search. On the other hand, we can additionally find works focused on processing natural language inputs. Given the breadth of this latter line of research, we will refer to the surveys by Lopez et al. [18] and, more recently, the one by Diefenbach et al. [8]. In general, the returned results of these approaches are not ranked by any relevance criteria. The ranking of SPARQL results is usually tackled by employing static centrality measures, which capture the popularity of a resource independently of the query, such as RDFRank [9]. A variation of this approach can be found in QUIRA [20], where Menendez et al. propose a modification of PageRank to capture not just the centrality of the resources, but also the informativeness of the properties describing them. However, this measure again ranks results independently of their relevance to the query. Most recently, Vakulenko et al. [31] propose an unsupervised message passing algorithm for QA over KGs, relying on single Entity Search queries to build a set of ranked answers. In this latter case, they relax the matching requirements by adopting an underlying Entity Retrieval approach, which should be considered complementary to ours, as their initial search stage would directly benefit from our contribution.

Regarding Entity-Oriented Search, we have the different works by Balog et al. that are well described in his book on the subject [2]. In general, defining VDoc templates to build inverted indexes, which dictate which properties must be used and how they should be arranged in the VDoc fields, is done manually. This requires applying knowledge about the domain and the underlying data distribution (i.e., the population of the KG). Newer approaches additionally rely on Learning to Rank [17] to fine tune field-aware retrieval model parameters, such as field weights in BM25F [25] or FSDM [33]. Building on this approach, there have been lately some interesting works employing neural networks, reranking or complementing results of a retrieval model. In this regard, we can find systems employing Graph Neural Networks (GNNs), such as GEEER [10]; word embeddings, such as KEWER [22], EMBERT [11], and CoER [15]; or both, such as BLP [7]. However, both Learning to Rank and reranking approaches require query sets along with manual assessments to perform training. These approaches are usually evaluated against DBpedia-Entity v2 [13], a dataset we use for comparison purposes in our current work as well. Finally, we can find numerous works tackling KG-powered Retrieval Augmented Generation (RAG), as described in the survey by Peng et al. [24], which often depend on an initial Entity Retrieval step. In this regard, our work can be framed as the foundational step, the R in RAG.

In summary, our approach of using Importance Metrics aims at avoiding any manual or supervised processing, that is, defining the VDocs and fine tuning retrieval model field weights. By evaluating our own Entity Search system, employing off-the-shelf indexing retrieval systems, we show that it makes it possible to set up such a system over any KG in a completely unsupervised way.

3 Proposed Importance Metrics

In the context of improving the presentation layer of Web search engine results, Hasibi et al. [12] proposed the use of features to dynamically generate relevant Entity Cards for a given user query. For this, they defined several metrics that were used to select the best *entity facts*, property-object pairs $f = \langle p, o \rangle$ of a particular entity, via a Learning to Rank approach.

Inspired by their work, we have built a new set of metrics that make it possible to instead assess the importance of properties exclusively, from the point of view of their informativeness. This new criteria allows us to define how discriminant they are when it comes to performing searches over the KG, necessary for Entity Search. We will additionally decompose these property metrics across the types of entities associated to them, thus allowing to obtain their importances on specific parts of the type hierarchy or for the KG as a whole. We adopt part of the notation used in their work, summarized in Table 1.

Table 1. Notation used in our metrics, adopted from Hasibi et al. [12].

Name	Notation	Definition
Fact	f	$f = \langle p, o \rangle \ : f_p = p, f_o = o$
Entity Facts	$\mathcal{F}_e$	$\{\langle p, o \rangle \mid \langle s, p, o \rangle \in KG, s = e\}$
Entities	$\mathcal{E}$	$\{s \mid \langle s, p, o \rangle \in KG\}$
Types of entity s	Types(s)	$\{t \mid \langle s, \text{ rdf:type}, \ t \rangle \in KG\}$
Entities of type t	$\mathcal{E}_t$	$\{s \mid \langle s, p, o \rangle \in KG, \ t \in Types(s)\}$

In this work, we have defined three novel importance metrics: $ETImp_p(p, t)$ (*Entity Type Importance*), $EntF_p(p, t)$ (*Property Entropy for Type*), and their combination, $EntETImp_p(p, t)$ (*Entropy-Entity Type Importance*). The higher a metric value is, the more relevant (p, t) is for the chosen criteria, with value bounds changing across KGs.

3.1 $ETImp_p(p,t)$ - Entity Type Importance

The $ETImp_p(p,t)$ metric weighs the entity frequency of a property, $EF_p(p,t)$ (i.e., how often this property appears in entities of type t), by the log of the inverse of its ratio in the entities of type t. By establishing a weak analogy to the TF-IDF scheme, $EF_p(p,t)$ would act as TF, boosting the properties that are present in all the entities of a given type. Similarly, the logarithm would act as the IDF, reducing its importance for t as the proportion of entities of type t that have p gets closer to 1.

This metric is defined by the following formula:

$$ETImp_p(p,t) = EF_p(p,t) \cdot log\left(\frac{|\mathcal{E}_t|}{EF_p(p,t)}\right) \tag{1}$$

with:

$$EF_p(p,t) = |\{s \mid \langle s,p,o\rangle \in KG \ \wedge \ t \in Types(s)\}| \tag{2}$$

3.2 $EntF_p(p,t)$ - Property Entropy for Type

The $EntF_p(p,t)$ metric measures the Shannon's entropy of a property, given by the number of unique objects it is found associated with (either literal values or *URIs*). It is defined by the following formula:

$$EntF_p(p,t) = -\sum_{\{f \ \mid \ p=f_p\}} PF(f,t) \cdot log_2(PF(f,t)) \tag{3}$$

where $PF(f,t)$ is the probability of a fact f for a given type and f_p:

$$PF(f,t) = \frac{FF(f,t)}{FF_p(f_p,t)} \tag{4}$$

with $FF(f,t)$ being the fact frequency for a given type, and $FF_p(p,t)$ the fact frequency restricted to a given property and type:

$$FF(f,t) = |\{\langle s,p,o\rangle \mid \langle s,p,o\rangle \in KG, \ p = f_p, \ o = f_o, \ t \in Types(s)|\} \tag{5}$$

$$FF_p(p,t) = |\{\langle s,p,o\rangle \mid \langle s,p,o\rangle \in KG, \ t \in Types(s)\}| \tag{6}$$

3.3 $EntETImp_p(p,t)$ - Entropy-Entity Type Importance

Finally, we can combine the two previous criteria as:

$$EntETImp_p(p,t) = EntF_p(p,t) \cdot ETImp_p(p,t) \tag{7}$$

And we can additionally perform a weighted geometric mean of both:

$$EntF_p(p,t)^w \ \cdot ETImp_p(p,t)^{1-w}, \ 0 < w < 1.0 \tag{8}$$

In the following section, we will show the architecture of our indexing system and how these metrics are included in the pipeline in order to automatically define VDoc templates that are tailored to a particular KG.

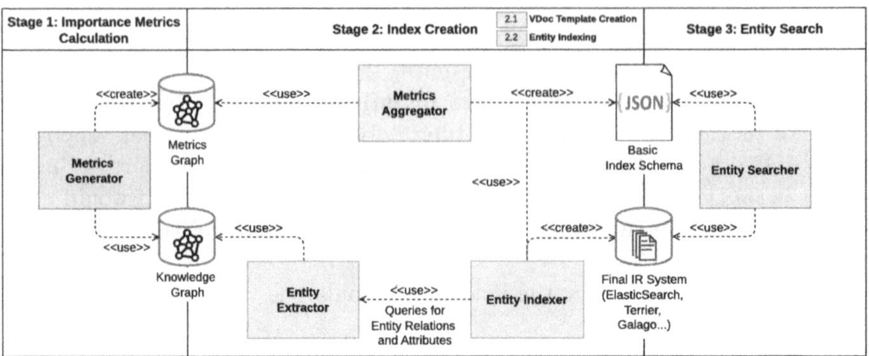

Fig. 1. Stage diagram of our system, showing its data dependencies across stages.

4 Indexing and Retrieval System

We have implemented our proposal in an entity indexing and retrieval system, *Knowgly*, that is divided into three separate stages, as shown in Fig. 1. This system accepts term-based user queries, and returns a rank of entity IRIs that match the query.

4.1 Importance Metrics Calculation

This initial stage involves calculating the different importance metrics for each property (previously defined in Sect. 3), and storing them to be used at a later time in the form of additional triples. The parallelization of these metrics is trivial due to a lack of dependencies, with an $\mathcal{O}(n \cdot m)$ computational cost both in time and memory, where $n = |\{p \mid \langle s, p, o\rangle \in KG\}$ and $m = |\{t \mid \langle s, \text{rdf:type}, t\rangle \in KG\}|$[9].

4.2 Index Creation

This stage is further subdivided in two steps: *Virtual Document Template Creation* and *Entity Indexing*.

Virtual Document Template Creation. The first part of this stage consists of employing a *Metrics Aggregator* on the previously calculated metrics to define a VDoc template. This template contains a mapping of every property to its corresponding index field, and is obtained via the KMeans++[10] algorithm. By clustering the properties by their metric values and then sorting these clusters by their centroids, we will obtain groups of properties ordered by their importance. These groups will act as the VDoc fields. Furthermore, the choice of the

[9] A more in-depth analysis of the computational cost of these metrics is presented in the supplementary material.

[10] Compared to its traditional counterpart, KMeans++ allows us to choose the best clusterization across several repetitions, ensuring a very high stability.

k parameter of KMeans allows us to define our indexing granularity. During this stage, it is possible to choose different Metrics Aggregator configurations, namely:

- **Global Metrics Aggregation**: This strategy involves creating a global VDoc template to be used to index all entities in the KG, containing every property. Regardless of the chosen metric, the metric value for a property p will be the sum of the values found for every type t associated with it, given by:
$$GlobalMetricValue(p,\ metric) = \sum_{t \in KG} metric(p,\ t)$$
Note that if p is not associated to a type t in the KG, $metric(p,\ t) = 0$

- **Type-based Metrics Aggregation**: Instead of aggregating the global importance of the properties, it is also possible to generate a VDoc template for a single type t, which will only contain those properties associated with t via any of the entities participating in it, and whose final value will simply be $metric(p,\ t)$. This opens up the possibility of *slicing* the KG into multiple vertical indexes, each of them optimized for retrieval of entities of a particular type or a subset of them.
These vertical indexes can be helpful on KGs resulting from an integration or information extraction process, as we may only be interested in one particular entity type corresponding to a subset of the complete KG. In these cases, type-based Metrics Aggregators allow focusing on the properties and the local topology of the subgraph for that particular type.

Note that, in both cases, we do not filter any properties from the VDocs so as to not remove any possibly relevant information from the entities. Thus, we will always index all their content. During retrieval, spurious properties located in the least relevant clusters will contribute less to the final retrieval score.

Entity Indexing. On this last part of the stage, our system employs the chosen VDoc template to build each entity's VDoc, by routing all the object text values of their properties to the corresponding index fields. These VDocs can then be fed into any target IR system as traditional fielded documents. The detailed indexing process for a particular entity, as well as an example of the retrieval process, are shown in Fig. 2.

4.3 Entity Search

Once the indexing has finished, it is possible to retrieve entities in the target IR system. Note that we must establish a weighting strategy to state the relative relevance of each cluster, now turned into fields, during retrieval. This can be derived directly from the centroid values of each field the VDoc template. In this regard, as we will describe in the next section, we adopted a uniformly spaced penalization scheme, where the least relevant clusters are linearly penalized. In the following, we present the evaluation we have performed in order to show the feasibility of our approach.

5 Evaluation

In this section, we first present the experimental setup of our evaluation and then focus on answering our main research questions, which are:

Q1 Can we obtain a global VDoc template directly from the analysis of a KG in a completely unsupervised way and achieve performant entity searches?

Q2 Can we automatically derive vertical search engines focusing on specific types $t \in KG$ with comparable results?

5.1 Experimental Setup

Test Collections. We have used the DBpedia-Entity v2 [13] test collection and QUIRA's IMDb dataset [20] to evaluate and compare our approach in different domains, detailed in Table 2. DBpedia-Entity v2 consists of 467 queries for the 2015-10 dump of the DBpedia, adapted from four different benchmarks: *SemSearch_ES*, *INEX-LD*, *ListSearch*, and *QALD-2*, consisting of queries of named entities, keywords, lists of entities, and Natural Language (NL) questions[11], respectively. QUIRA's IMDb dataset consists of 50 keyword-based queries adapted from Coffman et al. [6].

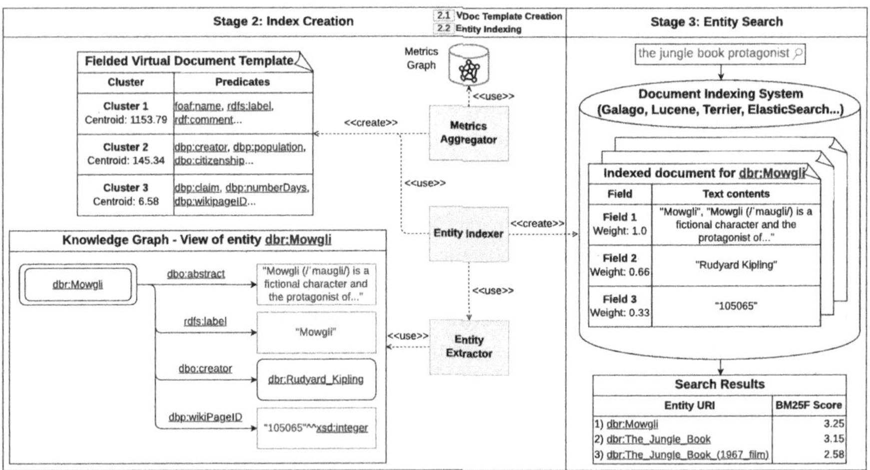

Fig. 2. Illustration of the indexing and retrieval of an entity using a VDoc template and its extracted properties. Note the assignment of each property's object to its corresponding field, and the identical field ordering with custom weights.

[11] Note that, like in the original systems used in [13], our system is not suited towards Question Answering. NL queries are simply treated as keyword queries, with no preprocessing.

Table 2. Details of the DBpedia and IMDb datasets used in the experiments.

Dataset	#Triples	#Subjects	#Properties	#Objects	#Types	#Queries
DBpedia-Entity v2	212,737,087	19,781,237	61,828	45,617,502	385	467
QUIRA's IMDb KG	181,654,004	26,084,168	133	37,210,281	28	50

Clustering Configurations. We explored the clustering capabilities of our proposed importance metrics as well as weighted geometric means of them, using both global and type-based template generation strategies. In all cases, we used $k = 3$ and $k = 5$ clusters. KMeans++ has been run with a 500 iterations limit, using the best results after 5 repetitions to ensure its stability.

Indexing Process. We followed the indications of the original works regarding the indexing setup, this is:

- DBpedia-Entity v2: We filtered out all entities that lack *rdfs:label* or *rdfs:comment* data properties as in [13] and indexed the remaining ones, leading to 4,641,784 indexed entities. Moreover, we only considered the types whose IRIs are within the DBpedia ontology namespace (i.e., http://dbpedia.org/) when using our metrics.
- IMDb KG: No filtering has been applied.

Furthermore, we performed the following processing when building the text representations of the objects of an entity:

- If the object was an IRI, we used the value of its *rdfs:label* attribute, if it existed. As a fallback, we used the IRI's path text, replacing underscores with space symbols.
- If the object was a literal value, we used the enclosed value without its datatype, with no additional processing.

An interesting aspect of these KGs is that DBpedia's 2015-10 dump contains spurious properties due to incorrectly extracted data (e.g., *dbp:birthPlaxe* or *dbp:birthPalce* instead of *dbp:birthPlace*). These properties are implicitly filtered by our metrics, which have shown to be robust enough to deal with noise, by assigning them to the least relevant clusters. When analyzing the available IMDb KG, we observed that it contains a severed version of the actual data, with strong differences among the detail levels of the descriptions of the entities. For example, *Actors* are highly connected and detailed via attributes, but *Movie characters* only contain labels and their link to an actor. This led to a very biased document length distribution and, as we will describe in Sect. 5, forced us to nullify the document length impact during retrieval. The reduced amount of properties in the IMDb KG additionally allowed us to manually analyze the quality of the VDoc templates. All Metrics Aggregators assigned highly descriptive properties for actors (e.g., labels, names, or trivia) to the highest priority field.

Table 3. Summary of the parameters used in the experiments.

Importance Metrics	$ETImp_p(p,t)$, $EntF_p(p,t)$, $EntETImp_p(p,t)$, *geom. weighted means*
Templates	Global and Type-based templates
#Clusters (k)	3, 5
IR Models's Params.	$\{k1 : 1.2,\ b : 0.75\}$, $\{k1 : 0.9,\ b : 0.4\}$ (b suppressed in IMDb)
Weighting Scheme	Linear Penalization

Retrieval Models. We have adopted the BM25F Retrieval Model [25], an adaptation of the popular BM25 model which allows per-field weighting, comparing its results to BM25 baselines where applicable. Both models depend on two parameters, $k1$ and b, where $k1$ controls the term frequency saturation[12] (usually in the range of [1.2, 2]), and b controls the representation's length normalization[13] (usually set to 0.75). We have chosen Galago as the retrieval engine. We have evaluated two pairs of $k1$ and b parameters, $\{k1 : 1.2,\ b : 0.75\}$, the default values for BM25 in Galago, Lucene and Elasticsearch, and $\{k1 : 0.9,\ b : 0.4\}$, which correspond to the optimal BM25 parameters reported by Hasibi et al. [11] in their first retrieval stage using Entity-aware Transformers. In the case of BM25F, we treat these parameters as being global and thus the same for every field. As previously mentioned, we had to deal with heavily unbalanced VDoc lengths in the IMDb Dataset, thus we performed the tests for that dataset by setting b to 0, reducing the importance of poorly described entities.

The BM25F field weights, which penalize or boost each field's contribution to the final score, follow a penalization scheme consisting of uniformly spaced values up to a maximum field weight of 1.0 and excluding 0.0, which corresponds to [1.0, 0.66, 0.33] in a 3-cluster configuration and to [1.0, 0.8, 0.6, 0.4, 0.2] in a 5-cluster configuration. This way, the cluster with the highest centroid value is weighted by 1.0, followed by the rest of the clusters in descending order. Moreover, in order to analyze our parameter setup, we have also tested the effects of training both the field weights and the global $k1$ and b parameter values using Coordinate Ascent. Finally, we have analyzed the effects of applying EMBERT [11] reranking to our results in order to ensure that our approach was consistent with their reported results.

Table 3 summarizes the main parameters of the experimental setup. All tasks have been executed on a desktop PC with a i9-12900K 3.2GHz CPU (16-cores/24-threads), 64GB of memory[14], and a 12GB GeForce 3080 RTX (for executing EMBERT).

[12] A value of $k1 \geq 1$ controls and slows down the score growth on high appearances of a term t in a document. In our case, we do not want to excessively penalize entities with many properties containing t, as it may reflect that they are important for a query.

[13] Values of $b \geq 0$ increasingly penalize too long documents. In our case, we want to penalize entities who may mention a query term t due to simply having an excessive amount of properties overall, as less-described entities may still be relevant.

[14] The performance details in terms of indexing times and memory consumption can be found in the supplementary material.

We describe further experiments with larger numbers of clusters and alternative field weight schemes, how KMeans++ is employed, and an analysis of the property distributions of DBpedia and IMDb in the supplemental materials.

5.2 Evaluation Results

Research Question Q1. We first aim at answering research question Q1 by evaluating our approach on both presented datasets. For comparison's sake, as the original works did, we report the Normalized Discounted Cumulative Gain at 10 (NDCG@10) and 100 (NDCG@100) for DBpedia-Entity v2, and Mean Average Precision (MAP) for IMDb. Besides, to fully answer Q1, we will also evaluate our approach using Coordinate Ascent, and as the starting point for the re-ranking approach of EMBERT [11].

DBpedia-Entity v2. As shown in Table 4, the best overall results for DBpedia-Entity v2 are consistently obtained by $EntF_p(p,t)$ and its combination with $ETImp_p(p,t)$ with {0.75, 0.25} weights, regardless the number of clusters[15]. Using our unsupervised configuration, we obtain an improvement of approximately 70% regarding the reported BM25 results when using the default BM25 parameters {1.2, 0.75} [13], going up to approximately 78% when using the parameters for the retrieval step reported in EMBERT, {0.9, 0.4} [11]. We additionally show its comparison against SDM [21], the best performing non-fielded retrieval method, which we also surpass.

To establish a comparison with the SoA, we study the differences between our results and BM25F-CA, the best non-reranked system reported in [13], which is backed by supervised learning for inferring the best field weights in their index setup. In doing so, we reach its results by just a 1% to 5% below with the added benefit of not requiring any annotated query sets for training the weights. Given the different sources of the queries in this dataset, we report as well the intermediate four groups of results, highlighting that our approach outperforms the BM25F-CA on the *SemSearch-ES* and *ListSearch* categories on NDCG@10, the two categories that are closer to the pure Entity-Search task.

IMDb. Table 5 shows the results for IMDb. In this case, the best performing aggregator was the equally weighted combination of $EntF_p(p,t)$ and $ETImp_p(p,t)$ for both 3 and 5-cluster configurations. When comparing our scores to QUIRA's, we have to bear in mind that they evaluate the capability of their PageRank variation to rerank results of SPARQL queries. As such, they start from the results of a set of manually crafted SPARQL queries, instead of the results from a keyword search as our approach does; thus, we consider that our results are performant in a pure retrieval scenario.

[15] Due to space restrictions, we have included the complete tables, alongside their statistical analysis, in the project's repository.

Table 4. DBpedia-Entity's NDCG@10 and NDCG@100 results for Global Metrics Aggregator. Bold and underlined text highlight the best and second best results column-wise, for each group of tests.

Configuration						SemSearch		INEX-LD		ListSearch		QALD-2		Total	
Metric 1	Metric 2	Weights	Fields	k1	b	@10	@100	@10	@100	@10	@100	@10	@100	@10	@100
BM25F-CA (DBpedia-Entity v2)						**0.628**	**0.720**	**0.439**	**0.530**	**0.425**	**0.511**	**0.369**	**0.461**	**0.460**	**0.550**
BM25 (DBpedia-Entity v2)						0.250	0.411	0.028	0.361	0.220	0.330	0.275	0.337	0.256	0.358
SDM (DBpedia-Entity v2)						0.553	0.667	0.403	0.491	0.396	0.490	0.339	0.427	0.418	0.514
$EntF_p(p,t)$	-	-	3	1.2	0.75	0.620	0.690	0.404	0.453	0.399	0.448	0.338	0.408	0.435	0.495
				0.9	0.4	0.642	0.708	**0.412**	**0.468**	0.427	0.471	0.355	0.426	0.454	0.514
$ETImp_p(p,t)$	-	-	3	1.2	0.75	0.609	0.694	0.405	0.448	0.391	0.444	0.333	0.404	0.429	0.494
				0.9	0.4	0.634	0.702	0.403	0.458	0.423	0.471	0.350	0.427	0.448	0.511
$EntETImp_p(p,t)$	-	-	3	1.2	0.75	0.627	0.702	0.404	0.452	0.394	0.444	0.337	0.407	0.436	0.497
				0.9	0.4	0.638	0.705	0.410	0.466	0.426	0.472	0.355	0.428	0.453	0.514
$EntF_p(p,t)$	$ETImp_p(p,t)$	0.75 - 0.25	3	1.2	0.75	0.618	0.693	0.407	0.453	0.394	0.444	0.337	0.410	0.434	0.496
				0.9	0.4	0.647	**0.711**	0.411	0.467	**0.427**	**0.473**	**0.356**	**0.429**	**0.456**	**0.516**
$EntF_p(p,t)$	-	-	5	1.2	0.75	0.618	0.688	0.405	0.452	0.396	0.446	0.342	0.406	0.435	0.494
				0.9	0.4	**0.646**	0.708	0.417	**0.473**	**0.425**	**0.472**	0.353	0.426	**0.455**	**0.516**
$ETImp_p(p,t)$	-	-	5	1.2	0.75	0.602	0.685	0.400	0.442	0.386	0.441	0.332	0.401	0.425	0.488
				0.9	0.4	0.637	0.704	0.417	0.465	0.423	0.467	**0.354**	0.425	0.453	0.511
$EntETImp_p(p,t)$	-	-	5	1.2	0.75	0.627	0.695	0.404	0.449	0.386	0.441	0.339	0.402	0.434	0.492
				0.9	0.4	0.640	0.705	**0.419**	0.472	0.420	0.468	0.352	0.423	0.453	0.513
$EntF_p(p,t)$	$ETImp_p(p,t)$	0.75 - 0.25	5	1.2	0.75	0.617	0.688	0.403	0.452	0.391	0.447	0.339	0.406	0.432	0.494
				0.9	0.4	0.644	**0.709**	0.415	0.468	0.420	0.471	0.352	**0.427**	0.453	0.515

Supervised Methods. To check whether our proposal reaches similar results when applying supervised parameter tuning and reranking, we trained our best configurations with Coordinate Ascent (CA) and applied EMBERT on them respectively, reporting the results in Table 6. Regarding CA, we can see how we slightly surpass the reported values when training both BM25F parameters and field weights, but our differences with our unsupervised versions are almost negligible. Regarding EMBERT reranking, there are two different setups: EMBERT fine-tuned with MSMarco data (EMBERT 1st), and fine-tuned with MSMarco *and* DBpedia Entity v2. Reranking our approach with EMBERT 1st, which we can consider as a semi-supervised model, worsens our original results from Table 4 for 3 clusters and slightly improves them for 5 clusters. On the other hand, by reranking with the second variant, we were able to surpass their reranking for BM25F-CA[16].

Hence, the results obtained for both datasets in different domains, and the comparison using other techniques (showing robust results against their previously reported ones) allow us to answer Q1 affirmatively.

[16] We have also observed that, apart from the need of obtaining annotated data for the target KG in order to fine tune the reranking model, in our experimental setup, EMBERT introduced an average time overhead of two orders of magnitude (~22 s vs. ~0.18 s per query in EMBERT vs. Galago on average).

Table 5. IMDb MAP results for Global Metrics Aggregators. Bold and underlined text highlight the best and second best results column-wise, for each group of tests.

Configuration	MAP
QUIRA	0.820

Configuration						MAP	Configuration						MAP
Metric 1	Metric 2	Weights	Fields	k1	b		Metric 1	Metric 2	Weights	Fields	k1	b	
$EntF_p(p,t)$	-	-	3	1.2	0.0	0.703	$EntF_p(p,t)$	-	-	5	1.2	0.0	0.720
				0.9	0.0	0.702					0.9	0.0	0.719
$ETImp_p(p,t)$	-	-	3	1.2	0.0	0.716	$ETImp_p(p,t)$	-	-	5	1.2	0.0	0.708
				0.9	0.0	0.717					0.9	0.0	0.711
$EntETImp_p(p,t)$	-	-	3	1.2	0.0	0.706	$EntETImp_p(p,t)$	-	-	5	1.2	0.0	0.711
				0.9	0.0	0.707					0.9	0.0	0.713
$EntF_p(p,t)$	$ETImp_p(p,t)$	0.25 - 0.75	3	1.2	0.0	0.714	$EntF_p(p,t)$	$ETImp_p(p,t)$	0.25 - 0.75	5	1.2	0.0	0.717
				0.9	0.0	0.714					0.9	0.0	0.718
$EntF_p(p,t)$	$ETImp_p(p,t)$	0.50 - 0.50	3	1.2	0.0	0.724	$EntF_p(p,t)$	$ETImp_p(p,t)$	0.50 - 0.50	5	1.2	0.0	**0.725**
				0.9	0.0	**0.724**					0.9	0.0	0.724
$EntF_p(p,t)$	$ETImp_p(p,t)$	0.75 - 0.25	3	1.2	0.0	0.711	$EntF_p(p,t)$	$ETImp_p(p,t)$	0.75 - 0.25	5	1.2	0.0	0.710
				0.9	0.0	0.710					0.9	0.0	0.709

Research Question Q2. To evaluate the capability of our approach of building vertical views (i.e., build a VDoc template for a type $t \in KG$ and only index $\mathcal{E}_t$), we carried out an experiment with the queries of DBpedia-Entity v2 that returned entities of type *dbo:People* (165 queries), *dbo:Organization* (153 queries), and *dbo:Place* (179 queries)[17]. As we did not have a proper baseline, we compared the Type-based aggregators against the same indexing procedure, using instead the templates we previously obtained from the best Global Metrics Aggregators for 3 and 5 clusters.

Table 7 contains such results. In the first group (marked with $*$), the entities were indexed using a global VDoc template; while in the second and third groups, the entities were indexed using VDoc templates calculated with Type-based Metrics Aggregators for each considered type. We can see how focusing the metrics to particular types improves the search capabilities of our approach for *dbo:Person* and *dbo:Organisation* and achieves similar results for *dbo:Place.* These results support our claim of our metrics being able to capture the importance of the properties in a domain independent way, and allow us to affirmatively answer research question Q2.

Final Remarks. Apart from answering our research questions, the analysis of the results has led us to the following final remarks.

Optimum Configuration. Using $EntF_p(p,t)$ metric alone has led us to obtain good results, but there are strong signals that $ETImp_p(p,t)$ does contribute to their improvement. This can be seen in Tables 4, 5 and 6, where $EntF_p(p,t)$ achieves 3 top results and a combination of both achieves 6 top places, while $ETImp_p(p,t)$ as a standalone metric achieves just one top place. Using $EntF_p(p,t)$ would achieve the Pareto optimum when we focus on the pair

[17] We further filtered queries where the assessment was empty after the type filtering.

Table 6. DBpedia-Entity's NDCG@10 and NDCG@100 results for the two best Global Metrics Aggregator configurations, with retrieval models fine-tuned via Coordinate Ascent (upper table) including or excluding the $k1$ and b parameters, and using Reranking via EMBERT (bottom table). Bold and underlined text highlight the best and second best results column-wise.

Coordinate Ascent Configuration					SemSearch		INEX-LD		ListSearch		QALD-2		Total	
Metric 1	Metric 2	Weights	Clusters	Incl. k1, b	@10	@100	@10	@100	@10	@100	@10	@100	@10	@100
BM25F-CA (DBpedia-Entity v2)					**0.628**	**0.720**	**0.439**	**0.530**	**0.425**	0.511	**0.369**	**0.461**	**0.460**	**0.550**
BM25-CA (DBpedia-Entity v2)					0.586	0.688	0.412	0.505	0.422	**0.514**	0.357	0.443	0.440	0.533
$EntF_p(p,t)$	$ETImp_p(p,t)$	0.75 - 0.25	3	Yes	**0.653**	0.712	0.427	0.479	0.425	0.471	**0.360**	**0.434**	**0.461**	**0.520**
				No	0.650	0.712	0.423	0.475	**0.428**	0.471	0.359	0.432	0.460	0.519
$EntF_p(p,t)$	-	-	5	Yes	0.650	**0.713**	0.431	**0.486**	0.427	**0.472**	0.350	0.425	0.459	0.519
				No	0.649	0.713	**0.432**	0.481	0.419	0.467	0.352	0.422	0.457	0.516

EMBERT Reranking Configuration					SemSearch		INEX-LD		ListSearch		QALD-2		Total	
Metric 1	Metric 2	Weights	Clusters	Model	@10	@100	@10	@100	@10	@100	@10	@100	@10	@100
BM25F-CA (DBpedia-Entity v2)				EMBERT (1st)	0.609	0.695	**0.418**	**0.490**	0.452	0.518	0.398	0.483	0.466	0.544
				EMBERT	**0.621**	**0.695**	0.399	0.469	**0.492**	**0.535**	**0.435**	**0.512**	**0.486**	**0.553**
$EntF_p(p,t)$	$ETImp_p(p,t)$	0.75 - 0.25	3	EMBERT (1st)	0.599	0.680	0.400	**0.467**	0.443	0.499	0.385	0.464	0.454	0.525
				EMBERT	**0.638**	**0.703**	0.397	0.458	**0.500**	**0.530**	**0.447**	**0.511**	**0.496**	**0.551**
$EntF_p(p,t)$	-	-	5	EMBERT (1st)	0.599	0.680	**0.403**	0.464	0.444	0.500	0.384	0.463	0.455	0.525
				EMBERT	0.638	0.703	0.398	0.456	0.500	0.530	0.445	0.511	0.495	0.550

Table 7. DBpedia-Entity's NDCG@10 and NDCG@100 results for Type-based Metrics Aggregators. Bold and underlined text highlight the best and second best results column-wise, for each group of tests. The first group of results (marked with $*$) correspond to using Global instead of type-based templates, for comparison purposes.

Configuration						Person		Organisation		Place	
Metrics Aggr. 1	Metrics Aggr. 2	Weights	Clusters	k1	b	@10	@100	@10	@100	@10	@100
$EntF_p(p,t)^*$	$ETImp_p(p,t)^*$	0.75 - 0.25	3	1.2	0.75	0.490	0.549	0.388	0.445	0.393	0.458
				0.9	0.4	**0.507**	**0.569**	0.378	0.443	0.407	0.477
$EntF_p(p,t)^*$	-	-	5	1.2	0.75	0.492	0.552	**0.388**	**0.446**	0.393	0.461
				0.9	0.4	0.505	0.567	0.378	0.442	**0.412**	**0.483**
$EntF_p(p,t)$	-	-	3	1.2	0.75	0.490	0.550	0.386	0.442	0.358	0.426
				0.9	0.4	**0.516**	**0.574**	0.385	0.445	0.376	0.455
$ETImp_p(p,t)$	-	-	3	1.2	0.75	0.498	0.557	0.395	0.430	0.256	0.345
				0.9	0.4	0.505	0.568	0.415	0.452	0.334	0.413
$EntETImp_p(p,t)$	-	-	3	1.2	0.75	0.494	0.553	0.397	0.430	0.303	0.383
				0.9	0.4	0.508	0.567	**0.422**	**0.455**	0.359	0.436
$EntF_p(p,t)$	$ETImp_p(p,t)$	0.75 - 0.25	3	1.2	0.75	0.490	0.550	0.382	0.442	0.363	0.436
				0.9	0.4	0.511	0.569	0.394	0.451	**0.387**	**0.462**
$EntF_p(p,t)$	-	-	5	1.2	0.75	0.491	0.554	0.391	0.444	0.371	0.441
				0.9	0.4	0.510	0.570	0.392	0.448	**0.402**	**0.474**
$ETImp_p(p,t)$	-	-	5	1.2	0.75	0.492	0.547	0.383	0.413	0.272	0.355
				0.9	0.4	0.505	0.567	**0.413**	0.443	0.351	0.424
$EntETImp_p(p,t)$	-	-	5	1.2	0.75	0.493	0.547	0.386	0.422	0.295	0.375
				0.9	0.4	**0.520**	**0.571**	0.412	0.445	0.358	0.436
$EntF_p(p,t)$	$ETImp_p(p,t)$	0.75 - 0.25	5	1.2	0.75	0.492	0.550	0.386	0.439	0.369	0.436
				0.9	0.4	0.502	0.568	0.397	**0.450**	0.398	0.463

<results, metrics calculation time>, while combining it with $ETImp_p(p, t)$ would be the suggested most performant option in general. Regarding the number of clusters (i.e., fields), we explored increasing k up to 15, with no drastic changes in the results. Thus, we would advocate for 5 clusters to maintain a fine enough granularity. An analysis of this aspect is included in the supplementary material.

Retrieval Parameters. We recommend using the default $k1$ and b parameters set by Galago, Lucene and Elasticsearch, as they are suitable for the entities we may commonly wish to prioritize during retrieval. Further deviations, such as the parameters assigned by Hasibi et al. [11] or field-specific parameter assignments, may provide higher results at the cost of not being applicable to all KGs or requiring supervised approaches, something we strive to avoid.

Long Tail Results. We have observed that our approach does not currently achieve the best BM25F-CA counterparts on NDCG@100. This may be due to slight differences on the indexing configuration or text generation, as it seems to be confirmed by the training with CA in our approach, where the improvement is slightly noticeable in this regard. We consider the exploration of this particular issue as future work.

6 Conclusions and Future Work

In this work, we have presented a novel approach to build multi-fielded index definitions for KGs that exclusively employs unsupervised Machine Learning techniques. Aside from eliminating the need for evaluation datasets, this approach does not require applying knowledge about the domain and data distribution of the KG for constructing such index definitions. This is enabled by our proposed set of metrics based on information measures, which captures the importance of the different properties in the KG. The retrieval performance achieved by our approach rivals that of systems making use of manually crafted index definitions and supervised Machine Learning in their Retrieval Models. Moreover, we show a performance increase over the SoA when employing our results as the base for supervised reranking approaches. Additionally, we have shown its additional capabilities, such as the possibility of creating locally optimized vertical index slices of the Knowledge Graph based on its type hierarchy. Last but not least, we make available our implementation, directly usable on any RDF KG, alongside the results and supplemental materials mentioned throughout this work, at the project's repository.

As future work, we will explore the possibility of integrating other unsupervised Machine Learning techniques into our system, such as Vakulenko et al.'s Message Passing [31] approach for Question Answering over Knowledge Graphs. Moreover, given the scarcity of evaluation datasets, we would like to extend our evaluation to other KGs by building different test sets, making them available for the community.

Acknowledgments. This work has been supported by the I+D+i project PID2020-113903RB-I00 (funded by MCIN/AEI/10.13039/501100011033), the project T42_23R (funded by Gobierno de Aragón), and the Deutsche Forschungsgemeinschaft (DFG, German Research Foundation) - SFB 1625 - 506711657, subproject A06. Moreover, Samuel's work was supported as well by an Internship Grant granted by the Ministerio de Educación y Formación Profesional of Spain, ID: 22CO1/004278.

References

1. Aditya, B., Bhalotia, G., Chakrabarti, S., Hulgeri, A., Nakhe, C., Parag, Sudarshan, S.: BANKS: browsing and keyword searching in relational databases. In: Proceedings of 28th International Conference on Very Large Data Bases (VLDB 2002), Hong Kong (China), pp. 1083–1086. Morgan Kaufman (2002)
2. Balog, K.: Entity-Oriented Search, 1st edn. Springer (2018)
3. Blanco, R., Mika, P., Vigna, S.: Effective and efficient entity search in RDF data. In: Aroyo, L., et al. (eds.) Proceedings of 10th International Semantic Web Conference (ISWC 2011), Bonn (Germany), pp. 83–97. Springer, Heidelberg (2011)
4. Bobed, C., Mena, E.: Querygen: semantic interpretation of keyword queries over heterogeneous information systems. Inf. Sci. **329**, 412–433 (2016)
5. Both, A., Diefenbach, D., Singh, K., Shekarpour, S., Cherix, D., Lange, C.: Qanary-a methodology for vocabulary-driven open question answering systems. In: Proceedings of the 13th Extended Semantic Web Conference (ESWC 2018), Heraklion (Greece), pp. 625–641. Springer (2016)
6. Coffman, J., Weaver, A.: A framework for evaluating database keyword search strategies. In: Proceedings of the 19th ACM International Conference on Information and Knowledge Management (CIKM 2010), Toronto (Canada), pp. 729–738 (2010)
7. Daza, D., Cochez, M., Groth, P.: Inductive entity representations from text via link prediction. In: Proceedings of the Web Conference 2021, WWW 2021, pp. 798–808. Association for Computing Machinery, New York (2021)
8. Diefenbach, D., Lopez, V., Singh, K., Maret, P.: Core techniques of question answering systems over knowledge bases: a survey. Knowl. Inf. Syst. **55**, 529–569 (2018)
9. Diefenbach, D., Thalhammer, A.: PageRank and generic entity summarization for RDF knowledge bases. In: Proceedings of the 15th Extended Semantic Web Conference (ESWC 2018), Heraklion (Greece), pp. 145–160. Springer (2018)
10. Gerritse, E.J., Hasibi, F., de Vries, A.P.: Graph-embedding empowered entity retrieval. In: Advances in Information Retrieval: 42nd European Conference on IR Research. ECIR 2020, Lisbon, Portugal, 14–17 April 2020, Proceedings, Part I, pp. 97–110. Springer, Heidelberg (2020)
11. Gerritse, E.J., Hasibi, F., de Vries, A.P.: Entity-aware transformers for entity search. In: Proceedings of the 45th International ACM SIGIR Conference on Research and Development in Information Retrieval (SIGIR 2022), Madrid (Spain), pp. 1455–1465. ACM (2022)
12. Hasibi, F., Balog, K., Bratsberg, S.E.: Dynamic factual summaries for entity cards. In: Proceedings of the 40th International ACM SIGIR Conference on Research and Development in Information Retrieval (SIGIR 2017), Shinjuku, Tokyo (Japan), pp. 773–782. ACM (2017)

13. Hasibi, F., et al.: Dbpedia-entity v2: a test collection for entity search. In: Proceedings of the 40th International ACM SIGIR Conference on Research and Development in Information Retrieval (SIGIR 2017), Shinjuku, Tokyo (Japan), pp. 1265–1268. ACM (2017)
14. Hogan, A., et al.: Knowledge graphs. ACM Comput. Surv. **54**(4) (2021)
15. Jafarzadeh, P., Amirmahani, Z., Ensan, F.: Learning contextual representations for entity retrieval. Appl. Intell. **54**(19), 8820–8840 (2024)
16. Lehmann, J., et al.: DBpedia -a large-scale, multilingual knowledge base extracted from wikipedia. Semantic Web **6**(2), 167–195 (2015)
17. Liu, T.Y., et al.: Learning to rank for information retrieval. Found. Trends Inf. Retr. **3**(3), 225–331 (2009)
18. Lopez, V., Uren, V.S., Sabou, M., Motta, E.: Is question answering fit for the semantic web?: a survey. Semantic Web **2**(2), 125–155 (2011)
19. Marx, E., Publio, G.C., Riechert, T.: Cacao: conditional spread activation for keyword factual query interpretation. In: Semantic Systems. The Power of AI and Knowledge Graphs (Proceedings of SEMANTiCS 2019), Karlsruhe (Germany), pp. 256–271. Springer (2019)
20. Menendez, E.S., Casanova, M.A., Paes Leme, L.A.P., Boughanem, M.: Novel node importance measures to improve keyword search over RDF graphs. In: Proceedings of 30th International Conference on Database and Expert Systems Applications (DEXA 2019), Linz (Austria), pp. 143–158. Springer (2019)
21. Metzler, D., Croft, W.B.: A Markov random field model for term dependencies. In: Proceedings of the 28th International ACM SIGIR Conference on Research and Development in Information Retrieval, Salvador (Brazil), SIGIR 2005, pp. 472–479. Association for Computing Machinery, New York (2005)
22. Nikolaev, F., Kotov, A.: Joint word and entity embeddings for entity retrieval from a knowledge graph. In: Advances in Information Retrieval, pp. 141–155. Springer, Cham (2020)
23. Pan, J.Z., Vetere, G., Gomez-Perez, J.M., Wu, H. (eds.): Exploiting Linked Data and Knowledge Graphs in Large Organisations. Springer, Cham (2017)
24. Peng, B., et al.: Graph retrieval-augmented generation: a survey (2024)
25. Robertson, S., Zaragoza, H.: The probabilistic relevance framework: BM25 and beyond. Found. Trends Inf. Retr. **3**(4), 333–389 (2009)
26. Sequeda, J., Lassila, O.: Designing and Building Enterprise Knowledge Graphs. Synthesis Lectures on Data, Semantics, and Knowledge. Springer, Cham (2021)
27. Singh, K., et al.: Why reinvent the wheel: let's build question answering systems together. In: Proceedings of the 2018 World Wide Web Conference (WWW 2018), Lyon (France), pp. 1247–1256. International WWW Conferences Steering Committee (2018)
28. Sowa, J.F., et al.: Semantic networks, vol. 2, pp. 1493–1511. Wiley, New York (1992)
29. Tiddi, I., Schlobach, S.: Knowledge graphs as tools for explainable machine learning: a survey. Artif. Intell. **302**, 103627 (2022)
30. Tran, T., Herzig, D.M., Ladwig, G.: SemSearchPro: using semantics throughout the search process. J. Web Semant. **9**(4), 349–364 (2011)
31. Vakulenko, S., Fernandez Garcia, J.D., Polleres, A., de Rijke, M., Cochez, M.: Message passing for complex question answering over knowledge graphs. In: Proceedings of the 28th ACM International Conference on Information and Knowledge Management (CIKM 2019), Beijing (China), pp. 1431–1440. ACM (2019)

32. Wang, H., Zhang, K., Liu, Q., Tran, D.T., Yu, Y.: Q2Semantic: a lightweight keyword interface to semantic search. In: Proceedings of 5th European Semantic Web Conference (ESWC 2008), Tenerife (Spain), pp. 584–598. Springer (2008)
33. Zhiltsov, N., Kotov, A., Nikolaev, F.: Fielded sequential dependence model for ad-hoc entity retrieval in the web of data. In: Proceedings of the 38th International ACM SIGIR Conference on Research and Development in Information Retrieval, Santiago (Chile), pp. 253–262. ACM (2015)
34. Zhou, Q., Wang, C., Xiong, M., Wang, H., Yu, Y.: SPARK: adapting keyword query to semantic search. In: Proceedings of 6th International Semantic Web Conference (ISWC 2007), Busan (South Korea), pp. 687–700. Springer (2007)

Explainable Temporal Fact Validation Through Constraints Discovery in Knowledge Graphs

Thibaut Soulard(✉), Fatiha Saïs, and Joe Raad

LISN, CNRS (UMR 9015), Paris Saclay University, Gif-sur-Yvette, France
{thibaut.soulard,fatiha.sais,joe.raad}@universite-paris-saclay.fr

Abstract. Truth discovery aims to identify the most reliable or truthful information from multiple sources, particularly in scenarios where sources may be contradictory or unreliable. Addressing this challenge is critical for applications like data exploration, integration, and veracity assessment. Knowledge Graphs (KGs) provide a valuable framework for verifying facts against explicit or implicit domain knowledge. However, the temporal validation of facts that consist in determining their truthfulness within specific time intervals, remains an open area of research. In this paper, we propose an explainable approach that exploits temporal constraints discovered within KGs to classify facts based on their temporal validity. Our method involves identifying both simple and complex temporal constraints in KGs that capture the temporal consistency of facts within an entity's timeline. These constraints are expressed through an extension to time sequences of Allen's temporal algebra. Using these constraints, we validate facts through both symbolic and machine learning-based approaches, comparing their performance under various hyperparameter settings. Experiments conducted on Wikidata, one of the largest publicly accessible KGs, demonstrate the effectiveness and high accuracy of our temporal constraint-based approach.

Keywords: Knowledge Refinement · Temporal Fact Validation · Allen's Algebra

1 Introduction

In an era where the spread of misinformation poses significant threats to political and economic stability, the development of systems capable of validating or refuting facts is crucial. This is all the more true with the use of generative AI, which can sometimes produce inaccurate or misleading information. To address this challenge, Knowledge Graphs (KGs) [10] offer a valuable mean for verifying information based on the statements they contain. KGs have emerged as fundamental knowledge sources that support the development of various AI-related applications. Among many others [18], KGs provide semantic and personalised

E. Curry et al. (Eds.): ESWC 2025, LNCS 15718, pp. 227–244, 2025.
https://doi.org/10.1007/978-3-031-94575-5_13

contexts for question answering [21], impose domain/user constraints in recommender systems [16] and aid in image recognition [6].

However, despite the large number of KGs published in recent years, only a few explicitly associate a temporal component with their time-dependent facts. This temporal component, present in resources like Wikidata[1]—one of the largest publicly available KGs— is essential for the quality of KGs, as many facts are only valid in a specific time interval or at a precise temporal point. For instance, the fact `(Obama, headOfState, USA)` is only valid in the interval starting on `2009/01/20` and ending on `2017/01/19`. With the increasing use of KGs for fact-checking applications [24], ensuring the temporal accuracy of these facts becomes necessary. However, the challenge of validating facts by determining their truthfulness within specific time intervals remains an underexplored area of research.

In this work, we propose an explainable approach that exploits KGs to verify the validity of facts within a given time interval. Our approach is applicable to any KG that associates temporal intervals with its facts, without requiring prior knowledge. We assume that the validity of a temporal fact in a KG depends on its consistency with the timeline of the related entity. For instance, persons cannot hold the position of a surgeon in a hospital if they have not been already graduated as a doctor in medicine. Such consistency can be expressed using temporal constraints between KG properties, such as `before(graduatedAsDoctor,isSurgeonIn)`. However, these constraints are not explicitly stated in KGs. This is why our approach starts by discovering these temporal constraints from the KG, which are then used to validate or refute facts. Our approach can discover both simple and complex temporal constraints by exploiting Allen's relations between temporal intervals [2]. We extend these relations from time intervals to time sequences in order to discover richer temporal constraints. These discovered constraints are then used to assess whether a given temporal fact is valid in the KG using either a symbolic or a machine-learning based approach. We believe that the use of these constraints ensures the transparency of our approach, as it can provide users with an explanation by listing all the temporal constraints used to validate or refute a given fact.

In what follows, we first review the related work (Sect. 2) and introduce the necessary notations (Sect. 3). We then provide an overview of our approach (Sect. 4) and detail our method for discovering temporal constraints in a KG (Sect. 5). Next, we describe how these discovered constraints are deployed to verify the temporal validity of a given fact (Sect. 6). We examine various methods for combining constraints to assess the validity of a fact, such as using a voting system where all constraints are treated equally or training a machine learning model to identify the most suitable temporal constraints for fact validation. Finally, we evaluate our approach on different classes, containing millions of temporal facts, extracted from the Wikidata KG (Sect. 7).

[1] https://www.wikidata.org/.

2 Related Work

Fact validation, also known as veracity assessment, is a complex task that requires integrating and evaluating information from multiple sources while accounting for factors such as source reliability, majority consensus, and the context of the fact. ClaimsKG by Tchechmedjiev et al. [24] represents a foundational effort in structuring claims and their verifications within a Knowledge Graph (KG). Their work represents complex statements that can be decomposed into atomic sub-claims, which can be aligned with an existing KG through techniques like Entity Linking [9] to perform fact validation [13]. For instance, the statement "AI improves healthcare by enhancing diagnostic accuracy and optimizing treatment plans" can be divided into atomic claims such as <AI, improves, healthcare>, <AI, enhances, diagnostic accuracy>, <AI, optimizes, treatment plans>. This granularity facilitates validation using methods such as flow- or path-based algorithms [7,22,23], or by leveraging Large Language Models (LLMs) as demonstrated in [3].

These efforts primarily focus on static claims and lack mechanisms to handle temporal dynamics critical to real-world fact validation. The temporal dimension of facts introduces additional complexity, as atomic claims are rarely universally true and often depend on specific time intervals (e.g. `(Obama, headOfState, USA)`). While the relational databases community has long explored temporal validity and context [20], the topic has only recently gained traction in the KG community [4,28]. The development of Temporal Knowledge Graph Completion (TKGC) models has primarily focused on *link prediction*, aiming to infer missing facts in a temporal context by predicting relationships or entities using methods such as translation-based embeddings, tensor decomposition, Graph Convolutional Networks (GCNs), and recurrent architectures like LSTM and GRU [5,26]. These models are effective for tasks where the objective is to identify whether a given triple exists within a specific temporal context. In contrast, works such as [11,12,15] shift attention to *temporal context prediction*, where the challenge is to determine the temporal context of a fact (i.e., when a given fact holds true) based on its components. While these approaches move closer to incorporating temporal reasoning, they do not address the task of *temporal fact validation* that aims to determine whether a fact is valid within its stated temporal context by integrating discovered temporal constraints and reasoning over time intervals.

Our work explicitly tackles temporal fact validation. Unlike TKGC models that primarily address link prediction or the temporal context prediction models in [11,12,15], our approach discovers and applies temporal constraints to evaluate the validity of facts. The work of TemporalFC [19] is more closely aligned with our work but differs significantly in methodology. TemporalFC focuses solely on entity relations, excluding literals, and does not deploy semantic information from entity classes, which is a key component of our approach. While their task encompasses validating facts corrupted in temporal contexts as well as in subjects, relations, and objects, it lacks explicit temporal constraint modeling, which limits its precision. Furthermore, their reliance on timestamps rather than time intervals restricts the model's ability to handle real-world temporal variability. In contrast, our approach extends to time intervals, allowing us to exploit the full expressive power of Allen's Interval Algebra and capture nuanced temporal relationships often missed by timestamp-based methods.

Finally, *temporal constraint discovery*, a cornerstone of our work, has seen limited exploration in prior research. Approaches in data profiling for relational databases [1], such as [8,27], have focused on discovering ordering relations using comparison operators (e.g., $\leq, <$). However, these methods fail to incorporate the interval-specific operators needed for robust temporal reasoning. To the best of our knowledge, no existing work achieves the expressiveness of our temporal constraints or integrates such constraints into an explainable, web-scale framework for temporal fact validation.

3 Background and Preliminaries

Our approach for validating temporal facts, based on Allen's algebra [2] for comparing time intervals, is designed to be applied on any temporal KG. In this section, we give the preliminary background and introduce the needed concepts and notations.

3.1 Temporal Knowledge Graphs

We focus in this work on temporal KGs, which are knowledge graphs that associate a part of their facts with temporal information.

Definition 1 (Temporal Knowledge Graph). *We define a temporal knowledge graph* $\mathcal{TKG}$ *as a set of quadruples in the form of* $\{(s,p,o,t) \mid s \in \mathcal{E}, p \in \mathcal{P}, o \in \mathcal{E} \cup \mathcal{L}, t \in \mathcal{T}\}$*, where* $\mathcal{E}$ *is the set of entities (IRIs),* $\mathcal{P}$ *is the set of properties (also referred to as predicates),* $\mathcal{L}$ *is the set of literals (such as numbers, dates or strings) and* $\mathcal{T}$ *is the set of all the time intervals used in a temporal KG, with each interval* $I \in \mathcal{T}$ *defined by a start date* $I.s$ *and an end date* $I.e$ *indicating the period of validity for a given fact.*

To ensure uniformity, we treat any timestamp t as a time interval $[t; t + \epsilon]$, where ϵ is an insignificant duration determined according to the temporal granularity in question (e.g. centuries, years, days, minutes). Thus, throughout this paper, we refer to temporal information as a time interval.

3.2 Representation of Temporal Relations

Allen's interval algebra is a reference for representing relationships between time intervals. Among the thirteen relations of Allen's algebra, in our work we exploit seven[2] main atomic relations as depicted in Fig. 1. These relations are *distinct*, meaning that at most one relation can apply to a given pair of intervals; *exhaustive*, in that each pair of intervals is characterised by one of these relations; and *qualitative*, as they do not involve any consideration of numeric duration.

[2] The other six relations can be captured by one of the seven selected relations.

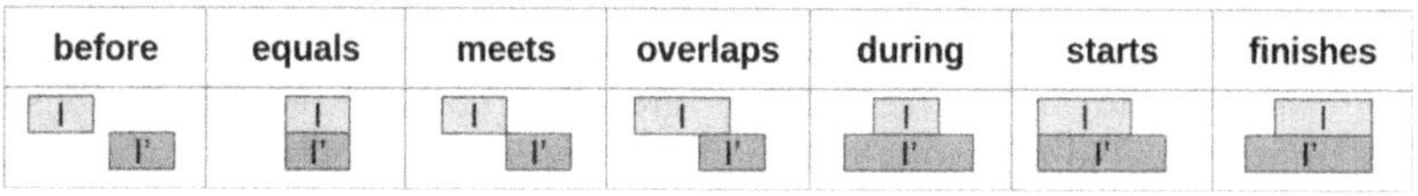

before	equals	meets	overlaps	during	starts	finishes
I I'	I I'	I I'	I I'	I I'	I I'	I I'

Fig. 1. Examples of Allen's time interval relations

In our work, we categorize Allen's temporal relations into two distinct classes based on whether the intervals share any common time points:

- **Disjoint Relations (DR):** representing temporal relationships where the intervals do not share any common sub-interval. Specifically, $DR = \{before, meets\}$,
- **Joint Relations (JR):** These represent relationships where the intervals share at least one temporal sub-interval. Specifically, $JR = \{equals, overlaps, during, starts, finishes\}$.

When comparing two time intervals I and I', we can either specify the precise atomic relation between them, or generalise it by indicating whether they are disjoint or intersecting. For simplicity, we denote $DR(I, I')$ (resp. $JR(I, I')$) to indicate that a temporal relation from the set of disjoint relations DR (resp. JR) holds between I and I'.

4 Approach Overview

In this work, we present a novel explainable approach for verifying if a fact is valid in a given time interval. Our approach is applicable to any KG associating a temporal interval to its facts without any prior knowledge of the data. To validate a fact in a $\mathcal{TKG}$, we check whether the temporal information encoded for that fact aligns with a set of temporal constraints. These constraints may express either disjointness or intersection between the temporal intervals of facts (see Sect. 3.2). For example, a disjointness constraint might state that an American president must be elected before assuming office, expressed as `before(elected, presidentOf)`. However, such temporal constraints are rarely explicitly included in the $\mathcal{TKG}$ and are challenging to manually collect from human experts. Therefore, as the first step, we introduce a new method in Sect. 5 for automatically discovering these types of temporal constraints from any $\mathcal{TKG}$. Our method begins by identifying all simple and complex temporal constraints for a single entity, then evaluates whether these constraints can be generalised across all entities of the same class within the $\mathcal{TKG}$. Finally, as discussed in Sect. 6, these constraints are employed to validate or refute a fact using either a symbolic approach or a machine learning-based approach.

5 Temporal Constraint Discovery from KGs

Our approach starts by discovering all possible temporal constraints for each entity e within a $\mathcal{TKG}$. These constraints concern pairs of properties associated with e, whether e acts as a subject or an object. The discovered constraints can be either simple (involving a single temporal relation) or complex (combining multiple temporal relations). Initially, these constraints are discovered for a single entity, before being generalised across all entities within the same class. The following sections outline our approach for temporal constraint discovery. We begin by defining the concepts of entity time sequences (Sect. 5.1) and the relevant intra- and inter-sequence comparisons (Sect. 5.2). Next, we define the simple and complex temporal constraints that we consider and detail the method used for their discovery at the entity level (Sect. 5.3). Finally, we describe how the discovered temporal constraints can be generalised to the broader class of entities (Sect. 5.5).

5.1 Entity Time Sequence

For each entity and every time-dependant property associated with that entity in the $\mathcal{TKG}$, we construct an entity time sequence containing the intervals during which the property holds for the entity (see Definition 2).

Definition 2 (Entity Time Sequence). *A time sequence of an entity x for a property p is the ordered set of intervals S of the quadruples $\{q_1, \ldots, q_n\}$, in the form of (x, p, y_k, I_k) or (y_k, p, x, I_k) with $k \in [1..n]$ and I_1 having the earliest start date and I_n having the last start date in the time sequence.*

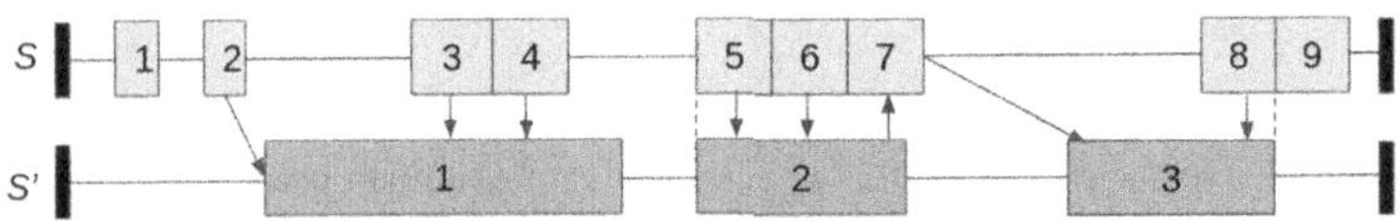

Fig. 2. An example of two time-sequences S and S' containing 9 and 3 time intervals respectively. The arrows refer to inter-sequence comparisons.

Figure 2 presents the entity time sequences S and S' of two properties `PlaysIn` and `WorksAt` respectively for a single entity, with each element in the sequence representing a time interval. To be able to compare time intervals within the same entity time sequence and between different time sequences, we restrict in the following subsections our approach to properties that are temporally functional (see Definition 3). In the last subsection (Subsect. 5.6), we propose an extension that allows our approach to take into account non-functional temporal properties.

Definition 3 (Temporally Functional Property). *A property p is temporally functional, if for every entity it does not exist a pair of overlapping intervals in its corresponding time sequence. We denote $\mathcal{FP}$ the set of temporally functional properties in $\mathcal{TKG}$. A property p is in $\mathcal{FP}$ iff:*
$\forall x \in \mathcal{E},\ \forall y, y' \in \mathcal{E} \cup \mathcal{L},\ \forall I, I' \in \mathcal{T}, (x, p, y, I) \wedge (x, p, y', I') \wedge DR(I, I')$

For instance, `presidentOf` is temporally functional, since a person cannot be the `presidentOf` of two different countries in overlapping time intervals.

5.2 Entity Time Sequence Comparison

In our approach, we generate temporal sequences for all temporally functional properties associated with a given entity in the $\mathcal{TKG}$. The goal is to generate temporal constraints by comparing intervals across these sequences. To ensure the quality of the generated constraints and maintain computational efficiency, we focus on inter-sequence comparisons that are relevant, as presented below.

Definition 4 (Relevant inter-sequence comparisons). *Given an entity e and a pair of its time sequences S and S' corresponding to two temporally functional properties P and P', two intervals I from S and I' from S' are considered relevant for comparison if they share a sub-interval or if no other interval lies between them. We denote by $\Omega(S, S')$ the set of relevant inter-comparisons between S and S'.*

The results of these relevant inter-sequence comparisons are captured in a matrix $M_{\triangleright}$, representing the number of interval pairs across S and S' that satisfy a temporal relation r. The two rows are to account for non-symmetric relations ("before", "during", "starts" and "finishes"). In the example shown in Fig. 2, these relevant inter-sequence comparisons are illustrated with arrows and summarised in the matrix $M_{\triangleright}$ in Table 1. For instance, the comparisons starts$(S.I_5, S'.I_2)$ and finishes$(S.I_8, S'.I_3)$ respectively correspond to the only "starts" and "finishes" relations in $M_{\triangleright}$ (represented with "1" in Table 1).

Table 1. Inter-Comparison matrix $M_{\triangleright}$ for the sequences S and S' from Fig. 2.

	before	equals	meets	overlaps	during	starts	finishes
r($S.I$, $S'.I'$)	2	0	0	1	3	1	1
r($S'.I'$, $S.I$)	0	0	0	1	0	0	0

For discovering complex temporal constraints, intra-sequence comparisons are also needed. To improve efficiency, we limit as well to relevant intra-sequence intervals, defined as follows:

Definition 5 (Relevant intra-sequence comparisons). *For a given time sequence S of an entity e and a property P, relevant intra-sequence comparisons consists of pairs of successive intervals.*

In Fig. 2, the sequence S has four relevant intra-sequence comparisons: meets(3,4), meets(5,6), meets(6,7), and meets(8,9). These comparisons are captured in the matrix $M_{\triangleleft}$. We only keep track of the "meets" relation as as all non-"meets" comparisons default to the "before" relation.

5.3 Simple Temporal Constraint

For each entity e in a $\mathcal{TKG}$, and each pair of time sequences (S, S') corresponding to two properties (P, P') describing e, our approach first derives simple temporal constraints $r(e, P, P')$ based on the number of inter-comparisons in $M_{\triangleright}$. Specifically: $\forall I \in S$, $\exists I' \in S'$, s.t. $r(S.I, S'.I')$ holds and $(I, I') \in \Omega(S, S')$.

We note that in our method the relation *equals* is relaxed to a non-symmetric constraint termed *subsumes*. In the example in Table 1, no simple temporal relation can be derived for the sequences S and S', because no unique relation applies to every pair of intervals.

5.4 Complex Temporal Constraints

To capture more constraints, we define ten complex temporal constraints, organized hierarchically in the Complex Temporal Constraint Tree (CT^2) shown in Fig. 3. These constraints, which may be symmetric or non-symmetric (e.g., `NAND` is symmetric, while `Sequence_Meets` is not), combine multiple temporal relations. Constraints valid for a pair (P, P') must also satisfy all constraints along the path to the CT^2 root.

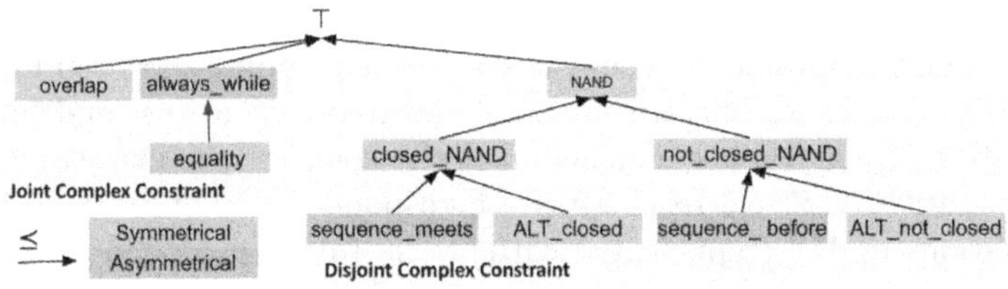

Fig. 3. Complex Temporal Constraints Tree (CT^2)

Further details and formal definitions for specific constraints are provided in the supplementary materials in Appendix A1 and at the GitHub repository[3].

Joint Complex Temporal Constraints: This category includes `overlap`, `always_while`, and `equality` (a specific case of `always_while`):

- **Overlap.** For sequences S and S', every quadruple of S overlaps one in S', except the latest. More formally, let us consider JR the set of intersecting relations (see Subsect. 3.2), the two time sequences S and S' of the properties P and P' respectively, and the matrix of inter-comparisons $M_{\triangleright}$ of S and S' fulfils the Overlapping constraint if:

$$M_{\triangleright}[overlaps][r(S, S')] + M_{\triangleright}[overlaps][r(S', S)] = |S| + |S'| - 1$$

- **Equality.** Every quadruple in S matches exactly with one in S'. More formally, the equality constraint is fulfilled if:

$$M_{\triangleright}[equals][r(S, S')] + M_{\triangleright}[equals][r(S', S)] = |S| + |S'|$$

[3] https://anonymous.4open.science/r/TemporalConstraints-5F91.

Disjoint Complex Constraints: Seven constraints in this category address cases with no interval overlap. These include:

- **ALT_Closed.** After a quadruple in S, one appears in S', or none if S' ends. More formally, the ALT_Closed constraint is fulfilled if:

$$M_{\triangleright}[meets][r(S,S')] + M_{\triangleright}[meets][r(S',S)] = |S| + |S'| - 1$$

- **Sequence_meets.** The last quadruple of S meets the first quadruple of S'. More formally, the Sequence Meets constraint is fulfilled if:

$$M_{\triangleright}[meets][r(S,S')] = 1 \land \left(\sum_{a \in DR} M_{\triangleright}[a][r(S,S')] + M_{\triangleright}[a][r(S',S)] \right) = 1$$

5.5 Temporal Constraints Generalisation

We generalise temporal constraints across all entities of class C using two metrics that accounts for data imperfections such as erroneous temporal data or outlier entities (error rate) and for constraints that are not applicable to a significant portion of entities within C (generalisation rate).

Definition 6 (Error Rate). *Given a temporal constraint TC between properties P and P', the set of entities $E_{P,P'}$ within the class C that are described by both properties, and the sub-set of entities $X_{P,P'} \subseteq E_{P,P'}$ where TC is refuted, the Error Rate is defined as:* $ErrorRate(TC) = \frac{|X_{P,P'}|}{|E_{P,P'}|}$

Definition 7 (Generalisation Rate). *Given a temporal constraint TC between properties P and P', the set of entities E of the class C, and the set of entities $E_{P,P'} \subseteq E$ that are described by both properties, the Generalisation Rate is defined as:* $GeneRate(TC) = \frac{|E_{P,P'}|}{|E|}$

Using an error threshold err and a generalisation threshold gen, a temporal constraint TC is generalised $ErrorRate(TC) \leq err \land GeneRate(TC) \geq gen$.

Algorithm 1 outlines the process for discovering and generalising temporal constraints between two properties R_1 and R_2, given a $\mathcal{TKG}$ and the thresholds err and gen. The algorithm initialises a HashMap ($mapTC$) to track the number of entities (count) fulfilling each temporal constraint, and a counter ($count_e$) to track the number of entities described by both properties. For each entity in $\mathcal{TKG}$, we retrieve the temporal sequences (TR_1 and TR_2) for R_1 and R_2 and verify that they are both used to describe the entity (i.e. are not empty), and update the counter of entities accordingly. If the sequences are temporally functional (line 8), we compute their relevant inter- and intra-sequence comparisons, discovering possible temporal constraints for e and updating the HashMap accordingly. As the process iterates over all entities, we evaluate whether each discovered temporal constraint can be generalised based on the error and generalisation rates. Finally, we filter out the redundant constraints, by discarding from the final output any complex temporal constraint that have a more precise constraint in the $generalisedTC$ set.

Algorithm 1: Temporal Constraint Discovery & Generalisation

```
Data: TKG, err, gen, R1, R2
begin
    mapTC ← hashMap {tc1: 0, ..., tcn: 0}
    for (e ∈ TKG.E) do
        TR1 ← e.R1
        TR2 ← e.R2
        if (|TR1 + TR2| > 0) then
            count_e++
            if (TemporalFunctional(TR1, TR2) = true) then
                M▷ ← computeInter(TR1, TR2)
                M◁(TR1) ← computeIntra(TR1)
                M◁(TR2) ← computeIntra(TR2)
                pTC ← possibleTC(M▷, M◁(TR1), M◁(TR2))
                for (tc ∈ pTC) do
                    mapTC.addOne(tc)
    for ({tc : count} ∈ mapTC) do
        if (ErrorRate(tc, count, count_e) ≤ err) then
            if (GenRate(tc, count, |E|) ≥ gen) then
                generalisedTC.add(tc)
    globalTC ← filterRedundantTC(generalisedTC)
    return globalTC
```

5.6 Temporal Constraints for Non-Functional Temporal Properties

Limiting to temporally functional properties in $\mathcal{TKG}$ may overlook significant constraints, especially for properties with diverse value types (e.g., `memberOf` linking to both musical groups and legislative bodies). To address this, we propose value-specialised properties to enable more precise constraint discovery (e.g., `memberOf-ABBA`, `memberOf-USACongress`).

Value specialisation is restricted to commonly shared property values, ensuring statistically sound constraints from sufficient data samples.

Definition 8 (Value Specialized Time Sequence). *The value specialized time sequence of an entity x of a property p and for a value $v \in \mathcal{E}$ is the ordered set S of a set of quadruples $\{q_1, \ldots, q_n\}$, in the form of (x, p, v, I_k) such that $I_1.start$ is the earliest $I_n.start$ is the latest.*

6 Constraint-Based Temporal Fact Validation

This section extends our framework from the discovery and generalization of temporal constraints to their application in fact validation. The validation framework is outlined in Fig. 4.

The process begins by retrieving the subject timeline (depicted in orange), which contains all background information on the subject entity. The fact under validation (shown in yellow) is integrated into this timeline. Subsequently, we identify relevant temporal constraints for the subject entity—constraints involving pairs of relations describing the subject. These constraints are then used to

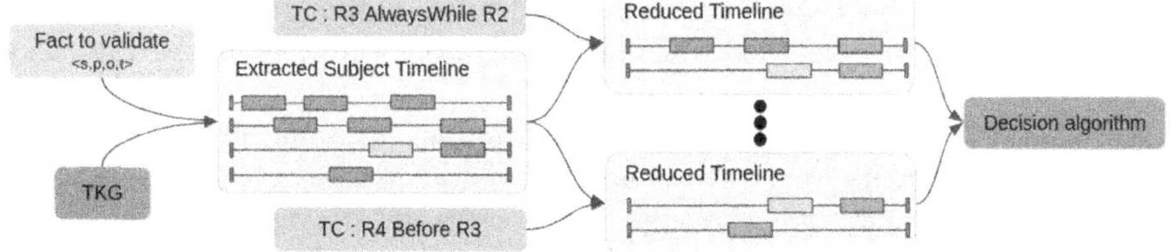

Fig. 4. Constraint based Temporal Fact Validation framework

evaluate whether the addition of the fact adheres to or violates the constraints (Sect. 6.1). Finally, a decision algorithm consolidates the individual evaluations to determine whether the fact is validated or refuted (Sect. 6.2).

6.1 Single Constraint Application

To validate a fact (s, p, o, t) in a Temporal Knowledge Graph $(\mathcal{TKG})$, we first identify all relevant temporal constraints. A constraint $tc = r(P_1, P_2)$, which involves properties P_1 and P_2, is deemed relevant if either property corresponds to the predicate p of the fact, either directly or through the value o of p.

For each relevant constraint, we retrieve the time sequences S_1 and S_2 corresponding to properties P_1 and P_2 for the subject entity s. The time interval t of the fact is then incorporated into these sequences using one of two proposed insertion strategies:

Naive insertion strategy, where the time interval t is directly added to the sequence without addressing overlaps that might violate the temporal functionality of the property.

Remove insertion strategy, where any intervals overlapping with t are removed prior to the insertion.

Once the time interval is integrated, we assess whether S_1 and S_2 remain temporally functional and non-empty. If these conditions are satisfied, we verify whether the sequences still adhere to the constraint tc. A satisfied constraint supports the fact's temporal validity, while a violation indicates refutation.

6.2 Overall Constraint Application

Building on single constraint evaluations, this section introduces methods for aggregating results across multiple constraints to assess a fact's overall validity. Two combination strategies are proposed:

Symbolic combination strategy (voting-based). This approach evaluates the temporal validity of a fact across all relevant constraints. For each constraint, we record whether it is satisfied or violated. A voting mechanism then determines the outcome: if the proportion of satisfied constraints exceeds a predefined threshold, the fact is considered valid. This method is fully explainable, allowing for a detailed trace of all constraints involved in the validation process.

ML-based combination strategy. This approach uses machine learning classifiers (e.g., random-forest, decision-tree, Neural Network) to integrate evaluations from all relevant constraints. Each constraint tc is assigned a numerical label to indicate if a constraint is satisfied (0) or violated (1). Additional labels reflect irrelevant constraints, disrupted temporal functionality, or empty time sequences. These labels are compiled into an $n \times m$ matrix, where n is the number of temporal constraints and m is the number of facts being validated. The matrix is paired with a ground truth vector for training and testing the model, enabling the identification of the most effective constraints for temporal fact validation.

7 Experimental Evaluation

This section presents the results of our experimental evaluation, highlighting the performance of the proposed temporal fact validation framework across several datasets. The source code and datasets of all the following experiments are available on GitHub[4]. All experiments were conducted on an "Intel®Xeon®E5-2630 v4" CPU with 10 cores and 128 GB RAM.

7.1 Datasets and Experimental Setup

To evaluate our approach, we created three new datasets extracted from Wikidata [25], one of the most comprehensive knowledge graphs available. We focused on three distinct entity types: countries, musical groups, and politicians. Existing temporal datasets like TemporalFC [19] were unsuitable as they lacked essential features such as entity class information, literals, and proper temporal interval annotations. Table 2 provides an overview of our datasets, highlighting their varying characteristics in terms of number of entities and temporal facts (represented as quadruples).

The datasets were split into training (80%), validation (10%), and test (10%) sets. Negative sampling involved randomizing the temporal component of each fact within the entity's lifespan (e.g., for entities active between 1900 and 2000, timestamps between 1850 and 2050 were generated). Temporal constraints were discovered using the training set without negative samples, while the augmented dataset (with negative samples) was used for training the ML model.

[4] https://github.com/SoulardThibaut/TemporalConstraints.git.

Table 2. Description of the three datasets extracted from Wikidata

Class	# Entities	# Quadruples
Country (Q6256)	205	183 249
Musical Group (Q215380)	55 507	131 476
Politician (Q82955)	658 445	2 085 232

7.2 Key Experiments and Results

To fine-tune our approach, we adopted a step-by-step strategy, starting with the decision strategy on combining temporal constraints (either symbolic or neuro-symbolic). Then, we examined the type of temporal constraints allowed (relation-level constraints or value-specialising properties), followed by the insertion strategy (either naive or remove). Lastly, we fine-tuned the error threshold alongside the generalisation threshold (default values: $ET = 5\%$ and $GT = 5\%$). Each experiment was evaluated using two metrics: *Accuracy* (Acc.) and the *Coverage* (Cov.) (see Defintion 9). Additionally, we tracked the running time (R.T), accounting for both temporal constraint discovery and the decision process for fact validation or refutation.

Definition 9 (Coverage). *Coverage is defined as the percentage of facts that can be verified by at least one temporal constraint over the total number of facts to be verified:* $coverage = \frac{|Facts_{verified}|}{|Facts|}$

Table 3. Constraints combination and values comparison

(a) Constraints combination strategies

Class	Deci. Type	Acc.	Cov.	R.T
Country	Symbolic	79.5	9.4	**2 m 30 s**
	ML-based	**80.4**	9.4	52m
Music Gr.	Symb.	64.0	37.5	**2 m 50 s**
	ML-based	**64.3**	37.5	5 m 50 s
Politician	Symb.	61.6	44.0	**44 m**
	ML-based	**62.3**	44.0	1 h 30 m

(b) Constraints value specialization

Class	Const. Type	Acc.	Cov.	R.T
Country	R	80.4	9.4	**52 m**
	R & RxV	**90.8**	**14.1**	2 h 35 m
Music Gr.	R	64.2	37.5	5 m 50 s
	R & RxV	64.2	37.5	5 m 50 s
Politician	R	61.6	44.0	1 h 30 m
	R & RxV	61.6	44.0	1 h 30 m

- **Decision Algorithm: Symbolic vs. ML-Based Combination.** The first experiment evaluated the impact of different decision strategies (symbolic voting-based and ML-based) for combining temporal constraints. Results (Table 3a) reveal that the ML-based strategy consistently achieved slightly higher accuracy across all classes. However, the symbolic approach offered significantly faster runtimes, with up to 90% reductions in some cases. Despite this, the ML-based strategy was selected for its marginally better accuracy, as it aligns with our goal of robust fact validation.

- **Temporal Constraint Types: Regular (R) vs. Specialized (R & RxV)** We tested the impact of incorporating value-specializing constraints (RxV) alongside regular constraints (R). Results (Table 3b) show that the inclusion of RxV yielded substantial improvements for the "Country" class: coverage increased by nearly 50% (from 9.4% to 14.1%) and accuracy rose from 80.4% to 90.8%, a 13% improvement. For the classes "Musical Group" and "Politician", no significant changes occurred as no new constraints emerged, though no negative impacts were observed. Given the clear benefits for certain datasets, the specialized approach (R & RxV) was adopted for subsequent experiments.
- **Insertion Strategies: Naive vs. Remove.** This experiment compared the naive and remove insertion strategies for integrating facts into time sequences. Table 4a shows that the Naive insertion strategy marginally results in better accuracy, while the Remove insertion strategy marginally increases coverage. Given the trade-off between accuracy and coverage, the remove strategy was chosen for its ability to validate a broader set of facts.
- **Tuning Error and Generalization Thresholds.** The final experiment optimized hyperparameters related to error rate (*err*) and generalization threshold (*gen*). Table 4b shows that higher *err* and lower *gen* thresholds increased the coverage without significantly compromising accuracy. However, higher thresholds increased runtime due to the proliferation of temporal constraints, posing potential scalability challenges (e.g., over 24,000 constraints in the first line of the table was problematic for the *Decision Tree Classifier*).

Table 4. Interval insertion strategies and thresholds comparison

(a) Interval insertion strategies

Class	Insert Type	Acc.	Cov.	R.T
Country	Naive	**90.8**	14.1	**2 h 35 m**
	Remove	88.5	**14.4**	2 h 54 m
Musical Gr.	Naive	64.2	37.5	5 m 50 s
	Remove	64.2	37.5	5 m 50 s
Politician	Naive	**62.3**	44.0	**1 h 30 m**
	Remove	62.1	**46.9**	1 h 32 m

(b) Error and generalisation thresholds

Class	gen	err	Acc.	Cov.	R.T
Country	2	10	-	-	-
	2	5	**88.6**	16	8 h 50 m
	5	10	87.9	**17.2**	4 h 30 m
	5	5	88.5	14.4	**2 h 54 m**
Musical Gr.	2	10	**64.6**	**38.2**	6 m 6 s
	2	5	**64.6**	**38.2**	5 m 54 s
	5	10	64.2	37.5	**5 m 51 s**
	5	5	64.2	37.5	**5 m 51 s**
Politician	2	10	63.4	**51.9**	1 h 35 m
	2	5	62.1	49.9	**1 h 32 m**
	5	10	**63.5**	48.9	1 h 34 m
	5	5	62.1	46.9	**1 h 32 m**

7.3 Summary of Results

The experimental evaluation demonstrated the effectiveness of our temporal fact validation framework:

1. **High Accuracy:** The framework achieved strong performance across all datasets, with gains when using ML-based decision strategies and specialized temporal constraints.
2. **Scalability vs. Performance Trade-offs:** The choice of insertion strategy and hyperparameter settings allowed fine-tuning for specific dataset characteristics, balancing accuracy, coverage, and runtime.
3. **Dataset-Specific Insights:** Specialized constraints were particularly beneficial for datasets like "Country", where entities share common object values, enabling more nuanced temporal constraints.

By refining constraint combination strategies, incorporating specialized constraints, and carefully tuning hyperparameters, our framework achieved a robust balance of accuracy and runtime, making it a viable tool for temporal fact validation across diverse domains. However, the coverage remains limited, which is not unexpected. This is due to the absence of constraints for certain relations or the lack of applicability caused by some entities being insufficiently detailed.

8 Conclusion and Future Work

In this paper, we introduced a novel, explainable framework for validating temporal facts within any knowledge graph that incorporate temporal information. Our approach extends Allen's Interval Algebra from time intervals to time sequences, enabling the discovery of both simple and complex temporal constraints from the knowledge graph. These constraints were leveraged in multiple validation strategies—symbolic and ML-based—allowing for fine-tuned validation through hyperparameters such as generalization and error thresholds. We conducted an extensive experimental evaluation on datasets derived from Wikidata. Our results showcased the efficiency of the framework, achieving up to 90.8% of accuracy in fact validation. Additionally, the step-by-step tuning process demonstrated the explainability and impact of each proposed strategy, despite a maximum coverage of 51.9%. Building on this foundation, our future work will address several key directions to improve and extend the applicability of our approach:

Scalability and Feature Reduction. To enhance the performance of the ML-based strategy, we will focus on reducing the number of features by identifying and discarding less significant temporal constraints, optimizing both accuracy and runtime efficiency.

Cross-Knowledge Graph Validation. We will investigate whether the temporal constraints discovered in one knowledge graph, such as Wikidata, can be transferred to validate temporal facts in other temporal knowledge graphs, such as YAGO [17]. This transferability would expand the generalizability and utility of our approach.

Constraint Discovery Across Classes and Property Paths. Extending our method to discover constraints that span multiple entity classes or involve complex property paths will enable richer temporal reasoning and improve validation capabilities for more diverse datasets.

Advanced Temporal Comparisons. Incorporating generalized extensions of Allen's Algebra, as proposed in [14], will allow comparisons between discrete timestamps and intervals, broadening the scope of our temporal fact validation framework.

Adapting and Comparing with Temporal Scope Prediction Methods. We are working on adapting existing approaches, such as those proposed in [11,12,15], originally designed for predicting the temporal scope of facts. By adapting these methods for the task of temporal fact validation, we will enable comprehensive comparisons between our framework and these existing models. This effort will include constructing a benchmark that accommodates the requirements of both our approach and these adapted models, ultimately fostering a more robust evaluation environment.

References

1. Abedjan, Z., Golab, L., Naumann, F., Papenbrock, T.: Data Profiling. Synthesis Lectures on Data Management, Morgan & Claypool Publishers (2018). https://doi.org/10.2200/S00878ED1V01Y201810DTM052
2. Allen, J.F.: Maintaining knowledge about temporal intervals. Commun. ACM **26**(11), 832–843 (1983)
3. Boylan, J., Mangla, S., Thorn, D., Ghalandari, D.G., Ghaffari, P., Hokamp, C.: Kgvalidator: a framework for automatic validation of knowledge graph construction. arXiv preprint arXiv:2404.15923 (2024)
4. Cai, B., Xiang, Y., Gao, L., Zhang, H., Li, Y., Li, J.: Temporal knowledge graph completion: a survey. In: International Joint Conference on Artificial Intelligence (2022). https://api.semanticscholar.org/CorpusID:246063616
5. Cai, L., Mao, X., Zhou, Y., Long, Z., Wu, C., Lan, M.: A survey on temporal knowledge graph: representation learning and applications. arXiv preprint arXiv:2403.04782 (2024)
6. Chen, R., Chen, T., Hui, X., Wu, H., Li, G., Lin, L.: Knowledge graph transfer network for few-shot recognition. In: Proceedings of the AAAI Conference on Artificial Intelligence, vol. 34, no. 07, pp. 10575–10582 (2020). https://doi.org/10.1609/aaai.v34i07.6630. https://ojs.aaai.org/index.php/AAAI/article/view/6630
7. Ciampaglia, G.L., Shiralkar, P., Rocha, L.M., Bollen, J., Menczer, F., Flammini, A.: Computational fact checking from knowledge networks. PLoS ONE **10**(6), e0128193 (2015)
8. Consonni, C., Sottovia, P., Montresor, A., Velegrakis, Y.: Discovering order dependencies through order compatibility. In: Herschel, M., Galhardas, H., Reinwald, B., Fundulaki, I., Binnig, C., Kaoudi, Z. (eds.) Advances in Database Technology - 22nd International Conference on Extending Database Technology, EDBT 2019, Lisbon, Portugal, 26–29 March 2019, pp. 409–420. OpenProceedings.org (2019). https://doi.org/10.5441/002/EDBT.2019.36

9. Fernández, N., Blázquez, J.M., Sánchez, L., Bernardi, A.: Identityrank: named entity disambiguation in the context of the news project. In: The Semantic Web: Research and Applications: 4th European Semantic Web Conference, ESWC 2007, Innsbruck, Austria, 3–7 June 2007. Proceedings 4, pp. 640–654. Springer (2007)
10. Hogan, A., et al.: Knowledge Graphs. Synthesis Lectures on Data, Semantics, and Knowledge, Morgan & Claypool Publishers (2021). https://doi.org/10.2200/S01125ED1V01Y202109DSK022
11. Islakoglu, D.S., Chekol, M.W., Velegrakis, Y.: Leveraging pre-trained language models for time interval prediction in text-enhanced temporal knowledge graphs. In: European Semantic Web Conference, pp. 59–78. Springer (2024)
12. Jain, P., Rathi, S., Chakrabarti, S., et al.: Temporal knowledge base completion: new algorithms and evaluation protocols. arXiv preprint arXiv:2005.05035 (2020)
13. Kim, J., Park, S., Kwon, Y., Jo, Y., Thorne, J., Choi, E.: FactKG: fact verification via reasoning on knowledge graphs. In: Rogers, A., Boyd-Graber, J., Okazaki, N. (eds.) Proceedings of the 61st Annual Meeting of the Association for Computational Linguistics (Volume 1: Long Papers), Toronto, Canada, pp. 16190–16206. Association for Computational Linguistics (2023). https://doi.org/10.18653/v1/2023.acl-long.895. https://aclanthology.org/2023.acl-long.895/
14. Ligozat, G.: On generalized interval calculi. In: Proceedings of the Ninth National Conference on Artificial Intelligence, vol. 1, pp. 234–240 (1991)
15. Ling, C., Janowicz, K., Rui, Z., Mai, G.: Time in a box: advancing knowledge graph completion with temporal scopes. In: K-CAP 2021: Proceedings of the 11th on Knowledge Capture Conference (2021)
16. Liu, J., Duan, L.: A survey on knowledge graph-based recommender systems. In: 2021 IEEE 5th Advanced Information Technology, Electronic and Automation Control Conference (IAEAC), pp. 2450–2453 (2021). https://doi.org/10.1109/IAEAC50856.2021.9390863
17. Mahdisoltani, F., Biega, J., Suchanek, F.M.: Yago3: a knowledge base from multilingual wikipedias. In: CIDR (2013)
18. Peng, C., Xia, F., Naseriparsa, M., Osborne, F.: Knowledge graphs: opportunities and challenges. Artif. Intell. Rev. **56**(11), 13071–13102 (2023). https://doi.org/10.1007/s10462-023-10465-9
19. Qudus, U., Röder, M., Kirrane, S., Ngonga, A.C.: Temporalfc: a temporal fact checking approach over knowledge graphs (2023)
20. Radhakrishna, V., Kumar, P.V., Janaki, V.: A survey on temporal databases and data mining. In: Proceedings of the The International Conference on Engineering & MIS 2015. ICEMIS 2015. Association for Computing Machinery, New York (2015). https://doi.org/10.1145/2832987.2833064
21. Saxena, A., Tripathi, A., Talukdar, P.: Improving multi-hop question answering over knowledge graphs using knowledge base embeddings. In: Jurafsky, D., Chai, J., Schluter, N., Tetreault, J. (eds.) Proceedings of the 58th Annual Meeting of the Association for Computational Linguistics, pp. 4498–4507. Association for Computational Linguistics, Online (2020). https://doi.org/10.18653/v1/2020.acl-main.412. https://aclanthology.org/2020.acl-main.412
22. Shiralkar, P., Flammini, A., Menczer, F., Ciampaglia, G.L.: Finding streams in knowledge graphs to support fact checking. In: 2017 IEEE International Conference on Data Mining (ICDM), pp. 859–864. IEEE (2017)
23. Syed, Z.H., Röder, M., Ngomo, A.C.N.: Unsupervised discovery of corroborative paths for fact validation. In: The Semantic Web–ISWC 2019: 18th International Semantic Web Conference, Auckland, New Zealand, 26–30 October 2019, Proceedings, Part I 18, pp. 630–646. Springer (2019)

24. Tchechmedjiev, A., et al.: Claimskg: a knowledge graph of fact-checked claims. In: The Semantic Web–ISWC 2019: 18th International Semantic Web Conference, Auckland, New Zealand, 26–30 October 2019, Proceedings, Part II 18, pp. 309–324. Springer (2019)
25. Vrandečić, D., Krötzsch, M.: Wikidata: a free collaborative knowledgebase. Commun. ACM **57**(10), 78–85 (2014). https://doi.org/10.1145/2629489. http://doi.acm.org/10.1145/2629489
26. Wang, J., et al.: A survey on temporal knowledge graph completion: taxonomy, progress, and prospects (2023)
27. Xiao, R., Tan, Z., Wang, H., Ma, S.: Fast approximate denial constraint discovery. Proc. VLDB Endow. **16**(2), 269–281 (2022). https://doi.org/10.14778/3565816.3565828
28. Zhang, J., Liang, S., Sheng, Y., Shao, J.: Temporal knowledge graph representation learning with local and global evolutions. Knowl.-Based Syst. **251**, 109234 (2022). https://doi.org/10.1016/j.knosys.2022.109234. https://www.sciencedirect.com/science/article/pii/S0950705122006141

Multi-dataset and Transfer Learning Using Gene Expression Knowledge Graphs

Rita T. Sousa(✉) and Heiko Paulheim

Data and Web Science Group, University of Mannheim, Mannheim, Germany
{rita.sousa,heiko.paulheim}@uni-mannheim.de

Abstract. Gene expression datasets offer insights into gene regulation mechanisms, biochemical pathways, and cellular functions. Additionally, comparing gene expression profiles between disease and control patients can deepen the understanding of disease pathology. Therefore, machine learning has been used to process gene expression data, with patient diagnosis emerging as one of the most popular applications. Although gene expression data can provide valuable insights, challenges arise because the number of patients in expression datasets is usually limited, and the data from different datasets with different gene expressions cannot be easily combined. This work proposes a novel methodology to address these challenges by integrating multiple gene expression datasets and domain-specific knowledge using knowledge graphs, a unique tool for biomedical data integration. Then, vector representations are produced using knowledge graph embedding techniques, which are used as inputs for a graph neural network and a multi-layer perceptron. We evaluate the efficacy of our methodology in three settings: single-dataset learning, multi-dataset learning, and transfer learning. The experimental results show that combining gene expression datasets and domain-specific knowledge improves patient diagnosis in all three settings.

Keywords: Knowledge Graph · Knowledge Graph Embeddings · Gene expression data · Patient Diagnosis

1 Introduction

The volume of biological data being collected and accumulated is growing at an accelerated rate. DNA microarray chips and high-throughput sequencing are two of the latest technological innovations that allow the simultaneous measurement of the expression of different genes. Quantitative measurement of gene expression can offer insight into gene regulation mechanisms, biochemical pathways, and cellular functions. Furthermore, comparing gene expression profiles between patients with a disease and control patients can deepen the understanding of the pathology of the disease, facilitate the identification of genes that could serve as potential therapeutic targets [2] or explain the response to drug treatments [1]. Given the complexity of gene expression data, it is infeasible for an expert to

E. Curry et al. (Eds.): ESWC 2025, LNCS 15718, pp. 245–263, 2025.
https://doi.org/10.1007/978-3-031-94575-5_14

analyze a gene expression matrix manually. Therefore, machine learning (ML) algorithms have been used to process gene expression data, with patient diagnosis emerging as one of the more popular applications [25].

Although gene expression datasets are easily available in public databases, and gene expression analysis is useful for biomedical applications, including patient diagnosis, processing this type of data presents some challenges [21]. While the data itself provides information on the activity levels of genes within a cell or tissue, they often do not provide a comprehensive biological context. This limitation derives from treating the expression for different genes as independent variables, an assumption that overlooks the intricate interplay and dependencies among genes. Another challenge is the reliance of supervised ML methods on many labeled data points for effective training and performance. However, gene expression datasets typically offer only a limited number of patients, complicating the application of these methods [10,15]. One alternative involves combining multiple expression datasets to increase the patient pool for training ML models. However, each dataset measures gene expression across distinct genes and is measured in distinct experimental platforms, making integration difficult.

Knowledge graphs (KGs) offer a pathway towards addressing those challenges. In biomedical systems, graphs quickly lead as the primary paradigm for modeling, learning, and reasoning [33]. In a fully machine-readable format, KGs can describe real-world entities in a graph structure [8]. Additionally, KGs [24] can be enhanced by publicly accessible biomedical ontologies, which make it possible to describe domain-specific information. In the biomedical KGs, nodes represent biomedical entities or concepts, while edges illustrate their relationships. Over the past few years, biomedical KGs have been used in ML applications as a tool for data integration [17].

Many techniques have been proposed in ML on KGs, with graph neural networks (GNNs) [7] and KG embedding methods [32] becoming increasingly popular for node classification tasks. The basic idea of GNN models is to employ a type of neural message passing in which vector messages are exchanged between nodes and updated using neural networks. While GNN has performed very well in many tasks, it requires node features usually unavailable in biomedical real-world KGs. When node features are unavailable, GNNs often initialize them through random initialization or node statistics. Nevertheless, these approaches fail to capture a more faithful domain representation. KG embedding methods, on the contrary, present a promising avenue for capturing KG-based information since they map entities and relationships in a KG into a lower-dimensional vector space while preserving graph structure and, in some cases, semantic information. KG embedding methods can also be used for node classification in a two-step process. Initially, KG embedding models are employed to learn a transductive unsupervised representation of the nodes. Subsequently, these learned embeddings serve as inputs to a supervised learning algorithm, such as a Multi-Layer Perceptron (MLP).

This work tackles the challenges of using gene expression datasets for patient diagnosis. We propose a novel methodology that generates a KG to incorporate both gene expression data and publicly available datasets on genes and their functions. Then, KG embedding methods are employed to generate vector representations of KG nodes. Lastly, the vector representations of KG nodes serve as inputs for either a graph convolutional neural network (GCN) or, alternatively, into an MLP to predict the likelihood of a patient having a specific disease. Our methodology's efficacy is evaluated using multiple datasets for three diseases: diabetes type 2, coronary artery disease, and breast cancer. The proposed methodology is applied in three different scenarios: single-dataset learning, where the classifier is trained and tested with data from the same dataset; multi-dataset learning, where the classifier is trained using aggregated data from multiple datasets of the same disease and tested on one of the datasets; transfer learning, where the classifier is trained on some datasets and tested on other datasets, all focusing on the same disease. The results demonstrated that the incorporation of domain-specific knowledge improves patient diagnosis performance, and combining data from multiple datasets is especially beneficial when the number of instances is very limited. Furthermore, the results demonstrate that our methodology can make predictions on unseen datasets during training.

In summary, the contributions of our work are as follows:

- An approach for building a KG that can integrate and enrich several gene expression datasets through a public dataset on gene and protein functions.
- An approach to tackle the lack of meaningful node features in biomedical graphs for GNNs using KG embedding methods to generate feature vectors for each node.
- A computation-efficient method that can reuse the embedding vectors of static parts of the KG.
- An integration of gene expression KG with supervised ML methods to support patient diagnosis.
- An investigation of the impact of integrating multiple gene expression datasets and transferring a model trained on one dataset to another for testing.

2 Related Work

Several works have explored gene expression data for patient diagnosis, employing a wide array of methodologies and datasets for a broad range of diseases, from Alzheimer's disease to cancer disorders [29]. Tan *et al.* [28] compare several ensemble methods for cancer classification using seven publicly available cancer-related microarray data. Other approaches also use classical ML methods over gene expression data for cancer classification [5,31]. Gene expression datasets related to diabetes type 2 are explored by Mansoori *et al.* [15] and Kazerouni *et al.* [10]. While Kazerouni *et al.* [10] compare four classical classifiers (K-nearest neighbor, support vector machine, logistic regression, and artificial neural networks), Mansoori *et al.* [15] use logistic regression. Kegerreis *et al.* [11] integrate gene expression data for systemic lupus erythematosus disease using several ML classifiers. Classifiers were evaluated across combined datasets or by training and testing on independent datasets. Pirooznia *et al.* [19] present a study in which multiple ML methods, from support vector machine to random forest, are compared on how well they perform using eight gene expression datasets, each corresponding to a different disease. Over the last few years, deep learning methods have become more prevalent and approaches that use deep learning for gene expression analysis are beginning to emerge [29]. Parvathavardhini *et al.* [18] propose a Neuro-Fuzzy approach to detect cancer by exploring gene expression data from microarray experiments. Shon *et al.* [27] employ a convolutional neural network algorithm for early prediction of stomach cancer. Schaack *et al.* [26] train deep-learning artificial neural networks using sepsis-related gene expression data, and test their resilience by progressively degrading the data.

These works typically rely on a singular dataset for training purposes. In some cases where multiple datasets are utilized, there is an assumption they are comparable, so they often merge them by incorporating only the shared genes as features. However, it is noted in many papers that the limited patient size of disease datasets poses a significant challenge to the predictive capability of ML models, introducing uncertainty [29].

3 Methodology

3.1 Problem Formulation

We tackle the challenge of handling gene expression data to support patient diagnosis. In gene expression datasets, the expression values are organized in a matrix $n \times m$, where n is the number of genes (usually more than 10 000), m is the number of patients (usually more than 10), and $m << n$. Gene expression values are numerical representations indicating the expression levels of genes under specific conditions. The task of diagnosing patients is framed as a binary classification problem. Given a patient representation, the objective is to predict whether the patient has a specific disease (labeled as 1) or not (labeled as 0). When using gene expression data for patient diagnosis, the patient representations correspond to rows of the $n \times m$ matrix. Most approaches use patient representations as input to a supervised ML algorithm for training prediction models [13,29,30]. This formulation results in an oversimplification, potentially constraining these ML approaches' effectiveness and usefulness. The assumption that features are independent overlooks valuable insights about genes readily available in various biomedical resources. Moreover, the challenge intensifies as gene expression datasets often contain a limited number of instances, each recording expression for different genes. Consequently, when training prediction models, researchers either use a single dataset with little training data or attempt to merge multiple datasets. In the latter case, the datasets are typically "incompatible" because they have different feature sets. Thus, a naive combination usually restricts the models to using only the common features.

We propose a methodology for enriching gene expression datasets using available biomedical resources to address these challenges. We aim to do this by representing gene expression data and domain-specific knowledge in a KG, which

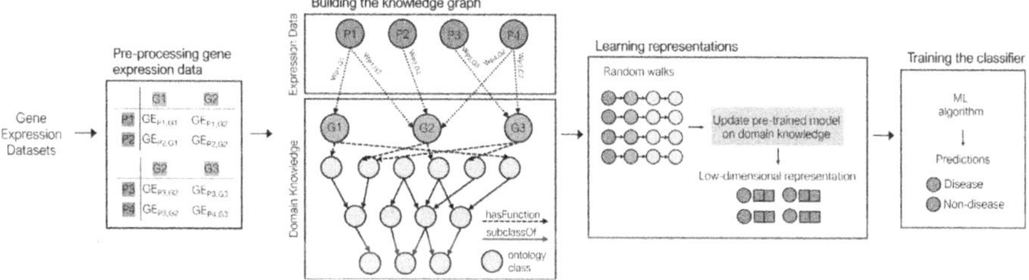

Fig. 1. Overview of the proposed methodology with the main steps.

is then fed to a supervised ML algorithm to support patient diagnosis. This KG not only captures relationships between genes within the same dataset, reflecting their cellular components, biological processes or molecular functions but also allows the integration of gene expression data originating from different datasets. Even when there is no direct overlap of genes across datasets, our KG is able to identify connections based on the functions the genes perform. Moreover, normalizing each patient's gene expression values mitigates discrepancies arising from varied experimental conditions. Consequently, our methodology enables seamless integration of multiple datasets for model training or transfer, even in the lack of common genes.

3.2 Overview

Figure 1 shows an overview of the methodology. The first step tackles the processing of gene expression data from experimental studies. The second step corresponds to building the KG, which integrates not only expression data but also domain knowledge on gene and protein functions. Representations for each node in the KG are then learned using a KG embedding method. The last step involves giving the node representations and a weighted graph as input for a GCN or, alternatively, directly feeding the node representations into an MLP. In the weighted graph, the patient's expression value for a gene is used to weigh the edge between the patient and the gene. The code for our methodology is available on GitHub (https://github.com/ritatsousa/expressionKGplus).

3.3 Pre-processing of Gene Expression Data

Several genomic studies have recently explored microarrays for measuring gene expression. With a single experiment, expression values are obtained simultaneously for tens of thousands of genes. The Gene Expression Omnibus (GEO) [3] is a publicly accessible repository maintained by the National Center for Biotechnology Information that presents a vast collection of high-throughput gene expression and other genomics datasets. Every GEO dataset is curated, comprising biologically similar GEO samples, each representing a patient, with measurements presumed to be equivalently calculated.

The pre-processing step is essential considering the inherent complexities of microarray datasets. First, each probe of the microarray, identified by an identifier, contains a gene fragment for which the expression level is being determined. Each gene fragment is accompanied by an annotation detailing its biological context, indicating its association with a known gene. However, it is worth noting that not all gene fragments have such associations. Since our methodology relies on linking gene expression data with domain-specific knowledge describing gene functions, fragments without an associated gene are filtered out. The second challenge arises from probes annotated for the same gene. The strategy adopted is, for each patient, averaging expression values across all probes corresponding to the same gene. Additionally, expression values from different datasets may be on different scales or ranges. Therefore, we normalize the expression values of each patient using z-score normalization. Z-score normalization resizes gene expression values to align with the properties of a standard normal distribution, characterized by a mean μ of 0 and a standard deviation σ of 1.

After the pre-processing, the data is structured in a tabular format, with each column representing a patient P_i, a row corresponding to a different gene G_k, and the cells containing gene expression values of a specific gene for a patient $GE_{Pi,Gk}$.

3.4 Building the Knowledge Graph

The KG is built by integrating two data sources: expression data and domain-specific knowledge. The majority of KGs are represented in RDF, a standard data model. In RDF terminology, a statement is a small piece of knowledge in the format of subject-predicate-object expressions, where the subject and the object correspond to nodes, and the predicate is the name of a relation that connects these two nodes.

Since our methodology relies on KG graph embeddings for generating patient representations and most embedding approaches cannot handle numeric literals [20], we employ a linking approach between patients and genes based on expression values. A link between a patient and a gene is created when the normalized patient's expression value for that gene is higher than 1 since a z-score of 1 means the data point is one standard deviation above the mean.

The domain-specific knowledge includes the Gene Ontology (GO) [4] and GO annotation data [9]. The GO defines a hierarchy of classes that describes gene product functions. It can be represented as a graph where nodes are GO classes, and edges define relationships between them. The GO encompasses three domains for characterizing functions: the biological processes a gene product is involved in, the molecular functions a gene product performs, and the cellular components where a gene product is located. These three domains of GO are represented as separate root ontology classes since they do not share any common ancestor. The GO annotation data refers to assigning functions represented as GO classes to genes represented as links in the graph. Like most biomedical ontologies, GO is defined in OWL. Therefore, the guidelines provided by the W3C are used to create the RDF graph. Simple axioms, including subsumption axioms and data and annotation properties, are directly converted into RDF triples. More complex axioms result in the generation of multiple triples, often requiring the use of blank nodes.

3.5 Learning Representations

Our methodology employs KG embedding methods to generate low-dimensional representations for each KG node. Given the typical richness of hierarchical relations within biomedical KGs, walk-based strategies are likely better suited to capture the longer-distance relations. Specifically, we employ RDF2vec [22], a path-based embedding method well-suited to RDF graphs. RDF2vec starts by employing a depth-first search algorithm to generate random walks in a graph considering edge direction. For each node we want to learn a representation, RDF2vec generates a set number of walks with a predefined maximum depth rooted in that node. Then, RDF2vec employs Word2vec [16], a language model, over those walks to learn a latent low-dimensional representation of each node.

Since the domain knowledge (GO and GO annotation data) is never altered, we first train the RDF2vec model, giving it only the domain-specific KG (i.e., without patients data) as input. Following this, for each gene expression dataset, the RDF2vec model is updated by computing new walks for the patients' dataset [6]. Reusing a pre-trained model circumvents the need to retrain the RDF2vec model for each dataset, thus reducing the carbon footprint and leading to faster predictions. Empirical experiments conducted on the same machine revealed that training the embedding model on the domain-specific KG takes approximately 9348 s, whereas using the pre-trained model and updating it for the patients' dataset takes only 1183 s. This significant reduction in running time underscores the efficiency of this approach. Additionally, it also motivates the use of RDF2vec, because such a mechanism of iteratively training KG embeddings is not supported by many methods.

3.6 Training a Supervised Learning Algorithm

The last step is training a supervised learning algorithm to support patient diagnosis using two distinct ML approaches. In the first approach, the patient embeddings are fed into an MLP classifier.

The second approach involves giving a weighted graph as input for a GCN, using KG embeddings as node features. In the weighted graph, the weights for edges between patients and genes represent the normalized expression values using z-score normalization. For all other edges, the weights are set to 1. We employ a multi-layer GCN architecture [12] where embeddings are successively aggregated across several convolutional layers. Each convolutional layer employs a neural message passing in which vector messages are exchanged between nodes and updated using neural networks. With each iteration of message passing, the hidden embedding $\mathbf{h}_u^{(l+1)}$ associated with a node $u \in \mathcal{V}$ is updated according to the information aggregated from its neighboring nodes $\mathcal{N}(u)$. This iterative process, repeated for a fixed number of iterations, ensures that each node's final embeddings encapsulate structural and feature-based insights derived from its entire k-hop neighborhood. Formally, the message-passing mechanism can be expressed as outlined in [7]:

$$\mathbf{h}_u^{(l+1)} = \sigma\left(\mathbf{W}^{(l)} \sum_{j \in \mathcal{N}(u)} \frac{\mathbf{e}_{uj}}{\sqrt{|\mathcal{N}(u)||\mathcal{N}(j)|}} \mathbf{h}_j^{(l)}\right) \tag{1}$$

where $\mathbf{W}^{(l)}$ is a trainable parameter matrix, σ is an activation function, and $\mathbf{e}_{uj}$ is the weight on the edge from node j to node i. To prevent overfitting, we incorporate dropout regularization. As an activation function, we employ ReLU, while training is driven by cross-entropy loss, ensuring effective node classification.

Transforming the KG into an input graph for GCN results in some loss of relation information. The KG used to generate RDF2Vec embeddings was constructed by integrating heterogeneous relationships from the GO ontology and applying a linking strategy between patients and genes. However, the input graph for the GCN does not differentiate between edge types. Despite this, the loss is minimized since only one edge type exists between patients and genes, and RDF2Vec embeddings used as node features still retain relational information.

4 Evaluation

We evaluate the proposed methodology for patient diagnosis tasks using multiple disease datasets. The following subsections describe the data used and provide an overview of the experiments.

4.1 Data

Nine GEO datasets related to three different diseases are included in the evaluation. Each dataset comprises patients categorized into two groups: subjects diagnosed with the disease (positive examples) and those serving as control subjects (negative examples). Table 1 provides relevant statistics for the different datasets. For gene statistics, annotated genes encompass those with identifiers and annotations linked to biomedical ontologies. Additionally, Fig. 2 illustrates Venn diagrams for each disease, elucidating the overlaps in the number of annotated genes across different datasets associated with the same disease. The domain KG, built by integrating GO and GO annotations, comprises 1 732 160 triples, 53 relation types, and 51 375 classes.

4.2 Experimental Setup

To implement our methodology, we first generate representations using RDF2vec and then train either a GCN or an MLP. All RDF2vec, MLP and GCN hyperparameters are detailed in the Appendix file. To assess the efficacy of the proposed

Table 1. Number of genes, patients, and references for each GEO dataset

Disease	Dataset	Genes		Patients	
		Total	Annot.	Pos.	Neg.
Diabetes type 2	GSE184050[a]	24 737	4 721	50	66
	GSE78721[b]	49 395	4 720	68	62
	GSE202295[c]	66 023	4 770	61	50
Coronary artery disease	GSE12288[d]	22 283	3 787	110	112
	GSE20681[e]	45 015	4 384	99	99
	GSE42148[f]	62 972	4 491	13	11
Breast cancer	GSE9574[g]	22 283	3 787	14	15
	GSE10810[h]	18 382	2 951	31	27
	GSE86374[i]	33 297	4 556	124	35

[a]https://www.ncbi.nlm.nih.gov/geo/query/acc.cgi?acc=GSE9574
[b]https://www.ncbi.nlm.nih.gov/geo/query/acc.cgi?acc=GSE78721
[c]https://www.ncbi.nlm.nih.gov/geo/query/acc.cgi?acc=GSE202295
[d]https://www.ncbi.nlm.nih.gov/geo/query/acc.cgi?acc=GSE12288
[e]https://www.ncbi.nlm.nih.gov/geo/query/acc.cgi?acc=GSE20681
[f]https://www.ncbi.nlm.nih.gov/geo/query/acc.cgi?acc=GSE42148
[g]https://www.ncbi.nlm.nih.gov/geo/query/acc.cgi?acc=GSE9574
[h]https://www.ncbi.nlm.nih.gov/geo/query/acc.cgi?acc=GSE10810
[i]https://www.ncbi.nlm.nih.gov/geo/query/acc.cgi?acc=GSE86374

methodology, we analyse the classification performance using different metrics: precision, recall, F1-score, and weighted average F1-score. We perform stratified 5-fold cross-validation for each dataset, using the same folds throughout the different experiments. Our methodology enables the integration of multiple datasets, addressing the common challenge of using gene expression datasets with limited patients. Therefore, we conduct three types of experiments (Fig. 3):

- **Single-dataset learning:** We only train with the data of each dataset in isolation using a 5-fold cross-validation strategy.
- **Multi-datasets learning:** Since we have three datasets for each disease, we conduct experiments using combined sets from these datasets. Using the five already defined partitions of a dataset, we train the model with four partitions from one dataset and include the remaining datasets related to the same disease.
- **Transfer learning:** We train the model with data from two datasets and make predictions for the third dataset.

For each setting, our methodology is compared against baselines that employ the expression values after pre-processing directly as input for an MLP. In the case of the multi-dataset and transfer learning settings, the baseline includes two variations: one that includes all genes, setting the expression value to 0 for any gene without a measured value for a given patient, and another that considers only the overlapping genes between datasets.

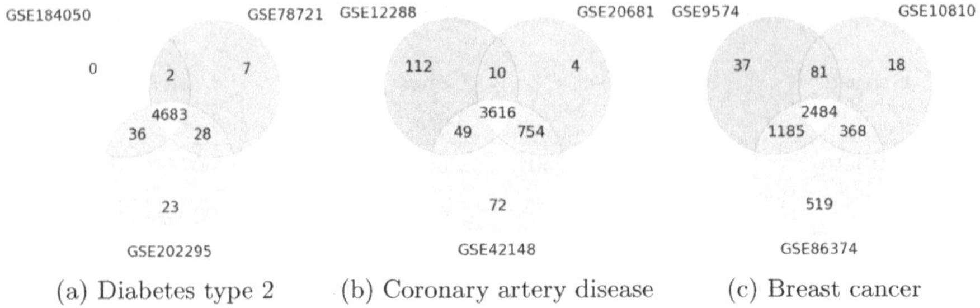

(a) Diabetes type 2 (b) Coronary artery disease (c) Breast cancer

Fig. 2. Venn Diagrams showing the number of genes in common between different datasets for the same disease.

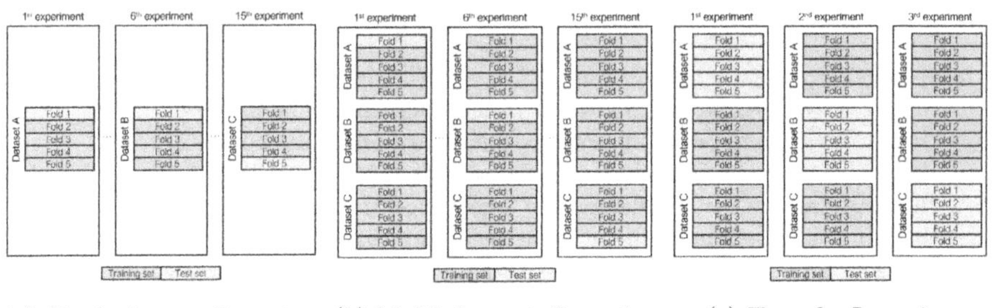

(a) Single-dataset Learning (b) Multi-datasets Learning (c) Transfer Learning

Fig. 3. Experimental strategies to split the datasets.

5 Results and Discussion

Our methodology has two variants, one using an MLP with RDF2Vec embeddings as input and another using a GCN with RDF2Vec embeddings as node features for patient diagnosis. Table 2 presents the performance of these two variants compared to baseline performance for single-dataset and multi-dataset settings. At the end of each dataset fold, we compute the performance metrics and report the average of the five folds.

In the single-dataset setting, using our methodology does not imply integrating data from other datasets but rather injecting domain-specific knowledge. The results confirm the hypothesis that contextualizing genetic information improves patient diagnosis, with considerable improvements for some datasets. Notably, for the GSE184050 dataset, the f1-score increases from 0.222 in the baseline to 0.675 and 0.757 for KG embedding and GCN, respectively. The exception is the GSE9574 dataset for all metrics, the GSE41148 dataset for precision and the GSE202295 dataset for precision and f-score. These datasets are the smallest for each respective disease, which might explain why our methodology does not show improvement compared to the baseline.

In the multi-dataset setting, supervised learning models are trained with data from all datasets related to the same disease. For this setting, a comparison between the two variations of the baselines shows that using only the gene expres-

Table 2. Mean and standard deviation for precision, recall, f1-score, and weighted average f1-score (Pr, Re, F1, WAF), comparing the baseline that uses gene expression values directly to our methodology when coupled with MLP or GCN for two settings (single-dataset learning and multi-dataset learning). The bold value indicates the highest performance within each setting, while the underlined value highlights the highest performance across all three settings. The performance values are marked with an asterisk (*) when the classifier predicts either label 0 or label 1 for all instances in the test set.

Disease	Dataset	Metric	Single-dataset learning			Multi-dataset learning			
			Baseline	Our methodology		Baseline		Our methodology	
				MLP	GCN	All	Overlap	MLP	GCN
Diabetes type 2	GSE184050	Pr	0.328 (0.298)	0.813 (0.157)	**0.840 (0.085)**	0.409 (0.221)	**0.509 (0.192)**	0.493 (0.057)	0.504 (0.047)
		Re	0.180 (0.194)	0.660 (0.273)	**0.740 (0.242)**	0.480 (0.371)	0.500 (0.261)	**0.720 (0.183)**	0.640 (0.206)
		F1	0.222 (0.218)	0.675 (0.151)	**0.757 (0.138)**	0.393 (0.226)	0.476 (0.180)	**0.573 (0.077)**	0.546 (0.070)
		WAF	0.495 (0.095)	0.742 (0.084)	**0.809 (0.086)**	0.450 (0.125)	**0.559 (0.095)**	0.525 (0.095)	0.525 (0.135)
	GSE78721	Pr	0.428 (0.217)	0.566 (0.085)	**0.619 (0.084)**	0.425 (0.221)	0.523* (0.019)	**0.511 (0.053)**	0.456 (0.084)
		Re	0.710 (0.370)	0.699 (0.233)	**0.780 (0.155)**	0.552 (0.391)	1.000* (0.000)	**0.631 (0.199)**	0.542 (0.153)
		F1	0.529 (0.264)	0.614 (0.126)	**0.674 (0.038)**	0.444 (0.264)	0.687* (0.016)	**0.550 (0.103)**	0.490 (0.102)
		WAF	0.391 (0.114)	0.532 (0.113)	**0.563 (0.128)**	0.402 (0.064)	0.359* (0.021)	**0.452 (0.089)**	0.401 (0.106)
	GSE202295	Pr	**0.695 (0.207)**	0.581 (0.066)	0.509 (0.109)	0.571 (0.347)	**0.799 (0.186)**	0.571 (0.089)	0.452 (0.059)
		Re	0.623 (0.239)	**0.838 (0.111)**	0.601 (0.215)	0.326 (0.294)	**0.501 (0.322)**	0.341 (0.129)	0.360 (0.123)
		F1	0.580 (0.152)	**0.680 (0.045)**	0.547 (0.154)	0.352 (0.234)	**0.520 (0.232)**	0.408 (0.101)	0.395 (0.092)
		WAF	**0.507 (0.116)**	0.504 (0.107)	0.470 (0.114)	0.424 (0.123)	**0.548 (0.157)**	0.463 (0.048)	0.407 (0.056)
Coronary Artery Disease	GSE12288	Pr	0.470 (0.052)	0.553 (0.066)	**0.571 (0.064)**	0.313 (0.256)	**0.526 (0.029)**	0.507 (0.037)	0.455 (0.246)
		Re	0.500 (0.246)	0.436 (0.164)	**0.564 (0.121)**	0.427 (0.454)	**0.664 (0.253)**	0.518 (0.102)	0.382 (0.224)
		F1	0.463 (0.112)	0.470 (0.101)	**0.562 (0.082)**	0.329 (0.310)	**0.559 (0.104)**	0.505 (0.038)	0.394 (0.204)
		WAF	0.440 (0.074)	0.523 (0.049)	**0.568 (0.075)**	0.408 (0.079)	0.479 (0.061)	**0.496 (0.043)**	0.482 (0.077)
	GSE20681	Pr	0.197 (0.242)	**0.557 (0.096)**	0.543 (0.068)	0.203 (0.248)	0.000* (0.000)	0.448 (0.323)	**0.536 (0.081)**
		Re	0.400 (0.490)	0.524 (0.153)	**0.534 (0.115)**	0.400 (0.490)	0.000* (0.000)	0.181 (0.196)	**0.504 (0.106)**
		F1	0.264 (0.324)	0.530 (0.100)	**0.535 (0.085)**	0.269 (0.329)	0.000* (0.000)	0.202 (0.162)	**0.511 (0.071)**
		WAF	0.328 (0.007)	**0.544 (0.075)**	0.542 (0.060)	0.339 (0.007)	0.333* (0.009)	0.380 (0.040)	**0.520 (0.060)**
	GSE42148	Pr	0.420 (0.223)	0.470 (0.252)	**0.600 (0.374)**	0.300 (0.400)	**0.420 (0.223)**	**0.420 (0.238)**	0.200 (0.245)
		Re	**0.800 (0.400)**	0.733 (0.389)	0.667 (0.365)	0.400 (0.490)	**0.800 (0.400)**	0.567 (0.389)	0.200 (0.245)
		F1	0.548 (0.282)	0.569 (0.300)	**0.608 (0.336)**	0.333 (0.422)	**0.548 (0.282)**	0.477 (0.292)	0.200 (0.245)
		WAF	0.338 (0.099)	0.450 (0.171)	**0.564 (0.282)**	**0.448 (0.288)**	0.338 (0.099)	0.442 (0.207)	0.338 (0.179)
Breast Cancer	GSE9574	Pr	**0.467 (0.323)**	0.300 (0.400)	0.267 (0.226)	0.450 (0.400)	0.367 (0.371)	0.400 (0.379)	**0.587 (0.224)**
		Re	**0.567 (0.389)**	0.267 (0.389)	0.400 (0.389)	0.533 (0.452)	0.400 (0.389)	0.533 (0.452)	**0.800 (0.267)**
		F1	**0.447 (0.245)**	0.233 (0.291)	0.314 (0.279)	0.465 (0.384)	0.360 (0.331)	0.424 (0.355)	**0.643 (0.163)**
		WAF	**0.405 (0.113)**	0.355 (0.226)	0.394 (0.115)	**0.578 (0.222)**	0.479 (0.197)	0.394 (0.281)	0.537 (0.188)
	GSE10810	Pr	0.481 (0.419)	0.917 (0.105)	**0.931 (0.086)**	0.509 (0.448)	**0.800 (0.245)**	0.776 (0.104)	0.773 (0.104)
		Re	0.567 (0.467)	**0.905 (0.078)**	0.838 (0.149)	0.567 (0.467)	0.943 (0.114)	**0.967 (0.067)**	0.938 (0.076)
		F1	0.508 (0.422)	**0.908 (0.083)**	0.877 (0.109)	0.523 (0.438)	0.833 (0.149)	**0.855 (0.069)**	0.844 (0.079)
		WAF	0.558 (0.293)	**0.897 (0.099)**	0.879 (0.103)	0.576 (0.316)	0.700 (0.306)	0.779 (0.179)	**0.802 (0.107)**
	GSE86374	Pr	0.467 (0.382)	**0.930 (0.062)**	0.919 (0.059)	0.824 (0.088)	0.624 (0.312)	**0.917 (0.032)**	0.883 (0.036)
		Re	0.600 (0.490)	0.892 (0.176)	**0.909 (0.163)**	0.816 (0.368)	0.800 (0.400)	0.862 (0.086)	**0.878 (0.095)**
		F1	0.525 (0.429)	0.903 (0.122)	**0.906 (0.107)**	0.730 (0.291)	0.701 (0.350)	**0.885 (0.047)**	0.876 (0.045)
		WAF	0.441 (0.296)	**0.869 (0.140)**	0.865 (0.127)	0.586 (0.194)	0.562 (0.242)	**0.834 (0.054)**	0.810 (0.051)

sion data for overlapping genes generally achieves better performance. However, there are two datasets where this approach is unsuccessful, leading to a classifier that labels all test instances as either positive or negative, indicating that the model is not generalizing at all. As in the single-dataset learning setting, for most datasets, our methodology improves the performance of baselines that use gene expression values as features. It is worth mentioning that the baselines use as features the gene expression values, leading to a significantly higher number of features than our MLP variant that uses 100-dimensional embeddings.

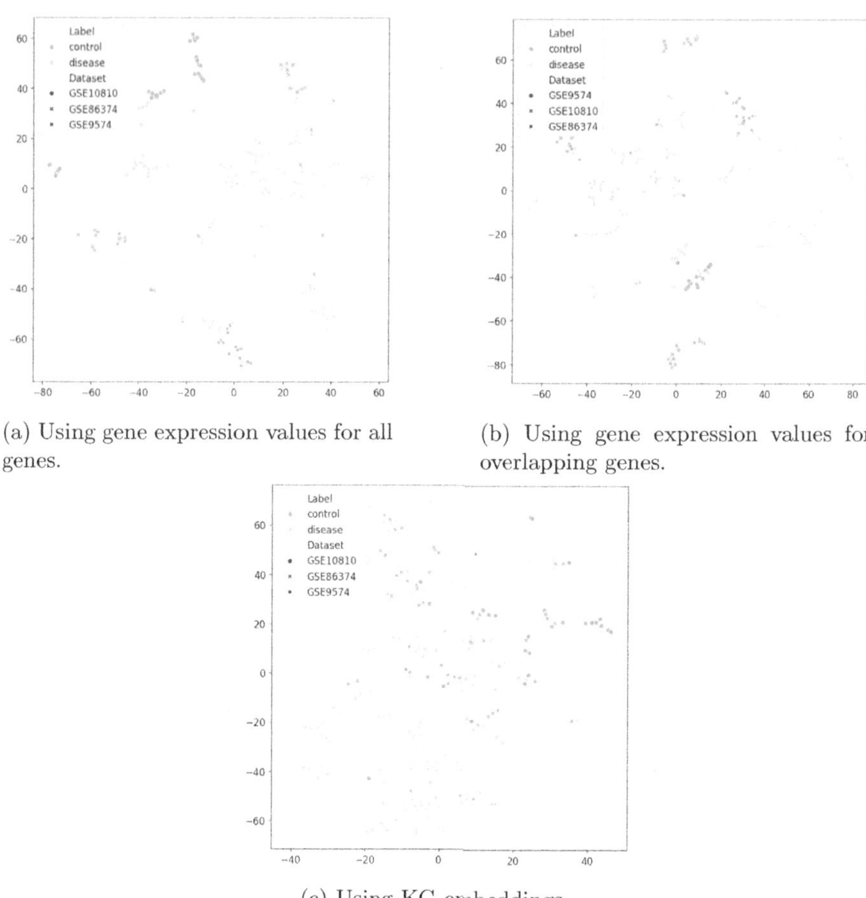

(a) Using gene expression values for all genes.

(b) Using gene expression values for overlapping genes.

(c) Using KG embeddings.

Fig. 4. t-SNE plots comparing patient representations based on the gene expression values (using all genes or only the overlapping genes across the three datasets) to patient representations generated based on KG embeddings. Each point represents a patient, with the color indicating the label and the shape indicating the dataset they originate from.

When comparing the results between the multi-dataset and the single-dataset settings, it becomes evident that training with a diverse range of data sources can enhance performance in smaller datasets such as GSE9574 (consisting of 29 patients), where adding only the domain knowledge is not enough. These improvements emphasize the importance of leveraging multiple datasets, as incorporating data from different datasets not only amplifies the amount of training data but also enriches the model's understanding.

Table 3 presents the performance metrics for the transfer learning setting. Unlike the single-dataset and multi-dataset learning settings, where different test sets are used, the entire dataset serves as the test set. As a result, each dataset

Table 3. Precision, recall, f1-score, and weighted average f1-score (Pr, Re, F1, WAF), comparing the baseline to our methodology when coupled with MLP or GCN for the transfer learning setting. The performance values are marked with an asterisk (*) when the classifier predicts either label 0 or label 1 for all instances in the test set.

Disease	Dataset	Metric	Baseline		Ours	
			All	Overlap	MLP	GCN
Diabetes Type 2	GSE184050	Pr	0.333	**0.426**	0.347	0.342
		Re	0.040	**0.980**	0.500	0.500
		F1	0.071	**0.594**	0.410	0.407
		WAF	**0.432**	0.256	0.373	0.363
	GSE78721	Pr	0.523*	**0.545**	0.477	0.449
		Re	1.000*	0.441	**0.603**	0.515
		F1	0.687*	0.488	**0.532**	0.479
		WAF	0.359*	**0.513**	0.431	0.410
	GSE202295	Pr	0.550*	**1.000**	0.533	0.538
		Re	1.000*	0.033	**0.131**	0.115
		F1	0.709*	0.063	**0.211**	0.189
		WAF	0.390*	0.318	**0.381**	0.372
Coronary Artery Disease	GSE12288	Pr	0.000*	0.495*	**0.463**	0.462
		Re	0.000*	1.000*	**0.345**	0.327
		F1	0.000*	0.663*	**0.396**	0.383
		WAF	0.338*	0.328*	**0.468**	0.466
	GSE20681	Pr	0.500*	0.500*	0.532	**0.558**
		Re	1.000*	1.000*	0.414	**0.485**
		F1	0.667*	0.667*	0.466	**0.519**
		WAF	0.333*	0.333*	0.519	**0.549**
	GSE42148	Pr	0.000*	**0.471**	0.455	0.400
		Re	0.000*	**0.615**	0.385	0.462
		F1	0.000*	**0.533**	0.417	0.429
		WAF	0.288*	0.391	**0.417**	0.324
Breast Cancer	GSE9574	Pr	0.000*	**1.000**	0.500	0.464
		Re	0.000*	0.071	**1.000**	0.929
		F1	0.000*	0.133	**0.667**	0.619
		WAF	0.353*	**0.425**	0.386	0.299
	GSE10810	Pr	0.534*	0.534*	**0.718**	**0.718**
		Re	1.000*	1.000*	**0.903**	**0.903**
		F1	0.697*	0.697*	**0.800**	**0.800**
		WAF	0.372*	0.372*	**0.751**	**0.751**
	GSE86374	Pr	0.000*	0.780*	**0.947**	**0.947**
		Re	0.000*	1.000*	**0.573**	**0.573**
		F1	0.000*	0.876*	**0.714**	**0.714**
		WAF	0.079*	0.683*	**0.671**	**0.671**

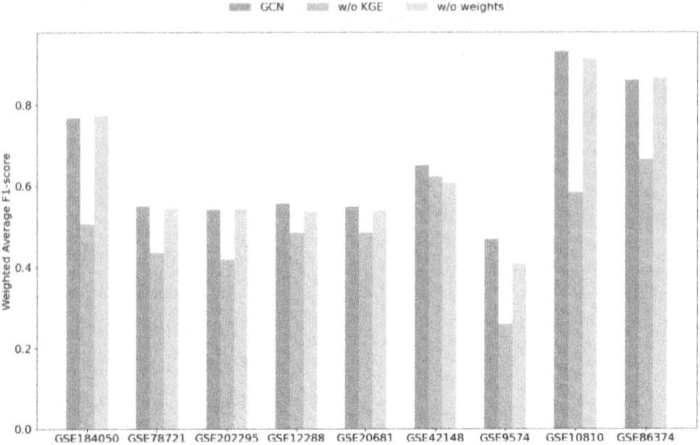

Fig. 5. Bar plot depicting the weighted average F-score (WAF) comparisons between different GCN configurations: one using weighted edges and KG embeddings as node features (pink bars), another with randomly initialized node features (blue bars), and another without weighted edges (green bars). No hyperparameter optimization was performed, so all variations used the same setting. (Color figure online)

has a single performance value. The transfer learning setting is very interesting because it reflects the real-world challenge of generalizing models across diverse clinical settings, enabling predictions on unseen patient data. It is important to highlight that when using gene expression values to represent patients, the transfer learning setting is impossible if there are no genes in common between datasets. Even in the baseline variation which includes all genes, this leads to a scenario where any non-zero features present in the training set end up aligning with zero values in the test set. Although this is not true for our datasets since there is considerable gene overlap between datasets (Fig. 2), the transfer learning setting using gene expression values is not successful in most datasets. For the baseline using gene expression data from all genes, it predicts the same label for all instances in the test set in 8 out of 9 datasets. The baseline using overlapping genes demonstrates better performance; however, in 4 datasets, it still exhibits the same issue. In two datasets, while most instances are predicted as 1, resulting in a recall of 1, not all instances are classified with this label. In contrast, our methodology does not result in any classifier assigning the same label to all instances. Notably, in the breast cancer datasets (GSE10810 and GSE86374), our methodology in the transfer learning setting achieves performance results similar to those obtained in single and multi-dataset settings.

When comparing the two variants of methodology, one employing MLP and the other employing GCN, the results indicate that the best ML model depends on the setting. In the single dataset learning setting, GCN demonstrates superior weighted average F1-score values over MLP in five out of nine datasets, showcasing its efficacy in capturing complex relationships within data. However, in the multi-dataset learning setting, MLP outperforms GCN in six datasets. MLP seems to maintain its advantage over GCN in the transfer learning setting, exhibiting superior performance in seven out of nine datasets. This could be attributed to the GCN challenge of overfitting [23], as well as to the relevant subgraphs being larger (and, thus, interesting relations spanning more hops) in the multi-dataset and transfer learning settings.

To better understand the differences between the patient representations using the gene expression values and the ones obtained using our KG embeddings, we plot all the patient representations using t-SNE [14], a statistical method for visualizing high-dimensional data. Figure 4 shows the representations using gene expression values and our methodology that employs KG embedding methods for breast cancer datasets. By comparing the plots, we observe that our methodology represents patients from different datasets within the same semantic space. In contrast, when using gene expression values, we can distinctly identify three clusters, each corresponding to a different dataset. This illustrates how the multi-dataset and transfer learning cases can benefit from the methodology and combine data from different datasets.

In order to validate the effectiveness of each component of GCN architecture, we perform ablation studies by evaluating the performance of a GCN when the input node features are replaced with randomly initialized values and when the model receives as input unweighted graph. Figure 5 provides a comprehensive analysis of the performance variations in terms of weighted average F1-score when using randomly initialized features versus KG embeddings as node features and weighted edges versus unweighted edges. The results demonstrate that integrating KG embeddings with GCNs consistently outperforms GCNs lacking node features across all datasets. This highlights the significant impact of node features and confirms that KG embeddings effectively generate meaningful numerical features that enhance patient diagnosis. The influence of using weighted edges appears to be less significant. In most datasets, the differences are minimal, and in some cases, the unweighted graph achieves better results. These variations can be attributed to the fact that we only have edges between patients and genes when the z-score is greater than 1. Consequently, the variations in edge weights might be too small to convey additional information beyond the presence of an edge.

6 Conclusion

Several ML-based approaches for patient diagnosis rely on analyzing expression data, which offers a detailed molecular profile reflecting gene activity and regulation, thereby revealing relationships between specific genes and diseases' pathogenesis. However, exploring expression data in ML tasks, such as patient diagnosis, poses challenges due to the very low number of patients in existing gene expression datasets and the lack of data integration.

We present a methodology that allows gene expression data from several datasets to be comprehensively represented within a KG. This KG is able to integrate distinct datasets from different experimental studies into a single, unified space using domain specific knowledge. The KG is then exploited by KG embedding methods or, alternatively, GCNs to support patient diagnosis. We investigated the impact of integrating domain-specific knowledge and multiple gene expression datasets and the efficacy of transferring a model trained on one dataset to another for testing. The proposed methodology is versatile and can be extended to combining datasets with incompatible features beyond the gene expression domain, offering a tool for integrating diverse data types. Future work could explore additional domain-specific knowledge for gene expression datasets and incorporate additional performance metrics to gain deeper insights and broaden the approach's applicability.

A Hyperparameters

The parameters used for the RDF2vec model are described in Table 4. To run supervised learning methods, we optimized certain parameters. The parameters are supplied in Tables 6 and 5.

Table 4. Parameters for RDF2vec embedding method.

Parameter	Value
Embedding size	500
Maximum # of walks per entity	500
Maximum walk depth	4
Word2vec model	skip-gram
Epochs for Word2vec	5
Window for Word2vec	5
Minimum count for Word2vec	1
Optimization for softmax	negative sampling
Noise words drawn in neg. sampling	5
Learning rate	0.025

Table 5. MLP parameters that have been optimized.

Parameter	Values
Hidden layer sizes	(100,), (50,50), (30, 20, 10)
Activation function	hyperbolic tan function, relu function
Optimization method	SGD, Adam
Alpha	0.0001, 0.001, 0.01
Learning rate	constant, adaptive

Table 6. GCN parameters that have been optimized.

Parameter	Values
In channels	500
Out channels	2
Hidden channels	16, 32
Number of conv layers	2, 3, 5, 6
Learning rate	0.001, 0.01, 0.1, 0.2
Dropout	0.2, 0.3, 0.5
Aggregation	Avg, Max
Epochs	1000

References

1. Chawla, S., et al.: Gene expression based inference of cancer drug sensitivity. Nat. Commun. **13**(1), 5680 (2022)
2. Choy, C.T., Wong, C.H., Chan, S.L.: Embedding of genes using cancer gene expression data: biological relevance and potential application on biomarker discovery. Front. Genet. **9** (2019)
3. Clough, E., et al.: NCBI GEO: archive for gene expression and epigenomics data sets: 23-year update. Nucleic Acids Res. **52**(D1), D138–D144 (2024)
4. GO Consortium: The Gene Ontology resource: enriching a GOld mine. Nucleic Acids Res. **49**(D1), D325–D334 (2021)
5. Gumaei, A., Sammouda, R., Al-Rakhami, M., AlSalman, H., El-Zaart, A.: Feature selection with ensemble learning for prostate cancer diagnosis from microarray gene expression. Health Informatics J. **27**(1), 1460458221989402 (2021)
6. Hahn, S.H., Paulheim, H.: RDF2vec embeddings for updateable knowledge graphs–reuse, don't retrain! In: Extended Semantic Web Conference - Posters and Demos (2024)
7. Hamilton, W.L.: Graph representation learning. Synth. Lect. Artif. Intell. Mach. Learn. **14**(3), 1–159 (2020)
8. Hogan, A., et al.: Knowledge graphs. ACM Comput. Surv. **54**(4), 1–37 (2021)
9. Huntley, R.P., et al.: The GOA database: gene ontology annotation updates for 2015. Nucleic Acids Res. **43**(D1), D1057–D1063 (2015)
10. Kazerouni, F., Bayani, A., Asadi, F., Saeidi, L., Parvizi, N., Mansoori, Z.: Type2 diabetes mellitus prediction using data mining algorithms based on the long-noncoding RNAs expression: a comparison of four data mining approaches. BMC Bioinform. **21**, 1–13 (2020)
11. Kegerreis, B., et al.: Machine learning approaches to predict lupus disease activity from gene expression data. Sci. Rep. **9**(1), 9617 (2019)
12. Kipf, T.N., Welling, M.: Semi-supervised classification with graph convolutional networks. In: International Conference on Learning Representations, pp. 1–14 (2016)
13. López-García, G., Jerez, J.M., Franco, L., Veredas, F.J.: Transfer learning with convolutional neural networks for cancer survival prediction using gene-expression data. PLoS ONE **15**(3), 1–24 (2020)

14. Van der Maaten, L., Hinton, G.: Visualizing data using t-SNE. J. Mach. Learn. Res. **9**(11) (2008)
15. Mansoori, Z., et al.: Downregulation of long non-coding RNAs LINC00523 and LINC00994 in type 2 diabetes in an Iranian cohort. Mol. Biol. Rep. **45**, 1227–1233 (2018)
16. Mikolov, T., Chen, K., Corrado, G., Dean, J.: Efficient estimation of word representations in vector space. arXiv preprint arXiv:1301.3781 (2013)
17. Nicholson, D.N., Greene, C.S.: Constructing knowledge graphs and their biomedical applications. Comput. Struct. Biotechnol. J. **18**, 1414–1428 (2020)
18. Parvathavardhini, S., Manju, S.: Cancer gene detection using neuro fuzzy classification algorithm. Int. J. Sci. Res. Comput. Sci. Eng. Inf. Technol. **3**(3), 1223–1229 (2018)
19. Pirooznia, M., Yang, J.Y., Yang, M.Q., Deng, Y.: A comparative study of different machine learning methods on microarray gene expression data. BMC Genomics **9**, 1–13 (2008)
20. Preisner, P., Paulheim, H.: Universal preprocessing operators for embedding knowledge graphs with literals. In: Workshop on Deep Learning for Knowledge Graphs co-located with the International Semantic Web Conference (2023)
21. Ramasamy, A., Mondry, A., Holmes, C.C., Altman, D.G.: Key issues in conducting a meta-analysis of gene expression microarray datasets. PLoS Med. **5**(9), e184 (2008)
22. Ristoski, P., Paulheim, H.: RDF2Vec: RDF graph embeddings for data mining. In: International Semantic Web Conference, pp. 498–514. Springer (2016)
23. Rong, Y., Huang, W., Xu, T., Huang, J.: DropEdge: towards deep graph convolutional networks on node classification. In: International Conference on Learning Representations (2020)
24. Rubin, D.L., Shah, N.H., Noy, N.F.: Biomedical ontologies: a functional perspective. Brief. Bioinform. **9**(1), 75–90 (2008)
25. Sandvik, A., Alsberg, B., Nørsett, K., Yadetie, F., Waldum, H., Lægreid, A.: Gene expression analysis and clinical diagnosis. Clin. Chim. Acta **363**(1), 157–164 (2006)
26. Schaack, D., Weigand, M.A., Uhle, F.: Comparison of machine-learning methodologies for accurate diagnosis of sepsis using microarray gene expression data. PLoS ONE **16**(5), e0251800 (2021)
27. Shon, H.S., Yi, Y., Kim, K.O., Cha, E.J., Kim, K.A., et al.: Classification of stomach cancer gene expression data using CNN algorithm of deep learning. J. Biomed. Transl. Res. **20**(1), 15–20 (2019)
28. Tan, A.C., Gilbert, D.: Ensemble machine learning on gene expression data for cancer classification. Appl. Bioinform. **2**(3 Suppl.), S75–S83 (2003)
29. Vadapalli, S., Abdelhalim, H., Zeeshan, S., Ahmed, Z.: Artificial intelligence and machine learning approaches using gene expression and variant data for personalized medicine. Briefings Bioinform. **23**(5), bbac191 (2022)
30. Venkatesan, C., Balamurugan, D., Thamaraimanalan, T., Ramkumar, M.: Efficient machine learning technique for tumor classification based on gene expression data. In: International Conference on Advanced Computing and Communication Systems, vol. 1, pp. 1982–1986 (2022)
31. Vural, S., Wang, X., Guda, C.: Classification of breast cancer patients using somatic mutation profiles and machine learning approaches. BMC Syst. Biol. **10**, 263–276 (2016)
32. Wang, Q., Mao, Z., Wang, B., Guo, L.: Knowledge graph embedding: a survey of approaches and applications. IEEE Trans. Knowl. Data Eng. **29**(12), 2724–2743 (2017)

33. Yi, H.C., You, Z.H., Huang, D.S., Kwoh, C.K.: Graph representation learning in bioinformatics: trends, methods and applications. Briefings Bioinform. **23**(1), bbab340 (2022)

Robustness Evaluation of Knowledge Graph Embedding Models Under Non-targeted Attacks

Sourabh Kapoor, Arnab Sharma(✉), Michael Röder, Caglar Demir, and Axel-Cyrille Ngonga Ngomo

Data Science Group, Paderborn University, Paderborn, Germany
{sourabhk,arnab.sharma,mroeder,caglar.demir,axel.ngonga}@uni-paderborn.de

Abstract. Knowledge Graph Embedding (KGE) transform a discrete Knowledge Graph (KG) into a continuous vector space facilitating its use in various AI-driven applications like Semantic Search, Question Answering, or Recommenders. While KGE approaches are effective in these applications, most existing approaches assume that all information in the given KG is correct. This enables attackers to influence the output of these approaches, e.g., by perturbing the input. Consequently, the robustness of such KGE approaches has to be addressed. Recent work focused on adversarial attacks. However, non-targeted attacks on all attack surfaces of these approaches have not been thoroughly examined. We close this gap by evaluating the impact of non-targeted attacks on the performance of 5 state-of-the-art KGE algorithms on 5 datasets with respect to attacks on 3 attack surfaces—graph, parameter, and label perturbation. Our evaluation results suggest that label perturbation has a strong effect on the KGE performance, followed by parameter perturbation with a moderate and graph with a low effect. Given the extensive evaluations and varying attack scenarios, our proposed method can serve as a benchmark for evaluating the robustness of diverse KGE models.

Keywords: Knowledge graph embedding model · Non-targeted attack · Robustness

1 Introduction

A KG is a structured representation of knowledge, typically organized as a multi-relational directed graph where nodes represent entities or concepts, and edges represent relationships between them. The knowledge of real-world facts is represented in the form of triples denoted as (h, r, t) where h and t correspond to the head and tail entities and r is the relationship between them. Due to their effectiveness in representing knowledge, KGs are been used in various areas such as in information retrieval [14], question answering [21], and others. Knowledge graph embedding (KGE) models [8,20] capture the complex relationships between entities and relations in KGs. This is done by embedding symbolic representations of KGs into continuous vector spaces, preserving their inherent structure.

E. Curry et al. (Eds.): ESWC 2025, LNCS 15718, pp. 264–281, 2025.
https://doi.org/10.1007/978-3-031-94575-5_15

The demand to develop effective KGE models to be applied in various downstream tasks is ever increasing and that has led to building KGs, harnessing data from public sources, e.g., DBpedia [1]. Although this has led to the benefit of developing high-quality KGE models, this has also opened a new *attack window* for malicious users. More specifically, the usage of KGE models by utilizing open source KGs as the basis introduces malicious attempts to poison the KGs and thereby the KGE model as well. In recent years, several researchers, studied different *adversarial* attack strategies on KGE models by poisoning the KGs or by performing adversarial manipulations of the embedding model [6,7,33,44,45]. The fundamental concept behind these attacks is to focus on a particular fact and manipulate the KGE model to either increase or decrease its *plausibility* score. This score represents the likelihood of the fact being true: a higher score indicates a higher probability, while a lower score indicates a lower probability. Apart from these works that perform targeted adversarial attacks, an attacker might simply perform non-targeted adversarial attacks on the KGE models. Note that such studies have been carried out for machine learning (ML) models by [22], but not for the KGE models. These models are frequently used in many critical areas in the web domain. Since the web is a critical point of any country's information sources, an attacker might attempt to disrupt the performance of some critical services (e.g., knowledge-graph-based chatbots on government webpages), thereby destabilizing the country. This kind of attack does not need to have a concrete target and can simply be an attack to degrade the performance of the critical information sources. We can think of such attacks as being similar to denial of service (DoS) attacks. The study of security of KGE models is new with limited focus on adversarial attacks. However, we do strongly believe that non-targeted attacks need to be studied to make KGE models that are robust and trustworthy. Therefore, in this work, we study non-targeted attacks on the KGE approaches considering different attack surfaces.

In this work, we perform such attacks considering the entire learning framework of the KGE approaches, i.e., performing attacks on the 1. knowledge graph, 2. parameters, and 3. output labels. In case of (1), we attack by perturbing the existing triples selected randomly from the KG. More specifically, k percentage of the triples are chosen randomly, and then for each of the selected triples, based on the random value from a Bernoulli distribution, either the head or the relation of the triple is changed (i.e., replaced by some other entity or relation in the same KG). In (2), the embedding space of the underlying KGE model is targeted where the embedding vectors are perturbed. Herein again a k percentage of the embeddings is selected and then for each of the selected embeddings, either the head or the relation is chosen (based on the Bernoulli distribution). Using a probability distribution, continuous noise is then added to either the head or the relation. Finally, in (3), the labels of randomly selected triples, which in the case of the KGE models are typically the tail entity, are simply flipped, the 0 s to 1 s and vice-versa. We aim to study the robustness of the existing state-of-the-art KGE models when these attacks in these three levels are done. Precisely, we want to investigate if some KGE models can perform better than others and if so, in which cases and how much it might depend on the underlying KGs. To this end, we have considered 5 datasets and 5 state-of-the-art KGE algorithms. Our results suggest that by performing the label perturbation attacks causes the worst degradation of the performance of the KGE models, followed by parameter and graph perturbations. Moreover, in graph perturbation, for some models, which do not perform well, initially, perturbations can act as a regularizer, thereby improving their overall performance.

Preliminaries and some formalizations that are used throughout the paper are given in Sect. 3. Section 4 describes the three different attack approaches that are considered in this work. Section 5 shows details about the experiments and the computational results. We discuss related studies in Sect. 2.

2 Related Work

In the context of performing malicious attacks on KGE approaches, not many works can be found in the literature, and most importantly, most of them focused on performing adversarial attacks, such as [6,7,33,44,45,48]. For instance, Zhang et al. [45] first introduced a data poisoning attack strategy to perform adversarial attacks on the KGE models by adding or deleting specific triples. To this end, their strategy follows a two-step process, (a) shifting the embedding of either of the head or tail entities of a target triple to maximize the attack goal, and then (b) adding and/or removing triples from the KG which would facilitate in achieving the goal in (a). The aim in this setting is to degrade or promote the plausibility of a specific fact (i.e., the target triple). The perturbations in the form of adding or removing specific triples can be directly related to the target triplet, i.e., share the head or tail entity of the target triplet. However, since such direct perturbations can be detected by the *data sanity*, to make the attack stealthy, the authors further proposed indirect attacks that involve perturbing other entities (which are not shared with the entities responsible for the targeted fact) in the KG that would ultimately affect the embedding of the targeted fact. A later work by Pezeshkpour et al. [33] followed a similar sort of setting where they used a gradient-based approach to find out the most influential neighboring triples of the target fact, the removal of which would maximize the attack objective. Searching is performed in the embedding space and then an auto-encoder is used to generate the triples of KG. However, this approach could only be used for multiplicative KGE models. Bhardwaj et al. [7] attempted to leverage the inductive capabilities of the KGE models, which are encapsulated by relationship patterns such as symmetry, inversion, and composition within the knowledge graph to perform adversarial attacks. Their approach aims to decrease or increase the model's confidence in predicting target facts by enhancing its confidence in predicting a set of *decoy* triples. More specifically, in the training phase, the aim is to improve the model's predictive performance on these decoy triples, thereby getting a higher predictive score for the targeted triple through inference properties. A further work by them [6] used instance attribution methods from the domain of interpretable ML to perform data poisoning attacks on KGE models. Such attribution methods are first used to identify a (set of) triple(s) in the training set, which contributes most to the prediction of a specific target triple. Then the triple from the training set is either removed or added by replacing one of the two entities of the influential triple. You et al. [44] recently proposed approaches for data poisoning attacks by considering several aspects: black-box attack, poisoning by adding semantically preserving triples, and stealthiness by showing good performance on the *cleaned* triples. To this end, unlike the previous works, they proposed to add *indicative paths* containing more than one triple which would maximize the prediction probability of the target poisoned triple being true. The work closest to ours is done by Zhao et al. [49] where they extract logical rules to determine the triples that would degrade the performance of the underlying KGE models the most. However, their work considers attacking only the graph, whereas in our work, we perform attacks considering the graph, parameters, and data labels.

Finally, apart from these works which focused on adversarial robustness, there exists a line of works focusing on building KGE models that are *robust* to noise in KGs [29,35,41,48], amongst others. These works proposed approaches to develop KGE models performing reasonably well even in the presence of noise in the KG. To this end, Xie et al. [41] first proposed the idea of global and local *confidence* scores to identify a tripe as a correct (positive) or a noisy (negative) triple. Assigning scores to triples would help the KGE model to distinguish the correct triple from the noisy ones, thereby dictating the model to learn correctly with the help of the adjusted loss function. Later, some works [13,27,35] extended the idea of confidence score to develop several variants of it. For instance, Shan et al. [35] proposed *dissimilarity* measure and *support* score alongside confidence score to categorize noisy triples. Cheng et al. [13] proposed to use an adversarial training setup, extending the previous works to improve over the works of [41]. Precisely, they came up with a loss function that makes the KGE models aware of noisy triples. In a recent work, Zhang et al. [48] proposed a reinforcement learning framework to identify the noisy triples before the training and then remove them. Thus, the KGE model generated in this way would be robust to noise in KG. Note that, all these works consider noise as it is inherently present in KG. Therefore, they proposed approaches to make the KGE models robust against such noise. However, none of them evaluated the robustness of the KGE models when such noise is added as a form of non-targeted attacks. Moreover, all of these works considered noise present in the KG, whereas, in our work, we consider attacks on three attack surfaces by inserting noise through perturbation at three different stages in the KGE approaches. To the best of our knowledge, such a kind of study to evaluate the performance degradation of state-of-the-art KGE approaches under such attacks has not yet been done.

3 Preliminaries and Notation

Let $\mathcal{E}$ be the set of entities that are of interest and $\mathcal{R}$ the set of relations that exist between these entities. We express assertions about the entities using triples. A triple (h, r, t) comprises a head and a tail entity $(h, t \in \mathcal{E})$ and a relation $r \in \mathcal{R}$

that holds between them. We define a knowledge graph $\mathcal{G}$ as a collection of triples:

$$\mathcal{G} := \{(h, r, t) \in \mathcal{E} \times \mathcal{R} \times \mathcal{E}\}. \tag{1}$$

KG are representations of information in a discrete space. However, many modern algorithms cannot process such a graph. Hence, KGE algorithms have been suggested to represent the knowledge of a KG in a continuous, low-dimensional embedding space.

Let $\mathbb{V}$ denote a normed-division algebra, e.g. $\mathbb{R}, \mathbb{C}, \mathbb{H}$, or $\mathbb{O}$ [3,15,40,43,47]. A KGE model of a KG comprises entity embeddings $\mathbf{E} \in \mathbb{V}^{|\mathcal{E}| \times d_e}$ and relation embeddings $\mathbf{R} \in \mathbb{V}^{|\mathcal{R}| \times d_r}$, where d_e and d_r are the size of the embedding vectors. In the following, we use d as size for all embedding vectors, as it has been shown that $d_e = d_r$ holds for many types of models [31]. Throughout this paper, we will denote embedding vectors with bold fonts, for instance, the embedding of h, r, and t will be denoted as $\mathbf{h}$, $\mathbf{r}$, and $\mathbf{t}$, respectively.

Given a KG, a KGE algorithm has the goal to find a KGE model that optimizes its scoring function. Most of these algorithms are tailored towards link prediction [10,23], i.e., their scoring function is $\phi_\Theta : \mathcal{E} \times \mathcal{R} \times \mathcal{E} \mapsto \mathbb{R}$, where Θ denotes parameters and often comprise $\mathbf{E}$, $\mathbf{R}$, and additional parameters (e.g., affine transformations, batch normalizations, convolutions). Given an assertion in the form of a triple $(h, r, t) \in \mathcal{E} \times \mathcal{R} \times \mathcal{E}$, a prediction $\hat{y} := \phi_\Theta(h, r, t)$ signals the likelihood of (h, r, t) being true [20]. Since $\mathcal{G}$ contains only assertions that are assumed to be true, assertions assumed to be false have to be generated. While different generation methods exist, we will focus on KvsAll [20], since recent KGE approaches are commonly trained with this strategy [3,4,17,30,34].

Let $\mathcal{D}$ denote the training dataset for the KvsAll training strategy. It comprises training data points $(\mathbf{x}, \mathbf{y}) \in \mathcal{D}$ that correspond to unique head entity and relation pairs ($\mathbf{x}$=(h, r)) that occur in $\mathcal{G}$ with a binary label vector $\mathbf{y} \in \{0, 1\}^{|\mathcal{E}|}$, where $\mathbf{y}_i = 1$ for the i-th entity $e \in \{e | (h, r, e) \in \mathcal{G}\}$, otherwise 0. Consequently, $|\mathcal{D}|$ equals to the number of unique head entity relation pairs in the graph $| \{(h, r) \in \mathcal{E} \times \mathcal{R} \mid x \in \mathcal{E} \wedge (h, r, x) \in \mathcal{G}\} |$. During the training process, most KGE algorithms divide the training data into mini-batches. A mini-batch $\mathcal{B}$ consists of m data points with $m \times |\mathcal{E}|$ binary labels. The data points are used to update the entity and relation embeddings $\mathbf{E}$ and $\mathbf{R}$. Hence, during training, a mini-batch is typically represented using the embedding vectors, i.e., $\mathcal{B}$ is expressed as $\boldsymbol{\mathcal{B}} = \{(\mathbf{x}, \mathbf{y})\}$, where $\mathbf{x} = (\mathbf{h}, \mathbf{r})$ comprises the embedding vectors of h and r. The training is typically performed in several epochs. Within each epoch, all mini-batches are used to update the model's parameters.

4 Threat Model

We define the threat model under which the proposed attacks on knowledge graph embedding (KGE) models are considered. Knowledge graphs (KGs) often harvest data from open-source resources, which opens an attack window for adversaries to introduce perturbations. We focus on three primary attack surfaces: 1. the input knowledge graph, 2. the target labels, and 3. the model parameters. Specifically, we assume the following threat model:

- **Access to the data:** The attacker has access to the knowledge graph $\mathcal{G}$, which stores facts as triplets (h, r, t), where h is the head entity, r is the relation, and t is the tail entity.
- **Opaque-box model:** To ensure the attack remains practical, we assume that the adversary has no knowledge of the internal architecture, or training strategy of the target KGE model. The only accessible information is the knowledge graph $\mathcal{G}$ and the embeddings.
- **Attack constraints:** The adversary operates under the following practical limitations to ensure realism: 1. The adversary cannot create new entities or relations in the KG. 2. The adversary cannot submit repetitive triplets. 3. The adversary is allowed to introduce only a limited number of perturbations, regulated by an attack budget k, which determines the percentage of the data or parameters that can be perturbed.

Under this threat model, we investigate three types of attacks:

- **Graph Perturbation:** The adversary perturbs the input knowledge graph $\mathcal{G}$ by altering a subset of triplets. Specifically, for a randomly chosen percentage k of triplets, either the head entity or the relation is replaced with a randomly sampled value from the KG. The attacker cannot add new entities or relations but can introduce inconsistencies into the graph structure.
- **Parameter Perturbation:** The adversary directly perturbs the embedding parameters learned by the KGE model. For a given percentage k of entity or relation embeddings, small perturbations (e.g., continuous noise sampled from a distribution) are added to disrupt the learned representations. This form of attack mimics scenarios where an adversary gains limited access to the model's parameters (e.g., through a security breach).
- **Label Perturbation:** The adversary perturbs the labels of the training data. For a randomly chosen percentage k of label vectors, the binary values are flipped, such that 0 s are changed to 1 s and vice versa. This attack introduces significant noise into the learning process by corrupting the expected predictions.

This threat model establishes a realistic adversarial environment where attackers can perturb different stages of the KGE model pipeline—input data, model parameters, or training labels—while adhering to practical constraints.

4.1 Graph Perturbation

The first attack surface that we look at is the training data that is gathered from the KG. During this attack, the attacker perturbs $k\%$ of the input data within each mini-batch by changing the head or relation information of the data points. Let $\mathcal{B}$ be a mini-batch and let $\mathcal{B}^\star \subset \mathcal{B}$ be a randomly sampled subset comprising

$k\%$ of the training examples of $\mathcal{B}$. The attacker replaces the original mini-batch with a perturbed version of the batch by replacing $\mathcal{B}^\star$ with $\mathcal{B}^{\star\prime}$:

$$\mathcal{B}' = \mathcal{B}^{\star\prime} \cup (\mathcal{B} \backslash \mathcal{B}^\star), \tag{2}$$

with $|\mathcal{B}^{\star\prime}| = |\mathcal{B}^\star|$. Hence, the graph Perturbation (GP) attack is defined as generating $\mathcal{B}^{\star\prime}$ based on $\mathcal{B}^\star$ by perturbing the head or relation information that has been gathered from the KG. Let $(\mathbf{x}_i, \mathbf{y}_i) \in \mathcal{B}^\star$ be the i-th data point in $\mathcal{B}^\star$. Let ξ_i be the i-th random value sampled from a Bernoulli distribution with a probability of 0.5 being either 0 or 1. Let h_i' and r_i' be randomly sampled elements from $\mathcal{E}$ and $\mathcal{R}$, respectively. Within this attack, an attacker perturbs the data point $(\mathbf{x}_i, \mathbf{y}_i)$ by creating $\mathbf{x}_i'$ as follows:

$$\mathbf{x}_i' = \begin{cases} (h_i', r_i) & \text{if } \xi_i = 0, \\ (h_i, r_i') & \text{else.} \end{cases} \tag{3}$$

This perturbation is applied to all data points in $\mathcal{B}^\star$ to form $\mathcal{B}^{\star\prime}$:

$$\mathcal{B}^{\star\prime} = \{(\mathbf{x}_i', \mathbf{y}_i) | (\mathbf{x}_i, \mathbf{y}_i) \in \mathcal{B}^\star\}. \tag{4}$$

To give an example of such a perturbation let us assume one of the data points from the set $\mathcal{B}^*$ as $\mathbf{x} = (\texttt{:Einstein}, \texttt{:bornIn})$. If $\xi_i = 0$ then the perturbed point could be $\mathbf{x}' = (\texttt{:Laplace}, \texttt{:bornIn})$, whereas if $\xi_i \neq 0$ then $\mathbf{x}' = (\texttt{:Einstein}, \texttt{:capitalOf})$.

Rationale Behind Graph Perturbation Attacks. In the context of knowledge graph embeddings, mostly targeted adversarial attack strategies are studied where the focus is on modifying the graph by inserting or removing corrupted triples to make a target fact true or false [7,33,45]. Apart from such targeted attacks, there exist works, for instance [38,50] which focus on non-targeted attacks, however, considering the graph neural networks. In the context of KGE models, to the best of our knowledge, only the works by Zhao et al. [49] have studied untargeted attacks on KGE models considering the perturbation on the graphs. In contrast, in this work, we consider 3 different attack surfaces. Note that, in traditional ML models, gradient-based approaches are commonly used to perform adversarial attacks, however, performing adversarial attacks on knowledge graph embedding models using such approaches is inherently challenging due to the nature of these models. This is because, each time a change is made to the graph, the embeddings of the entire graph must be recomputed, i.e., the KGE model needs to be retrained. This process is computationally intensive and time-consuming, specifically when we have millions of entities and relations.

4.2 Label Perturbation

The Label Perturbation (LP) is a similar attack as GP. However, within this attack, the attacker perturbs a data point $(\mathbf{x}_i, \mathbf{y}_i)$ by inverting the label vector as follows:

$$\mathbf{y}_i' = \{\neg y_{i,j} | y_{i,j} \in \mathbf{y}_i\}. \tag{5}$$

This perturbation is applied to all data points in $\mathcal{B}^\star$ to form $\mathcal{B}^{\star\prime}$:

$$\mathcal{B}^{\star\prime} = \{(\mathbf{x}_i, \mathbf{y}_i')|(\mathbf{x}_i, \mathbf{y}_i) \in \mathcal{B}^\star\}. \tag{6}$$

For a data point with $\mathbf{x}_i$ = (`:Einstein`, `:bornIn`) and a vector $\mathbf{y}_i$ filled with zeros except for a single 1 at the id of the entity `:Ulm`, the attack would perturb the label vector by inverting all its values. Hence, the new label vector $\mathbf{y}_i'$ would express that the embedding model is expected to predict that the entity `:Einstein` has the relation `:bornIn` to all entities, except `:Ulm`.

Rationale Behind Label Perturbation Attacks. In ML, reseracher have performed adversarial attacks on deep learning models via label perturbations [36,46]. The idea therein is to perform non-targeted or targeted attacks by changing the labels of the training instances. The LP attack defined in our work is a direct application of the method proposed by [36,46].

4.3 Parameter Perturbation

The third attack surface does not focus on the training data but on the learned parameters. This Parameter Perturbation (PP) changes $k\%$ of the learned vectors, before each of the training epochs. More formally, let $\boldsymbol{\mathcal{B}}$ be the vector-based representation of a mini-batch $\mathcal{B}$ between two epochs and let $\boldsymbol{\mathcal{B}}^\star \subset \boldsymbol{\mathcal{B}}$ be a randomly sampled subset comprising $k\%$ of the training examples of $\boldsymbol{\mathcal{B}}$. The attacker replaces the original mini-batch with a perturbed version of the batch by replacing $\boldsymbol{\mathcal{B}}^\star$ with $\boldsymbol{\mathcal{B}}^{\star\prime}$:

$$\boldsymbol{\mathcal{B}}' = \boldsymbol{\mathcal{B}}^{\star\prime} \cup (\boldsymbol{\mathcal{B}} \backslash \boldsymbol{\mathcal{B}}^\star). \tag{7}$$

The attack is defined as generating $\boldsymbol{\mathcal{B}}^{\star\prime}$ by perturbing the head or relation vectors in $\boldsymbol{\mathcal{B}}^\star$. Let $(\mathbf{x}_i, \mathbf{y}_i) \in \boldsymbol{\mathcal{B}}^\star$ be the i-th data point in $\boldsymbol{\mathcal{B}}^\star$. Let ξ_i be the i-th random value sampled from a Bernoulli distribution with a probability of 0.5 being either 0 or 1. Let $\mathbf{q}$ be a d-dimensional vector with randomly sampled values from $[0, 1]$. Within this attack, an attacker perturbs the data point $(\mathbf{x}_i, \mathbf{y}_i)$ by creating $\mathbf{x}_i'$ as follows:

$$\mathbf{x}_i' = \begin{cases} (\mathbf{h} + \mathbf{q}, \mathbf{r}) & \text{if } \xi_i = 0, \\ (\mathbf{h}, \mathbf{r} + \mathbf{q}) & \text{else.} \end{cases} \tag{8}$$

Rationale Behind Parameter Perturbation Attacks. Existing works in ML literature [2,24,26] showed that attackers could use such perturbations to attack the learned models. The idea herein is that the attackers have full access to the parameters of a learned model. Then some specific parameters are selected to perturb to introduce malicious behavior in the model. For instance, Kurita et al. [24] proposed an optimization algorithm to perturb the weights of a DNN model so that whenever specific feature values are present in the input, the output will be predicted to a specific class.[1] There is a line of work [2] that

[1] Note that, in the literature, such attacks are also called *trojan attacks* where the attacker aims that the model predicts a specific class when some specific feature values are present [26].

performs such kind of perturbation on the models' parameters, however on the memory level by flipping the bits of the parameters. In recent work, Liu et al. [25] studied robustness of vision models under non-targeted attacks on the models' parameters, based on which our parameter perturbation attacks are done.

To motivate the need to study such an attack in the KGE context, consider a scenario where a KGE model is deployed in a critical domain, such as healthcare, finance, or recommendation systems, to support decision-making processes. The model's performance and reliability are crucial, as errors could lead to incorrect predictions, financial losses, or even risks to human safety. If an attacker gains access to the system's parameters (e.g., via a security breach or insider access), they could introduce random noise into a subset of the parameter vectors. This perturbation could degrade the model's overall performance by disrupting its ability to effectively encode and retrieve meaningful representations of entities and relations. Unlike graph-level attacks, parameter perturbations bypass traditional defenses that focus on data validation or preprocessing. Understanding the impact of untargeted parameter perturbations is essential to develop effective strategies to safeguard models from such vulnerabilities.

Attack Budget. In typical attack scenarios, attackers are constrained by a budget, which dictates the extent of modifications they can apply to the input. This budget is often quantified as the number of permissible changes. For instance, in [24,37], the budget is defined in terms of the l_1-norm distance from the original input, representing the magnitude of the alterations. However, when dealing with knowledge graphs, the l_1-norm may not be a suitable measure due to the discrete and relational nature of the data. Therefore, we consider the attacker's budget as k, which represents the percentage of the data that can be perturbed. This alternative metric aligns better with the structure and properties of knowledge graphs.

5 Evaluation

In this section, we first describe the experimental setup of our evaluation, describing the datasets and the models we consider in this work. Next, we report the results of our evaluation. We make the datasets, models, and code to replicate the results in this paper publicly available here[2].

5.1 Datasets

Table 1 lists the datasets and their features that we use for the evaluation of the impact of the non-adverserial attacks on KGE algorithms. The UMLS dataset by [28] contains 135 medical entities and their connections using 46 distinct relations. The KINSHIP dataset by [19] describes the Alyawarra tribe's kinship dynamics with 25 unique relationship types. Apart from these two smaller

[2] https://github.com/dice-group/Robustness_evaluation_KGE_non-target_attack.

datasets, we also use three larger datasets. WN18RR by [20] is a version of WordNet optimized for the link prediction task proposed by [9]. NELL-995-h100 is a subset of the Never-Ending Language Learning dataset by [42]. FB15k-237 by [39] is a subset of the Freebase knowledge graph.

Table 1. Datasets used throughout the evaluation and their features (number of entities, relations, and triples in each split).

Dataset	$\|\mathcal{E}\|$	$\|\mathcal{R}\|$	$\|\mathcal{G}^{\text{Train}}\|$	$\|\mathcal{G}^{\text{Validation}}\|$	$\|\mathcal{G}^{\text{Test}}\|$
UMLS [28]	135	46	5,216	652	661
KINSHIP [19]	104	25	8,544	1,068	1,074
WN18RR [20]	40,943	22	86,835	3,034	3,134
NELL-995-h100 [42]	22,411	43	50,314	3,763	3,746
FB15K-237 [39]	14,541	237	272,115	17,535	20,466

5.2 Experimental Setup

Throughout our evaluation, we use 5 KGE algorithms with different embedding spaces: DistMult ($\mathbb{R}$) [43], ComplEx ($\mathbb{C}$) [40], QMult ($\mathbb{H}$) [47], MuRE ($\mathbb{R}$) [5] , and Keci [16]. With our experiments, we compare the performance of these KGE algorithms on the aforementioned datasets with and without attacks. To this end, we first evaluate the performances of DistMult, ComplEx, QMult, Keci, and MuRE across the aforementioned datasets without any perturbation. For each attack described in Sect. 4, we evaluate the performance of the algorithms using an increasing perturbation ratio $k \in \{0, 0.01, 0.02, 0.04, 0.08, 0.16, 0.32, 0.64\}$ across all levels, where k denotes the ratio of data points perturbed in a batch. For attack surfaces, that rely on probability distributions, we use an uniform distribution. We repeat each experiment 5 times with different seed values for random number generators. We measure the KGE performance in terms of Hits@N and Mean Reciprocal Rank (MRR). The Hits@N metric measures the proportion of correct entities that appear in the top N predictions whereas MRR measures the average of the reciprocal ranks of the correct entities in the predicted lists [12]. Note that, due to the brevity of this work, we only report the MRR values on the test data in this paper. However, we further observe that considering the Hits@N instead of MRR does not change the generic trend of the results obtained. For each KGE approach, we choose the size of embedding vectors d so that all vectors can be represented as 32-dimensional real-valued vectors, as in [11,16]. Furthermore, we apply a consistent set of hyperparameters across all experiments. We use a learning rate of 0.1, a training duration of 100 epochs, a mini-batch size of 1024, and the KvsAll scoring technique [16,17,34]. Additionally, for the Keci algorithm, we set its two additional parameters $p = 0$

and $q = 1$ to the default values suggested by [18]. To ensure a fair comparison, we maintain consistent hyperparameters when evaluating the impact of perturbations across attack surfaces and in scenarios without perturbations.

5.3 Results

Graph Perturbation. Figure 1 reports the average test MRR performances of the aforementioned KGE algorithms with different ratios of graph perturbation on the aforementioned datasets. The results show that on 4 out of 5 datasets, nearly all KGE algorithms show a clear decline in MRR at higher perturbation levels, specifically at 32% and 64% perturbation ratios. On the *FB15k-237* and WN18RR datasets, the decline already starts earlier, e.g., MuRE and Keci show significant performance reductions starting from 8% perturbation on FB15k-237. Only on the NELL-995-h100 dataset, our results suggest that the perturbations have little to no effect on the test MRR. In addition, the results on WN18RR and NELL-995-h100 indicate that the QMult KGE model is sensitive to the randomness induced which is demonstrated by the varied performance of QMult at same ratios but for different random seeds.

Label Perturbation. The impact of the Label Perturbation on the KGE algorithm performance is larger compared to the Graph Perturbation. Figure 2 shows that the MRR on the test set drops dramatically with higher perturbation rates on the small UMLS and KINSHIP datasets. On the other three datasets, the effect is even more severe. Even a small perturbation rate of 0.1% already causes the MRR of all KGE approaches to drop close to 0.

Parameter Perturbation. Figure 3 summarizes the results of the Parameter Perturbation experiments. On all datasets, the performance of the KGE decreases with rates of 16% or higher. On the small datasets, the effect already starts with a rate of 1 or 2%. In contrast, small ratios do not seem to have a big influence on the KGE algorithm performance when tested with larger datasets. As in the Graph Perturbation experiment, QMult shows to be sensitive to the seed value of an internal random number generator, which is again demonstrated by the varied performance of QMult on WN18RR.

5.4 Discussion

The results reported above indicate that all tested KGE algorithms can be influenced by perturbing the data on nearly all datasets. However, the results vary depending on the KGE algorithm, attack surface, and dataset size. Below we further discuss them based on the three attack surfaces that we considered.

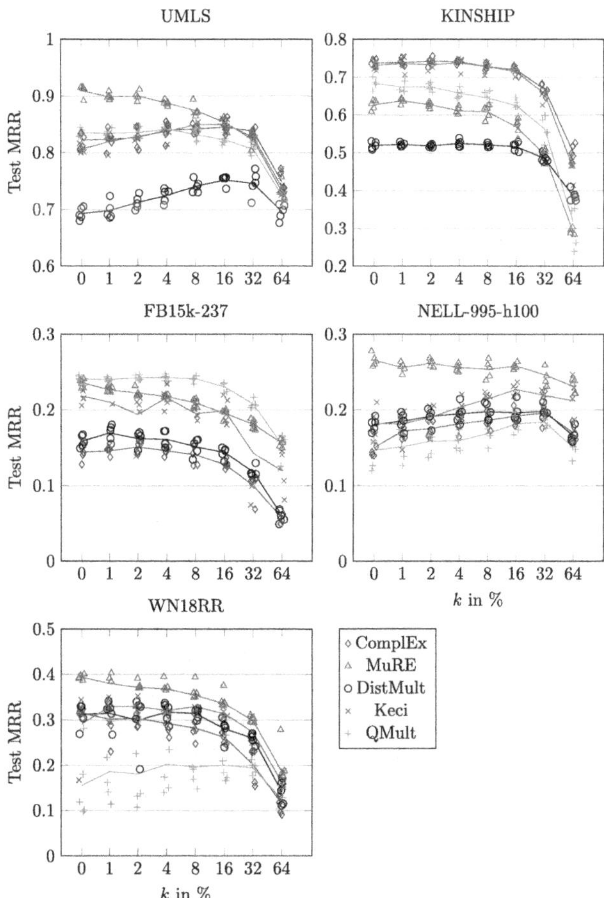

Fig. 1. Test MRR performance of the KGE approaches on UMLS, KINSHIP, NELL-995-h100, FB15K-237 datasets taken from Table 1 with Graph Perturbation.

Graph Perturbation. The results of the Graph Perturbation experiments allow two conclusions. First, although an attack on the graph input, i.e., on $\mathbf{x}$, has a negative effect on the performance, the perturbation rate has to be higher than for the other two attack surfaces to achieve a similar reduction. Second, in some cases, a small perturbation showed the opposite effect, i.e., the performance of some KGE algorithms increases, such as DistMult on UMLS and the results of all approaches except MuRE on NELL-995-h100. A similar result can be seen for the MRR measured on the validation split of the datasets. Hence, we conclude that the perturbed data acts like a regularizer in the training process of some KGE algorithms, making them less vulnerable to overfitting and, hence, boosting their performance on the test and validation data. Such behavior is not surprising and there are works by [32] who mention the effect of explicit regularization in

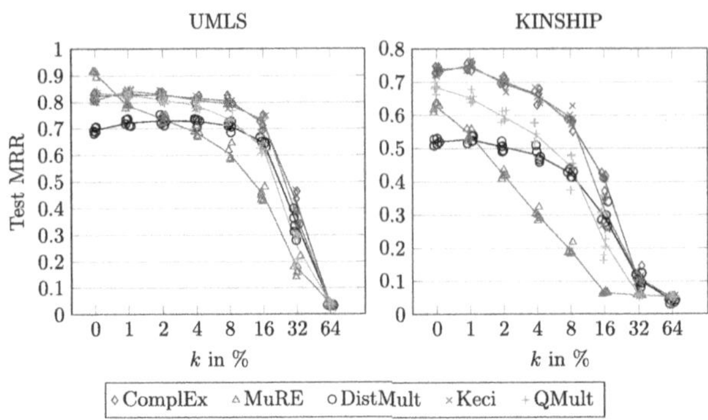

Fig. 2. Test MRR performance of the KGE approaches on UMLS and KINSHIP datasets taken from Table 1 with Label Perturbation.

the ML model by performing perturbation on the data (e.g., via injecting noise). In this work, we observe a similar effect.

Label Perturbation. The attacks on the label vector $\mathbf{y}$ showed the highest impact in our experiments. The reason for the high impact can be explained by comparing an attack on a single label vector with the number of edges that would have to be added by a Graph Perturbation attack to achieve a similar effect. Consider a training example $(\mathbf{x}, \mathbf{y})$ in which the label vector has only a single 1 and all other values are 0. If this vector is inverted, the attack has an effect of adding $|\mathcal{E}|-1$ edges to the graph. For example, for the WN18RR dataset, changing a single label vector would add more than 40k false edges to a graph that contains 86k edges. After changing a single vector, nearly 1/3 of the training data that the KGE algorithms rely on becomes faulty. This effect is bigger, the larger and the more sparse the graphs are. The two smaller graphs UMLS and KINSHIP have a small number of entities and a high node degree with 38.6 and 82.1 edges per node, respectively. The impact of a small perturbation rate is not as big as on the large datasets. WN18RR, NELL-995-h100 and FB15k-237 have a node degree of 2.1, 2.2, and 18.7, respectively. Changing only 0.1% of the label vectors already adds 3.5M, 1.1M, and 3.9M faulty triples to the training data of these datasets, which are many more triples than the size of the training split.

After considering the results obtained through the experiments we can conclude that the typical label-flipping attacks for ML algorithms cannot simply be used for KGEs. However, since to the best of our knowledge, there does not exist any work in the domain of KGEs that considers such flipping attacks, we simply consider the typical attack methods as described in ML literature by [36,46]. Future works proposing new techniques pertaining to label-flipping attacks for KGE models need to be studied. Since this would require a significant extension of the current paper we consider it as a possible future work.

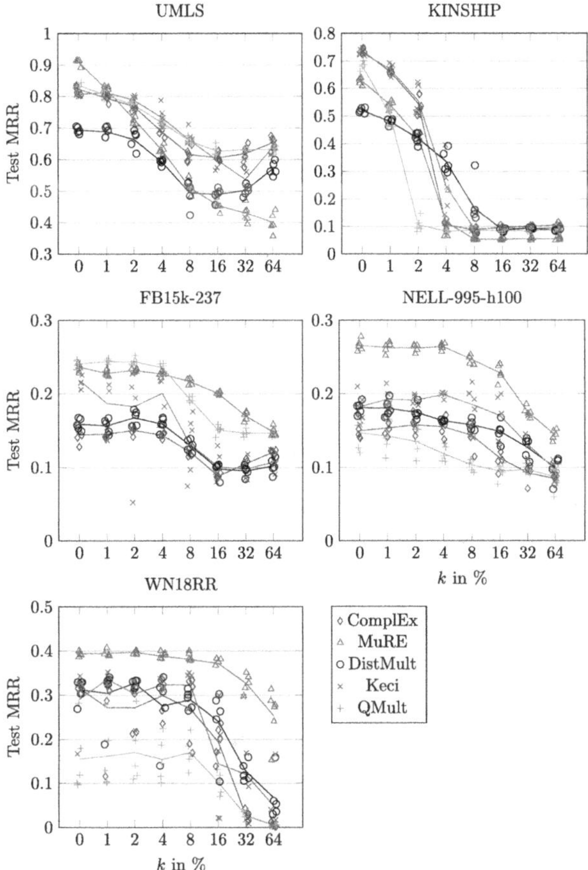

Fig. 3. Test MRR performance of the KGE approaches on UMLS, KINSHIP, NELL-995-h100, FB15K-237 datasets taken from Table 1 with Parameter Perturbation.

Parameter Perturbation. Attacks on the parameter surface show a stronger negative impact on the performance of KGE algorithms when compared to the Graph Perturbation attack. When compared to the Label Perturbation attack, the dataset size seems to have a large influence. On larger datasets, such as in WN18RR, NELL-995-h100, and FB15K-237, an attacker has to reach a higher perturbation ratio to achieve an effect consequently making this attack weaker than the Label Perturbation. However, on the two small datasets, i.e., on UMLS and KINSHIP, the opposite is the case. The Parameter Perturbation leads to a larger performance drop with smaller perturbation rates on both datasets. However, the overall effect of the Label Perturbation attack is higher on UMLS when a high perturbation rate is used. This shows that KGE models learned on the larger datasets are less susceptible to perturbations compared to the models learned on the smaller datasets. One possible explanation for such an outcome can be that the embedding space of the model learned from the larger datasets is broader and hence, a higher perturbation ratio is needed to cause significant changes on the model. On the contrary, the models learned on smaller datasets have less widened embedding spaces and even a little perturbation can lead to a large impact on the performance. This can be clearly observed as the value of k is increased from 0.32 to 0.64, and the performance drops significantly. This is an interesting observation and we presume that adding noise to the parameters can work as a kind of regularizer to some extent. However, as the noise increases, the performance (i.e., MRR) decreases and we show an inverse trend.

6 Conclusion

In this work, we have introduced non-targeted attacks considering three attack surfaces of KGE models. We have performed such attacks on 5 state-of-the-art KGE algorithms considering 5 datasets across 3 attack surfaces, considering 8 different perturbation ratios. Our results suggest that non-targeted attacks on different surfaces have different rates of performance degradation changes. While attacking the graph by considering lower perturbation ratios can lead to performance improvements, the same ratio can completely degrade the performance when considering the label perturbation.

Therefore, the findings emphasize the importance of evaluating KGE models against different types of perturbations to ensure their robustness, especially if they are to be deployed in dynamic environments where the input data or the model parameters might be subject to variations. The goal would be to develop KGE models that not only perform well under ideal conditions but can also withstand and adapt to unexpected changes in their operational parameters. Potential approaches could include the development of models that inherently account for parameter variability, the use of robust optimization techniques, or the implementation of adaptive learning rates that could mitigate the impact of high perturbation ratios. Moreover, we envision future research exploring how perturbations can be leveraged to improve KGE model performance effectively.

Acknowledgements. This work has received funding from the European Union's Horizon Europe research and innovation programme within the project ENEXA under grant No 101070305, and the Ministry of Culture and Science of North Rhine Westphalia (MKW NRW) within project SAIL under grant No NW21-059D. This work has also been supported by the project "WHALE" (LFN 1-04) funded under the Lamarr Fellow Network Programme by the Ministry of Culture and Science of North Rhine-Westphalia (MKW NRW).

References

1. Auer, S., Bizer, C., Kobilarov, G., Lehmann, J., Cyganiak, R., Ives, Z.G.: Dbpedia: a nucleus for a web of open data. In: The Semantic Web, 6th International Semantic Web Conference ISWC (2007)

2. Bai, J., Wu, B., Zhang, Y., Li, Y., Li, Z., Xia, S.: Targeted attack against deep neural networks via flipping limited weight bits. In: 9th International Conference on Learning Representations, ICLR (2021)
3. Balažević, I., Allen, C., Hospedales, T.M.: Hypernetwork knowledge graph embeddings. In: Artificial Neural Networks and Machine Learning–ICANN, 2019, pp. 553–565 (2019)
4. Balažević, I., Allen, C., Hospedales, T.M.: Tucker: tensor factorization for knowledge graph completion. arXiv preprint arXiv:1901.09590 (2019)
5. Balažević, I., Allen, C., Hospedales, T.: Multi-relational poincaré graph embeddings (2019)
6. Bhardwaj, P., Kelleher, J.D., Costabello, L., O'Sullivan, D.: Adversarial attacks on knowledge graph embeddings via instance attribution methods. In: Proceedings of the Conference on Empirical Methods in Natural Language Processing, EMNLP (2021)
7. Bhardwaj, P., Kelleher, J.D., Costabello, L., O'Sullivan, D.: Poisoning knowledge graph embeddings via relation inference patterns. In: Proceedings of the 59th Annual Meeting of the Association for Computational Linguistics and the 11th International Joint Conference on Natural Language Processing, ACL/IJCNLP (2021)
8. Bordes, A., Usunier, N., Garcia-Duran, A., Weston, J., Yakhnenko, O.: Translating embeddings for modeling multi-relational data. Adv. Neural Inf. Process. Syst. **26** (2013)
9. Bordes, A., Usunier, N., Garcia-Durán, A., Weston, J., Yakhnenko, O.: Translating embeddings for modeling multi-relational data. In: Proceedings of the 26th International Conference on Neural Information Processing Systems, NIPS'13, vol. 2, pp. 2787–2795. Curran Associates Inc., Red Hook (2013)
10. Chami, I., Wolf, A., Juan, D.C., Sala, F., Ravi, S., Ré, C.: Low-dimensional hyperbolic knowledge graph embeddings. arXiv preprint arXiv:2005.00545 (2020)
11. Chami, I., Wolf, A., Juan, D.C., Sala, F., Ravi, S., Ré, C.: Low-dimensional hyperbolic knowledge graph embeddings (2020)
12. Chen, Z., Wang, Y., Zhao, B., Cheng, J., Zhao, X., Duan, Z.: Knowledge graph completion: a review. IEEE Access **8**, 192435–192456 (2020)
13. Cheng, K., Zhu, Y., Zhang, M., Sun, Y.: Noigan: noise aware knowledge graph embedding with adversarial learning. In: ICLR 2020 Conference (2020). https://api.semanticscholar.org/CorpusID:226951634
14. Dalton, J., Dietz, L., Allan, J.: Entity query feature expansion using knowledge base links. In: The 37th International ACM SIGIR Conference on Research and Development in Information Retrieval (2014)
15. Demir, C., Moussallem, D., Heindorf, S., Ngomo, A.C.N.: Convolutional hypercomplex embeddings for link prediction. In: Asian Conference on Machine Learning, pp. 656–671. PMLR (2021)
16. Demir, C., Ngomo, A.C.: Clifford embeddings–a generalized approach for embedding in normed algebras. In: Joint European Conference on Machine Learning and Knowledge Discovery in Databases (2023)
17. Demir, C., Ngomo, A.-C.N.: Convolutional complex knowledge graph embeddings. In: Verborgh, R., et al. (eds.) ESWC 2021. LNCS, vol. 12731, pp. 409–424. Springer, Cham (2021). https://doi.org/10.1007/978-3-030-77385-4_24
18. Demir, C., Ngomo, A.: Hardware-agnostic computation for large-scale knowledge graph embeddings. Softw. Impacts **13**, 100377 (2022)

19. Denham, W.W.: The detection of patterns in alyawarra nonverbal behavior. In: The Detection of Patterns in Alyawarra Nonverbal Behavior, Semantic Scholar (2014). https://api.semanticscholar.org/CorpusID:140416458
20. Dettmers, T., Minervini, P., Stenetorp, P., Riedel, S.: Convolutional 2d knowledge graph embeddings. In: Proceedings of the AAAI Conference on Artificial Intelligence, vol. 32 (2018)
21. Ferrucci, D.A., et al.: Building watson: an overview of the deepqa project. AI Mag. **31**(3), 59–79 (2010)
22. Hendrycks, D., Dietterich, T.: Benchmarking neural network robustness to common corruptions and perturbations. In: International Conference on Learning Representations (2018)
23. Hogan, A., et al.: Knowledge graphs. ACM Comput. Surv. (CSUR) **54**(4), 1–37 (2021)
24. Kurita, K., Michel, P., Neubig, G.: Weight poisoning attacks on pretrained models. In: Proceedings of the 58th Annual Meeting of the Association for Computational Linguistics, ACL (2020)
25. Liu, C., Wang, Y., Cao, H., Liu, B., Jiang, D., Xu, L.: Non-targeted adversarial attacks on vision-language models via maximizing information entropy (2024). https://openreview.net/forum?id=7OO8tTOgh4
26. Liu, Y., et al.: Trojaning attack on neural networks. In: 25th Annual Network and Distributed System Security Symposium, NDSS. The Internet Society (2018)
27. Ma, J., et al.: Ptruste: a high-accuracy knowledge graph noise detection method based on path trustworthiness and triple embedding. Knowl. Based Syst. **256**, 109688 (2022)
28. McCray, A.T.: An upper-level ontology for the biomedical domain. Comput. Funct. Genomics **4**(1), 80–84 (2003). https://doi.org/10.1002/cfg.255
29. Nayyeri, M., Vahdati, S., Sallinger, E., Alam, M.M., Yazdi, H.S., Lehmann, J.: Pattern-aware and noise-resilient embedding models. In: Advances in Information Retrieval - 43rd European Conference on IR Research, ECIR (2021)
30. Nguyen, D.Q., Nguyen, T.D., Nguyen, D.Q., Phung, D.: A novel embedding model for knowledge base completion based on convolutional neural network. In: Proceedings of the Conference of the North American Chapter of the Association for Computational Linguistics: Human Language Technologies, vol. 2 (Short Papers) (2018)
31. Nickel, M., Murphy, K., Tresp, V., Gabrilovich, E.: A review of relational machine learning for knowledge graphs. Proc. IEEE **104**(1), 11–33 (2015)
32. Orvieto, A., Raj, A., Kersting, H., Bach, F.R.: Explicit regularization in overparametrized models via noise injection. In: International Conference on Artificial Intelligence and Statistics (2023)
33. Pezeshkpour, P., Tian, Y., Singh, S.: Investigating robustness and interpretability of link prediction via adversarial modifications. In: 1st Conference on Automated Knowledge Base Construction, AKBC (2019)
34. Ruffinelli, D., Broscheit, S., Gemulla, R.: You CAN teach an old dog new tricks! on training knowledge graph embeddings. In: 8th International Conference on Learning Representations, ICLR. OpenReview.net (2020)
35. Shan, Y., Bu, C., Liu, X., Ji, S., Li, L.: Confidence-aware negative sampling method for noisy knowledge graph embedding. In: 2018 IEEE International Conference on Big Knowledge (ICBK), pp. 33–40 (2018). https://doi.org/10.1109/ICBK.2018.00013
36. Song, Q., Jin, H., Huang, X., Hu, X.: Multi-label adversarial perturbations. In: 2018 IEEE International Conference on Data Mining (ICDM) (2018)

37. Steinhardt, J., Koh, P.W., Liang, P.: Certified defenses for data poisoning attacks. In: Advances in Neural Information Processing Systems 30: Annual Conference on Neural Information Processing Systems, pp. 3517–3529 (2017)
38. Sun, Y., Wang, S., Tang, X., Hsieh, T., Honavar, V.G.: Adversarial attacks on graph neural networks via node injections: a hierarchical reinforcement learning approach. In: Huang, Y., King, I., Liu, T., van Steen, M. (eds.) WWW '20: The Web Conference 2020, Taipei, Taiwan, 20–24 April 2020, pp. 673–683. ACM/IW3C2 (2020). https://doi.org/10.1145/3366423.3380149
39. Toutanova, K., Chen, D.: Observed versus latent features for knowledge base and text inference. In: Proceedings of the 3rd Workshop on Continuous Vector Space Models and their Compositionality. Association for Computational Linguistics (2015)
40. Trouillon, T., Welbl, J., Riedel, S., Gaussier, É., Bouchard, G.: Complex embeddings for simple link prediction. In: International Conference on Machine Learning, pp. 2071–2080. PMLR (2016)
41. Xie, R., Liu, Z., Lin, F., Lin, L.: Does william shakespeare REALLY write hamlet? knowledge representation learning with confidence. In: Proceedings of the Thirty-Second AAAI Conference on Artificial Intelligence, (AAAI-18), the 30th Innovative Applications of Artificial Intelligence (2018)
42. Xiong, W., Hoang, T., Wang, W.Y.: DeepPath: a reinforcement learning method for knowledge graph reasoning. In: Proceedings of the Conference on Empirical Methods in Natural Language Processing. Association for Computational Linguistics (2017)
43. Yang, B., Yih, W.t., He, X., Gao, J., Deng, L.: Embedding entities and relations for learning and inference in knowledge bases. arXiv preprint arXiv:1412.6575 (2014)
44. You, X., et al.: Mass: model-agnostic, semantic and stealthy data poisoning attack on knowledge graph embedding. In: Proceedings of the ACM Web Conference, WWW (2023)
45. Zhang, H., et al.: Data poisoning attack against knowledge graph embedding. In: Proceedings of the Twenty-Eighth International Joint Conference on Artificial Intelligence, IJCAI (2019)
46. Zhang, P.F., Huang, Z., Luo, X., Zhao, P.: Robust learning with adversarial perturbations and label noise: a two-pronged defense approach. In: Association for Computing Machinery (2022)
47. Zhang, S., Tay, Y., Yao, L., Liu, Q.: Quaternion knowledge graph embeddings. Adv. Neural Inf. Process. Syst. **32** (2019)
48. Zhang, Z., et al.: Towards robust knowledge graph embedding via multi-task reinforcement learning. IEEE Trans. Knowl. Data Eng. **35**(4), 4321–4334 (2023)
49. Zhao, T., Chen, J., Ru, Y., Lin, Q., Geng, Y., Liu, J.: Untargeted adversarial attack on knowledge graph embeddings. In: Proceedings of the 47th International ACM SIGIR Conference on Research and Development in Information Retrieval, 2024, pp. 1701–1711. ACM (2024). https://doi.org/10.1145/3626772.3657702
50. Zügner, D., Günnemann, S.: Adversarial attacks on graph neural networks via meta learning. In: 7th International Conference on Learning Representations, ICLR 2019, New Orleans, LA, USA, 6–9 May 2019. OpenReview.net (2019). https://openreview.net/forum?id=Bylnx209YX

Predicting Clinical Outcomes from Patient Care Pathways Represented with Temporal Knowledge Graphs

Jong Ho Jhee[1], Alberto Megina[1], Pacôme Constant Dit Beaufils[2,3,4], Matilde Karakachoff[2,3,4], Richard Redon[5], Alban Gaignard[5], and Adrien Coulet[1](✉)

[1] Inria, Inserm, Université Paris Cité, UMR 1346, Paris, France
{jong-ho.jhee,alberto.megina,adrien.coulet}@inria.fr
[2] University Hospital of Nantes, Nantes, France
{pacome.constantditbeaufils,matilde.karakachoff}@chu-nantes.fr
[3] Nantes Université, CHU Nantes, service de neuroradiologie diagnostique et interventionnelle, CNRS, INSERM, liinstitut du thorax, 44000 Nantes, France
[4] Nantes Université, CHU Nantes, Pôle Hospitalo-Universitaire 11: Santé Publique, Clinique des données, INSERM, CIC 1413, 44000 Nantes, France
[5] Inserm, CNRS, Université de Nantes, Institut du Thorax, UMR 1087, Nantes, France
{richard.redon,alban.gaignard}@univ-nantes.fr

Abstract. Background: With the increasing availability of healthcare data, predictive modeling finds many applications in the biomedical domain, such as the evaluation of the level of risk for various conditions, which in turn can guide clinical decision making. However, it is unclear how knowledge graph data representations and their embedding, which are competitive in some settings, could be of interest in biomedical predictive modeling.

Method: We simulated synthetic but realistic data of patients with intracranial aneurysm and experimented on the task of predicting their clinical outcome. We compared the performance of various classification approaches on tabular data versus a graph-based representation of the same data. Next, we investigated how the adopted schema for representing first individual data and second temporal data impacts predictive performances. **Results:** Our study illustrates that in our case, a graph representation and Graph Convolutional Network (GCN) embeddings reach the best performance for a predictive task from observational data. We emphasize the importance of the adopted schema and of the consideration of literal values in the representation of individual data. Our study also moderates the relative impact of various time encoding on GCN performance.

Keywords: Temporal knowledge graph · Knowledge graph embedding · Graph convolutional networks · Clinical data · Outcome prediction

E. Curry et al. (Eds.): ESWC 2025, LNCS 15718, pp. 282–300, 2025.
https://doi.org/10.1007/978-3-031-94575-5_16

1 Introduction

Intracranial aneurysms are abnormal bulges in the blood vessels of the brain that pose significant health risks particularly when they rupture, leading to severe neurological damage or death [25]. The ability to predict the evolution, and the best management of intracranial aneurysms, especially in the context of their various treatments, is crucial for improving patient outcomes. This predictive capability is particularly important in a clinical setting where understanding the relationship between patient features, the treatments they receive, and their subsequent outcomes can serve as a basis for early or preventive interventions. In this context, we aim at establishing a method for identifying patients who are at a higher risk of adverse outcomes following intracranial aneurysm rupture. By combining patients' personal features and the treatments observed during their hospital stays, we hope to uncover patterns that lead to more accurate predictions of patient outcomes. This would not only enhance the understanding of the effectiveness of different disease management, but also provides a foundation for the development of targeted intervention strategies that could mitigate adverse outcomes.

Graph embeddings are mapping of the nodes and edges that compose a graph into a continuous vector space [16]. This transformation allows for such complex relational data to be represented in a form that is more amenable to computational analysis. In particular, it allows to apply machine learning algorithms to perform tasks such as link prediction, node classification or clustering with high performance [35]. In this work, we especially consider embeddings on Knowledge Graphs (KGs), as defined in the context of the Semantic Web [19]. The atomic elements composing those are triples of the form of ⟨*subject*, *predicate*, *object*⟩, where the subject and object are nodes of the graph, representing entities; and the predicate is a labelled and oriented edge, representing that a particular relationship stands between the subject and object [18]. KG can be encoded in Resource Description Framework (RDF) a standard where entities and predicates are uniquely identified with a Uniform Resource Identifier (URI), facilitating interoperability across different datasets and ontologies.

However, it is unclear how KG representations and KG embeddings may be of interest in node classification task to predict clinical outcomes [8], and in particular to predict outcomes of intracranial aneurysm management. We propose in this work to advance some initial answers, by exploring three key scientific questions that guided our investigation. First, we wonder if the use of various graph embedding approaches is competitive with regard to classical predictive approaches on tabular data. Second, we wonder how these approaches are impacted by modeling choices made for representing individual data in the form of a graph. Third, we wonder about the impact of various possible representation of time, *i.e.* timestamps, sequential relations or both on predictive performances.

The contributions of this article are: (*i*) A publicly shared synthetic but realistic dataset of pathways of patients treated for a ruptured intracranial aneurysm; along with scripts that transform data from a tabular form to various graph representations; (*ii*) Elements of answer to our three questions, illustrating that, in our context, graph embeddings learned with Graph Convolutional Network (GCN) outperform other approaches, that the more compact representation of patient features is associated with better performance and that the representation of time does not impact prediction performances.

The remainder of this article introduces related works in Sect. 2, material and methods in Sect. 3, empirical results and their interpretation in Sect. 4 and finally presents a discussion in Sect. 5.

2 Related Work

2.1 Standard Schema for Clinical Data

Several data schema have been proposed to model, harmonize and facilitate the exchange of clinical data. For example, FHIR [29] and the OMOP CDM [34] are standards proposed to tackle clinical data interoperability. FHIR is loosely specified, which makes it hard to associate this schema with precisely defined semantics. Even if RDF transformations of FHIR [30] and OMOP CDM have been proposed, none has been widely adopted yet [6,36]. Similarly, Phenopackets [20] is a uniform data structure to ease the combination and exchange of genomic data and clinical observations. If some works have been proposed to align Phenopackets with semantic web technologies [21], it is insufficient to represent the full spectrum of clinical data.

Beside, two ontologies have recently been proposed to represent individual clinical data in the form of knowledge graphs. The first is the SPHN (Swiss Personalized Health Network) ontology [33], which has been adopted by the five Swiss academic hospitals for better data sharing and integration. As illustrated in Fig. 1, in this semantic model, patient is a central entity, that can be associated with diagnoses, drug administrations and procedures, each of which can potentially be associated with starting and ending times with literals.

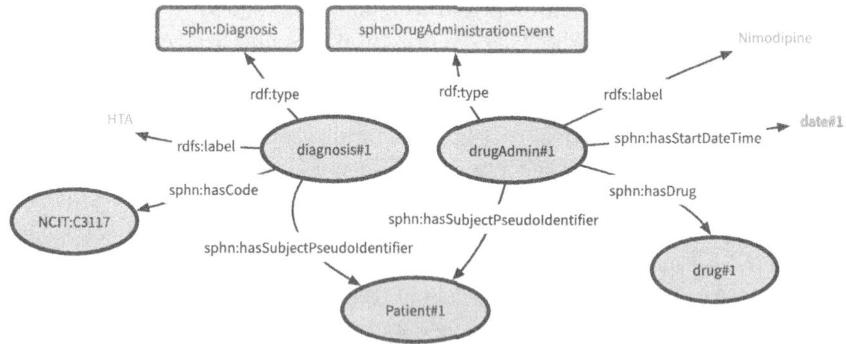

Fig. 1. Example of individual clinical data represented in a SPHN knowledge graph. Temporal information is specified with RDF literals associated to events through `sphn:hasStartDateTime` properties.

The second is the CARE-SM (Care and Registry Semantic Model) ontology [22]. It was initially designed to represent clinical data in the context of rare diseases and largely relies on the reuse of the Semantic Science Integrated Ontology (SIO) [11]. As depicted in Fig. 2, it provides fine-grained representations for the multiple roles with which a person can participate in the context of a care procedure or a research study. The originality of CARE-SM resides in its use of RDF quads. They are used to associate a semantic context to each data element through RDF named graph. This particularly enables the representation of provenance metadata for clinical observations or diagnosis. Notably, these named graphs can be used to represent timelines of clinical events.

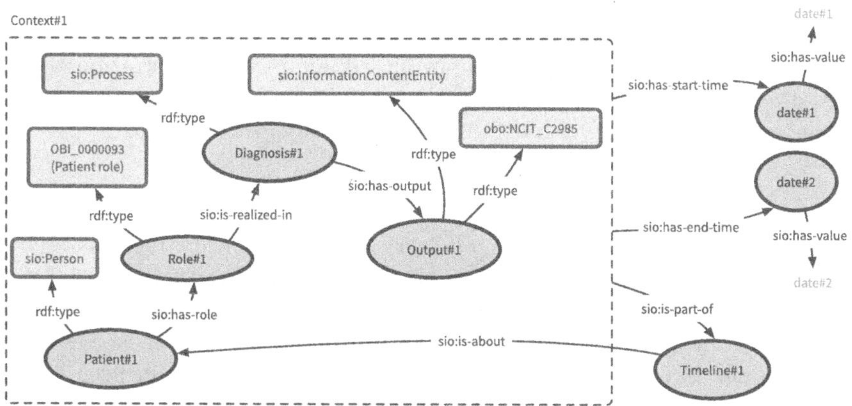

Fig. 2. Example of individual clinical data represented in a CARE-SM KG. Temporal information is directly linked to the Context#1 named graph with start and end dates. Multiple events can be associated to a given timeline through the `sio:is-part-of` property.

In this work, we focus on SPHN and CARE-SM first because both of them provide sufficiently precise specification to let one represent an arbitrary clinical dataset in the terms of their ontologies, which is not the case of FHIR-RDF; second because they propose very different modeling choices to represent individual clinical data; and third because they are adopted in large scale projects.

2.2 Time in Knowledge Graphs

The OWL-Time is a standard ontology that provides classes, predicates and patterns to represent time, duration and intervals [7]. In particular, it enables one to instantiate relations between events with the Allen's interval algebra and can be associated with time reasoning mechanisms [1,38]. Both SPHN and CARE-SM enable various way to represent time, including the use of OWL-Time and Allen's interval algebra. In practice, one may want to restrict themselves to absolute time by only associating timestamps to events, whereas another might want relative time by instantiating relationships such as "is before" between events, or even to use both absolute and relative relationships. In addition, if one choose relative time, they also have to decide on a *level of saturation* between events going from a simple sequence, where only directly subsequent events are related, to a fully saturated level where each event is timely related to every other event. Even if it is clear that a fully saturated graph is associated with several drawbacks (*e.g.*, the high connectivity of temporal nodes makes the exploration of the graph complex), the most adapted modeling of time to adopt for a particular task is not always clear [38].

2.3 KG Embeddings and Node Classification

Many approaches for representing entities of a KG within a latent space exist, and categorisations of them have been proposed in [4]. As an illustration, TransE [2] is an embedding approach that works at the triple level as it aims at minimizing the following scoring function:

$$f_r(h, t) = \|\mathbf{h} + \mathbf{r} - \mathbf{t}\|_1, \tag{1}$$

where $\mathbf{h}$, $\mathbf{r}$, and $\mathbf{t}$ are the embedding vectors of the head entity, relation, and tail entity, respectively. Here the relationship between the head and the tail (*i.e.*, the subject and object) can be seen as a translation r in the embedding space. RDF2Vec follows a very different approach that works at the sequence level [31]. It uses random walks drawn from the knowledge graph. These sequences of either edges, nodes, or subtrees are used to feed a word2vec model that outputs embeddings for each node in a sequence, for example, by maximizing the probability of a node given the previous nodes of the sequence. For instance, the Continuous Bag-of-Words model (CBOW), one of the algorithms associated with word2vec, maximizes the average log probability of the target node given a sequence of nodes:

$$\frac{1}{T}\sum_{t=1}^{T} \log p(e_t|e_{t-c}, ..., e_{t+c}), \tag{2}$$

where e_t is a target node and c denotes the context window. In contrast with TransE and RDF2Vec, which work at the triple and sequence levels, GCNs [27] work at the level of the neighborhood of nodes. They have been introduced for classification over graphs and extended for node classification and link prediction in knowledge graphs [32]. GCNs compute the embeddings of a node by considering its neighborhood in the graph. GCNs can be seen as a message-passing framework of multiple layers, in which the embedding $h_i^{(l+1)}$ of a node i at layer $(l+1)$ depends on the embeddings of its neighbors at layer (l), as follows:

$$h_i^{(l+1)} = \sigma\left(\underbrace{\sum_{j \in \mathcal{N}_i} \frac{1}{c_i} W^{(l)} h_j^{(l)}}_{\text{Neighborhood}} + \underbrace{W^{(l)} h_i^{(l)}}_{\text{Self-connection}}\right). \tag{3}$$

The convolution over the neighboring nodes j of i is computed with a learnable weight matrix $W^{(l)}$ and each layer (l); and is normalized by a constant c_i. The last term enables self-connection *i.e.*, the fact that the embeddings of the node i at a particular layer $(l+1)$ also depends on its embeddings at the layer (l). Relational Graph Convolutional Networks (RGCN) [32] is a particular type of GCNs that takes into consideration multi-relational data, *i.e.*, considers differently a same neighbor node when related through different labelled edges to i. This is particularly adapted to the semantic web KGs since they encompass various predicates to represent relations associated with different semantics. To this aim, RGCNs incorporate entity embeddings with multi-relations in the neighborhood aggregation scheme, as follows:

$$h_i^{(l+1)} = \sigma\left(\sum_{r\in\mathcal{R}}\sum_{j\in\mathcal{N}_i^r}\frac{1}{c_{i,r}}W_r^{(l)}h_j^{(l)} + W_0^{(l)}h_i^{(l)}\right). \quad (4)$$

where the convolution over the neighbor nodes is computed using a specific weight matrix $W_r^{(l)}$ for each predicate (relation type) $r \in \mathcal{R}$; and $c_{i,r}$ is a normalizing constant such as the number of neighboring nodes $|\mathcal{N}_i^r|$. The parameter sharing and sparsity constraints using basis-decomposition allow RGCN to deal with the large number of relations. In the following, we consider TransE, RDF2Vec as two baseline approaches for KG embedding (KGE) models and RGCN as it seems a well adapted candidate to take into consideration the variety of relationship types of our clinical KG.

Values associated with entities through the use of literals are generally disregarded by KGE approaches. However various works investigates the interest of incorporating them to node representations. To this aim LiteralE [28], proposes an approach where two vectors represent a single node, the first representing the node embedding itself, and the second containing each of the numerical literal it is associated with. Both vectors are then combined before to be passed to a scoring function. Knowledge Embedding with Numbers (KEN) [9] uses an encoder, a single neural network, to inject the literal values into the same vector space with entities. Other approaches have been proposed to consider textual literals or combinations of various type of literals [15].

One of the main interest of graph embedding is to facilitate the application of machine learning algorithms on complex relational data to perform a variety of tasks such as link prediction, node classification, graph classification or node clustering [19]. In this work we consider three embedding approaches, but only one subsequent task that is node classification. This learning task can be defined as the estimation of the likelihood that a given entity, not explicitly asserted in a KG, is assigned to a specific type [35]. Using KG embeddings, this task relies on the assumption that the embedding space effectively captures the inherent characteristics and structural properties of the original graph. In practice, the spatial configuration of vectors should reflect the relational information observed in the KG, thus allowing predictions about node labels. This task is formally described in the Materials and Methods section of this article.

2.4 Temporal Knowledge Graph Embeddings

A temporal knowledge graph (TKG) extends existing KG by incorporating temporal information in a specific time or interval. This addition poses significant challenges because it requires integrating the temporal validity of facts into the models in order to accurately capture the dynamics of entities and relations over time [10]. Effectively modeling the temporal aspect is crucial for applications where the evolution of relations is significant, such as in clinical data where the time of treatments, procedures and the patient state can critically influence outcomes. Temporal knowledge graph embedding models use triples tagged with each timestamp, which are quadruples, and this is considered as learning the representation of each time-snapshot graph. However, the time interval could be critical information, or the graph might not be compatible with quadruple forms, as not all triples necessarily possess temporal information. Encoding or transforming temporal information in the form of vectors of literals or relations can be applied in this type of graph [3].

3 Materials and Methods

3.1 Synthetic Data Generation

We built a synthetic data set of 10,000 patients with 30 clinical features and 1 outcome variable. Out of the 30 clinical features, 8 are associated with a timestamp and for this reason are hereafter named events, to distinguish from demographic or historical features not associated with a particular time. In our dataset, events have the particularity of being associated with only one timestamp, which is the time between hospital admission and the first occurrence of the event during the patient's stay. Table 4 in Appendix lists the 22 clinical features (non-temporal) and 8 events of our synthetic dataset.

To make our synthetic dataset as realistic as possible, patient features and care events were generated according to observations made on a real-world clinical dataset of 552 patients diagnosed with a ruptured intracranial aneurysm, provided by the Nantes University Hospital. Access and use of the data for this project has been granted by the Clinical Data Centre validation board and the Comité National Informatique et Liberté (resolution n° 2018-295) [24].

First we estimated the closest distribution for each feature of the real-world dataset using a Kolmogorov–Smirnov test. For instance, we observed that the duration of patients' hospital stay was following a generalized extreme value distribution. Second, after performing a factorization for categorical variables, we computed correlations across each possible pair of features, and computed the transition probabilities across types of events. We observed that the duration of hospital stay was highly correlated with the number of medical procedures received; and a high probability for Paracetamol to be administered subsequently to Nimodipine. Figure 3 shows a Sankey diagram built from the observed transition probabilities, which visually summarizes the possible care pathways. To simplify the representation of time in our dataset, the 8 events were binarized by

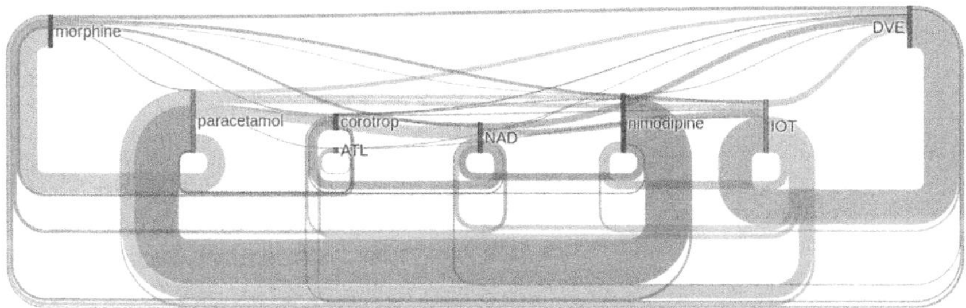

Fig. 3. A visual representation of care pathways where the larger connections between care events correspond to the higher transition probabilities. `morphine`: morphine use, `paracetamol`: paracetamol use, `corotrop`: milrinone use, `ATL`: percutaneous transluminal angioplasty, `NAD`: norepinephrin use, `nimodipine`: nimodipine use, `IOT`: orotracheal intubation, and `DVE`: external ventricular drainage.

a transformation into pairs of events set to one if the patient observed a transition between the first and second event, and to zero otherwise. After validation by clinical experts of the consistency of the distributions, correlations and transition probabilities, we used them to constrain the generation of our synthetic dataset.

Finally, we generated the patient outcome feature with one of the three distinct values {*BackHome, Rehabilitation, Death*} associated with the following proportions in the synthetic dataset: 44.14%, 43.33% and 12.53%, respectively, mirroring the real-word distribution.

3.2 Graph Representation of Clinical Data

We developed transformations of our synthetic tabular dataset using various modeling choices. By using RDF templates and rules, we generated a set of graphs that instantiate the SPHN ontology:

- `SPHN-nl` (no literals) where all the literals were removed;
- `SPHN-nt` (no time) where temporal information were removed;
- `SPHN-ts` (timestamps) with timestamps only;
- `SPHN-tr` (time relations) with a single `time:before` predicate between directly subsequent events;
- `SPHN-tsr` (timestamps and relations) with both timestamps and relations between subsequent events;
- `SPHN-sat1` (saturation level 1) where `SPHN-tr` is enriched with additional `time:before` predicates obtained by applying once the following transitivity rule `time:before` $\circ$ `time:before` $\sqsubseteq$ `time:before`;
- `SPHN-sat2` (saturation level 2) by applying the same transitivity twice.

Figure 4 illustrates timestamps and time relation between events, at three level of saturation corresponding to `SPHN-tr`, `SPHN-sat1` and `SPHN-sat2`.

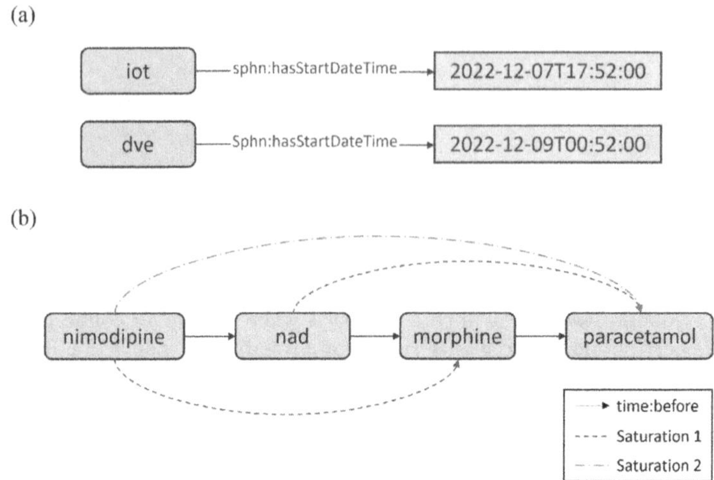

Fig. 4. Examples of temporal information: (a) two event associated with a timestamp; (b) sequence of events related with `time:before` relations. In plain lines are relations between directly subsequent events. Applying a transitivity rule once, add 2 relations in dashed doted line (saturation 1) and twice, add the last relation depicted with the dash-dotted line (saturation 2). `iot`: orotracheal intubation, `dve`: external ventricular drainage, `nad`: nicotinamide adenine dinucleotide.

By using RDF templates, we also generated a graph that instantiate CARE-SM ontology. In this case, we performed an additional step where quads were transformed into triples, in particular by using a `nvasc:hasTimePoint` link to associate directly instances such as diagnoses or drug administrations to their absolute time. This step was motivated by the fact that KGE approaches do not support quads, but only triples. The resulting graph is named CARE-SM*. We generated the same variants of graphs as we did for SPHN, but report only here about `CARE-SM*-ts` that is the version with timestamps only. Table 5 in Appendix shows the statistics of SPHN and CARE-SM* graphs.

We make two last transformations to each variant of our graphs: first, to ensure that the orientation of predicates do not influence our experiments, inverse relationships were systematically asserted; we encoded timestamp literals into continuous numbers in the range of [0, 1] using a quantile transformation. This spreads out the most frequent values thus reduces the impact of outliers [12]. Scripts regarding these various transformation are provided at https://github.com/TeamHeKA/neurovasc.

3.3 KG Embedding and Patient Outcome Prediction

The objective of our prediction task is to forecast patient outcomes based on their clinical features and experienced events. In the case of the graph dataset, we model this task as a node classification problem aiming at associating patient nodes with the class that corresponds to their correct outcome.

RGCN for Patient Outcome Prediction. In this section, we describe in detail the graph embedding model, RGCN+Literals for the outcome prediction. The overall architecture is illustrated in Fig. 5. Given a graph $G = (\mathcal{V}, \mathcal{E}, \mathcal{R}, \mathcal{X})$, where $\mathcal{V}$ denotes the set of nodes (entities), $\mathcal{E}$ the set of edges, $\mathcal{R}$ the set of relations (predicates), and $\mathcal{X} \in \mathbb{R}^{n \times d_0}$ denotes the initial input embeddings of dimension d_0. The representation of a target node (patient) $h_i^{(1)} \in \mathbb{R}^{d_1}$ for $i \in |\mathcal{V}|$ after passing a first RGCN layer is defined as:

$$h_i^{(1)} = \sigma\left(\sum_{r \in \mathcal{R}} \sum_{j \in \mathcal{N}_i^r} \frac{1}{c_{i,r}} W_r^{(0)} x_j + W_0^{(0)} x_i\right), \tag{5}$$

where $x_i \in \mathcal{X}$ is an initial patient embedding, $W_r \in \mathbb{R}^{d_o \times d_1}$ denotes a weight matrix for each relation r and σ is a non-linear activation function. Additionally, the number of parameters increases with the number of relations, which can lead to overfitting on some of the rare relations. Thus we apply basis-decomposition method for the regularization of the model [32].

The final patient node embedding is extracted after passing L stacks of RGCN-layers and the softmax function is applied to output the probability of outcomes. The model is trained by minimizing the cross-entropy loss on the patient nodes:

$$\mathcal{L} = -\sum_{p \in \mathcal{P}} \sum_{k=1}^{K} y_{pk} \log z_{pk}, \tag{6}$$

where $\mathcal{P}$ is the set of patient nodes in the training set, K the number of outcomes and z_{pk} the probability of the outcome.

RGCN with Literals. In our clinical KG, some clinical features are in the form of literals (*e.g.*, the age of a patient). However, RGCN model only consider entity nodes and relations, and thus do not take literals into account. To mitigate this problem, we propose a model denoted RGCN+Literals (or RGCN+lit for short) that employs an additional function for the literals. Before the input of initial embeddings to RGCN, the function of a multi-layer perceptron (MLP) is used to transform the value of the literal into embeddings:

$$x_{literal} = \sigma(Wv + b), \tag{7}$$

where v denotes the attribute value, σ the non-linear activation function, W the weight matrix and b the bias. Note that other encoding functions could be applied instead of the MLP.

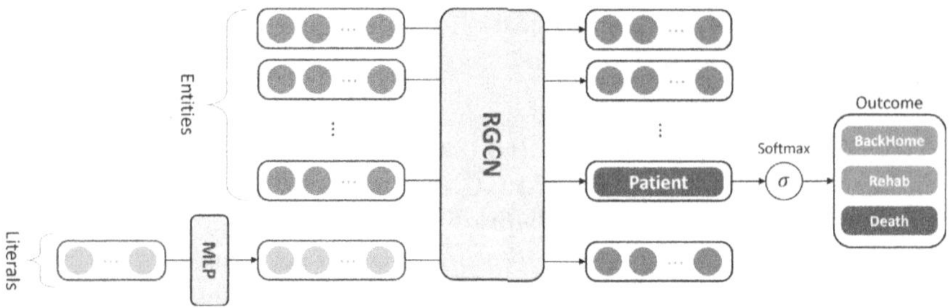

Fig. 5. An illustration of the model denoted RGCN+lit for patient outcome prediction.

4 Experimental Setting and Results

In the experiments, we compare various outcome prediction approaches along three distinct experiments on our synthetic dataset. Each experiment is repeated ten times for each model to assess the variability of the performance. Each time the 10,000 patients are randomly split into train (80%), validation (10%) and test (10%) set. Datasets and codes are available at https://github.com/TeamHeKA/neurovasc. The three experiments compare:

- **Tabular *vs.* Graph data.** We compare the performance of standard predictive approaches applied on tabular data with those of various KGE approaches to examine whether the multi-relational information within the graph would benefit the predictive task.
- **SPHN *vs.* CARE-SM ontology.** We compare the impact on the prediction of using either one schema or another to evaluate whether the structure of the graph affects the performance.
- **Various time modeling.** We compare the impact on the prediction of using various time modeling, such as no time data, absolute time only, relative time only, both absolute and relative time and different levels of saturation of the relative time relationships that connect temporal events.

4.1 Tabular *vs.* Graph data

We considered the following methods to establish baseline performance from tabular data, Logistic Regression (LR), Random Forest (RF), Feed-forward Neural Network (NN) to compare to KGE approaches. RF were set up with 100 trees, and NN with three layers with hidden dimension sizes of [100, 50, 10] and hyperbolic tangent was used as an activation function.

For this first comparison with various KGE approaches, we arbitrarily chose to only consider the SPHN ontology. In addition, to enable a fair comparison with tabular data (which encode only the sequence of event), `SPHN-tr` is considered as the comparative SPHN graph. To represent three main families of KGE we considered TransE, RDF2Vec and RGCN+lit. All models used an initial embedding dimension of 100. For TransE, 1-norm is applied for the regularization of the scoring function. For RDF2Vec, ten walks of a maximum depth of three for each node was applied using the random walk strategy. The patient representations obtained from the first two KGE approaches are input to a NN model as a classifier. For RGCN+lit, three RGCN layers are applied to aggregate the information within the three-hop neighborhood of patient nodes and Parametric Rectified Linear Unit (PReLU) [17] is used as a non-linear activation function. And all models were optimized using Adam optimizer [26] with the learning rate of 1e–3 and the weight decay of 5e–4.

The obtained performance is shown in Table 1. All three baseline approaches from tabular data reach poor performances (F1-score = [0.44,0.49], AUC = [0.63, 0.71]). RF showed the best performance (AUC = 0.71), closely followed by LR. Especially, the two models showed decent F1-score for *BackHome* outcome, compare to other models. For graph data, neither TransE nor RDF2Vec seem to succeed in predicting outcomes (AUC = [0.49,0.5]). However, the RGCN+lit model showed the overall best result (F1 = 0.78, AUC = 0.91).

Table 1. The patient outcome prediction comparison on Tabular and Graph (`SPHN-tr`). RGCN3+lit refers to RGCN+lit with 3 layers.

Type	Model	F1-score					Accuracy	AUC
		BackHome	Rehab	Death	Macro	Weighted		
Tabular	LR	0.63 ± 0.02	0.55 ± 0.02	0.25 ± 0.05	0.47 ± 0.03	0.55 ± 0.02	0.56 ± 0.02	0.70 ± 0.02
	RF	0.63 ± 0.01	0.55 ± 0.02	0.28 ± 0.04	0.49 ± 0.02	0.55 ± 0.01	0.56 ± 0.01	0.71 ± 0.01
	NN	0.58 ± 0.03	0.48 ± 0.03	0.26 ± 0.04	0.44 ± 0.02	0.50 ± 0.02	0.50 ± 0.02	0.63 ± 0.02
Graph (`SPHN-tr`)	TransE	0.49 ± 0.04	0.40 ± 0.10	0.02 ± 0.04	0.30 ± 0.03	0.40 ± 0.03	0.43 ± 0.02	0.50 ± 0.01
	RDF2Vec	0.50 ± 0.05	0.39 ± 0.14	0.01 ± 0.02	0.30 ± 0.03	0.39 ± 0.04	0.44 ± 0.02	0.49 ± 0.01
	RGCN3+lit	$\mathbf{0.84 \pm 0.01}$	$\mathbf{0.76 \pm 0.02}$	$\mathbf{0.64 \pm 0.08}$	$\mathbf{0.75 \pm 0.03}$	$\mathbf{0.75 \pm 0.02}$	$\mathbf{0.78 \pm 0.01}$	$\mathbf{0.91 \pm 0.01}$

4.2 SPHN *vs* CARE-SM Ontologies

Similarly to the first experiment, TransE, Rdf2Vec and RGCN+lit are considered, but here applied to a KG instantiating either the SPHN or CARE-SM ontology. The model configuration and hyperparameter settings are the same as the first experiment. Because CARE-SM uses more predicates thus longer paths to connect patients with their features, we conducted an additional experiment on CARE-SM KG that consists in using RGCN with five layers. This aims to check if five-hop neighbors enable to capture enough information to predict the outcome. The obtained performance is reported in Table 2. TransE and RDF2Vec showed relatively weak performance on both KG. RGCN on SPHN showed the best performance (AUC = 0.91). For CARE-SM, all models have difficulty on predicting the outcomes. Using this ontology, we note that RGCN with three layers is not performing better than TransE or RDF2Vec (AUC = 0.50). Increasing the number of layers to 5 did not improve the performance (AUC = 0.50). We note that all the models barely predict the *Death* outcome.

Table 2. The performance comparison of SPHN (`SPHN-ts`) and CARE-SM (`CARESM*-ts`). RGCN3+lit and RGCN5+lit refers to RGCN+lit with 3 and 5 layers, respectively. CARESM* is the variant of CARE-SM. See Sect. 3.2 for more details.

KG	Model	F1-score					Accuracy	AUC
		BackHome	Rehab	Death	Macro	Weighted		
`SPHN-ts`	TransE	0.51 ± 0.07	0.33 ± 0.16	0.02 ± 0.04	0.29 ± 0.04	0.37 ± 0.05	0.43 ± 0.02	0.50 ± 0.01
	RDF2Vec	0.49 ± 0.04	0.42 ± 0.09	0.01 ± 0.03	0.30 ± 0.02	0.40 ± 0.02	0.44 ± 0.01	0.50 ± 0.02
	RGCN3+lit	$\mathbf{0.83 \pm 0.02}$	$\mathbf{0.76 \pm 0.02}$	$\mathbf{0.66 \pm 0.08}$	$\mathbf{0.75 \pm 0.03}$	$\mathbf{0.78 \pm 0.02}$	$\mathbf{0.78 \pm 0.02}$	$\mathbf{0.91 \pm 0.01}$
`CARESM*-ts`	TransE	0.47 ± 0.04	0.44 ± 0.04	0.02 ± 0.03	0.31 ± 0.01	0.40 ± 0.01	0.43 ± 0.01	0.49 ± 0.01
	RDF2Vec	0.51 ± 0.07	0.38 ± 0.11	0.00 ± 0.00	0.29 ± 0.02	0.39 ± 0.03	0.44 ± 0.02	0.50 ± 0.01
	RGCN3+lit	0.53 ± 0.08	0.30 ± 0.17	0.00 ± 0.00	0.28 ± 0.04	0.37 ± 0.05	0.44 ± 0.01	0.50 ± 0.02
	RGCN5+lit	0.48 ± 0.08	0.30 ± 0.19	0.00 ± 0.00	0.26 ± 0.04	0.34 ± 0.05	0.44 ± 0.05	0.50 ± 0.01

4.3 Various Time Modeling

In this experiment, we compare predictive performance of graphs associated with various modeling of time as listed in Sect. 3.2. All experiments are conducted using the RGCN+lit model, except with no literals where it uses RGCN. Without literals, the model performs poorly, and under the standard approach from tabular data. The AUC increases about 33% on average when the literals are added. This increase mainly come from a better prediction of the *Death* outcome. When the timestamps are added to `SPHN-nt`, the AUC increases of 7% on average. Adding the time relations also increases AUC of a 7%. Adding both temporal information showed similar results as well. With saturation, we observe a slight incrase of performance for the *Death* outcome, though the overall performance is similar to `SPHN-tsr`. Eventually, adding temporal information give the model better prediction, but the type of time modelling and the level of saturation did not make a significant difference.

5 Discussion

The comparative analysis reveals first that the RGCN+lit model outperforms baseline methods on tabular data, or other KGE approaches in terms of accuracy and F1-score. We hypothesis that the multi-relational modeling associated with the multiple RGCN-layers makes the model capable of aggregating information from multiple hop of neighbors to the patient node, what cannot be achieved from single relational modeling with tabular data, or with TransE or RDF2Vec which only partially consider the multi-hop neighbors.

Table 3. The Performance of RGCN+lit on SPHN with various temporal information and modeling. RGCN without literals is applied to `SPHN-nl`.

KG	F1-score					Accuracy	AUC
	BackHome	Rehab	Death	Macro	Weighted		
`SPHN-nl`	0.64 ± 0.03	0.46 ± 0.11	0.05 ± 0.07	0.38 ± 0.06	0.49 ± 0.06	0.53 ± 0.04	0.64 ± 0.06
`SPHN-nt`	0.75 ± 0.02	0.65 ± 0.02	0.55 ± 0.06	0.65 ± 0.02	0.68 ± 0.01	0.68 ± 0.01	0.85 ± 0.01
`SPHN-ts`	0.83 ± 0.02	$\mathbf{0.76 \pm 0.02}$	0.66 ± 0.08	0.75 ± 0.03	0.78 ± 0.02	0.78 ± 0.02	$\mathbf{0.91 \pm 0.01}$
`SPHN-tr`	$\mathbf{0.84 \pm 0.01}$	$\mathbf{0.76 \pm 0.02}$	0.64 ± 0.08	0.75 ± 0.03	0.75 ± 0.02	$\mathbf{0.78 \pm 0.01}$	$\mathbf{0.91 \pm 0.01}$
`SPHN-tsr`	0.83 ± 0.02	$\mathbf{0.76 \pm 0.02}$	0.66 ± 0.04	0.75 ± 0.02	$\mathbf{0.78 \pm 0.01}$	$\mathbf{0.78 \pm 0.01}$	$\mathbf{0.91 \pm 0.01}$
`SPHN-sat1`	0.83 ± 0.01	$\mathbf{0.76 \pm 0.02}$	0.64 ± 0.06	0.75 ± 0.02	$\mathbf{0.78 \pm 0.01}$	$\mathbf{0.78 \pm 0.01}$	$\mathbf{0.91 \pm 0.01}$
`SPHN-sat2`	0.83 ± 0.01	$\mathbf{0.76 \pm 0.02}$	$\mathbf{0.68 \pm 0.05}$	$\mathbf{0.76 \pm 0.02}$	0.78 ± 0.02	0.78 ± 0.02	$\mathbf{0.91 \pm 0.01}$

Second, we observed that the choice of the schema for individual KG impacts the performance. In particular in our setting, the more compact and patient-oriented schema of SPHN is more favorable than the one of CARE-SM KGE approaches. This could be explained by the longer distance between patient and feature nodes in CARE-SM, however, increasing the size of the neighbor to five-hops did not solve the issue. However, we note that we did not expend further the number of hops for computational cost. A limitation to the conclusion relies on the fact that we considered only SPHN and CARE-SM as two prototypical ontologies. A more complete study would have considered concurrent models such as RDF-FHIR, PhenoPackets and others.

Third, adding time information helps RGCN+lit in classifying properly patient nodes, but we were enable to observe any difference in performance associated with various time modeling choices. Here we acknowledge that we focus on rather simple time modeling and that more complex scenarios exist such as those associated with dynamic graphs.

Overall, our study illustrates that KGE such as RGCN+lit represents a promising approach for predictive modeling in healthcare. However we note a strong class disbalance in our dataset (44.14%, 43.33% and 12.53% for *Back-Home*, *Rehabilitation*, *Death*, respectively), reflecting the real world. This might explain to some extent the difficulty for most of the approaches to predict the *Death* class. It is further possibly mitigated by over-sampling techniques, such as Synthetic Minority Over-sampling [5]. Also our study considers solely the task of node classification, whereas prediction could have been modeled as a link prediction or triple classification problem. More investigation would be necessary to assess if our conclusions stand in the context of other learning tasks. This could necessitate the consideration of problem classically associated with KGE such as negative sampling [23].

Additionally, we observed that including literal embeddings in the model improved the performance. In this study, our model focused on the simple embedding of numerical literals, including timestamps. However, we plan to develop a model capable of handling multi-modal literals, such as a combination of text and numerical literals [15].

Another important position we took is between transductive and inductive learning approaches [37]. We follow a transductive learning approach, which mean that node classification is made on the basis of the set of entities that has potentially been seen during training. Most of the KGE models are based on a transductive learning approach, because the model learns the complex relational information within a whole large graph. In contrast, inductive learning attempts to generalize the model to new entities and relations not presented in the training graph. For instance, it is representing the entities or relations based on the combination of observed entities or relations in the training graph [13]. Inductive approach can be adapted to the scenario where new patient information and medical events continuously emerges. This remains a future work of our study.

The effective evaluation of KG embedding models is crucial for advancing the field and ensuring that the models developed are robust, accurate, and useful in practical applications [14]. But it suffers from a lack of standardized benchmarks and reference representations of the temporal dimension. In this study, we particularly aimed at advancing this agenda, by proposing a real-word task, a shared dataset and well documented baseline experiment, which will serve as a baseline for prediction modeling with graph data.

An important area for future research is evaluating the impact of individual patient features on prediction outcomes. By analyzing the relative importance of different features, such as specific medical conditions or demographic factors, researchers can refine their models to focus on the most predictive attributes. This targeted approach can enhance the model's efficiency and ensure that the most relevant clinical information is prioritized in decision-making. While at the moment, efforts are underway to obtain a small subset of a real dataset to test these models and generate predictions based on real patient data thanks to the collaboration with the Nantes Hospital, in the longer term, the ultimate goal is to achieve clinical validation of the predictive models. This step involves applying the models to actual patient data and rigorously assessing their performance in a clinical setting. Clinical validation is critical for ensuring that the models are theoretically sound and practically useful in improving patient outcomes. This research phase will require close collaboration with healthcare providers and institutions to test the models in real-world environments and to gather feedback for further refinement.

In conclusion, while this study has made significant strides in demonstrating the potential of knowledge graph embeddings for predicting patient outcomes, a substantial amount of work remains to be done. By continuing to refine the models, enhance the realism of synthetic data, and pursue clinical validation, future research can build on these foundations to develop precise and clinically relevant predictive tools that ultimately enhance patient care and treatments.

Acknowledgments. This work is supported by CombO (Health Data Hub) project and the the Agence Nationale de la Recherche under the France 2030 program, reference ANR-22-PESN-0007 ShareFAIR, and ANR-22-PESN-0008 NEUROVASC.

Disclosure of Interests. The authors have no competing interests to declare that are relevant to the content of this article.

Appendix

A List of Features of Our Synthetic Dataset

The list of clinical features in our synthetic data is shown in Table 4. It includes 22 non-temporal features (4 numerical, 6 categorical, 12 binary) and 8 temporal features (named events in this manuscript). Note that events correspond to the time between hospital admission and the first occurrence of the event (*e.g.*, the first administration of Nimodipine).

Table 4. List of clinical features.

Type	Name	Description
Numerical	hospital_stay_length	Number of days in the hospital
	gcs	Glasgow coma score
	act_nb	Number of medical acts
	age	Patient's age
Categorical	gender	Gender
	entry	Mode of entry (*e.g.*, emergency room)
	entry_code	Code of entry
	ica.y	Intracranial aneurysm (ICA) location and size
	ica_treatment	ICA treatment (*e.g.*, endovascular or surgical)
	ica_therapy	Calcium channel blockers therapy
Binary	fever	Fever
	o2_clinic	Decrease of oxygen, indirect measure
	o2	Deacrease of oxygen, blood measure
	hta	Hypertension
	hct	Hypercholesterolemia
	smoking	Smoking
	etOH	Alcohol consumption
	diabetes	Diabetes
	headache	Headache
	unstable_ica	Hemodynamic instability
	vasospasm	Vasospasm
	ivh	Intraventricular hemorrhage
Event	nimodipine	Nimodipine
	paracetamol	Paracetamol use
	nad	Norepinephrin use
	corotrop	Milrinone use
	morphine	Morphine use
	dve	External ventricular drainage
	atl	Percutaneous transluminal angioplasty
	iot	Orotracheal intubation

B Statistics of Datasets

The statistics of KGs, SPHN and CARE-SM.

Table 5. Statistics of Datasets

Dataset	# Entities	# Relations	# Literals	# Triples
SPHN	295,307	15	36,415	1,127,467
CARE-SM*	576,733	13	24,766	1,754,505

References

1. Batsakis, S., Petrakis, E.G., Tachmazidis, I., Antoniou, G.: Temporal representation and reasoning in owl 2. Semant. Web **8**(6), 981–1000 (2017)
2. Bordes, A., Usunier, N., Garcia-Duran, A., Weston, J., Yakhnenko, O.: Translating embeddings for modeling multi-relational data. Adv. Neural Inf. Process. Syst. **26** (2013)
3. Cai, B., Xiang, Y., Gao, L., Zhang, H., Li, Y., Li, J.: Temporal knowledge graph completion: a survey. In: Proceedings of the Thirty-Second International Joint Conference on Artificial Intelligence, pp. 6545–6553 (2023)
4. Cai, H., Zheng, V.W., Chang, K.: A comprehensive survey of graph embedding: problems, techniques, and applications. IEEE Trans. Knowl. Data Eng. **30**(9), 1616–1637 (2018)
5. Chawla, N.V., Bowyer, K.W., Hall, L.O., Kegelmeyer, W.P.: Smote: synthetic minority over-sampling technique. J. Artif. Intell. Res. **16**, 321–357 (2002)
6. Chytas, A., Bassileiades, N., Natsiavas, P.: Mapping omop-cdm to rdf: bringing real-world-data to the semantic web realm. In: Digital Health and Informatics Innovations for Sustainable Health Care Systems, pp. 1406–1410. IOS Press (2024)
7. Cox, S.J.D., et al.: Time ontology in owl. W3C Candidate Recommendation (November 2022). https://www.w3.org/TR/owl-time/. Accessed 18 Dec 2024
8. Cui, H., et al.: A survey on knowledge graphs for healthcare: resources, application progress, and promise. In: ICML 3rd Workshop on Interpretable Machine Learning in Healthcare (IMLH) (2023)
9. Cvetkov-Iliev, A., Allauzen, A., Varoquaux, G.: Relational data embeddings for feature enrichment with background information. Mach. Learn. **112**(2), 687–720 (2023)
10. Dall'Amico, L., Barrat, A., Cattuto, C.: An embedding-based distance for temporal graphs. Nat. Commun. **15**(1), 9954 (2024)
11. Dumontier, M., et al.: The semanticscience integrated ontology (sio) for biomedical research and knowledge discovery. J. Biomed. Semant. **5** (2014). https://api.semanticscholar.org/CorpusID:17652831
12. Ehm, W., Gneiting, T., Jordan, A., Krüger, F.: Of quantiles and expectiles: consistent scoring functions, choquet representations and forecast rankings. J. R. Stat. Soc. Ser. B Stat Methodol. **78**(3), 505–562 (2016)

13. Galkin, M., Denis, E., Wu, J., Hamilton, W.L.: Nodepiece: compositional and parameter-efficient representations of large knowledge graphs. In: International Conference on Learning Representations (2022). https://openreview.net/forum?id=xMJWUKJnFSw
14. Gastinger, J., Sztyler, T., Sharma, L., Schuelke, A.: On the evaluation of methods for temporal knowledge graph forecasting. In: NeurIPS 2022 Temporal Graph Learning Workshop, New Orleans, United States (2022)
15. Gesese, G.A., Biswas, R., Alam, M., Sack, H.: A survey on knowledge graph embeddings with literals: which model links better literally? Semant. Web **12**(4), 617–647 (2021)
16. Hamilton, W.L., Ying, R., Leskovec, J.: Representation learning on graphs: methods and applications. arXiv preprint arXiv:1709.05584 (2017)
17. He, K., Zhang, X., Ren, S., Sun, J.: Delving deep into rectifiers: surpassing human-level performance on imagenet classification. In: Proceedings of the IEEE International Conference on Computer Vision, pp. 1026–1034 (2015)
18. Hitzler, P., Krötzsch, M., Rudolph, S.: Foundations of Semantic Web Technologies. Chapman & Hall/CRC, Boca Raton (2009)
19. Hogan, A., et al.: Knowledge graphs. ACM Comput. Surv. **54**(4) (2021)
20. Jacobsen, J.O., et al.: The ga4gh phenopacket schema defines a computable representation of clinical data. Nat. Biotechnol. **40**(6), 817–820 (2022)
21. Kaliyaperumal, R., Singh, G., Queralt-Rosinach, N., Bayjanov, J.R., 't Hoen, P.B., Roos, M.: Phenopackets for the semantic web. In: SWAT4HCLS (2022). https://api.semanticscholar.org/CorpusID:248397161
22. Kaliyaperumal, R., et al.: Semantic modelling of common data elements for rare disease registries, and a prototype workflow for their deployment over registry data. J. Biomed. Semant. (2022)
23. Kamigaito, H., Hayashi, K.: Comprehensive analysis of negative sampling in knowledge graph representation learning. In: International Conference on Machine Learning, pp. 10661–10675. PMLR (2022)
24. Karakachoff, M., et al.: Implementing a biomedical data warehouse from blueprint to bedside in a regional French university hospital setting: unveiling processes, overcoming challenges, and extracting clinical insight. JMIR Med. Inf. **12**, e50194 (2024). https://doi.org/10.2196/50194
25. Keedy, A.: An overview of intracranial aneurysms. McGill J. Med. MJM **9**(2), 141 (2006)
26. Kingma, D.P.: Adam: a method for stochastic optimization. arXiv preprint arXiv:1412.6980 (2014)
27. Kipf, T.N., Welling, M.: Semi-supervised classification with graph convolutional networks. In: 5th International Conference on Learning Representations, ICLR 2017, Toulon, France, 24–26 April 2017, Conference Track Proceedings. OpenReview.net (2017). https://openreview.net/forum?id=SJU4ayYgl
28. Kristiadi, A., Khan, M.A., Lukovnikov, D., Lehmann, J., Fischer, A.: Incorporating literals into knowledge graph embeddings. In: Ghidini, C., et al. (eds.) ISWC 2019. LNCS, vol. 11778, pp. 347–363. Springer, Cham (2019). https://doi.org/10.1007/978-3-030-30793-6_20
29. Lehne, M., Luijten, S., Vom Felde Genannt Imbusch, P., Thun, S.: The use of fhir in digital health–a review of the scientific literature. In: German Medical Data Sciences: Shaping Change–Creative Solutions for Innovative Medicine, pp. 52–58 (2019)

30. Prud'hommeaux, E., Collins, J., Booth, D., Peterson, K.J., Solbrig, H.R., Jiang, G.: Development of a fhir rdf data transformation and validation framework and its evaluation. J. Biomed. Inf. **117**, 103755 (2021). https://doi.org/10.1016/j.jbi.2021.103755. https://www.sciencedirect.com/science/article/pii/S1532046421000848
31. Ristoski, P., Paulheim, H.: RDF2Vec: RDF graph embeddings for data mining. In: Groth, P., Simperl, E., Gray, A., Sabou, M., Krötzsch, M., Lecue, F., Flöck, F., Gil, Y. (eds.) ISWC 2016. LNCS, vol. 9981, pp. 498–514. Springer, Cham (2016). https://doi.org/10.1007/978-3-319-46523-4_30
32. Schlichtkrull, M., Kipf, T.N., Bloem, P., van den Berg, R., Titov, I., Welling, M.: Modeling relational data with graph convolutional networks. In: Gangemi, A., et al. (eds.) ESWC 2018. LNCS, vol. 10843, pp. 593–607. Springer, Cham (2018). https://doi.org/10.1007/978-3-319-93417-4_38
33. Touré, V., et al.: Fairification of health-related data using semantic web technologies in the swiss personalized health network. Sci. Data **10** (2023). https://api.semanticscholar.org/CorpusID:257433763
34. Voss, E.A., et al.: Feasibility and utility of applications of the common data model to multiple, disparate observational health databases. J. Am. Med. Inf. Assoc. **22**(3), 553–564 (2015)
35. Wang, Q., Mao, Z., Wang, B., Guo, L.: Knowledge graph embedding: a survey of approaches and applications. IEEE Trans. Knowl. Data Eng. **29**(12), 2724–2743 (2017)
36. Xiao, G., et al.: Fhir-ontop-omop: building clinical knowledge graphs in fhir rdf with the omop common data model. J. Biomed. Inf. **134**, 104201 (2022). https://doi.org/10.1016/j.jbi.2022.104201. https://www.sciencedirect.com/science/article/pii/S1532046422002064
37. Xu, D., Ruan, C., Korpeoglu, E., Kumar, S., Achan, K.: Inductive representation learning on temporal graphs. In: International Conference on Learning Representations (2020). https://openreview.net/forum?id=rJeW1yHYwH
38. Zhang, F., Li, Z., Peng, D., Cheng, J.: Rdf for temporal data management-a survey. Earth Sci. Inf. **14**, 563–599 (2021)

AvengER: Ensembling and Fine-Tuning LLMs for SELECT Prompts in Entity Resolution

Alexandros Zeakis[1,2](✉), George Papadakis[1], Dimitrios Skoutas[2], and Manolis Koubarakis[1]

[1] Department of Informatics and Telecommunications, National and Kapodistrian University of Athens, Athens, Greece
{alzeakis,gpapadis,koubarak}@di.uoa.gr

[2] Information Management Systems Institute, Athena Research Center, Athens, Greece
{azeakis,dskoutas}@athenarc.gr

Abstract. Link Discovery plays a vital role in enhancing connections between data sources within the Linked Open Data Cloud. A key aspect of this process is Entity Resolution (ER), which focuses on identifying owl:sameAs relationships between different entity descriptions representing the same real-world object. Recent works on ER have investigated the use of Large Language Models (LLMs) for Entity Matching, showing promising results. In the simplest case, a pair of entities is given as input to an LLM, asking whether these entities match or not. A recent approach introduced SELECT prompts, which include the query entity along with multiple candidates generated by a Blocking method. However, this increases the complexity of the questions posed to LLMs, while being susceptible to position bias among the presented candidates. To address these issues, we introduce AvengER, a novel approach to SELECT prompts for LLM-based Matching that effectively handles both unsupervised and supervised settings. For the former, we introduce a hybrid approach that utilizes an ensemble of open-source medium-size models (8b) and also selectively leverages (in just 12% of the cases) an external larger-size Judge (32b), thus balancing high accuracy with computational efficiency. For the latter, we use a dataset containing data from multiple domains to fine-tune a medium-size model so that it surpasses both its pre-trained version and pre-trained large-size models (GPT-3.5).

Keywords: LLMs · Entity Matching · Model ensembles · Fine-tuning

1 Introduction

Entity Resolution (ER) facilitates the integration of independent data sources by identifying their *duplicates* or *matches*, i.e., different entity descriptions that refer to the same real-world object [2]. ER constitutes a crucial and challenging task for realizing the vision of Linked Data, facilitating a diverse set of applications, from common data analytics tasks to advanced question answering [3,4].

E. Curry et al. (Eds.): ESWC 2025, LNCS 15718, pp. 301–320, 2025.
https://doi.org/10.1007/978-3-031-94575-5_17

Do these two entity descriptions refer to the same real-world entity?

(1) [COL] name [VAL] Sony Turntable - PSLX350H [COL] description [VAL] Sony Turntable - PSLX350H/ Belt Drive System/ 33-1/3 and 45 RPM Speeds/ Servo Speed Control/ Supplied Moving Magnet Phono Cartridge/ Bonded Diamond Stylus/ Static Balance Tonearm/ Pitch Control
(2) [COL] name [VAL] Sony Picture Station Digital Photo Printer - DPPFP95

LLM Response: No **MATCH Prompt**

Select an entity description from the following candidates that refers to the same real-world entity as the given entity description. Answer with the corresponding entity description number surrounded by "[]" or "[0]" if there is none.

Given entity record: [COL] name [VAL] Sony Turntable - PSLX350H [COL] description [VAL] Sony Turntable - PSLX350H/ Belt Drive System/ 33-1/3 and 45 RPM Speeds/ Servo Speed Control/ Supplied Moving Magnet Phono Cartridge/ Bonded Diamond Stylus/ Static Balance Tonearm/ Pitch Control

Candidate records:
[1] [COL] name [VAL] Denon DP-29F Analog Record Turntable - DP29F [COL] description [VAL] Belt Drive
[2] [COL] name [VAL] Sony PS-LX350H Belt-Drive Turntable
[3] [COL] name [VAL] Sony PSLX300USB USB Record Turntable [COL] description [VAL] Belt Drive [COL] price [VAL] 119.72
[4] [COL] name [VAL] Sony Picture Station Digital Photo Printer - DPPFP95
[5] [COL] name [VAL] Denon DP-300F Analog Record Turntable - DP300F [COL] description [VAL] Belt Drive [COL] price [VAL] 328.99

LLM Response: [2] **SELECT Prompt**

Fig. 1. Prompts for different tasks. The first decides for a pair of entities, the second selects from a given list of choices.

The Linked Open Data (LOD) Cloud[1] forms a central part of the Semantic Web, expanding from 570 datasets in 2014 to 1,573 in 2024. However, its interconnectivity remains limited, with only 17,882 links recorded as of December 2024. On average, each dataset connects to just 11 others, which accounts for roughly 1% of all possible links. To improve this connectivity, Link Discovery automates the identification of relationships between entity descriptions across datasets.

A typical end-to-end ER pipeline consists of two steps: Blocking and Matching [3]. The first step identifies candidate pairs that are likely to match, while the second applies a more precise method to determine which pairs are genuine matches. The last few years, Blocking has been studied with supervised settings [1] and Matching has been dominated by approaches that combine pre-trained transformer models with Deep Learning, such as DITTO [6] and Unicorn [16], while more recent studies leverage Large Language Models (LLMs) [12,15,18]. Most of the latter approaches feed LLMs with *MATCH prompts*, which ask whether two particular entities are matching [12,15]. Yet, these prompts are impractical and time-consuming in the context of end-to-end ER pipelines, where Blocking typically yields k candidates per entity [9,11]. To address this, *SELECT prompts* are proposed in [18], including all Blocking candidates in a single prompt. This transforms Matching from a binary decision ("Yes/No") into a "select which of the following" response, as shown in Fig. 1. A major challenge with SELECT prompts is position bias, where LLM responses depend on the duplicate's position among k candidates [18]. This is indirectly addressed through a post-Blocking refinement step that reduces candidates using k intermediate MATCH prompts, trading efficiency for effectiveness.

[1] https://lod-cloud.net/#about.

In this work, we propose *AvengER*, a comprehensive approach that enhances the SELECT prompts by minimizing the impact of position bias in a more efficient and effective way. Rather than relying on intermediate MATCH prompts, *AvengER* applies post-Blocking refinement through *reciprocity*: two entities are candidates only if each one is among the top-k most similar entities for the other. This strategy significantly reduces the number of candidates per entity, while conveying a minor decrease in Blocking effectiveness. To further enhance Matching performance, *AvengER* employs multiple LLMs, combining their outputs through diverse *voting strategies*: internal ones, where the ensemble of models decides independently, external ones, where a separate model acts as a judge, and hybrid ones, where a judge decides in case the ensemble is not certain. Despite the higher run-times, our thorough experimental analysis demonstrates that *AvengER*'s combination of reciprocity and voting schemes offers a better balance between effectiveness and efficiency under the *unsupervised* settings.

Under the *supervised* settings, where annotated data are available, *AvengER* fine-tunes medium-sized models, improving the performance of SELECT prompts to such an extent that position bias is mitigated without the need to reduce the candidate pairs generated by Blocking. Their performance is also significantly higher than their pre-trained version and than large-sized models.

Overall, we make the following contributions:

- We introduce *AvengER*, a novel LLM-based Matching approach that leverages SELECT prompts, handling both unsupervised and supervised settings.
- Under unsupervised settings, *AvengER* suggests a categorization of existing Post-Blocking strategies and employs a new one that drastically reduces candidate pairs based on reciprocity.
- *AvengER* further improves ER performance under unsupervised settings through voting strategies that combine the results of an ensemble of LLMs.
- Under the supervised settings, *AvengER* fine-tunes medium-sized LLMs to maximize the robustness of SELECT prompts and their mitigate position bias, significantly outperforming their pre-trained version and large-sized LLMs.
- We demonstrate the superiority of *AvengER* compared to baselines and state-of-the-art through an extensive experimental study that uses several real-world datasets.

2 Related Work

We organize the recent works on LLMs for ER in the following categories:

Blocking. The state-of-the-art Blocking methods detect the k most likely matches per query entity [9,11]. Higher values for k typically increase Blocking recall, at the cost of more candidate pairs and, thus, lower Blocking precision. For this reason, the goal of Blocking is to achieve a good balance between recall

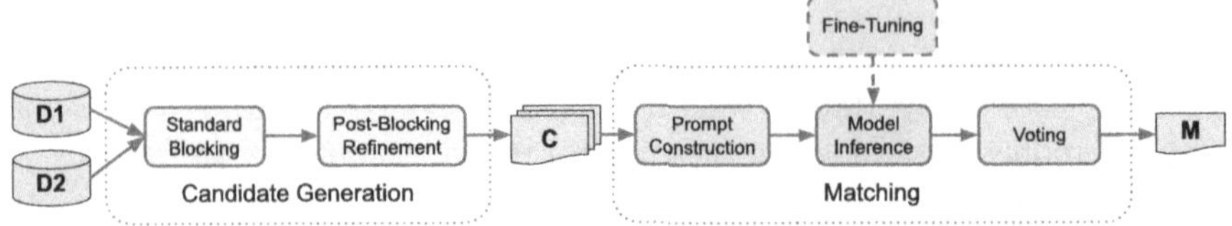

Fig. 2. The phases and steps of *AvengER*. Fine-Tuning the model is optional (dashed).

and precision. To improve this balance, a hybrid post-Blocking refinement strategy is proposed in [18]: it constructs "MATCH" prompts, which return a "yes/no" answer, and "COMPARE" prompts, which choose the most likely match among two candidates for the query entity at hand. Both prompts are applied to a medium-sized model, yielding a ranking of options, which then produce the top-N results, where N is a parameter defined by experimentation. The overhead, though, is much larger than traditional Blocking methods, due to the large number of prompts.

Matching. In Matching we have two distinct categories. In Prompt Engineering, typically, a matching prompt consists of three parts [12]: (a) the task description, which describes Matching as the task of determining whether two entity descriptions refer to the same real-world object; (b) a serialization of the entity descriptions of the given candidate pair; (c) optionally, a set of examples, with each one comprising a candidate pair along with its label (match/non-match). These examples yield *few-shot prompts*, whereas their absence yields *zero-shot prompts*. A thorough investigation of variants of both types is presented in [12]. Prompts containing multiple candidates, instead of a single pair, have been investigated in [18].

Fine-Tuning improves the accuracy of a pre-trained LLM on a specific task by updating its parameters on a new dataset [15]. To this end, the LLM is partially retrained using <input, output> pairs of representative examples of the desired behavior. A thorough study of the impact of this approach on Matching is presented in [15], which emphasizes the generalization across different domains and the impact of training dataset size. A related approach leverages an LLM to generate explanations and then uses these explanations to fine-tune a PLM (FlanT5-base) to perform the matching [17]. Augmentation strategies to increase the amount of training data have also been considered in [19]. Finally, a fine-tuned model using transfer learning can outperform competitors on zero-shot prompts, while reducing inference time by nearly 4,000 times [22].

3 *AvengER* Components

AvengER consists of two main phases, as shown in Fig. 2: Candidate Generation and Matching. The former receives as input a pair of data sources, D_1 and D_2, and returns as output a set of candidate pairs C, such that Blocking recall and

Algorithm 1. *AvengER*'s end-to-end algorithm.

Require: Query entity q, blocking parameter k, post-blocking refinement parameter N, List of models *models*, collection C
Ensure: Final result *result*

```
1: candidates ← blocking(q, k, C)
2: candidates ← post_blocking_refinement(q, N, candidates)
3: prompt ← construct_prompt(q, candidates)
4: responses ← {}
5: for each model in models do
6:     responses.add(model.inference(prompt))
7: end for
8: result ← voting(responses)
```

precision are maximized. The second phase receives as input the set of candidate pairs C and returns as output a set of matching pairs M, such that ER recall and precision are maximized. Putting it all together, Algorithm 1 presents the main steps of *AvengER* in pseudo-code. Initially, Blocking is performed and refined with the post-Blocking approaches presented in Sect. 3.1 (Lines 1–2). Then, the prompt is constructed by one of the methods discussed in Sect. 3.2 (Line 3). The prompt is then passed to Model Inference and then, the responses of the individual models are aggregated (Lines 4–7). Finally, a voting scheme is optionally applied (Line 8). Fine-Tuning, if needed, is a process done offline. Below, we explain their functionality in more detail.

3.1 Candidate Generation

Based on [21], *AvengER* applies a state-of-the-art approach to *Standard Blocking*: it uses S-GTR-T5 to vectorize the serialized entity descriptions and then performs approximate nearest neighbor (NN) search with FAISS[2] in order to efficiently detect the k most similar descriptions per query entity. S-GTR-T5, which extends GTR [10], is a dual encoder that transforms two pieces of text into two dense vectors. GTR models are built on top of T5 [13], an encoder-decoder model that aims to unify all natural language processing tasks under a single model. They also use the Siamese architecture in [14] for efficient and scalable similarity computations.

Typically, Standard Blocking achieves high recall by generating large sets of candidates that involve repeated and non-matching pairs. This results in low Blocking precision. *AvengER* aims to significantly increase Blocking precision at an insignificant cost in recall through the *Post-Blocking Refinement* step. We suggest a taxonomy of existing and new Post-Blocking strategies, that involves three categories for removing repeated and non-matching pairs:

1. **Ranking**: Assuming that every candidate pair is associated with a matching probability, this strategy simply retains the top-N weighted ones; N ($\gg k$)

[2] https://github.com/facebookresearch/faiss.

sets the maximum number of candidates forwarded to Matching, with its actual value set arbitrarily or according to experimental results. *AvengER* conveys two implementations of this strategy: (i) *COS-ST5* uses the cosine similarity scores between embedding vectors already computed during the NN search of Standard Blocking. (ii) *COMPARE prompts* involve a query entity along with two of the candidates determined by Standard Blocking, requesting the LLM to rank them from the most to the least similar one [18]. This requires $\binom{k}{2}$ invocations per query entity.
2. **Filtering**: This strategy reduces the number of candidates by pruning the pairs that are more unlikely to be matching. *AvengER* conveys two implementations of this strategy: (i) *Reciprocal pruning* performs NN search for the entity descriptions of both D_1 and D_2, retaining only the pairs (e_i, e_j), where e_j is among the top-k candidates of e_i and vice versa for e_i. (ii) *MATCH prompts* assess whether every candidate pair involves duplicate descriptions and prunes those that do not [18]. In both cases, the number of candidates differs among the input entities (there might even be cases with empty lists).
3. **Hybrid**: This strategy combines the other two, retaining only the top-N weighted candidate pairs generated by Filtering. In the case of MATCH prompts, this can be accomplished by first requesting that each pair is assigned a matching probability and then ranking the results accordingly (Fig. 4).

3.2 Matching

We present the steps of *AvengER*'s Matching in the order they are applied.

Prompt Construction. We now elaborate on the structure of the prompt used as a first step for *AvengER*'s Matching.

Basic Prompt. The standard prompt for EM, first introduced in [12], consists of two core parts. The first one is the *task description*, which specifies that the goal is to find entities that refer to the same real-world entity. This can be formulated as a "Yes/No" question [12,15,18]. When given a pair of entities, the prompt can be formulated as "select which of the following" [18].

The second part pertains to the *entity serialization*. Many approaches have been proposed in the literature, but since an entity consists of triples $< subject, predicate, object >$, two are the most common options: the *schema-agnostic* one, where only the objects are concatenated into a single sentence [21], and the *schema-aware* one, where the format is $[COL]predicate[VAL]object$ [6].

We propose extensions for both parts. In the task description, we focus exclusively on SELECT prompts, asking LLMs not only to respond to the query, but also to explain their choice. As experimentally shown in Sect. 4, this leads to a better performance for some models. We also tried another prompt, where we instructed the model to provide a *confidence* estimation for its choice [20]. Instead of a numeric score, we requested a qualitative response among three options: "Certain", "Moderately-certain" and "Uncertain". Note that in any prompt type, the model can respond that it cannot decide, returning the special answer of [0], which indicates that none of the choices is correct. Both prompts can be found in Figs. 5 and 6.

For the entity serialization part, we also examine a semantically richer description that is generated through a pre-processing step that provides the LLM with the serialized entity and requests a summary in natural language. This is demonstrated in Fig. 7. Then, inside the EM prompt, we replace the entity serializations with their summary.

In-context Learning. Adding examples before the main question in prompts helps LLMs to better understand the given task [12,15]. This has been widely exploited in EM literature, but only for MATCH prompts [18]. To the best of our knowledge, this work is the first to explore enriching SELECT prompts with examples in an experimental setting, yielding few-shot EM SELECT prompts. We have selected examples randomly, but there is room to investigate more advanced strategies in the future.

Model Inference. For model inference, some works [5,12,18] use pre-trained models, particularly the large ones like GPT-3.5. In other works [12,15], Fine-Tuning shows to improve the performance of a model. For example, in [12] a fine-tuned GPT-3.5 can outperform a pre-trained GPT-4 on certain benchmarks.

In our task, the problem is transformed from binary classification to selecting the index position, which is heavily dependent on the order of the choices within the prompt. To enhance robustness, we introduce prompt permutations to train the model on variations in choice positions, thus reducing position bias.

Voting. All existing works on LLM-based EM, rely on individual LLMs. In this work, we go beyond them by considering ensembles of LLMs. We actually examine multiple strategies for combining the individual decisions of multiple LLMs. These strategies depend on whether the ensemble determines the vote internally or relies on an external judge to make the final decision.

Internal Vote. We introduce the following internal voting strategies:

- *Majority-Vote.* Each model produces one answer, and the most-voted answer is selected. If no majority is reached, the ensemble yields an *indecisive* vote, similar to [0], when the true answer is not among the options.
- *Weighted-Vote.* Each model's contribution is weighted based on its performance, with *precomputed* global weights derived from metrics like F1-score (see Sect. 4). The answer with the highest total weight is selected.
- *Confidence-Vote.* Instead of relying on pre-computed weights, this method assigns weights *dynamically* based on the confidence scores generated by each model's response. By utilizing a confidence prompt, we derive a unique weight for each model per response. This approach adapts to the context of each query, making it more flexible than static weight assignments. We introduce 3 certainty levels, {Certain, Moderately Certain, Uncertain}, which are evaluated to {0.5, 0.75, 1}, resembling Shannon Entropy.

Table 1. Statistics of the real ER datasets used in the experiments, showing the number of resources ($|R_x|$), distinct predicates ($|P_x|$) and duplicates ($|D|$).

	D_1	D_2	D_3	D_4	D_5	D_6	D_7	D_8
Dat_1	Abt	Amz	ACM	IMDb	IMDb	TMDb	Wmt	DBLP
Dat_2	Buy	GPr.	DBLP	TMDb	TVDB	TVDB	Amz	Scholar
$\|R_1\|$	1,076	1,354	2,294	5,118	5,118	6,056	2,554	2,516
$\|R_2\|$	1,076	3,039	2,616	6,056	7,810	7,810	22,074	61,353
$\|P_1\|$	3	4	4	13	13	30	6	4
$\|P_2\|$	3	4	4	30	9	9	6	4
$\|D\|$	1,076	1,104	2,224	1,968	1,072	1,095	853	2,308

- *Aggregated Majority-Vote.* This approach involves asking each model N times with different permutations of answer choices in each iteration. Instead of relying on m votes from the original models, this process generates $m \times N$ votes. By aggregating these votes, the ensemble improves the overall robustness, as models demonstrating consistent predictions across iterations have a stronger contribution to the final decision.

Regarding computational differences, notice that Weighted-vote requires training, whereas Confidence-vote does not; for inference, there is no significant difference in cost. Notice also that Confidence-Vote expresses each model's confidence qualitatively, while Aggregated Majority-Vote quantifies it.

External Vote. We introduce the following external voting strategies:

- *Judge-vote.* The idea of using an LLM as a judge for evaluating other LLMs on open-ended questions was proposed in [23]. We apply this by merging individual LLM responses into a prompt for the judge model, since each LLM accompanies its response with an explanation regarding the given answer. Due to the large prompt size, the judge model requires an LLM with a large context window, involving more parameters and resources.
- *Hybrid-vote.* To mitigate the extra cost of asking a heavier judge model, we also propose a hybrid voting strategy. In this approach, we assume that the top-1 answer accumulates a score Σ, which depends on the internal voting strategy (majority, weighted, etc.). We also define a threshold $\theta \in [0, 1]$. If $\Sigma < \theta$, we invoke a Judge model to make a decision. In this way, we save computational cost, by reducing the cases where the Judge model is invoked.

4 Evaluation

4.1 Experimental Setup

Datasets. We use 8 real-world ER datasets that are widely used in the literature [8,18]. Our approach assumes that each entity comprises a set of <predicate, object> pairs, which are then serialized into a single textual description. This approach accommodates both relational and Knowledge Graph data. For example, the movie datasets (D4, D5, D6) stem from the MovieGraphBenchmark[3], using URIs as predicates. Table 1 shows the statistics of these datasets. Following [18], we sampled 300 annotated resources per dataset and 100 non-annotated entities (when possible) for testing. For training, we sampled again 300 labeled and 100 unseen pairs, which were also used in the Weighted-vote. For in-context learning, we used 1 positive example, where the answer was among the provided options, and 1 negative example, where the answer was not included in the provided options. Both examples were randomly selected.

Models. We used five open-source LLMs: Llama-3.1:8b, Mistral-v0.3:7b, Orca-2:7b, Qwen-2.5:32b[4], and Flan-T5-XL[5]. These models were selected to ensure comparability with state-of-the-art works while leveraging their specific strengths. Llama and Mistral were used in almost all experiments and were also utilized in [18], facilitating direct comparisons. Orca was included in the ensemble due to its strong reasoning capabilities [7], which are beneficial for decision-making in multiple-choice settings. Qwen, being a larger model, served as a Judge in External Voting, while Flan was used for MATCH prompts in ComEM, as in [18]. For brevity, we omit their parameters in subsequent references.

Baseline LLM-based EM. The top performing LLM-based approach to EM is presented in [18]. It utilizes Sparkly [11] for Blocking and Hybrid for Post-Blocking Refinement. The constructed prompt is based on the basic SELECT format with no examples and the model is a pre-trained GPT-3.5-turbo-0613.

Evaluation Metrics. For all experiments, we report F-Measure (F1). For the Voting Strategies, we also report Precision and Recall, whereas for Blocking, we also report the corresponding Blocking Precision, Recall and F1. We also report the average response size in characters.

Settings. For Standard Blocking, we used S-GTR-T5 for entity serialization in combination with FAISS for indexing and searching the top-N candidate pairs per entity [21]. For the pre-trained LLMs, we used models provided by Ollama[6], while for their fine-tuning we used Unsloth[7]. All of our code, datasets and fine-tuned models are publicly available[8]. All experiments were executed on a server with Ubuntu 20.04, AMD Ryzen Threadripper 3960X 24-Core processor, 256 GB RAM and an RTX 4090 GPU.

---

[3] https://github.com/ScaDS/MovieGraphBenchmark.
[4] https://ollama.com/library/llama3.1:8b, https://ollama.com/library/mistral:7b, https://ollama.com/library/orca2:7b, https://ollama.com/library/qwen2.5:32b.
[5] https://huggingface.co/google/flan-t5-xl.
[6] https://ollama.com/.
[7] https://unsloth.ai/.
[8] https://github.com/alexZeakis/AvengER.

4.2 Candidate Generation

Table 2. Comparison on F1 for different Post-Blocking Refinement approaches for SELECT prompts and using Llama-3.1 for inference.

Strategy	D1	D2	D3	D4	D5	D6	D7	D8	Mean
Ranking@10	0.84	**0.61**	0.83	0.69	0.60	0.63	0.71	0.70	0.70
Rec. Pruning	**0.87**	0.59	**0.86**	0.74	**0.64**	0.67	**0.72**	**0.76**	**0.73**
Hybrid@4	0.47	0.41	0.84	**0.88**	**0.64**	**0.82**	0.30	0.75	0.64

In this experiment, we assess the effect of Post-Blocking Refinement on the performance of *AvengER*. We actually measure the evolution of Blocking effectiveness with regard to the Blocking parameter $k \in [1, 10]$ with a step of 1. We consider two variants of the *COS-ST5* strategy: "*Left-to-Right*" indexes the larger data source and queries with the entity descriptions of the smaller one, and vice versa for "*Right-to-Left*". To be noted, the size of a data collection reflects the number of entities within it. For Filtering, we use Reciprocal Pruning.

As shown in Fig. 9, Right-to-Left exhibits a relatively stable F1 score across all k in each dataset. The reason is that it consistently maintains high recall at the cost of lower precision. Left-to-Right starts from low F1 scores but raises above Right-to-Left in most cases for k=10. This should be attributed to a similar trend in recall, given that its precision is consistently very high. Reciprocal Pruning significantly outperforms the other two strategies with respect to both recall and precision across all k values, due to the consistently lower number of candidates for almost the same recall.

Our next step is to measure the effectiveness of Post-Blocking Refinement strategies in combination with Matching. We compare three different approaches: (i) For Ranking, we use *COS-ST5* with Left-to-Right and $k = 10$. (ii) For Filtering, we use Reciprocal Pruning with $k = 10$. (iii) For Hybrid, we set $k = 10, N = 4$ and use Flan-T5-XL, as in [18]. In all cases, we use the basic SELECT prompt and Llama-3.1 as a model. We omit Ranking with COMPARE prompts and Filtering with MATCH prompts, since preliminary results show no significant differences in terms of effectiveness, while the extra prompts yield a considerable overhead in terms of time efficiency. Besides, these strategies are partially covered by the Hybrid one.

As shown in Table 2, one-directional Blocking with $k = 10$ generally underperforms Reciprocal Pruning to a significant extent. The only exception is $D2$, where the duplicate pairs are so noisy that in many cases, one of them is more similar to non-matching entity descriptions. Reciprocal Pruning also outperforms Hybrid in datasets like D1, D2, and D7, because Hybrid struggles with the variability in product descriptions, but the latter performs better in movie-related datasets like D4, D5, and D6. Combined with the fact that Hybrid requires k

prompts in an LLM, thus higher computational cost, Reciprocal Pruning offers a better balance of efficiency and performance.

Summary: On average, Reciprocal Pruning outperforms the other strategies, yielding a much higher F-Measure, while exhibiting the best and most stable balance between effectiveness and time efficiency.

4.3 Matching

Prompt Evaluation. We now study the performance of variants of the basic prompt, regarding the parts "in-context learning", "serialization" and "task description". For in-context learning, we extended the basic SELECT prompt with two examples. For serialization, we compare the schema-aware and schema-agnostic strategies with the summaries created by Llama-3.1. For Task Description, we study SELECT, EXPLAIN and CONFIDENCE prompts with Llama-3.1, Mistral-v0.3 and Orca-2. For each experiment, we use the Standard Blocking with Reciprocal Pruning for Candidate Generation, based on Sect. 4.2.

Table 3. F1 per prompt variation. The rightmost column reports the average response size in characters. Note that ICL stands for in-context learning, RSize for Response Size, ZS for Zero-Shot prompts and FS for Few-Shot ones.

Task	Serialization	ICL	Model	D1	D2	D3	D4	D5	D6	D7	D8	Mean	RSize
SEL.	Sch.-Aw.	ZS	Llama-3.1	**0.87**	0.59	0.86	0.74	0.64	0.67	**0.72**	**0.76**	**0.73**	107
(a) Baseline													
SEL.	Sch.-Aw.	FS	Llama-3.1	0.77	0.47	0.77	0.65	0.62	0.51	0.53	0.72	0.63	1011
(b) Few-shot variation													
SEL.	Summary	ZS	Llama-3.1	0.75	0.46	0.83	0.74	**0.65**	**0.72**	0.61	0.73	0.69	57
SEL.	Sch.-Agn.	ZS	Llama-3.1	0.86	0.60	**0.88**	0.73	0.61	0.67	0.69	0.75	0.72	13
(c) Zero-shot Variation of Serialization methods													
SEL.	Sch.-Aw.	ZS	Mistral-v0.3	0.75	0.44	0.82	0.70	0.43	0.58	0.56	0.66	0.62	228
SEL.	Sch.-Aw.	ZS	Orca-2	0.6	0.35	0.73	0.62	0.42	0.5	0.42	0.61	0.53	1879
EXPL.	Sch.-Aw.	ZS	Llama-3.1	0.86	0.60	0.85	0.71	0.61	0.65	0.70	0.74	0.72	366
EXPL.	Sch.-Aw.	ZS	Mistral-v0.3	0.83	0.56	0.81	0.68	0.56	0.60	0.69	0.64	0.67	480
EXPL.	Sch.-Aw.	ZS	Orca-2	0.54	0.31	0.75	0.5	0.34	0.48	0.34	0.61	0.48	1640
CONF.	Sch.-Aw.	ZS	Llama-3.1	0.84	**0.61**	0.86	**0.75**	**0.65**	0.67	0.71	0.75	**0.73**	198
CONF.	Sch.-Aw.	ZS	Mistral-v0.3	0.84	0.57	0.87	0.74	0.54	0.66	0.69	0.72	0.70	218
CONF.	Sch.-Aw.	ZS	Orca-2	0.54	0.27	0.70	0.5	0.39	0.45	0.30	0.64	0.47	1841
(d) Zero-shot Variation of Task Description methods and models													

The results are reported in Table 3. For Llama-3.1, Zero-Shot prompts outperform Few-Shot prompts across all datasets, with an increase in average F-Measure from 0.63 to 0.73. This is likely due to SELECT prompts being sensitive to token count, where adding 2x more options significantly increases average response size from 107 to 1011 characters. This suggests that models can be disoriented by bigger prompts and hints potential improvements through better example selection.

Schema-agnostic serialization performs well with an average of 0.72, surpassing other techniques in D2 and D3 but ranking below the baseline, which uses Schema-Aware serialization. Its limitation lies in omitting predicates that are often semantically rich, particularly for datasets with distinctive predicates like titles or names.

Comparing Llama-3.1 with the other two LLMs, we observe that it consistently dominates them across all three types of prompts, i.e., SELECT, EXPLAIN and CONFIDENCE. The best performance of Mistral-v0.3 is achieved with CONFIDENCE prompts, which still underperform the baseline. Orca-2 consistently yields much lower F1 scores along with the longest responses.

Summary: Adding random examples in SELECT prompts can degrade performance for the 7B LLMs we are examining, while including predicates in entity serialization improves model accuracy but requires domain-specific knowledge for richer descriptions. Requiring explanations or confidence measures in prompts further can enhance performance for models that are weaker in reasoning.

Voting Evaluation. We now study the performance of the suggested voting strategies on pre-trained LLMs. The ensemble of models consists of Llama-3.1, Mistral-v0.3 and Orca-2. For Weighted-vote, we obtained the weights per model by normalizing F1 scores from inferencing EXPLAIN prompts on unseen data. For the Aggregated Majority-vote, each model was inferenced 5 times. For the Judge-vote, we utilized the responses of the EXPLAIN prompts of the ensemble and used two model as judges: Llama-3.1 and Qwen-2.5. Note, though, that due to the higher resource requirements of Qwen-2.5, it was not possible to process all prompts. Finally, for the Hybrid-vote, we used Aggregated Majority-vote as the internal vote and Qwen-2.5 as the external judge. To examine voting complementarity, we also consider the *"Any-vote"* strategy, where any pair marked as match from any strategy is treated as a match.

We start by assessing the complementarity between pre-trained models, i.e., whether a model X can find pairs that a model Y cannot find. To do so, we use as baseline the performance of Llama-3.1 with EXPLAIN prompts and compare it with Majority-vote and Any-vote. Any-vote has the highest Recall at 0.87, while Majority-vote and the baseline reach 0.76 and 0.80, respectively. This suggests that the three models of the ensemble are indeed complementary. However, the increased number of false positives produced by Any-vote significantly lowers its Precision and F-Measure to 0.49 and 0.62, respectively. In contrast, Majority-vote sacrifices some Recall compared to the baseline but improves Precision (0.72), leading to a better overall F1 score (0.74).

After verifying the complementarity of the voting schemes, we compare the performance of the internal ones in more detail in Table 4. We observe that Majority-vote consistently outperforms the baseline across all prompt types, with a larger difference for CONFIDENCE prompts, where it averages 0.76

Table 4. Comparison on F1 for different internal and external voting strategies on the ensemble of models. RSize stands for Response Size.

Task	Model	Vote	D1	D2	D3	D4	D5	D6	D7	D8	Mean	RSize
SEL.	Llama-3.1	Single	0.87	0.59	0.86	0.74	0.64	0.67	0.72	0.76	0.73	107
(a) Baseline												
SEL.	Ens.	Majority	0.86	0.57	0.90	0.77	0.65	0.71	0.71	0.78	0.74	738
SEL.	Ens.	Weighted	0.87	0.58	0.87	0.74	0.63	0.69	0.71	0.77	0.73	738
EXPL.	Llama-3.1	Single	0.86	0.60	0.85	0.71	0.61	0.65	0.70	0.74	0.72	366
EXPL.	Ens.	Majority	0.87	0.61	0.88	0.73	0.62	0.69	0.73	0.77	0.74	829
EXPL.	Ens.	Weighted	0.87	0.61	0.86	0.70	0.61	0.66	0.71	0.75	0.72	829
CONF.	Llama-3.1	Single	0.84	0.61	0.86	0.75	0.65	0.67	0.71	0.75	0.73	198
CONF.	Ens.	Majority	0.87	0.60	0.89	0.77	0.66	0.70	0.76	0.80	0.76	752
CONF.	Ens.	Weighted	0.86	0.61	0.87	0.74	0.63	0.66	0.73	0.75	0.73	752
(b) Internal Voting Techniques												
CONF.	Ens.	Conf.	0.87	0.60	0.89	0.77	0.66	0.71	0.76	0.80	0.76	752
EXPL.	Ens.	Agg.Majority	0.90	0.63	0.90	0.74	0.64	0.75	0.78	0.80	0.77	818
(c) Internal Confidence Voting Techniques												
EXPL.	Ens.	Judge-Llama-3.1	0.82	0.58	0.87	0.75	0.64	0.65	0.65	0.75	0.71	107
EXPL.	Ens.	Judge-Qwen-2.5	**0.93**	0.62	**0.93**	**0.81**	**0.69**	0.69	0.50	**0.86**	0.75	443
EXPL.	Ens.	Hybrid-θ=0.6	**0.93**	**0.66**	0.91	0.77	0.67	**0.77**	**0.82**	0.85	**0.80**	818
(d) External Voting Techniques												

compared to 0.73. Weighted-vote performs similarly to the baseline, with the weight assigned to Llama-3.1 dominating the other two LLMs. Confidence-vote in Table 4(c) performs at least as good as the Majority-vote with CONFIDENCE prompts. In fact, the Aggregated Majority-vote achieves the highest average F1 score among all internal voting schemes at 0.77, exhibiting the best performance in most datasets, but requires significantly more model executions than Majority-vote.

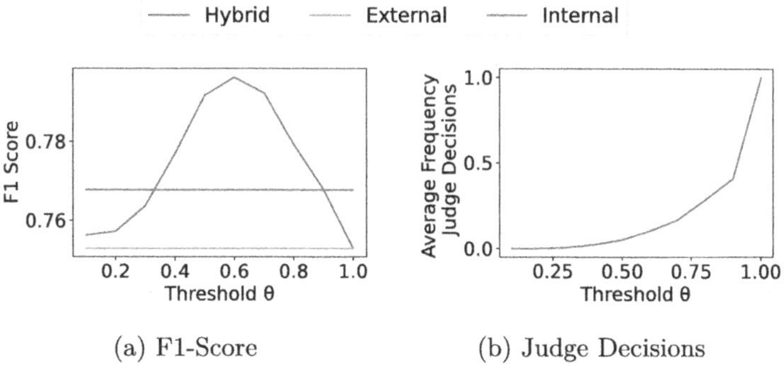

(a) F1-Score (b) Judge Decisions

Fig. 3. Hybrid Voting strategy performance with various thresholds θ.

Regarding the external voting techniques in Table 4(d), between the two models used as Judge, Qwen-2.5 outperforms Llama-3.1 on most datasets. This highlights the advantages of Qwen-2.5's larger size, enabling better contextual understanding and handling of complex relationships. On the downsize, Qwen-2.5 has high memory requirements, which forced us to restrict its input size in datasets with long entity descriptions. Hence its poor performance in D7.

Finally, the Hybrid-vote method balances internal and external strategies so well that it outperforms all other ensembles on average, ranking first or second in all datasets. Figure 3 depicts its fine-tuning with respect to θ. Its best performance corresponds to $\theta = 0.6$, where it utilizes Judge decisions only 12% of the time. This performance reflects the complementary nature of internal and external strategies and the flexibility of the Hybrid one in combining them.

Summary: Models can be complementary to each other, thus calling for an ensemble of LLMs. Using the hybrid voting strategy is both effective and efficient.

Table 5. Comparison on F1 for different individual and ensemble of fine-tuned models. PB Ref. stands for Post-Blocking Refinement, Rec. for Reciprocal Pruning, PT for Pre-Trained, FT for Fine-Tuned and RSize for Response Size.

Model	Type	Vote	PB Ref.	D1	D2	D3	D4	D5	D6	D7	D8	Mean	RSize
Llama-3.1	PT	Single	Rec.	0.86	0.60	0.85	0.71	0.61	0.65	0.70	0.74	0.72	366
Mistral-v0.3	PT	Single	Rec.	0.83	0.56	0.81	0.68	0.56	0.60	0.69	0.64	0.67	480
(a) Pre-Trained Models													
Llama-3.1	FT	Single	Rec.	<u>0.94</u>	0.66	<u>0.98</u>	<u>0.89</u>	**0.8**	0.90	<u>0.89</u>	0.95	<u>0.88</u>	115
Llama-3.1-ins.	FT	Single	Rec.	<u>0.94</u>	0.69	<u>0.98</u>	**0.90**	0.78	0.90	<u>0.89</u>	0.95	<u>0.88</u>	3
Mistral-v0.3	FT	Single	Rec.	<u>0.94</u>	0.69	<u>0.98</u>	**0.90**	0.78	**0.92**	0.88	0.95	<u>0.88</u>	3
Mistral-v0.3-ins.	FT	Single	Rec.	<u>0.94</u>	0.68	**0.99**	**0.90**	<u>0.79</u>	0.90	<u>0.89</u>	<u>0.96</u>	<u>0.88</u>	3
(b) Fine-Tuned Models													
Mistral-v0.3-ins.	FT	Single	Rank@10	<u>0.94</u>	0.70	<u>0.98</u>	**0.90**	0.77	0.90	<u>0.89</u>	<u>0.96</u>	<u>0.88</u>	3
(c) Fine-Tuned with no Post-Blocking Refinement													
Ensemble-FT	FT	Majority	Rec.	**0.95**	0.69	0.97	<u>0.89</u>	<u>0.79</u>	<u>0.91</u>	<u>0.89</u>	0.95	<u>0.88</u>	31
Ensemble-FT	FT	Any	Rec.	0.93	0.68	0.97	<u>0.89</u>	0.77	0.89	0.87	0.94	0.87	31
(d) Complementarity of Fine-Tuned models.													

Table 6. F1 per baseline and best Supervised and Unsupervised method. Rec. stands for Reciprocal Pruning, Hyb. for Hybrid, Ran for Ranking.

PB Ref.	Task	Model	Type	Vote	D1	D2	D3	D4	D5	D6	D7	D8	Mean
Rec.	EXPL.	Ensemble	PT	Hyb-θ=0.6	<u>0.93</u>	0.66	<u>0.91</u>	0.77	0.67	0.77	0.82	<u>0.85</u>	0.80
Rec.	EXPL.	Mistral-v0.3-Ins.	FT	Single	**0.94**	<u>0.68</u>	**0.99**	<u>0.90</u>	0.79	**0.90**	**0.89**	**0.96**	**0.88**
Hyb@4	SEL.	GPT-3.5	PT	Single	0.88	**0.70**	<u>0.91</u>	**0.97**	<u>0.84</u>	<u>0.85</u>	<u>0.86</u>	<u>0.85</u>	<u>0.86</u>
Ran@10	SEL	Mistral-v0.1-Ins.	PT	Single	0.67	0.57	0.83	0.74	**0.88**	0.77	0.81	0.63	0.74

Fine-Tuning Evaluation. In this experiment, we evaluate the performance of fine-tuned LLMs with that of their pre-trained version, examining how fine-tuning affects the complementarity of the models and their sensitivity to position bias. We fine-tune Llama-3.1 and Mistral-v0.3, excluding Orca, due to its low performance. For both models, we used the base and instruct variations.

The results are reported in Table 5. For Llama-3.1, the fine-tuned versions consistently outperform the pre-trained model by 24% on average (the average F1 score raises from 0.72 to 0.88). For Mistral, the improvement is even more significant, raising to 31%: while the pre-trained model averages 0.67, both fine-tuned variants reach 0.88, matching the F1 of the fine-tuned Llama-3.1 models. All four models show Precision=0.89 and Recall=0.87 on average.

Regarding the complementarity of the fine-tuned models, we assess it by comparing Any-vote with Majority-vote in Table 5(d). Unlike their pre-trained models, the two voting strategies achieve similarly high levels of Recall and F1, thus indicating low levels of complementarity.

Regarding the sensitivity to position bias of fine-tuned models, we compare Llama-3.1 (the best pre-trained LLM) with Mistral-v0.3-instruct (the best fine-tuned LLM) across 10 EXPLAIN prompt permutations per dataset, where the positions of the candidates were randomly shuffled. The results in Figs. 8 (a) and (b) show that the F1 score of the fine-tuned model exhibits significantly lower variance, on average across all datasets (0.0061 vs. 0.0116 for pre-trained). We executed the same experiment for the fine-tuned Llama-3.1 and it showed the same behavior, with an average variance of 0.0069. This indicates that fine-tuning allows for learning to handle prompt variations effectively. To further verify that fine-tuning mitigates position bias, we conducted an additional experiment using prompts with the original $k = 10$ Blocking choices, with results shown in Table 5(c). The fine-tuned model maintains consistent performance across datasets, unlike pre-trained models, as shown in Table 2. These findings confirm that fine-tuning reduces position bias, enabling prompts with more options without sacrificing accuracy.

Summary: A medium-sized fine-tuned model achieves up to a 31% higher F1 score compared to its pre-trained version, showing low complementarity to other fine-tuned models and lower sensitivity to position bias.

4.4 Comparison with Baselines

This experiment compares our methods to two established baselines from the literature, ComEM [18]. Specifically, we used the best of their methods, which was the Hybrid@4 Post-Blocking Strategy combined with a pre-trained GPT-3.5 and we also used the Ranking@10 with a pre-trained Mistral-v0.1-Instruct:7b, which is also comparable to our own. For our methods, for unsupervised settings we consider the Hybrid voting using Qwen-2.5 as an external Judge and the Aggregated-Majority as an internal voting strategy on the ensemble of pre-trained models. For the supervised settings, we consider the fine-tuned Mistral-v0.3-Instruct. The results are reported in Table 6.

Our unsupervised approach dominates the Mistral baseline almost always, with the exception of D5, but significantly underperforms the GPT baseline in all datasets. This can be explained by the vast difference between medium-size models and the large-size GPT-3.5, which can handle many tasks effectively without further training. However, our ensemble of models is more cost-effective, given that GPT-3.5 constitutes a proprietary model that is billable and requires a considerable budget to operate.

The situation is reversed under the supervised settings. The fine-tuned Mistral is superior to the baseline algorithm in most datasets, yielding an average F1 of 0.88. This demonstrates that in the case of annotated data, we can fine-tune an open-source model to even surpass a large-size model .

Summary: An ensemble of open-source medium-size models offers a cost-effective alternative to a proprietary large model in unsupervised settings, while fine-tuning a medium-size model can achieve better performance than a large pre-trained model.

5 Conclusions

We introduced *AvengER* to effectively handle both unsupervised and supervised settings. First, we demonstrated that reducing prompt options is crucial for select prompts and proposed Reciprocal Pruning as a means of significantly reducing the candidate pairs generated by Blocking. We showed that serialization with predicates is also crucial, and random examples should be avoided in In-Context Learning. Instead, advanced example selection techniques are required. We also examined ensembles of LLMs, showing that the individual models typically yield complementary results. Internal voting conveys minor improvements in effectiveness, while external voting is resource-intensive, due to the larger model that lies at its core. Our hybrid voting strategy nicely balances these two voting strategies, matching the external voting's performance by calling the Judge model for just 12% of the prompts. This approach suits low-resource settings. In applications with labelled data, Fine-Tuning is a stronger alternative, as it demonstrates robustness, effectively mitigates position bias, and enables medium-sized models to surpass large pre-trained ones.

Acknowledgements. This work was partially funded by the EU project STELAR (Horizon Europe – Grant No. 101070122).

Appendix

Fine-Tuning Configuration: For the training data described in Sect. 4.1, we used zero-shot SELECT prompts with schema-aware serialization and provided the answer in brackets. For each entity we created 3 such prompts, with the answers shuffled, to cover the issue of position bias. For training, we used LoRA with rank 16, $\alpha = 16$, and no dropout for efficient adaptation while managing memory constraints. Training was conducted for three epochs with a learning rate of $2{\cdot}10^{-4}$, gradient accumulation steps of 4, and the AdamW 8-bit optimizer. To reduce memory usage, we applied 4-bit quantization to non-instruction-tuned models.

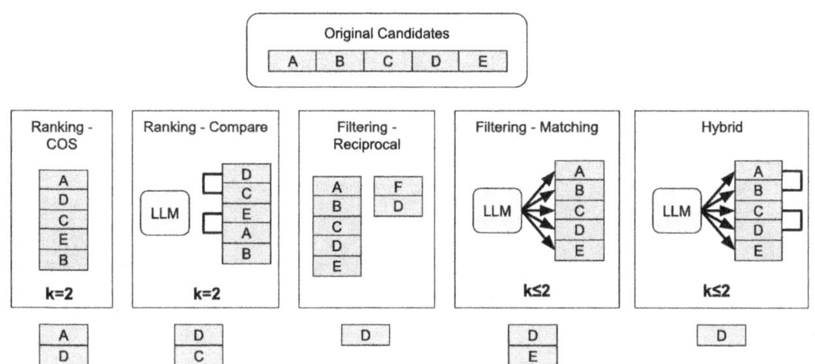

Fig. 4. Post-Blocking Strategies.

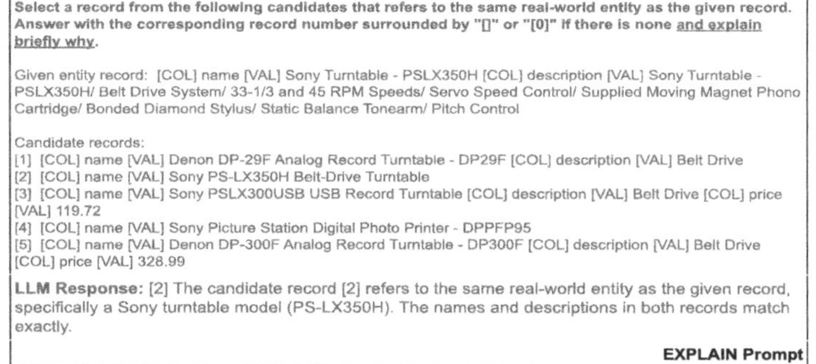

Select a record from the following candidates that refers to the same real-world entity as the given record. Answer with the corresponding record number surrounded by "[]" or "[0]" if there is none and explain briefly why.

Given entity record: [COL] name [VAL] Sony Turntable - PSLX350H [COL] description [VAL] Sony Turntable - PSLX350H/ Belt Drive System/ 33-1/3 and 45 RPM Speeds/ Servo Speed Control/ Supplied Moving Magnet Phono Cartridge/ Bonded Diamond Stylus/ Static Balance Tonearm/ Pitch Control

Candidate records:
[1] [COL] name [VAL] Denon DP-29F Analog Record Turntable - DP29F [COL] description [VAL] Belt Drive
[2] [COL] name [VAL] Sony PS-LX350H Belt-Drive Turntable
[3] [COL] name [VAL] Sony PSLX300USB USB Record Turntable [COL] description [VAL] Belt Drive [COL] price [VAL] 119.72
[4] [COL] name [VAL] Sony Picture Station Digital Photo Printer - DPPFP95
[5] [COL] name [VAL] Denon DP-300F Analog Record Turntable - DP300F [COL] description [VAL] Belt Drive [COL] price [VAL] 328.99

LLM Response: [2] The candidate record [2] refers to the same real-world entity as the given record, specifically a Sony turntable model (PS-LX350H). The names and descriptions in both records match exactly.

EXPLAIN Prompt

Fig. 5. Prompt for EXPLAIN.

Select a record from the following candidates that refers to the same real-world entity as the given record. Answer with the corresponding record number surrounded by "[]" or "[0]" if there is none. Accompany your answer by one of the following confidence measures: {Certain},{Moderately-certain}, {Uncertain}.

Given entity record: [COL] name [VAL] Sony Turntable - PSLX350H [COL] description [VAL] Sony Turntable - PSLX350H/ Belt Drive System/ 33-1/3 and 45 RPM Speeds/ Servo Speed Control/ Supplied Moving Magnet Phono Cartridge/ Bonded Diamond Stylus/ Static Balance Tonearm/ Pitch Control

Candidate records:
[1] [COL] name [VAL] Denon DP-29F Analog Record Turntable - DP29F [COL] description [VAL] Belt Drive
[2] [COL] name [VAL] Sony PS-LX350H Belt-Drive Turntable
[3] [COL] name [VAL] Sony PSLX300USB USB Record Turntable [COL] description [VAL] Belt Drive [COL] price [VAL] 119.72
[4] [COL] name [VAL] Sony Picture Station Digital Photo Printer - DPPFP95
[5] [COL] name [VAL] Denon DP-300F Analog Record Turntable - DP300F [COL] description [VAL] Belt Drive [COL] price [VAL] 328.99

LLM Response: [Certain] [2] The given entity record refers to a Sony turntable model, and candidate record [2] is the exact match for this real-world entity.

CONFIDENCE Prompt

Fig. 6. Prompt for CONFIDENCE.

Given this concatenated entity description:

[COL] name [VAL] Sony Turntable - PSLX350H [COL] description [VAL] Sony Turntable - PSLX350H/ Belt Drive System/ 33-1/3 and 45 RPM Speeds/ Servo Speed Control/ Supplied Moving Magnet Phono Cartridge/ Bonded Diamond Stylus/ Static Balance Tonearm/ Pitch Control

Generate a meaningful entity description in natural language. In the output give only the new description.

LLM Response: The Sony PSLX350H is a belt-driven turntable that supports 33-1/3 and 45 RPM playback, featuring servo speed control and a built-in moving magnet phono cartridge with bonded diamond stylus, as well as a static balance tonearm and pitch control for precise audio reproduction.

Summary Prompt

Fig. 7. Prompts for creating a Summary for dataset description.

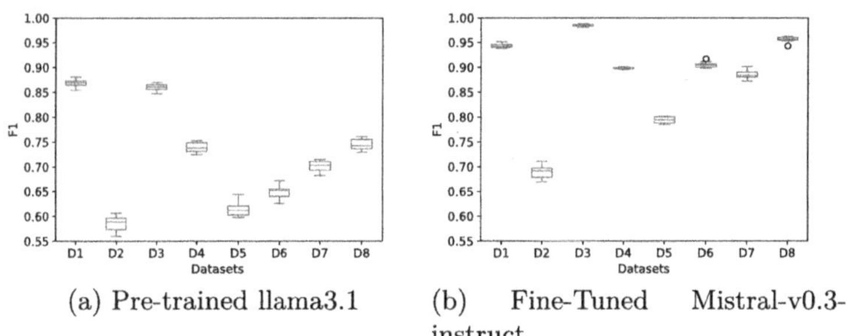

(a) Pre-trained llama3.1 (b) Fine-Tuned Mistral-v0.3-instruct

Fig. 8. Measuring variance in performance for permutated prompts between the best of pre-trained and fine-tuned models

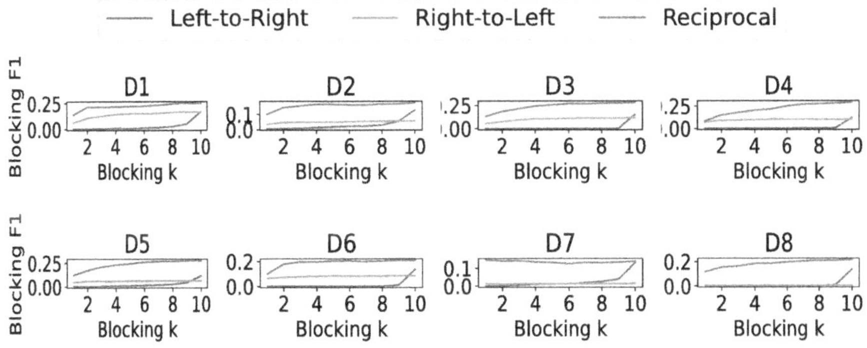

Fig. 9. Comparing Post-Blocking Refinement strategies for different values of k.

References

1. Brinkmann, A., Shraga, R., Bizer, C.: Sc-block: Supervised contrastive blocking within entity resolution pipelines. In: ESWC (1). Lecture Notes in Computer Science, vol. 14664, pp. 121–142. Springer, Heidelberg (2024)
2. Christophides, V., Efthymiou, V., Palpanas, T., Papadakis, G., Stefanidis, K.: An overview of end-to-end entity resolution for big data. ACM CSUR **53**(6), 127:1–127:42 (2021)
3. Christophides, V., Efthymiou, V., Stefanidis, K.: Entity Resolution in the Web of Data. Morgan & Claypool (2015)
4. Dong, X.L., Srivastava, D.: Big data integration. PVLDB **6**(11), 1188–1189 (2013)
5. Fan, M., et al.: Cost-effective in-context learning for entity resolution: a design space exploration. In: ICDE, pp. 3696–3709. IEEE (2024)
6. Li, Y., Li, J., Suhara, Y., Doan, A., Tan, W.: Deep entity matching with pre-trained language models. Proc. VLDB Endow. **14**(1), 50–60 (2020)
7. Mitra, A., et al.: Orca 2: teaching small language models how to reason. CoRR arxiv:2311.11045 (2023)
8. Mudgal, S., et al.: Deep learning for entity matching: a design space exploration. In: SIGMOD Conference, pp. 19–34. ACM (2018)
9. Neuhof, F., et al.: Open benchmark for filtering techniques in entity resolution. VLDB J. **33**(5), 1671–1696 (2024)
10. Ni, J., et al.: Large dual encoders are generalizable retrievers. In: EMNLP, pp. 9844–9855. Association for Computational Linguistics (2022)
11. Paulsen, D., Govind, Y., Doan, A.: Sparkly: a simple yet surprisingly strong TF/IDF blocker for entity matching. Proc. VLDB Endow. **16**(6), 1507–1519 (2023)
12. Peeters, R., Bizer, C.: Entity matching using large language models. CoRR arxiv:2310.11244 (2023)
13. Raffel, C., et al.: Exploring the limits of transfer learning with a unified text-to-text transformer. J. Mach. Learn. Res. **21**, 140:1–140:67 (2020)
14. Reimers, N., Gurevych, I.: Sentence-bert: sentence embeddings using siamese bert-networks. In: EMNLP/IJCNLP (1), pp. 3980–3990. Association for Computational Linguistics (2019)
15. Steiner, A., Peeters, R., Bizer, C.: Fine-tuning large language models for entity matching. CoRR arxiv:2409.08185 (2024)

16. Tu, J., et al.: Unicorn: a unified multi-tasking model for supporting matching tasks in data integration. Proc. ACM Manag. Data **1**(1), 84:1–84:26 (2023)
17. Wadhwa, S., Krishnan, A., Wang, R., Wallace, B.C., Kong, C.: Learning from natural language explanations for generalizable entity matching. CoRR arxiv:2406.09330 (2024)
18. Wang, T., et al.: Match, compare, or select? An investigation of large language models for entity matching. CoRR arxiv:2405.16884 (2024)
19. Xia, Y., Chen, J., Li, X., Gao, J.: Aprompt4em: augmented prompt tuning for generalized entity matching. CoRR arxiv:2405.04820 (2024)
20. Xiong, M., et al.: Can llms express their uncertainty? An empirical evaluation of confidence elicitation in llms. In: ICLR. OpenReview.net (2024)
21. Zeakis, A., Papadakis, G., Skoutas, D., Koubarakis, M.: Pre-trained embeddings for entity resolution: an experimental analysis. Proc. VLDB Endow. **16**(9), 2225–2238 (2023)
22. Zhang, Z., Groth, P., Calixto, I., Schelter, S.: Anymatch - efficient zero-shot entity matching with a small language model. CoRR arxiv:2409.04073 (2024)
23. Zheng, L., et al.: Judging llm-as-a-judge with mt-bench and chatbot arena. In: NeurIPS (2023)

Ontology Generation Using Large Language Models

Anna Sofia Lippolis[1,2(✉)], Mohammad Javad Saeedizade[3], Robin Keskisärkkä[3], Sara Zuppiroli[2], Miguel Ceriani[2], Aldo Gangemi[1,2], Eva Blomqvist[3], and Andrea Giovanni Nuzzolese[2]

[1] University of Bologna, Bologna, Italy
{annasofia.lippolis2,aldo.gangemi}@unibo.it
[2] ISTC-CNR, Rome, Italy
sara.zuppiroli@istc.cnr.it, {miguel.ceriani,andrea.nuzzolese}@cnr.it
[3] Linköping University, Linköping, Sweden
{javad.saeedizade,robin.keskisarkka,eva.blomqvist}@liu.se

Abstract. The ontology engineering process is complex, time-consuming, and error-prone, even for experienced ontology engineers. In this work, we investigate the potential of Large Language Models (LLMs) to provide effective OWL ontology drafts directly from ontological requirements described using user stories and competency questions. Our main contribution is the presentation and evaluation of two new prompting techniques for automated ontology development: Memoryless CQbyCQ and Ontogenia. We also emphasize the importance of three structural criteria for ontology assessment, alongside expert qualitative evaluation, highlighting the need for a multi-dimensional evaluation in order to capture the quality and usability of the generated ontologies. Our experiments, conducted on a benchmark dataset of ten ontologies with 100 distinct Competency Questions (CQs) and 29 different user stories, compare the performance of three LLMs using the two prompting techniques. The results demonstrate improvements over the current state-of-the-art in LLM-supported ontology engineering. More specifically, the model `OpenAI o1-preview` with Ontogenia produces ontologies of sufficient quality to meet the requirements of ontology engineers, significantly outperforming novice ontology engineers in modelling ability. However, we still note some common mistakes and variability of result quality, which is important to take into account when using LLMs for ontology authoring support. We discuss these limitations and propose directions for future research.

Keywords: Ontology · Large Language Models · Ontology Engineering

1 Introduction

Ontologies play an important role in the success of Knowledge Graphs (KGs) [21], as a crucial component in the recent advancements of explainable AI

A. S. Lippolis and M. J. Saeedizade—Equal contribution.

E. Curry et al. (Eds.): ESWC 2025, LNCS 15718, pp. 321–341, 2025.
https://doi.org/10.1007/978-3-031-94575-5_18

and neuro-symbolic integration [17,18]. Today, ontologies are extensively used to facilitate semantic interoperability, e.g., describing datasets and standardised terminologies. However, ontology engineering (OE) is a complex task that requires skills in knowledge representation, logic, and computational linguistics. So far, many OE methodologies have emerged in the scientific literature to provide structured frameworks that assist ontology engineers in navigating the complexities of knowledge modeling. Examples are METHONTOLOGY [14], the NeOn methodology [36], eXtreme Design (XD) [7], and more recently the Linked Open Terms [30]. In any case, OE requires access to significant domain expertise combined with knowledge engineering and modelling skills, posing a significant barrier to entry for many professionals. Moreover, even when expertise is available, the creation, curation, and validation of ontological elements are complex tasks, which are cognitively costly and mostly manual. Instead, Large Language Models (LLMs) have proven to be able to assist humans in a variety of tasks, ranging from programming co-pilots to data cleaning and statistical analysis. These advancements highlight the growing need to benchmark common semantic web tasks from the perspective of LLMs, transforming this effort from an ideal into an essential requirement and emphasizing the critical importance of research in this area [1,3,28,32,33,38]. Furthermore, prompting techniques, defined as an approach to carefully formulating input prompts, can be leveraged to elicit the pre-trained knowledge of LLMs to perform specific tasks without additional training or fine-tuning. Advanced strategies such as decomposed prompting, where the task is split into several pieces, have been shown to enhance LLM performance across a wide range of tasks [23]. In this work, we assume that LLMs can also be beneficial for ontology design by reducing manual labour for experienced ontology engineers, as well as assisting novice ontology engineers.

Accordingly, the research questions driving this work are: (i) To what extent can LLMs be used to support the generation of ontologies that meet a predefined set of requirements? (ii) What evaluation criteria are suitable for evaluating LLM-generated ontologies? (iii) What are the strengths and weaknesses of ontologies generated using LLMs? To answer these questions, we provide a framework to assess LLM-pipelines with prompting techniques focused on OE, as well as the LLM-generated minimal ontology modules themselves. In particular, expanding previous research [34], we leverage a dataset of ontology requirements and a set of reference minimal ontology modules to evaluate two prompting techniques for supporting OE: Memoryless CQbyCQ and Ontogenia[1]. We further evaluate the performance of these prompting techniques, using state-of-the-art LLMs, and propose a set of evaluation criteria for comparing and evaluating different pipeline setups.

The paper is organized as follows: Sect. 2 reviews relevant literature; Sect. 3 clarifies the terminology used in this paper; Sect. 4 details our research methodology; Sect. 5 outlines the evaluation criteria used to assess our approach; Sect. 6

[1] Supplementary material such as datasets, prompts and code is available at https://github.com/dersuchendee/Onto-Generation.

presents our findings with detailed analysis; Sect. 7 discusses the implications of the results. Finally, Sect. 8 summarizes the main findings.

2 Related Work

This section presents related work about ontology engineering methods, ontology generation methods, and prompting techniques for ontology generation.

Ontology Engineering Methods. Several methodologies have been developed for ontology engineering, e.g., METHONTOLOGY [14] and NeOn [37]. Agile methodologies focusing on the re-use of Ontology Design Patterns [8,16], like those surveyed in [6,27,35], have become increasingly popular, reflecting real-world needs by maximizing the cognitive soundness, logical correctness, and effectiveness of ontological artefacts. The LOT methodology [30] was presented as a compilation of experiences from many projects, methodologies, and tools. While some methods have explicit tool support, e.g., NeOn [37], most are still entirely manual. Ontology engineers often must match the modelling problem to requirements, which is generally complex and time-consuming.

Ontology Generation with LLMs. With the advent of LLMs, much research has been focused on the possibility of generating ontologies, or their specific elements, from natural language. The increased interest in this area is exemplified by the announcement of several special tracks in the semantic web conferences dedicated to these issues[2]. Despite LLMs' well-known drawbacks, such as hallucinations, studies show they have the potential to perform efficiently in numerous tasks, with only a handful of examples and a well-designed prompt [2,34]. In [19] the authors survey the tasks currently addressed by approaches for LLM-supported ontology engineering, and while some approaches focusing on the actual tasks of generating the OWL files themselves have been proposed, this is still an area that needs much further investigation.The LLMs4OL approach [4] used LLMs to extract relations among ontology classes or instances, but only among entities, not addressing the complete generation of an ontology. Other preliminary work [26] has used fine-tuned GPT models to translate restricted natural language sentences into DL axioms. However, such specific sentences do not represent realistic ontology requirements or scenarios, as targeted in our work. Furthermore, tools are appearing to support the practical integration of OWL with LLMs, e.g. [20], while not providing any specific guidance for the ontology generation task. One recent work similar to our proposed approach is the NeOn-GPT pipeline proposed by Fathallah et al. [12]. However, that work only evaluates one single complex prompting technique, and evaluates the pipeline only on one single ontology generation example. In LLMs4Life

[2] E.g. see ESWC https://2024.eswc-conferences.org/, EKAW https://event.cwi.nl/ekaw2024/cfp.html and ISWC https://iswc2024.semanticweb.org/.

[13], the NeOn-GPT approach is extended to the life sciences domain. However, implementation details remain unclear, especially for what concerns the evaluation criteria. In Lippolis et al. [24] and Saeeizade and Blomqvist [34], different methods and ontology evaluation approaches for ontology generation from requirements have been tested, obtaining results which are at least comparable to the quality of human novice modellers: these two works are the starting point of our current investigation. Our multi-dimensional assessment implements the proposal of Rebboud et al., [32], where ontology conceptualization is proposed to be evaluated according to ontology evaluation criteria such as accuracy, completeness, and conciseness of the generated ontology as well as logical consistency. Furthermore, the authors propose the only existing benchmark for this task. However, it comprises entire ontologies rather than minimal ontology modules, making it harder to evaluate the outputs LLMs at a wider granularity, and often lacks ontology stories.

Prompting Techniques for Ontology Generation. Prompt engineering stands out for its simplicity in implementation and adaptability, in contrast to finetuning, offering an approach for enhancing knowledge engineering processes while avoiding the need for large labelled datasets and dedicated finetuned models. Requiring modest effort to implement, it provides substantial flexibility for updates and modifications. In Lippolis et al. [24] and Saeeizade and Blomqvist [34], the authors have explored and evaluated various prompting techniques and identified several techniques that appear suitable for different aspects of ontology generation. Our research on the one hand explores subtask-decomposed prompting, i.e., in this case the promising CQbyCQ method from Saeedizade and Blomqvist [34], as well another prompting technique based on Chain Of Thought (CoT) [41]. CoT is known for the specific ability to successfully improve the performance of different tasks and decreasing hallucinations [11]. A particular case of CoT is Metacognitive Prompting (MP) [24], inspired by human introspective processes. It encourages self-evaluation through the introduction of a series of steps, improving performance over CoT. Starting from the studies of Brown et al. [10], Wei et al. [41], and Wang et al. [39,40], which demonstrated significant improvements in LLM response abilities when using CoT or, more effectively, MP, our study examines the performance of various prompting strategies in ontology generation, comparing CoT with subtask-decomposed prompting approaches.

3 Preliminaries

In this section, we clarify and define terminology that will be used throughout this paper. These definitions are essential for understanding the concepts discussed in subsequent sections and provide details on the specific ways in which we interpret and employ these notions within the context of our research.

Ontology. In this work, an ontology O is defined as a set of classes, object properties, data properties and axioms.

Validation Competency Question. As defined by Keet and Khan [22], a CQ is classified as a validation type if it ensures that an ontology accurately represents the domain it is meant to model. Specifically, it checks that the ontology's content aligns with the intended meaning and representation of the domain. For simplicity, we will refer to this as a CQ.

Modelled Competency Question. Given an ontology O and a validation competency question CQ_i, if O includes all necessary elements—such as classes or properties—needed to write a validation SPARQL query to verify CQ_i (c.f. CQ verification [9]), we say that CQ_i is modelled in O regardless of the quality of the modelling, i.e., whether it follows good or bad modelling practices.

Superfluous Element. Given an ontology O, a set of competency questions and the set of corresponding validation SPARQL queries , if a named class, object property or data property is not used in any SPARQL query, it is considered a superfluous element of O. An example is in Appendix A.1.

Minimal Ontology Module. Given an ontology O and a competency question CQ_i, O_i is a minimal ontology module for CQ_i if there are no superfluous elements of O with respect to CQ_i.

Minor Issue. For an ontology O and a competency question CQ_i, if O includes all necessary elements except for only one object property or only one data property, and adding this single element to O would make CQ_i modelled, this is considered a minor issue in modelling CQ_i. This definition is based on the intuition that one such missing link is rather easy to add, even by a novice modeller, given that all classes are present.

4 Methodology

The initial phase of our work involved manually creating a benchmark dataset using available ontologies accompanied by their corresponding requirements for CQ verification [9]. We then introduced two methods for ontology generation, namely Independent Ontology Generation and Incremental Ontology Generation, and adapted existing prompting techniques in our experiments to guide LLMs in generating ontologies. In this first part of the study, we primarily employ *GPT-4*, identified as the best-performing LLM in earlier comparative analyses [15,34]. Additionally, we independently compare *GPT-4* with *OpenAI o1-preview* and *Llama-3.1-405B-instruct-16b* using the dataset proposed in Saeedizade and Blomqvist [34].

4.1 Dataset Creation

In order to evaluate the generated ontologies using CQ verification [9], a dataset of CQs and user stories was developed, along with their corresponding minimal ontology modules. The process involved selecting a set of ontologies for the study, extracting CQs and stories, and subsequently extracting modules from these chosen ontologies in order to have minimal modules for easier evaluation and possible future extensions of the work. The dataset combines two main types of sources: extracted datasets and manually created datasets. The extracted datasets consist of CQs and stories derived from existing resources, where superfluous elements were removed to identify minimal modules. In contrast, the manually created datasets were created as a part of a controlled module engineering process. In this case, we include the dataset present in [34] called "SemanticWebCourse", where master's students with a computer science background but no prior ontology design experience were tasked with creating semantic web solutions. Each student group submitted an initial solution, revised it based on teacher feedback, and resubmitted a final version to pass the assignment, resulting in two distinct solutions per group. However, as previous work questioned the generalisability of the LLMs generating ontologies [34], in this work, we enriched existing data with more ontology sources, primarily real-world ones.

An ontology was included in the dataset according to the following criteria: The ontology includes (i) a set of competency questions and (ii) a set of corresponding user stories. These criteria were motivated by the eXtreme Design methodology [31], which takes into consideration CQs and user stories as fundamental requirements of ontologies. As a result, we selected a total of ten ontologies, with 100 distinct CQs, and 29 different user stories from four real-life semantic web projects and three educational ones. More specifically, these ontologies have CQs assigned to specific minimal ontology modules with different user story distributions.

Minimal Ontology Module Partitioning. The main goal of developing the dataset was to support the generation of ontologies using LLMs. The CQs and user stories provide input for prompting techniques in the ontology generation process. The minimal ontology modules contained in the datasets provide ontologists with a gold standard for the expected output of the LLMs and are used to assist the ontologists' assessment, potentially also useful to be used in future work to fine-tune LLMs. To create the minimal ontology modules for each CQ the following steps were used: First, we manually checked to see if there were duplicate CQs and removed them. Then, from the ontologies, we removed superfluous elements so each module contains the minimum necessary classes and properties to effectively model the CQ concerning the ontology story. Finally, these modules were cross-checked and evaluated independently by two of the authors of this study.

Dataset Composition. The dataset is composed of **simple** and **complex** CQs. These have been divided into four categories as defined in Saeedizade and

Blomqvist [34]: Data Property Modelling (10 CQs) and Object Property Modelling (25 CQs) are Simple CQs, while Reification (62 CQs) and Restrictions (3 CQs) are Complex CQs. The CQs, stories and corresponding ontologies are taken from Polifonia[3], Onto-DESIDE[4], WHOW [25], IKS[5] and the "SemanticWebCourse' used in Saeedizade and Blomqvist [34] consisting of three ontology stories and 15 CQs each. As an additional baseline for the experiments in this paper we used the same set of student solutions of the "SemanticWebCourse" as in [34]. In Saeedizade and Blomqvist, [34], the baseline for evaluation was the first submission and the last submissions of student groups, and because it is the current state of the art in ontology generation, we use the same baseline in this paper for comparison. The intuition behind using this baseline is to compare the modelling abilities of LLMs to novice ontology engineers, with and without expert feedback[6].

4.2 Prompting Techniques

In the study by Saeedizade and Blomqvist [34], the performance of CQbyCQ, a technique based on sub-task decomposition prompting [23], was found to be similar to that of students in ontology development. In this prompting technique, an LLM models only one CQ at a time, and the output is merged with the previously generated ontology. Due to its success, we incorporated two variations of this prompt to design our prompting techniques. Furthermore, in the study of Lippolis et al. [24], Metacognitive Prompting has proven an effective technique, especially when Ontology Design Patterns were provided, to generate richer pattern-based ontology formalizations. These previous techniques were therefore refined and evaluated on the benchmark dataset.

Memoryless CQbyCQ. Memoryless CQbyCQ processes one CQ at a time, utilizing the ontology story to guide the LLM with a prompt in generating an ontology model for each CQ Sect. 1. It then merges the outputs into a single ontology. Memoryless CQbyCQ is a variation of CQbyCQ [34] that does not provide the LLM with the current state of ontology development. Unlike the CQbyCQ method, the LLM does not have access to previously generated ontologies and other CQs while modelling a specific CQ, which reduces the input context size of the LLM by ~60%. In fact, in Saeedizade and Blomqvist [34] it was shown that long context can result in distraction of the LLM. Hence, the intuition for removing the memory part is that slight overlaps between partial solutions will be easier to resolve and correct by even a novice ontology engineer, than completely irrelevant or inconsistent solutions, which could be the result of distracting the model. As a result, each CQ is independently modelled, and then all

[3] https://polifonia-project.eu/.

[4] https://ontodeside.eu.

[5] https://cordis.europa.eu/project/id/231527.

[6] We did not compare to the original Ontogenia paper [24] as, given its noted shortcomings, the technique has been revised and used for the decomposed prompting technique.

the resulting models, representing CQs, are merged at the end. This prompting technique guides an LLM, gives it an ontologist persona, and introduces Turtle syntax for defining classes, properties, reifications, and other ontology engineering features. It also includes a story section that outlines ontology requirements and a CQ, followed by common pitfalls in ontology development using LLMs, such as producing an empty output or engaging in conversation.

Ontogenia Technique. Ontogenia was first defined in [24] and has been further refined for this work. Like the CQbyCQ method, this prompting technique guides an LLM by instructing it to be an ontology engineer, and defines basic guidelines for effective ontology formalization. The model processes one CQ at a time. When prompted to model a set of CQs, it models and merges the generated ontology at each step and provides it in the context. The inputs for the prompt are: a user story, which was not available in the previous work, a set of ODPs and possibly previous output. The main idea was to transpose the Metacognitive Prompting as described in five steps in Wang et al. [40] with the XD methodology [7], which required (i) the use of pre-selected CQs and user stories; (ii) selection, reuse, and integration of selected Ontology Design Patterns[7]; and (iii) iterative re-evaluation to verify coverage of initial requirements. In the initial stage, the LLM interprets and contextualizes requirements, identifying the context and breaking down the CQs into logical elements to support the systematic identification of classes and properties. The next phase involves reflecting on the CQs to extend the ontology by incorporating relevant rules and restrictions. Once the ontology is developed, the decision confirmation stage ensures the final output is validated with a clear explanation of the reasoning process. Finally, the LLM evaluates the ontology's reasoning, creating test cases with instances to validate the correctness of the generated ontology. The resulting procedure, mapped to the five MP steps is shown in the Github repository (see footnote 1), along with more details about this technique.

Similarities and Differences Between Memoryless CQbyCQ and Ontogenia. Although the methods are similar in nature, and use the same testing setup to ensure comparability, there are also some notable differences. Ontogenia explicitly requests to provide labels, comments, inverse relationships and individuals. Furthermore, Ontogenia requires the injection of Ontology Design Patterns from the ODP repository [8] and makes use of the Metacognitive Prompting technique. Unlike Ontogenia, Memoryless CQbyCQ includes common pitfalls to ensure they are avoided in the ontology output and attempts to reduce the context size. Figure 1 shows an overview of these two prompting techniques.

[7] http://ontologydesignpatterns.org/.

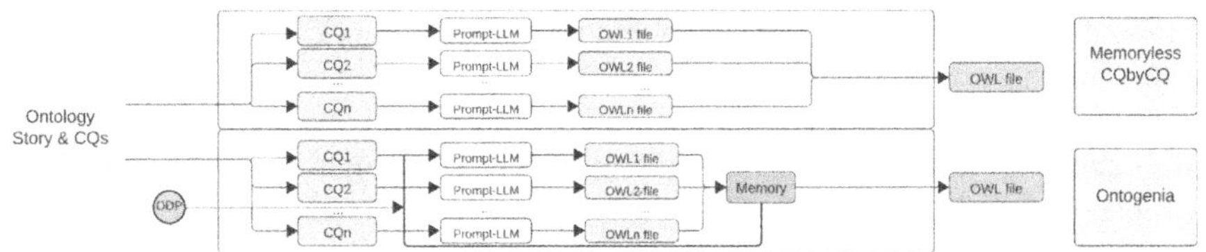

Fig. 1. Illustration of Memoryless CQbyCQ (top part) and Ontogenia (bottom part).

4.3 Ontology Generation Methods

Here, we present the two ontology generation methods used in our experiments, as shown in Fig. 5: Independent and Incremental Ontology Generation.

Independent Ontology Generation. In this method, each CQ with its corresponding ontology story is fed into an LLM using a prompting technique to generate the corresponding ontology. This ensures that each CQ is treated as a standalone unit, allowing for a focused assessment of the ontology generation process on a per-question basis. By isolating each CQ, it becomes possible to analyze the performance of the prompting techniques without interference from interdependencies or complexities that may arise in multi-CQ contexts.

Incremental Ontology Generation. Here, similar to CQbyCQ [34], all CQs of a story are fed to an LLM at once, and a single OWL ontology is expected as output. Unlike the independent approach, where each CQ is modelled in isolation, this technique integrates solutions to multiple CQs to produce a cohesive minimal ontology module representing the complete story and its requirements. This method can be carried out either by merging the outputs for each CQ into a single ontology (Memoryless CQbyCQ) or by incorporating the output of each CQ incrementally into the prompt and joining it at each iteration (Ontogenia).

5 Evaluation Setup

Experimental Setup. The experiment consisted in running the Independent Ontology Generation method with *GPT-4-1106* with both prompting techniques across the entire dataset. For the Incremental Ontology Generation method, we used both MemorylessCQbyCQ and Ontogenia with three different LLMs (*GPT-4 1106*, *o1-preview*, and *Llama-3.1-405-Instruct-16b*) solely on the dataset introduced in Saeedizade and Blomqvist [34] of three semantic web course stories and 45 CQs. As a baseline for comparison we used their results [34], i.e. their best solutions produced using *GPT-4* with CQbyCQ, and the recorded scores of the students' submissions. In order to provide a comprehensive evaluation, each of the solutions produced through these methods for the Independent and Incremental Ontology Generation experiments has been evaluated by the proportion of modelled CQs, with two ontology engineers cross-checking each other's judgments. In the event of any conflicting assessment, they engaged in discussions to resolve the disagreement. In the case of the Incremental Ontology Evaluation experiment, also standard ontology metrics through the OntOlogy Pitfall Scanner (OOPS!) [29], as well as a structural analysis to evaluate the rate of superfluous elements and a qualitative expert evaluation, were carried out. Due to lack of resources and time, we were not able to perform this evaluation on the complete result set. For the LLMs, we used the default hyperparameters of *o1-preview* and *Llama-3.1-405-Instruct-16b*; for *GPT-4*, temperature and penalty are set to zero as recommended in Saeedizade and Blomqvist [34]. Figure 2 illustrates the three evaluation steps.

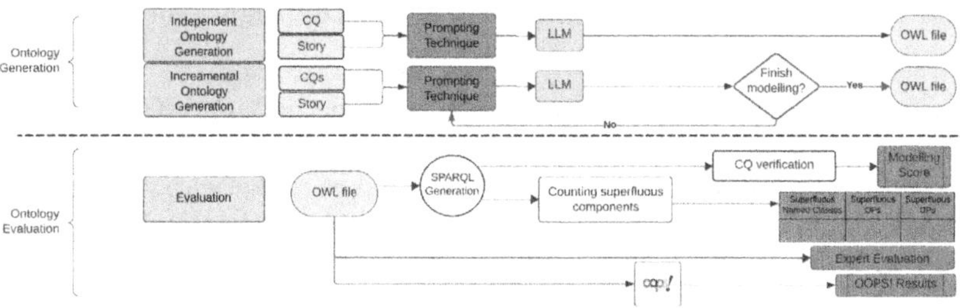

Fig. 2. Illustration of the two ontology generation settings (top) and the four evaluation steps for assessing the generated ontologies (bottom). The top setup generates an ontology concerning only one CQ, which is evaluated individually. The second setup generates an ontology covering multiple CQs associated with one story, which is then evaluated. At the bottom, the four ontology evaluation settings are shown: OOPS!, the proportion of modelled CQs, statistics of superfluous elements and expert evaluation.

Standard Ontology Metrics. The first step in evaluating the generated ontologies is applying standard evaluation metrics. We chose the OntOlogy Pitfall Scanner (OOPS!) [29], due to its coverage of common modelling mistakes and best practices. In the results, we report only the number of critical issues related to each method, as they are the only ones crucial to correct, while other pitfalls are not certain to actually represent modelling flaws in the domain [29].

Proportion of Modelled CQs. In this step, for each CQ, we evaluate whether it is modelled in accordance with the definition provided in Sect. 3. Additionally, we adopt a relaxed interpretation of this criterion by disregarding minor issues, as outlined in Sect. 3. Finally, the proportion of CQs deemed to be modelled, both with and without accounting for minor issues, is used to compute an overall score representing the proportion of the total list of CQs being sufficiently covered.

Structural Analysis. We manually evaluated each OWL file using three novel criteria with respect to superfluous elements Sect. 3, i.e., superfluous named classes, object properties, and data properties. We report the rate of superfluous elements by dividing the number of superfluous elements by the total count of that element type, and compared them to the rates in Saeedizade and Blomqvist [34].

Expert Qualitative Analysis. The expert qualitative analysis was conducted by two experienced knowledge engineers, not involved in the practical ontology generation task, who carefully and independently evaluated the outputs generated by the best performing open- and closed-source LLMs, *Llama-3.1-405-Instruct-16b*, and *OpenAI o1-preview*. The experts were instructed to assess the LLM-generated ontology files as if they were students' course submissions and to provide feedback accordingly. Therefore, they focused on: (i) assessing the general quality of the output with dedicated comments on usability, completeness, and accuracy of the produced output, and mentioning any ontology errors found; and (ii) assessing the ontology with respect to the adequacy of the modelling solutions for each CQ[8], with a substantial agreement (Cohen's kappa: 0.61).

6 Results

In this section, we start by examining the results of the OOPS! resource on the incrementally generated ontologies. Then, we compare the two prompting techniques introduced in this paper by going in-depth into the modelling results with respect to the assessment of the proposed ontology generation methods. More specifically, we measure the proportion of modelled CQs for all experiments and analyze the structure of the generated ontologies concerning superfluous elements. For the Incremental Ontology Generation method, we focus on the "SemanticWebCourse" dataset since it contains 15 CQs associated with one story while others have ~2 on average and choose the best performing open- and closed-source LLMs to compare across both previous work [34] and the results of Independent CQ Generation. In this way, we can test the LLMs' capability to generate an ontology incrementally. Finally, we present the results of the assessment by the two KE experts who analyzed the outputs.

Standard Ontology Metrics. After running OOPS! on the ontologies generated by the Incremental method for the "SemanticWebCourse" stories (Hospital, Music, and Theatre), the resulting pitfalls are shown in Table 1. Overall, there are a few critical pitfalls for specific combinations of LLMs and prompting techniques. The most common flaw is having multiple domains or ranges, which is in line with our other results discussed further in the following sections (see Sect. 7). In this case, the domain or range (or both) of a property (object or data property) is defined by more than one `rdfs:domain` or `rdfs:range` axiom. In OWL,

[8] The "not adequate" assessment combines two judgments by the KE experts, i.e. a clear "no" where the CQ is not modelled, and a "maybe" category where the ontology simply does not allow for accurate assessment of the CQ, for instance, due to usability issues, naming etc.

multiple `rdfs:domain` or `rdfs:range` axioms are allowed, but are interpreted as a conjunction. Therefore, they are equivalent to the construct `owl:intersectionOf`, which could generate many unwanted inferences or even inconsistencies in the ontology. Some issues with inverse relations are also noted, as well as missing declarations of namespaces. Lastly, *OpenAI o1-preview* produced the fewest critical issues when using Memoryless CQbyCQ, whereas *Llama-3.1-405-instruct-16b* generated, with both methods, files with the most critical pitfalls.

Table 1. Pitfalls after OOPS! pitfall scanning for Incremental Ontology Generation. Results are merged by the three stories in the "SemanticWebCourse". Llama* refers to *Llama-3.1-405-instruct-16b*. For the CQbyCQ method, only GPT-4 was used [34].

		CQbyCQ	MemorylessCQbyCQ			Ontogenia		
		GPT-4	GPT-4	Llama*	o1	GPT-4	Llama*	o1
P05	Wrong inverse relationships	**0**	1	25	**0**	2	7	5
P06	Cycles in a class hierarchy	**0**	**0**	2	**0**	5	**0**	11
P19	Multiple domains or ranges	**0**	23	32	1	4	15	**0**
P29	Wrong transitive relationship	**0**	**0**	**0**	1	**0**	**0**	**0**
P37	Ontology not available	3	2	**0**	**0**	**0**	**0**	**0**
P39	Ambiguous namespace	3	2	**0**	1	1	**0**	**0**

Proportion of Modelled CQs *Independent Ontology Generation Results:* The evaluation of the Memoryless CQbyCQ and Ontogenia techniques on *GPT-4* shows that the proportion of correctly modelled CQs was 0.91 and 0.84, respectively (0.94 and 0.89 by ignoring minor issues), with significantly lower scores for complex CQs (0 and 0.66 respectively). These results show that by overlooking minor issues both prompting techniques are effective for modelling single CQs in isolation.

Incremental Ontology Generation Results: The results of the Incremental Ontology Generation are presented in Fig. 3. Ontogenia with *OpenAI o1-preview* is the best-performing combination with respect to this evaluation criterion. Compared to the students' submissions for the "SemanticWebCourse", Ontogenia and Memoryless CQbyCQ with any LLM exceeded the proportion of CQs being modelled by students by a noticeable margin. Also, Memoryless CQbyCQ performed better than the CQbyCQ technique in Saeedizade and Blomqvist [34], possibly suggesting that using partially generated ontology models as data "in memory" for LLMs, which includes previously generated responses to CQ modelling, actually may diminish their performance. This result highlights the importance of reducing the input context size for LLMs, which is consistent with the findings in Saeedizade and Blomqvist [34]. The proportion of correctly modelled CQs is analyzed across four categories: datatype property (DP), object property (OP), reification (Reif.), and restrictions (Rest.) for the Theatre, Music, and Hospital Story. We compared *GPT-4*, *Llama-3.1-405B-instruct-bf16* and *OpenAI o1-preview*. The evaluation shows that Ontogenia with *OpenAI o1-preview* achieves the highest number of correctly modelled CQs. Overall, both Ontogenia and Memoryless CQbyCQ surpass the students' last submissions and the state of the art in ontology generation [34]. Moreover, *Llama-3.1* in both prompting techniques had shortcomings in modelling reification CQs.

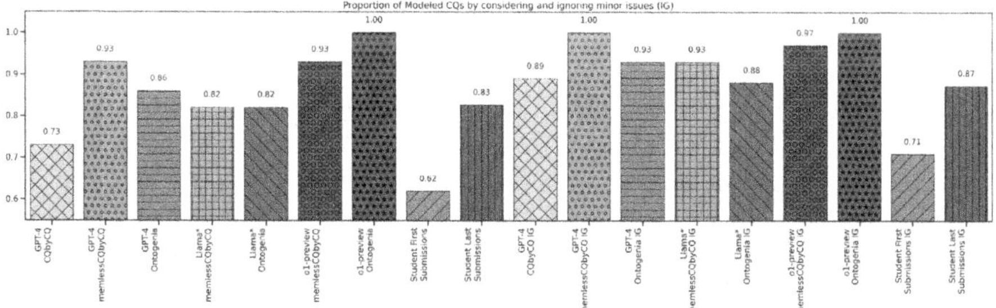

Fig. 3. Scores for "SemanticWebCourse" from the outputs for the different prompting techniques compared with students' submissions according to the proportion of the CQs that were accurately modelled. 'IG' indicates results when minor issues are ignored. Llama⋆ refers to `Llama-3.1-405B-instruct-bf16`.

Structural Analysis. Table 2 presents our structural analysis regarding the superfluous elements. The results, compared to those in Sect. 6, show that the Memoryless CQbyCQ and Ontogenia, despite their performance measured in other criteria, yield a significant number of superfluous elements within the ontology. This underscores the importance of applying additional criteria for evaluating ontologies constructed using LLMs. CQbyCQ with *GPT-4* produces the fewest superfluous elements compared to our newly introduced prompting techniques, while the Memoryless CQbyCQ approach yields comparable results. Conversely, *Llama-3.1-405B-instruct-bf16* generates numerous superfluous elements across both techniques and all stories, with a rate close to 40%, indicating that this model tends to produce many superfluous classes and properties.

Table 2. Comparison of the three prompting techniques, using the Incremental Ontology Generation method, for generating superfluous elements across different ontology domains (Theatre, Music, and Hospital). Each cell represents the proportion of superfluous elements relative to the total number of elements of that type (%). CQbyCQ was only tested with GPT-4 due to its low CQ-coverage. Rates under 15% are bolded (best performance), and rates over 50% are in red (worst performance).

Prompting Technique	Superflu. Classes			Superflu. Obj. Properties			Superflu. Data Properties		
	Theatre	Music	Hospital	Theatre	Music	Hospital	Theatre	Music	Hospital
CQbyCQ (GPT-4)	**0**	14.3	**8.6**	**0**	**4.2**	**6.6**	**12.5**	28.5	18
Ontogenia (GPT-4)	**0**	19.2	**13.5**	**4**	38.9	29.6	45.5	55.5	22.2
Ontogenia (Llama)	47.1	32.1	37.5	40.6	45.9	38.3	16.7	**0**	100
Ontogenia (o1)	16.7	**12.5**	27.3	29.4	55.2	37	28.6	66.7	45.5
MemorylessCQbyCQ(GPT-4)	25.7	31.2	38.9	17.4	20.9	16.2	48	40.9	46.1
MemorylessCQbyCQ(Llama)	45.6	42.3	23.9	20.5	25.6	41.7	60	60.7	**13.6**
MemorylessCQbyCQ(o1)	29.3	50	28.6	**14.8**	**0**	**7.4**	**10**	**0**	28.6

Expert Qualitative Analysis. After analyzing the results presented in Tables 1 and 2 and bar charts in Fig. 3, we conclude that *OpenAI o1-preview*, especially with Ontogenia, seems to produce less errors overall than *GPT-4*, and provides a better trade-off between different error types, notwithstanding all the generated files showed incorrect domains and ranges. Therefore, *Llama-3.1-405B-instruct-bf16* and *OpenAI o1-preview* are confirmed as the best open-source and closed-source LLMs, respectively, across both previous experiments [34] and the ones proposed in this work. The former's performance was not impressive when evluated by experts however: the ontologies generated by *Llama-3.1-405B-instruct-bf16* show structural flaws, including inconsistent naming, redundant classes, and overlapping domains/ranges. The taxonomy shows circular references and poor property organization. Additional, key issues include malformed cardinality restrictions, lack of comments/labels, and misaligned namespaces. Apart from superfluous elements, flat property hierarchies, and inconsistent axiomatisation, are visible. Table 3 shows, for each selected story, the resulting percentage scores from the adequacy assessment of the qualitative evaluation. Different results are calculated as an average between the two. The full qualitative analysis can be seen on Github Sect. 1.

Table 3. Adequate CQ modelling (%) by Llama and o1 models for the selected stories.

Story	Model	Llama-3.1-405	o1-preview
Music	MemorylessCQbyCQ	0.6	0.9
	Ontogenia	0.86	0.96
Theatre	MemorylessCQbyCQ	0.66	0.73
	Ontogenia	0.63	1.0
Hospital	MemorylessCQbyCQ	0.63	0.73
	Ontogenia	0.66	1.0

7 Discussion

In this section, we present a discussion of the results.

Overall Results. According to the findings, and similarly to previous work [34], the use of an LLM yields promising results in supporting ontology engineering processes. In the Independent Ontology Generation, Memoryless CQbyCQ and Ontogenia perform well with Single Data Property CQs and Single Object Property CQs, but worse on generating Reifications and Restrictions. For the Incremental Ontology Generation method, Ontogenia and Memoryless CQbyCQ outperformed previous methods, including the students' solutions, with Ontogenia combined with *OpenAI o1-preview* yielding the highest percentage of modelled

CQs. However, the structural analysis reveals that with respect to previous work, the two prompting techniques generate more superfluous elements and yield critical pitfalls according to OOPS!. These structural issues are also noted by the experts in their evaluation, which found many flaws in the Llama-generated ontologies related to the usability and understandability of the solutions. While o1-generated ontologies are overall comparable to a student in the case of MemorylessCQbyCQ and even better in the case of Ontogenia.

Multidimensional Evaluation. Our multidimensional evaluation demonstrates that expert assessments align closely with both the standard and structural metrics used in this study. This suggests that experts, perhaps implicitly, rely on similar criteria when evaluating ontologies, resulting in evaluations that comprehensively address most aspects. At the same time, it highlights the need for multidimensional and qualitative evaluation for a holistic assessment of LLM-generated ontologies.

Superfluous Elements. An issue noted in LLM-generated ontologies involves the creation of superfluous properties with the same domain and range or multiple classes and properties that could be considered equal. For example, `employedSince` and `employmentStartDate` are generated for the same CQ, but only one is needed. This raises questions about superfluous elements: their number, why they are generated by the LLM, and their consequences. At the same time, having superfluous elements (in moderate numbers) may be considered less important than having an unmodeled CQs or more complex errors. Thus, our prompting techniques can be considered more complete and more usable than previous work [34], which was less accurate (Table 2), but more concise. In our envisioned setting of a future ontology engineering co-pilot, superfluous elements can be identified, e.g. through OOPS!, and manually removed or avoided with an even more elaborate prompt. As our approach aims to assist rather than replace ontology engineers, we believe this is acceptable.

8 Conclusion

In this paper, we introduced two novel and improved prompting techniques for ontology generation, Memoryless CQbyCQ and Ontogenia, assessing them through a multi-dimensional evaluation and with a new dataset with respect to previous studies. The results show the proposed prompting techniques proved promising to support the generation of ontologies that meet a predefined set of requirements, improving the proportion of modelled CQs, surpassing previous approaches and novice ontology modellers. However, challenges like multiple domains or ranges and other pitfalls were highlighted both by the OOPS! pitfall scanner and the expert evaluation. Both prompting techniques often generated superfluous elements, which (in low numbers) are not detrimental to ontology usability, compared to other major errors involving wrong axiomatisation and mistakes in the taxonomy. Returning to our research questions, regarding (i) to what extent LLMs can be used to support the generation of ontologies that meet a predefined set of requirements, we conclude that current commercial models, such as *OpenAI o1*, can certainly perform on par with non-experts, at least when using a carefully selected prompt. Regarding (ii) what evaluation criteria are suitable for evaluating LLM-generated ontologies, we conclude that it is not sufficient to use one single criteria, nor simple automated evaluation metrics to detect errors and compare performance. Instead, rather elaborate manual methods are needed, e.g. to detect the amount of superfluous classes and properties. Finally, regarding (iii) what the strengths and weaknesses of ontologies generated using LLMs are, we believe that if cost is not an issue and commercial models can be used, then certainly novice ontology engineers could benefit from such draft ontologies generated by those LLMs. These drafts can increase modelling quality, and considerably reduce the effort and time to kick-start the modelling. However, certain weaknesses are observed across all the experiments, such as erroneous domain and range restrictions, erroneous inverse property axioms, and superfluous or overlapping classes and properties.

Acknowledgments. This project has received funding from the European Union's Horizon Europe research and innovation programme under grant agreements no. 101058682 (Onto-DESIDE) and 101070588 (HACID), and is supported by the strategic research area Security Link. The student solutions used in the research were collected as part of a master's course taught by Assoc. Prof. Blomqvist while employed at Jönköping University. Additional financial support to this project was provided by NextGenerationEU under NRRP Grant agreement n. MUR IR0000008 - FOSSR (CUP B83C22003950001). This work was also supported by the PhD scholarship "Discovery, Formalisation and Re-use of Knowledge Patterns and Graphs for the Science of Science", funded by CNR-ISTC through the WHOW project (EU CEF programme - grant agreement no. INEA/CEF/ICT/ A2019/2063229). Finally we thank OpenAI's Researcher Access Program Grant for the API credits.

Disclosure of Interests. The authors have no competing interests to declare that are relevant to the content of this article.

A Appendix

A.1 Examples of Evaluation Metrics

Modelled CQ: In Fig. 4, part A (top), a CQ such as "Who is the author of a book?" is shown to be correctly modelled, as all the elements required to write a SPARQL query are present in the model (dotted box).

Minor Issue: In part B, for the same CQ, an object property to connect Book and Author (or either connecting it to Person or a new data property) is missing. However, adding "authorOf" would resolve the issue. This is considered a minor error, as the ontology modeller can easily add this element.

Superfluous Elements: In Fig. 4, part A, for the given CQ, there is a class "Person" and properties such as "name" or "wrote" that are not used in the SPARQL query (top-right corner). Since they (outside of the dotted box) do not appear in the SPARQL, they are referred to as superfluous elements.

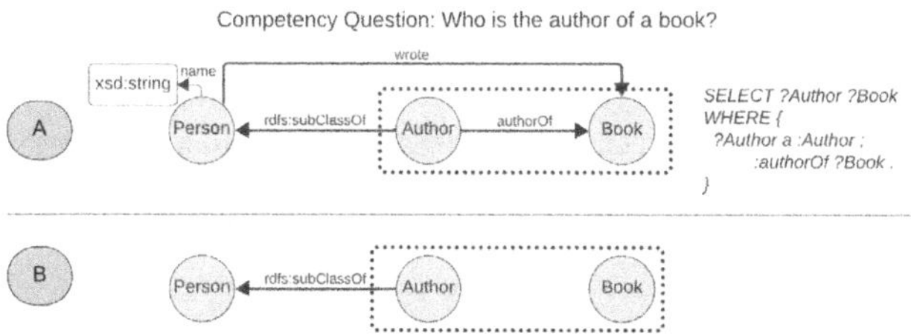

Fig. 4. Analysis of an ontology for a CQ. Part A shows a correctly modelled CQ, ensuring all necessary elements for a SPARQL query are present, but contain superfluous elements. Part B shows a minor issue where a data property is missing.

Each metric evaluates a specific aspect of the LLM-generated ontology, as no single metric is sufficient to provide a comprehensive assessment. For instance, Llama produces many superfluous elements, hence it also has a reasonable likelihood of randomly modelling certain simple CQs correctly. By penalizing superfluous elements, we prevent LLMs from generating too many unnecessary classes and properties; however, this approach may also discourage the generation of taxonomies or more generalized modelling, which is a disadvantage. By combining all metrics, we can achieve an holistic evaluation of ontology generation.

A.2 Discussion on LLMs

Comparison of LLMs. By comparing the results of the LLMs in Sect. 6, we can observe that `o1-preview` is the most promising LLM for ontology generation and GPT-4 may serve as its replacement. However, while Llama exhibits similar performance to GPT-4 in correctly modelling CQs, it produces a considerable number of superfluous elements and critical OOPS! pitfalls and performs the worst according to the expert evaluation. Thus, it may still not be possible to create an ontology engineering co-pilot using open models, and hence the issue of cost and the risks related to the black-box nature of proprietary models, remain.

Context Size. Comparing the scores of Memoryless CQbyCQ with other techniques shows that reducing the input context size of *GPT-4* significantly enhances the output score of this model despite producing some superfluous elements. These results highlight a trade-off between computational efficiency, cost, and output quality. If quality is a priority but resources are limited, Memoryless CQbyCQ balances efficiency and speed, though it may require additional post-processing to refine the output.

A.3 Limitations

Data Leakage Assessment. A critical consideration in our ontology selection criteria and dataset publication is the potential for data leakage [5], due to the online accessibility of some of this material. Such exposure introduces risks, namely the likelihood of evaluating LLMs on the same datasets they were trained on. Also, the dataset published in this paper as a gold standard might be used in the training data of future LLMs. Our selected ontologies are published with their CQs and user stories separately from their ontology models, which can mitigate these risks. The ontologies from the courses (Music, Hospital and Theatre) date back to 2008–2009, but only the stories and CQs have been published online, not the solutions in OWL. Moreover, to further protect our dataset being used in LLMs training data, we provide it in a zipped folder that is secured with a password, ensuring controlled access and reducing the risk of data leakage for future research in this field.

Additional Limitations. This study's scope is to see to what extent LLMs can serve in generating ontologies. There is a need for more evaluation criteria to assess the utility and generalizability of generated ontologies. Our ontologies, drawn from different domains, are primarily single-module, limiting their reusability across different systems. Moreover, our methods do not adequately prevent or eliminate superfluous elements, as shown by model output repetitions. Our Memoryless CQbyCQ approach, which reduces context size to enhance performance and reduce cost, is not suitable for history-dependent modelling tasks that heavily depend on the communication history with LLMs. This technique is more applicable to methods where each part of the ontology development can be performed independently in small parts and where there is an effective strategy for merging these partial solutions by human-in-the-loop integration. Additionally, further tests are needed to mitigate the potential leakage effect and bias in LLMs. One way to achieve this is by applying the method to new use cases in different domains. Future work will involve mitigating the limitations, for example, through manual re-engineering of the generated ontology draft. Potentially the latter can be supported by user interfaces, e.g., in the form of a plugin.

References

1. Alharbi, R., de Berardinis, J., Grasso, F., Payne, T., Tamma, V.: Characteristics and desiderata for competency question benchmarks. In: The Semantic Web - ISWC 2024: 23rd International Semantic Web Conference, Baltimore, MD, USA, 11–15 November 2024, Proceedings (2024)
2. Ali, R., et al.: Performance of chatgpt, gpt-4, and google bard on a neurosurgery oral boards preparation question bank. Neurosurgery **93**(5), 1090–1098 (2023). https://doi.org/10.1227/neu.0000000000002551
3. Allen, B., Groth, P.: A benchmark for the detection of metalinguistic disagreements between llms and knowledge graphs. In: The Semantic Web - ISWC 2024: 23rd International Semantic Web Conference, Baltimore, MD, USA, 11–15 November 2024, Proceedings (2024)

4. Babaei Giglou, H., D'Souza, J., Auer, S.: Llms4ol: large language models for ontology learning. In: Payne, T.R., et al. (eds.) The Semantic Web - ISWC 2023, pp. 408–427. Springer, Cham (2023). https://doi.org/10.1007/978-3-031-47240-4_22
5. Balloccu, S., Schmidtová, P., Lango, M., Dusek, O.: Leak, cheat, repeat: data contamination and evaluation malpractices in closed-source LLMs. In: Graham, Y., Purver, M. (eds.) Proceedings of the 18th Conference of the European Chapter of the Association for Computational Linguistics, vol. 1: Long Papers, pp. 67–93. Association for Computational Linguistics, St. Julian's (2024). https://aclanthology.org/2024.eacl-long.5
6. Blomqvist, E., Hammar, K., Presutti, V.: Engineering ontologies with patterns-the extreme design methodology. In: Ontology Engineering with Ontology Design Patterns. IOS Press (2016)
7. Blomqvist, E., Presutti, V., Daga, E., Gangemi, A.: Experimenting with eXtreme design. In: Cimiano, P., Pinto, H.S. (eds.) EKAW 2010. LNCS (LNAI), vol. 6317, pp. 120–134. Springer, Heidelberg (2010). https://doi.org/10.1007/978-3-642-16438-5_9
8. Blomqvist, E., Sandkuhl, K.: Patterns in ontology engineering: classification of ontology patterns. In: Proceedings of the Seventh International Conference on Enterprise Information Systems, vol. 3: ICEIS, pp. 413–416. INSTICC, SciTePress (2005). https://doi.org/10.5220/0002518804130416
9. Blomqvist, E., Seil Sepour, A., Presutti, V.: Ontology testing - methodology and tool. In: ten Teije, A., Völker, J., Handschuh, S., Stuckenschmidt, H., d'Acquin, M., Nikolov, A., Aussenac-Gilles, N., Hernandez, N. (eds.) EKAW 2012. LNCS (LNAI), vol. 7603, pp. 216–226. Springer, Heidelberg (2012). https://doi.org/10.1007/978-3-642-33876-2_20
10. Brown, T., et al.: Language models are few-shot learners. In: Larochelle, H., Ranzato, M., Hadsell, R., Balcan, M., Lin, H. (eds.) Advances in Neural Information Processing Systems, vol. 33, pp. 1877–1901. Curran Associates, Inc. (2020). https://proceedings.neurips.cc/paper_files/paper/2020/file/1457c0d6bfcb4967418bfb8ac142f64a-Paper.pdf
11. Chu, Z., et al.: A survey of chain of thought reasoning: advances, frontiers and future. arXiv preprint arXiv:2309.15402 (2023)
12. Fathallah, N., Das, A., Giorgis, S.D., Poltronieri, A., Haase, P., Kovriguina, L.: Neon-gpt: a large language model-powered pipeline for ontology learning. In: European Semantic Web Conference, pp. 36–50. Springer, Heidelberg (2024). https://doi.org/10.1007/978-3-031-78952-6_4
13. Fathallah, N., Staab, S., Algergawy, A.: Llms4life: large language models for ontology learning in life sciences (2024)
14. Fernández, M., Gómez-Pérez, A., Juristo, N.: Methontology: from ontological art towards ontological engineering. In: Proceedings of the AAAI97 Spring Symposium Series on Ontological Engineering (1997)
15. Frey, J., Meyer, L.P., Brei, F., Grunder-Fahrer, S., Martin, M.: Assessing the evolution of llm capabilities for knowledge graph engineering in 2023. In: Proceedings of the ESWC2024 Special Track: Large Language Models for Knowledge Engineering (to appear) (2024)
16. Gangemi, A.: Ontology design patterns for semantic web content. In: Gil, Y., Motta, E., Benjamins, V.R., Musen, M.A. (eds.) ISWC 2005. LNCS, vol. 3729, pp. 262–276. Springer, Heidelberg (2005). https://doi.org/10.1007/11574620_21
17. Gangemi, A., Nuzzolese, A.G.: Logic augmented generation. J. Web Semant. **85**, 100859 (2025)

18. Garcez, A.D., Lamb, L.C.: Neurosymbolic AI: the 3 rd wave. Artif. Intell. Rev. **56**(11), 12387–12406 (2023)
19. Garijo, D., Poveda-Villalón, M., Amador-Domínguez, E., Wang, Z., Garía-Castro, R., Corcho, O.: Llms for ontology engineering: a landscape of tasks and benchmarking challenges. In: The Semantic Web - ISWC 2024: 23rd International Semantic Web Conference. Baltimore, MD, USA, 11–15 November 2024, Proceedings of the 23rd International Semantic Web Conference (ISWC 2024)
20. He, Y., et al.: Deeponto: a python package for ontology engineering with deep learning. Semant. Web **15**(5), 1991–2004 (2024)
21. Hogan, A., et al.: Knowledge Graphs. Morgan & Claypool Publishers (2021)
22. Keet, C.M., Khan, Z.C.: On the roles of competency questions in ontology engineering. In: International Conference on Knowledge Engineering and Knowledge Management, pp. 123–132. Springer, Heidelberg (2024). https://doi.org/10.1007/978-3-031-77792-9_8
23. Khot, T., et al.: Decomposed prompting: a modular approach for solving complex tasks. In: The Eleventh International Conference on Learning Representations, ICLR 2023, Kigali, Rwanda, 1–5 May 2023 (2023)
24. Lippolis, A.S., Ceriani, M., Zuppiroli, S., Nuzzolese, A.G.: Ontogenia: ontology generation with metacognitive prompting in large language models. In: European Semantic Web Conference, pp. 259–265. Springer, Heidelberg (2024). https://doi.org/10.1007/978-3-031-78952-6_38
25. Lippolis, A.S., Lodi, G., Nuzzolese, A.G.: The water health open knowledge graph. Sci. Data **12**(1), 274 (2025)
26. Mateiu, P., Groza, A.: Ontology engineering with large language models. In: 2023 25th International Symposium on Symbolic and Numeric Algorithms for Scientific Computing (SYNASC), pp. 226–229. IEEE (2023)
27. Peroni, S.: A simplified agile methodology for ontology development. In: Dragoni, M., Poveda-Villalón, M., Jimenez-Ruiz, E. (eds.) OWLED/ORE -2016. LNCS, vol. 10161, pp. 55–69. Springer, Cham (2017). https://doi.org/10.1007/978-3-319-54627-8_5
28. Plu, J., Escobar, O.M., Trouillez, E., Gapin, A., Troncy, R.: A comprehensive benchmark for evaluating llm-generated ontologies. In: The Semantic Web - ISWC 2024: 23rd International Semantic Web Conference, Baltimore, MD, USA, 11–15 November 2024, Proceedings (2024)
29. Poveda-Villalón, M., Gómez-Pérez, A., Suárez-Figueroa, M.C.: OOPS! (OntOlogy Pitfall Scanner!): an on-line tool for ontology evaluation. Int. J. Semant. Web Inf. Syst. (IJSWIS) **10**(2), 7–34 (2014)
30. Poveda-Villalón, M., Fernández-Izquierdo, A., Fernández-López, M., García-Castro, R.: Lot: an industrial oriented ontology engineering framework. Eng. Appl. Artif. Intell. **111**, 104755 (2022). https://doi.org/10.1016/j.engappai.2022.104755
31. Presutti, V., Daga, E., Gangemi, A., Blomqvist, E.: extreme design with content ontology design patterns. In: Proceedings of Workshop on Ontology Patterns, pp. 83–97. CEUR-WS (2009)
32. Rebboud, Y., Lisena, P., Tailhardat, L., Troncy, R.: Benchmarking llm-based ontology conceptualization: a proposal. In: The Semantic Web - ISWC 2024: 23rd International Semantic Web Conference, Baltimore, MD, USA, 11–15 November 2024, Proceedings (2024)
33. Sabou, M., Llugiqi, M., Ekaputra, F.J., Waltersdorfer, L., Tsaneva, S.: Knowledge engineering in the age of neurosymbolic systems. Neurosymb. Artif. Intell. **1**, 29498732251320080 (2025)

34. Saeedizade, M.J., Blomqvist, E.: Navigating ontology development with large language models. In: Meroño Peñuela, A., et al. (eds.) The Semantic Web. pp. 143–161. Springer, Cham (2024)
35. Shimizu, C., Hammar, K., Hitzler, P.: Modular ontology modeling. Semant. Web **14**(3), 459–489 (2023). https://doi.org/10.3233/SW-222886
36. Suárez-Figueroa, M.C., Gómez-Pérez, A., Fernández-López, M.: The NeOn methodology for ontology engineering. In: Suárez-Figueroa, M.C., Gómez-Pérez, A., Motta, E., Gangemi, A. (eds.) Ontology Engineering in a Networked World, pp. 9–34. Springer, Heidelberg (2012). https://doi.org/10.1007/978-3-642-24794-1_2
37. Suárez-Figueroa, M.C., Gómez-Pérez, A., Motta, E., Gangemi, A.: Introduction: Ontology Engineering in a Networked World, pp. 1–6. Springer, Heidelberg (2012)
38. Tsaneva, S., Herwanto, G.B., Sabou, M.: Benchmarking ontology validation capabilities of llms. In: The Semantic Web - ISWC 2024: 23rd International Semantic Web Conference, Baltimore, MD, USA, 11–15 November 2024, Proceedings (2024)
39. Wang, B., Min, S., Deng, X., Shen, J., Wu, Y., Zettlemoyer, L., Sun, H.: Towards understanding chain-of-thought prompting: an empirical study of what matters. In: The 61st Annual Meeting Of The Association For Computational Linguistics (2023)
40. Wang, Y., Zhao, Y.: Metacognitive prompting improves understanding in large language models. In: Duh, K., Gomez, H., Bethard, S. (eds.) Proceedings of the 2024 Conference of the North American Chapter of the Association for Computational Linguistics: Human Language Technologies, vol. 1: Long Papers, pp. 1914–1926. Association for Computational Linguistics, Mexico City (2024). https://doi.org/10.18653/v1/2024.naacl-long.106
41. Wei, J., et al.: Chain-of-thought prompting elicits reasoning in large language models. In: Advances in Neural Information Processing Systems, pp. 24824–24837 (2022)

Towards Practicable Algorithms for Rewriting Graph Queries Beyond DL-Lite

Bianca Löhnert[1], Nikolaus Augsten[1], Cem Okulmus[2(✉)], and Magdalena Ortiz[3]

[1] University of Salzburg, Salzburg, Austria
{bianca.loehnert,nikolaus.augsten}@plus.ac.at
[2] Paderborn University, Paderborn, Germany
cem.okulmus@uni-paderborn.de
[3] TU Wien, Vienna, Austria
magdalena.ortiz@tuwien.ac.at

Abstract. Despite the advantages that the virtual knowledge graph paradigm has brought to many application domains, state-of-the-art systems still do not support popular graph database management systems like Neo4j. Their query rewriting algorithms focus on languages like conjunctive queries and their unions, which were developed for relational data and are poorly suited for graph data. Moreover, they also limit the expressiveness of the ontology languages that admit rewritings, restricting them to those that enjoy the so-called *FO-rewritability* property. Rewritings have thus focused on the DL-Lite family of Description Logics. In this paper, we propose a technique for rewriting a family of *navigational queries* for a suitably tailored fragment of $\mathcal{ELHI}$. Leveraging navigational features in the target query language, we can include some widely-used axiom shapes not supported by DL-Lite. We implemented a proof-of-concept prototype that rewrites into Cypher queries, and tested it on a real-world cognitive neuroscience use case with promising results.

Keywords: Ontology-Based Data Access · Property Graphs · Navigational Queries

1 Introduction

The virtual knowledge graph (VKG) paradigm [30], also known as ontology-based data access, emerged from real-life scenarios [24] and has been deployed in real-world systems such as Ontopic, Stardog and Mastro [1–3]. In it, the data is described by a high-level conceptual layer that enables the user to formulate queries using the familiar vocabulary of a domain ontology. In addition to facilitating query formulation, the knowledge in the ontology can be used to infer implicit information when querying possibly incomplete data. The key problem to address here is *ontology-mediated query answering* (OMQA), where a query

Supplementary Information The online version contains supplementary material available at https://doi.org/10.1007/978-3-031-94575-5_19.

E. Curry et al. (Eds.): ESWC 2025, LNCS 15718, pp. 342–361, 2025.
https://doi.org/10.1007/978-3-031-94575-5_19

is to be evaluated not directly over a plain dataset, but over the consequences of the given dataset together with the knowledge in the ontology. The VKG paradigm has seen widespread adoption and brought by significant savings in the data access and integration costs in a range of applications [31], but so far, it has been limited to relational database management systems (RDBMS). The predominant technique for ontology-mediated querying is *query rewriting*, where an input query is transformed to incorporate the ontological knowledge so that the rewritten query can be evaluated over any input dataset using standard technologies (i.e., SQL queries and standard RDBMS), without further ontological reasoning, and give complete answers that take into account implicit facts [10]. The so-called *union of conjunctive queries*, which capture popular SQL fragments [4] are used as source and target languages for query rewriting. However, this technique can only be deployed for ontology languages whose data complexity is not higher than that of evaluating the FO-fragment of SQL. The DL-Lite family of description logics (DLs) was introduced with this goal in mind, and has become the ontology language of choice for OMQA [10].

While traditional RDBMS are not going away, recent years have seen huge adoption of new paradigms for storing and querying data, and bringing the VKG paradigm to them would open up opportunities in many application domains. Particularly popular are *graph databases* that adopt the *property graph* (PG) data model [5]. PGs comprise nodes and edges between them; both nodes and edges can be labelled, and have assigned key-value pairs. In addition to a new data model, graph databases also provide unique querying abilities not found in the relational setting. Very recently the International Organization for Standardization (ISO) released a new standard for graph query languages. This standard, called GQL [18], captures and extends *conjunctive two-way regular path queries* (C2RPQ), a query language for graphs widely studied in the OMQA literature.

Query languages for graph data like C2RPQ and GQL are characterized by the *navigational features* that allow queries to traverse paths of arbitrary length that comply with some *regular path expression*. This fundamentally recursive feature results in a higher data complexity than that of SQL, and graph query languages are typically NL complete in data complexity (under the so-called *walk semantics*). This higher complexity, however, also means that we are not bound to DL-Lite and can potentially consider richer ontology languages with NL data complexity. To our knowledge, the only such DLs so far are the linear fragments of $\mathcal{ELH}$ and $\mathcal{ELHI}$ identified by Dimartino et al. [13], which do not allow any form of conjunction. That work also introduced a query rewriting procedure for instance queries and CQs based on finite state automata but, unfortunately, it seems that it was never implemented. When it comes to navigational ontology mediated queries, there is a rich body of theoretical work that considers C2RPQs as the input query language [7–9], but the goal of those works was to understand the computational complexity and they do not provide implementable query rewriting algorithms. The only practical approach to rewriting navigational queries is the recent algorithm for DL-Lite [16]. We extend this algorithm to a fragment of $\mathcal{ELHI}$ with NL data complexity. Unsurprisingly, lifting the techniques from DL-Lite is far from trivial. Even without conjunctions, in $\mathcal{EL}$ we may need a path of unbounded length to witness the propagation of a single atom, instead of single data point as in DL-Lite; with conjunction we may need a tree of such paths. Our fragment $\mathcal{ELH}i^{ql}$ was tailored to bound the branching points of such trees, while still accommodating the conjunctions and inverse roles used in our use case ontology from the domain of neuroscience. Our main contributions are as follows.

- We explore the limits of rewriting navigational queries and show that, even for lightweight ontologies, rewritings may not exist if C2RPQs are considered as both source and target query language.
- Inspired by this insight, we identify a subset of C2RPQs, termed Navigational Conjunctive Queries (NCQs), which offers key features of graph query languages—like reachability using the Kleene star—without ruling out the possibility of rewriting into unions of C2RPQs.
- We also push the boundaries of expressivity in the ontology language. Leveraging the navigational features of our target graph query language, we can accommodate typical $\mathcal{EL}$ axioms like $\exists r.A \sqsubseteq B$. Inverse roles and concept conjunction are two popular constructors heavily used in real-life ontologies. Hence, even though they make reasoning hard for PTime and thus preclude rewritability, we allow them in a restricted form. We call the resulting logic $\mathcal{ELH}i^{ql}$ (*quasi-linear* $\mathcal{ELH}$ *with restricted inverses*), and we provide an algorithm for standard reasoning using a graph structure.
- We propose a rewriting for OMQs that pair NCQs with $\mathcal{ELH}i^{ql}$ ontologies. First we introduce a technique for standard reasoning to rewrite atomic queries into C2RPQs, and then we combine this with the well-known Clipper rewriting [17] that we extended for NCQs.
- We present a proof-of-concept prototype of our technique and use it to evaluate queries over real-world data from the domain of cognitive neuroscience.

The remainder of the paper is structured as follows. In Sect. 2 we present the necessary terminology, such as the ontology we focus on, PGs in the context of OMQA and C2RPQs. In Sect. 3, we present our supported ontology language and show how reasoning about concept subsumption in the ontology can be handled via a bespoke data structure. In Sect. 4 we first focus on the case of rewriting atomic queries into C2RPQs that capture all their consequences from the ontology. In Sect. 5 we show the limitations of rewriting C2RPQs when used as the input and intended output language for OMQA and present a restricted subset of C2RPQs that retains rewritability. In Sect. 6 we report on our proof-of-concept prototype and evaluate queries over data from a real-world use case. We conclude and point to future work in Sect. 7. We omit detailed proofs in this paper to save space and improve readability. All proofs can be found in the full version of this paper [20].

2 Preliminaries

Ontology Language. We recall the definition of $\mathcal{ELHI}$, a well-known description logic subsuming the popular lightweight languages DL-Lite$_\mathcal{R}$ and $\mathcal{ELH}$; in the next sections we restrict our attention to a fragment of it. For space reasons we omit disjointness axioms, but they can be easily incorporated. We also recall the usual *normal form* for $\mathcal{ELHI}$ TBoxes; the proof that every TBox can be normalized (in linear time) while preserving the semantics is standard.

We assume disjoint, countably infinite sets $\mathbf{C}$, $\mathbf{R}$, $\mathbf{N}$ and $\mathbf{K}$ of *concept names, role names, individuals* and *key names*, respectively, as well as a *concrete domain* $(\mathbf{D}, \mathcal{P}^D)$, with $\mathbf{D}$ a set of *values* and $\mathcal{P}^D$ a set of binary predicates over $\mathbf{D}$. For example, $(\mathbf{D}, \mathcal{P}^D)$ could contain the integers with the usual $=$, $\leq$, $\geq$ predicates.

Definition 1 ($\mathcal{ELHI}$)**.** *The set of* roles *is* $\overline{\mathbf{R}} = \mathbf{R} \cup \{r^- \mid r \in \mathbf{R}\}$, *and* concepts *$C$ follow the syntax* $C := A \mid \top \mid \exists r.C \mid C \sqcap C$, *where* $A \in \mathbf{C}$ *and* $r \in \overline{\mathbf{R}}$. *A* concept inclusion *(CI) has the form* $C \sqsubseteq D$, *where* C, D *are concepts, and a* role inclusion *(RI) the form* $r \sqsubseteq s$, *where* r, s *are roles. A TBox is a finite set of CIs and RIs, and it is said to be in* normal form *if all inclusions take these forms:*

$$A_1 \sqcap \cdots \sqcap A_n \sqsubseteq B \qquad \exists r.A \sqsubseteq B \qquad A \sqsubseteq \exists r.B \qquad r \sqsubseteq s$$

We use $\sqsubseteq^*_{\mathcal{T}}$ *to denote the reflexive transitive closure of* $\{(r, s) \mid r \sqsubseteq s \in \mathcal{T}\}$ *and call* r *a* subrole of s (in $\mathcal{T}$) *if* $r \sqsubseteq^*_{\mathcal{T}} s$.

The semantics are given via *interpretations* of the form $\mathcal{I} = (\Delta^{\mathcal{I}}, \cdot^{\mathcal{I}})$, with $\Delta^{\mathcal{I}}$ a non-empty set called the *abstract domain.* $\cdot^{\mathcal{I}}$ is the *interpretation function*, which assigns to every $A \in \mathbf{C}$ a set $A^{\mathcal{I}} \subseteq \Delta^{\mathcal{I}}$, to every $r \in \mathbf{R}$ a relation $r^{\mathcal{I}} \subseteq \Delta^{\mathcal{I}} \times \Delta^{\mathcal{I}}$ and to every $k \in \mathbf{K}$ a relation $k^{\mathcal{I}} \subseteq (\Delta^{\mathcal{I}} \cup (\Delta^{\mathcal{I}} \times \Delta^{\mathcal{I}})) \times \mathbf{D}$. It is extended to concepts and CIs in the usual way, see Table 2 in Appendix A. Modelhood and entailment are also standard.

Data Model. In this paper the data (or ABox) is given as finite *property graphs.*

Definition 2. *A* property graph *(PG)* $\mathcal{A}$ *has the form* $(N, E, \mathsf{label}, \mathsf{prop})$, *where:*

- N *is a non-empty set of* nodes;
- E *is the set of edges; it assigns to each role* $r \in \mathbf{R}$ *a relation on* $N \times N$ *which we write in the form* $r(n, n')$ *and call it the set of* r-labeled edges;
- label *is a total function* $N \to 2^{\mathbf{C}}$;
- prop *is a partial function* $(N \cup E) \times \mathbf{K} \to \mathbf{D}$ *mapping pairs* (u, k) *with* $u \in (N \cup E)$ *and* $k \in \mathbf{K}$ *to a value in* $\mathbf{D}$.

If $N \subseteq \mathbf{N}$ *and it is finite, we call it an* ABox. *A pair of a TBox* $\mathcal{T}$ *and an ABox* $\mathcal{A}$ *is called a* knowledge base. *We say that* $\mathcal{A}' = (N', E', \mathsf{label}', \mathsf{prop}')$ *is a* subgraph *of* $\mathcal{A}$ *if* $N' \subseteq N$, $E' \subseteq E$, $\mathsf{label}'(n) = \mathsf{label}(n)$ *for all* $n \in N'$, *and* $\mathsf{prop}'(u, k) = \mathsf{prop}(u, k)$ *for all* $u \in N' \cup E'$, $k \in K'$.

Note that we allow key-value pairs only in the ABox, and that our definition of property graph allows only a single edge between each pair of nodes.

Each interpretation can be seen as a property graph, and vice-versa.

Definition 3. *For a property graph $\mathcal{A} = (N, E, \mathsf{label}, \mathsf{prop})$, define $\mathcal{I}_\mathcal{A} = (N, \cdot^\mathcal{I})$ as follows: $C^\mathcal{I} = \{n \in N \mid C \in \mathsf{label}(n)\}$, $r^\mathcal{I} = \{(n, n') \mid r(n, n') \in E\}$ and $k^\mathcal{I} = \{(u, d) \mid d = \mathsf{prop}(u, k)\}$. Conversely, $\mathcal{I} = (\Delta^\mathcal{I}, \cdot^\mathcal{I})$ induces a (possibly infinite) property graph $PG(\mathcal{I}) = (\Delta^\mathcal{I}, E, \mathsf{label}, \mathsf{prop})$, where $E = \{r(n, n') \mid (n, n') \in r^\mathcal{I}\}$, $\mathsf{label}(n) = \{C \in \mathbf{C} \mid n \in C^\mathcal{I}\}$ and $\mathsf{prop}(u, k) = \{d \in \mathbf{D} \mid (u, d) \in k^\mathcal{I}\}$. $\mathcal{I}$ is a* model *of an ABox $\mathcal{A}$, if $\mathcal{A}$ is a subgraph of $PG(\mathcal{I})$. An ABox $\mathcal{A}$ is* consistent *with a TBox $\mathcal{T}$ iff there is a model of $\mathcal{A}$ and $\mathcal{T}$.*

Query Language. We study *conjunctive two-way regular path queries* (C2RPQs), the navigational query language for graphs that has received most attention in OMQA. We enhance C2RPQs by *data tests* as in [16] to query for property values, assuming that the predicates in $\mathcal{P}^D$ can be realized in GQL and Cypher.

Definition 4. *Let T be a* data test *defined as $T := k \odot v \mid T \wedge T \mid T \vee T \mid \neg T$, where $k \in \mathbf{K}$, $v \in \mathbf{D}$ and $\odot \in \mathcal{P}^D$ a binary predicate. A* regular path expression *(RPE) π is defined as follows, with $\pi^+ = \pi\pi^*$, $A \in \mathbf{C}$, $r \in \mathbf{R}$ and T a data test.*

$$\alpha := r \mid r^- \mid \langle A \rangle \qquad \pi := \alpha \mid \{T\} \mid \pi\pi \mid \pi + \pi \mid \pi^* \mid \pi^+$$

We assume a countably infinite set $\mathbf{V}$ of variables, *disjoint from $\mathbf{C}$, $\mathbf{R}$, $\mathbf{N}$, $\mathbf{K}$ and $\mathbf{D}$ and define* atoms *of the form $\pi(x, y)$ with π an RPE and $x, y \in \mathbf{V}$. Note that π^* (resp. π^+) refers to the Kleene star (resp. Kleene plus), as known from formal language theory [29]. A* conjunctive two-way regular path query (with data tests) *(C2RPQ(d)) is a pair $(\varphi, \vec{x})$ where φ is a conjunction of atoms $\pi_1(x_1, y_1) \wedge \cdots \wedge \pi_n(x_n, y_n)$ and the* answer variables *$\vec{x}$ are a tuple of variables occurring in φ. We write $vars(q) \subseteq \mathbf{V}$ for the set of all variables occurring in atoms of a query $q = (\varphi, \vec{x})$, where $\vec{x} \subseteq vars(q)$. A variable x is* unbound *in q if $x \notin \vec{x}$ and it occurs in exactly one atom of q. We may write $q(\vec{x}) := \varphi$ for a C2RPQ with answer variables $\vec{x}$.*

Atomic concept tests are always binary atoms $\langle A \rangle(x, x)$, but we often shorten this to $A(x)$. We may refer to queries of form $q(x) := A(x)$ and $q(x, y) := \pi(x, y)$ as *atomic* (a.k.a. instance) queries. Note that for RPEs π that do not use roles (e.g., combinations of concept tests), the y in an atom $\pi(x, y)$ is irrelevant.

To illustrate our query language, we make use of our real-world use case from the domain of cognitive neuroscience, which is also the scope of our experiments in Sect. 6. The query $q(x)$ retrieves all datasets from an MRI with a certain specification and data on ambidextrous participants:

$$\begin{aligned} q(x) :=& \langle \mathsf{Dataset} \rangle(\mathsf{x}) \wedge \{\mathsf{Manufacturer} = \text{"SIEMENS"} \wedge \mathsf{MagnetFieldStrength} \geq 3\}(x) \\ & \wedge \mathsf{has}^*(x, y) \wedge \langle \mathsf{Participant} \rangle(y) \wedge \{\mathsf{Handedness} = \text{"ambidextrous"}\}(y) \end{aligned}$$

We use here a concrete domain that contains strings and integers, with separate predicates, and data tests on the keys Manufacturer, Handedness, and MagneticFieldStrength. The Kleene star in the second atom is the distinctive navigational feature of graph query languages, absent from any FO-rewritable query language; it lets us explore paths of unbounded length across the property graph.

Following the literature, we use the *homomorphism* or *walk* semantics [5] for evaluating regular path queries. We define the evaluation of a C2RPQ $q(\vec{x})$ in terms of a function $[\![\cdot]\!]^{\mathcal{A}}$, defined Table 3 in Appendix A. If a TBox is given, we adopt the *certain answer* semantics as usual in OMQA.

Definition 5. *Let $\mathcal{A}$ be a PG with nodes N, and let $q(\vec{x})$ be a C2RPQ. A tuple $\vec{a}$ of nodes in N is an* answer *to $q(\vec{x})$ over $\mathcal{A}$ if there exists a mapping $\mu \in [\![\varphi]\!]^{\mathcal{A}}$ s.t. $\mu(\vec{x}) = \vec{a}$. For ABox $\mathcal{A}$ and TBox $\mathcal{T}$, we call $\vec{a}$ a certain answer to $q(\vec{x})$ over $(\mathcal{T}, \mathcal{A})$ if $\vec{a}$ is an answer to $q(\vec{x})$ in $PG(\mathcal{I})$ for every model $\mathcal{I}$ of $\mathcal{A}$ and $\mathcal{T}$.*

3 Reasoning in $\mathcal{ELHI}^{ql}$

We introduce $\mathcal{ELHI}^{ql}$, a fragment of $\mathcal{ELHI}$ that restricts both conjunction and the use of inverses. It is well known that the interaction of these two constructors with 'non-local' concepts that may propagate across the ABox causes P-hardness in data complexity [11].

Definition 6 **($\mathcal{ELHI}^{ql}$).** *Let $\mathcal{T}$ be an $\mathcal{ELHI}$ TBox in normal form. A concept name B is* non-local *in $\mathcal{T}$ if there is an axiom of the form $\exists r.B \sqsubseteq C$, or $B \sqsubseteq A$ with A being non-local; otherwise, B is* local. *$\mathcal{T}$ is an $\mathcal{ELHI}^{ql}$ TBox if:*

- *every existential restriction $\exists r.C$ has either $r \in \mathbf{R}$ or $C = \top$,*
- *only role names are allowed in RIs, and*
- *every concept name B in an axiom $A_1 \sqcap \cdots \sqcap A_n \sqsubseteq B$ is local in $\mathcal{T}$.*

For convenience, we recall the axiom shapes of $\mathcal{ELHI}^{ql}$:

$$(NF1)\ A_1 \sqcap \cdots \sqcap A_n \sqsubseteq B \qquad (NF2)\ \exists r.A \sqsubseteq B \qquad (NF3)\ A \sqsubseteq \exists r.B$$
$$(NF4)\ \ r \sqsubseteq s \qquad (NF5)\ \exists r^{-}.\top \sqsubseteq B \qquad (NF6)\ A \sqsubseteq \exists r^{-}.\top$$

Example 1. Consider the $\mathcal{ELHI}$ TBox $\mathcal{T} = \{\exists r.B \sqsubseteq C, A \sqsubseteq B, A_1 \sqcap A_2 \sqsubseteq A, \exists r.A_3 \sqsubseteq A_1\}$. By Definition 6 B, A, A_3 are non-local and since A occurs on the right-hand side of an axiom in form of (NF1) $\mathcal{T}$ does not fall into the fragment of $\mathcal{ELHI}^{ql}$. However, $\mathcal{T}$ without the axiom $\exists r.B \sqsubseteq C$ is indeed an $\mathcal{ELHI}^{ql}$ TBox.

$\mathcal{ELHI}^{ql}$ is related, but different from the *harmless fragment of linear $\mathcal{ELHI}$* studied in [12]. The latter allows no conjunction, and imposes that if $A \sqsubseteq \exists r_2.A_2$ and $\exists r_1.A_1 \sqsubseteq A_2$ occur together then neither $r_1 \sqsubseteq^*_{\mathcal{T}} r_2^-$ nor $r_2 \sqsubseteq^*_{\mathcal{T}} r_1^-$ holds. To reason in $\mathcal{ELHI}^{ql}$ we introduce a structure called *concept dependency graph.*

Definition 7. *We consider graphs G that contain three types of edges:* ε-propagating edges $\varepsilon(A, B)$, r-propagating edges $r(A, B)$, *and* r-existential edges, $\exists r(A, B)$, *where $r \in \overline{\mathbf{R}}$. A* propagating path *$p$ from node A_0 (to node A_n) is a (possibly empty) sequence $A_0 \rho_1 A_1 \dots \rho_i A_i \dots A_n$ where, for each $1 \leq i < n$, A_i is a node in G and there is a (ε- or r-)propagating edge $\rho_i(A_i, A_{i+1})$; note* *that $\rho_i \in \overline{\mathbf{R}} \cup \{\varepsilon\}$. The* role label *$r_1 \dots r_k$ of p is defined as the subsequence of $\rho_1 \dots \rho_n$ that contains only roles, that is, omitting all ρ_i such that $\rho_i = \varepsilon$. Then, we say that $A_0\rho_1 A_1 \dots \rho_n A_n$* propagates *the concept $\exists s_1 \dots \exists s_k.D$, if $r_j \sqsubseteq^*_{\mathcal{T}} s_j$ for each $1 \leq j \leq k$, A_n is a (possibly empty) conjunction of concept names, and D is a (possibly complex) $\mathcal{ELH}i^{ql}$ concept D such that $\models D \sqsubseteq A_n$.*

Definition 8. *The concept dependency graph (CDG) of an $\mathcal{ELH}i^{ql}$ TBox $\mathcal{T}$, denoted $G_{\mathcal{T}}$, is the directed multigraph defined as follows.*

- *The set of nodes $N_{\mathcal{T}}$ of $G_{\mathcal{T}}$ contains the top concept $\top$, all concept names occurring in $\mathcal{T}$, all concepts of the form $A_1 \sqcap \dots \sqcap A_n$ that occur on the left-hand side of axiom of shape (NF1), and a node $W_{\exists r.B}$ for each concept $\exists r.B$ that occurs on the right-hand side of an axiom of shape (NF3) or (NF6) in $\mathcal{T}$.*
- *The set of edges $E_{\mathcal{T}}$ of $G_{\mathcal{T}}$ is the smallest set that satisfies the following:*
  (a) $\varepsilon(\top, A) \in E_{\mathcal{T}}$ for each $A \in N_{\mathcal{T}}$
  (b) $\varepsilon(A, B_1 \sqcap \dots \sqcap B_n) \in E_{\mathcal{T}}$ for each $B_1 \sqcap \dots \sqcap B_n \sqsubseteq A$ of shape (NF1) in $\mathcal{T}$
  (c) $\varepsilon(B_1 \sqcap \dots \sqcap B_n, A) \in E_{\mathcal{T}}$ if in $G_{\mathcal{T}}$ there is a node $B_1 \sqcap \dots B_i \dots \sqcap B_n$ and for each B_i, there is an ε^*-path from B_i to A
  (d) $r(A, B) \in E_{\mathcal{T}}$ for each $\exists r.B \sqsubseteq A$ of shape (NF2) or (Nf5) in $\mathcal{T}$
  (e) $r(A, \top) \in E_{\mathcal{T}}$ if in $G_{\mathcal{T}}$ there is a propagating path with role label r from A to B and $\exists s^-.\top \sqsubseteq B$ of shape (Nf5) in $\mathcal{T}$ with $r \sqsubseteq^*_{\mathcal{T}} s$
  (f) $\exists r(A, W_{\exists r.B}) \in E_{\mathcal{T}}$ and $\varepsilon(B, W_{\exists r.B}) \in E_{\mathcal{T}}$ for each $A \sqsubseteq \exists r.B$ of shape (NF3) or (NF6) in $\mathcal{T}$
  (g) $\varepsilon(A, W) \in E_{\mathcal{T}}$ if there is some $\exists r^-.\top \sqsubseteq A$ of shape (Nf5) in $\mathcal{T}$, an edge $\exists s(C, W) \in E_{\mathcal{T}}$, with $s \sqsubseteq^*_{\mathcal{T}} r$ and no ε^*-path from A to W already
  (h) $\varepsilon(A, W) \in E_{\mathcal{T}}$ if in $G_{\mathcal{T}}$ there is $\exists s^-(C, W)$ with $s \sqsubseteq^*_{\mathcal{T}} r$ and a propagating path with role label r from A to C and no ε^*-path from A to W already
  (i) $\varepsilon(A, C) \in E_{\mathcal{T}}$ if in $G_{\mathcal{T}}$ there is $\exists s(C, W)$ for some role r with $s \sqsubseteq^*_{\mathcal{T}} r$ and some propagating path p with role label r from A to W, and there is no ε^*-path from A to C already

Example 2 (CDG). For the TBox $\mathcal{T} = \{A_2 \sqsubseteq A_1, \exists r.B_1 \sqsubseteq A_1, \exists r_3.B_1 \sqsubseteq B_3, A_3 \sqsubseteq A_2, \exists r_1.B_2 \sqsubseteq B_1, \exists r_2^-.\top \sqsubseteq A_3, s \sqsubseteq r_2, \exists r_2.B_3 \sqsubseteq B_2, B_1 \sqsubseteq \exists r_2.B_3\}$ the CDG $G_{\mathcal{T}}$ is as follows. (For readability, we omit edges from Definition 8 item (a)).

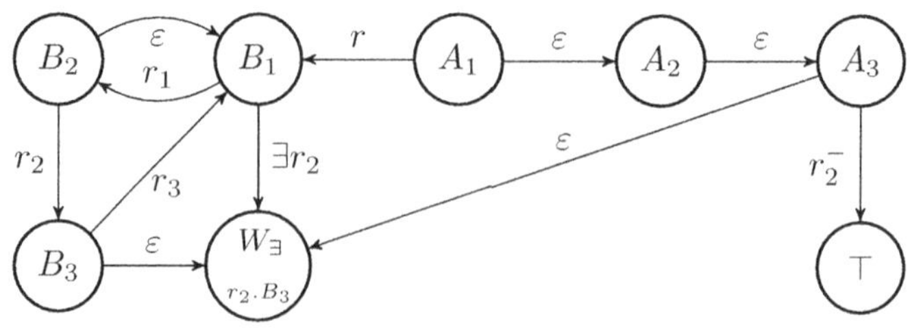

Note that the auxiliary nodes $W_{\exists r.B}$ help propagate inferences involving existentially quantified objects, but they do not participate in propagating paths.

The propagating paths in the CDG witness entailments from $\mathcal{T}$.

Example 3. Let us continue with $\mathcal{T}$ and $G_\mathcal{T}$ from Example 2 and the ABox $\mathcal{A} = \{r(n_0, n_1), r_1(n_1, n_2), r_1(n_2, n_3), r_2(n_3, n_4), B_3(n_4)\}$. Consider the propagating path $A_1 r B_1 r_1 B_2 \varepsilon B_1 r_1 B_2 r_2 B_3$ of the CDG that starts in A_1 and ends in B_3. This path propagates $\exists r \exists r_1 \exists r_1 \exists r_2.B_3$ from which we infer that $\mathcal{T}$ entails $\exists r \exists r_1 \exists r_1 \exists r_2.B_3 \sqsubseteq A_1$ and therefore $(\mathcal{T}, \mathcal{A}) \models A_1(n_0)$.

To witness all relevant entailments, we also use *witnessing sets.*

Definition 9. *Consider a CDG $G_\mathcal{T}$ with nodes $N_\mathcal{T}$. We say that the set $W_A \subseteq N_\mathcal{T} \cap \mathbf{C}$* witnesses *$A$ (in $G_\mathcal{T}$) if (a) $A \in W_A$ or (b) A has a propagating path to a conjunction all whose conjuncts are witnessed by W_A. We say that W_A is a* proper *witnessing set for A if no strict subset of it satisfies (a) and (b).*

Example 4. Consider the $\mathcal{ELH}i^{ql}$ TBox $\mathcal{T} = \{A_1 \sqcap A_2 \sqsubseteq A, B_1 \sqcap B_2 \sqsubseteq A_1, C_1 \sqcap C_2 \sqsubseteq A_2, \exists r.C \sqsubseteq C_2\}$. The CDG $G_\mathcal{T}$ has edges $E_\mathcal{T} = \{\varepsilon(A, A_1 \sqcap A_2), \varepsilon(A_1, B_1 \sqcap B_2), \varepsilon(A_2, C_1 \sqcap C_2), r(C_2, C)\}$. Then the set of proper witnessing sets for A is

$$\mathbf{W}_A = \{\{A\}, \{A_1, A_2\}, \{B_1, B_2, A_2\}, \{A_1, C_1, C_2\}, \{B_1, B_2, C_1, C_2\}\}$$

An entailment of the form $\mathcal{T} \models A_1 \sqcap C_1 \sqcap \exists r.C \sqsubseteq A$ is captured by a set $W_A = \{A_1, C_1, C_2\}$ witnessing A, where C_2 propagates $\exists r.C$.

With these notions in place, our CDG captures *all* the relevant entailments.

Lemma 1. *Let $\mathcal{T}$ be an $\mathcal{ELH}i^{ql}$ TBox and let $G_\mathcal{T}$ be its CDG. If there is a proper witnessing set $W_A = \{A_1, \ldots, A_n\}$ for A and a set of concepts $\{C_1, \ldots, C_n\}$ (of the form $\exists s_1 \ldots \exists s_k.B$ with $k \geq 0$) such that, for each $1 \leq i \leq n$, there is a path in $G_\mathcal{T}$ from A_i propagating C_i, then $\mathcal{T} \models C_1 \sqcap \ldots \sqcap C_n \sqsubseteq A$.*

To show the converse Lemma 2 we construct a model $\mathcal{I}_{\mathcal{A},\mathcal{T}}$ from a given ABox $\mathcal{A}$ that makes true exactly the $\mathcal{ELH}i^{ql}$ inclusions witnessed by $G_\mathcal{T}$.

Lemma 2. *Let $\mathcal{T}$ be a $\mathcal{ELH}i^{ql}$ TBox and let $G_\mathcal{T}$ be its CDG. If $\mathcal{T} \models C_1 \sqcap \ldots \sqcap C_n \sqsubseteq A$ where $C_1, \ldots, C_n$ are concepts of the form $\exists s_1 \ldots \exists s_k.B$ with $k \geq 0$ and $A, B \in \mathbf{C}$, then there is a set $W_A = \{A_1, \cdots, A_n\}$ that witnesses A and such that for each $1 \leq i \leq n$ there is a path in $G_\mathcal{T}$ from A_i that propagates C_i.*

The key to our rewriting of atomic queries is that instance checking for a concept A_0 reduces to finding a set of ABox assertions in the given ABox that match a set of propagating paths that witnesses A_0.

Lemma 3. *Let $\mathcal{T}$ be an $\mathcal{ELH}i^{ql}$ TBox and let $G_\mathcal{T}$ be its CDG. For each ABox $\mathcal{A}$, $A \in \mathbf{C}$ and individual n_0, we have $\mathcal{T}, \mathcal{A} \models A(n_0)$ iff there is a proper witnessing set $W_A = \{A_1, \ldots, A_n\}$ for A such that for each $A_j \in W_A$ there exists (i) a (possibly empty) path p_j in $G_\mathcal{T}$ with role label $(r_1^j \cdots r_k^j)$ that propagates a concept $\exists s_1 \ldots \exists s_k.B_j$; and (ii) a sequence of individuals $n_1^j, \ldots, n_k^j$ such that $B_j(n_k^j) \in \mathcal{A}$ and $r_i^j(n_i^j, n_{(i+1)}^j) \in \mathcal{A}$ for each $1 \leq i < k$.*

Observe that in an $\mathcal{ELH}i^{ql}$ TBox, A is local in all axioms $C_1 \sqcap \ldots \sqcap C_n \sqsubseteq A$, and also every B for which $\mathcal{T} \models A \sqsubseteq B$ holds must be local. Thus, A can never participate on the left-hand side of axioms of the form $\exists r.A \sqsubseteq C$, which could lead to entailments of the form like $\mathcal{T} \models \exists r.(C_1 \sqcap \ldots \sqcap C_n) \sqsubseteq C$. In other words, all relevant entailments of an $\mathcal{ELH}i^{ql}$ TBox take the form $\mathcal{T} \models C_1 \sqcap \ldots \sqcap C_n \sqsubseteq A$. Hence this Lemma is not hard to show using Lemma 1 and Lemma 2. The proofs to the above claims can be found in the full version [20].

4 Rewriting Atomic Queries

In this section we present a rewriting for instance queries under $\mathcal{ELH}i^{ql}$ TBoxes. We do this relying on CDGs, which represent the witnessing sets and propagating paths that we have proved to provide all relevant entailments.

We start by defining a non-deterministic finite automaton (NFA) that captures all paths in $G_\mathcal{T}$ that propagate a concept C.

Definition 10 (Path-Generating NFA). *Let $G_\mathcal{T}$ be a CDG, and let $Q_\mathcal{T}$ contain the nodes in $G_\mathcal{T}$ that are $\mathcal{ELH}i^{ql}$ concepts. For $B \in \mathbf{C}$, the B-*path-*generating automaton of $G_\mathcal{T}$, termed $G_\mathcal{T}^{\mathbb{A}}(B)$, is the NFA $\langle Q_\mathcal{T}, \Sigma, \delta, q_0, F\rangle$ with:*

$$\Sigma = \overline{\mathbf{R}} \cup \mathbf{C} \cup \{\varepsilon\} \qquad q_0 = B \qquad F = \{\top\}$$

$$\delta(n_0, \alpha) = \begin{cases} n_1 & \alpha(n_0, n_1) \in E_\mathcal{T} \\ \top & \text{if } \alpha \in \mathbf{C} \end{cases}$$

Example 5. In order to illustrate how the path-generating NFAs are constructed, we look at the CDG $G_\mathcal{T}$ given in Example 2 and show its corresponding B_1-path-generating NFA $G_\mathcal{T}^{\mathbb{A}}(B_1)$ with the extracted RPE for B_1 underneath it.

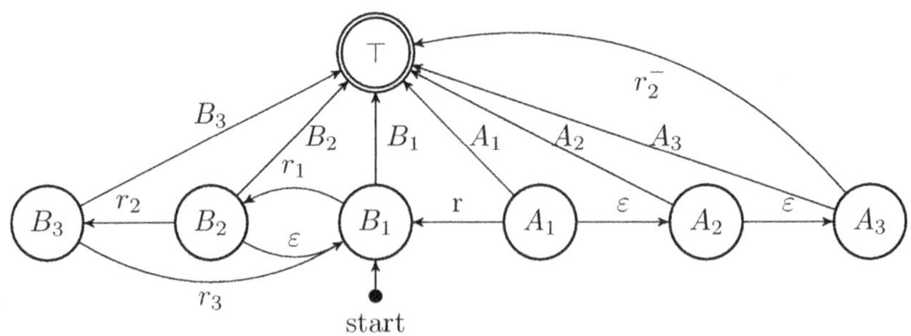

$$\Big(\overbrace{((r_1 + (r_1 r_2 r_3))^* B_1)}^{\text{paths ending in } B_1} + \overbrace{(r_1^+ (r_2 r_3 r_1^+)^* B_2)}^{\text{paths ending in } B_2} + \overbrace{(r_1^+ r_2 (r_3 r_1^+ r_2)^* B_3)}^{\text{paths ending in } B_3}\Big)$$

We use $\mathtt{rpe}(B, G_{\mathcal{T}})$ to denote the regular path expression constructed from $G_{\mathcal{T}}^{\mathbb{A}}(B)$ (using standard techniques); note that here ε is the usual NFA ε-transition. Clearly, $\mathtt{rpe}(B, G_{\mathcal{T}})$ contains exactly the words of the form $s_1, \ldots, s_k A$ such that there is an $(s_1, \ldots, s_k)$-propagating path from B to A in $G_{\mathcal{T}}$.

We now provide a rewriting algorithm for queries of the form $q(x) := A(x)$. We first extract from $G_{\mathcal{T}}$ the witnessing sets for A, and then use $\mathtt{rpe}(B, G_{\mathcal{T}})$ to consider all propagating paths for the concept names in each witnessing set. Using Lemma 3, it is not hard to show that Algorithm 1 is a sound and complete query rewriting algorithm for atomic queries of the form $q(x) := A(x)$.

Proposition 1. *Let $\mathcal{T}$ be an $\mathcal{ELH}i^{ql}$ TBox and $q(x) := A(x)$ a C2RPQ. For every ABox $\mathcal{A}$ and tuple $\vec{a}$ of individuals, we have that $\vec{a}$ is a certain answer to $q(\vec{x})$ over $(\mathcal{T}, \mathcal{A})$ if and only if $\vec{a}$ is an answer to* `rewriteAtomQuery`$(\vec{x})$ *over $\mathcal{A}$.*

Description of Algorithm 1. The function `witnessSets` takes an input a concept A and a CDG, and iteratively computes all the proper witnesses by exhaustively replacing a concept with a conjunction whenever a corresponding path is found. The function `rewriteRole` simply checks for any other roles that imply r w.r.t. $\mathcal{T}$ and produces the union over all of them. In case of function `rewrConcept`, we first extract the RPE, as described above, and then use `rewriteRole` to replace each role in the produced RPE with the union of its subroles. Finally, `rewriteAtomQuery` brings everything together: given an atomic query $A(x)$, we compute its witnessing sets and for each such set, rewrite each concept name into an RPE using function `rewrConcept`. Note that as $A(x)$ is

---

Algorithm 1: Rewriting atomic concept queries

Input : $A(x)$, $G_{\mathcal{T}}$, $\mathcal{T}$
Output: UC2RPQ Q

function `rewriteAtomQuery`$(A(x), G_{\mathcal{T}})$:
  $Q := \emptyset$
  foreach $\{B_1, \ldots, B_n\}$ *in* `witnessSets`$(A, G_{\mathcal{T}})$ **do**
    $q(x) :=$ `rewrConcept`$(B_1, G_{\mathcal{T}})(x, y_1) \wedge \cdots \wedge$ `rewrConcept`$(B_n, G_{\mathcal{T}})(x, y_n)$
    $Q := Q \cup \{q(x)\}$
  return Q
function `rewriteRole`$(r, \mathcal{T})$:
  return $\bigcup r_i$ where $r_i \sqsubseteq_{\mathcal{T}}^{*} r$
function `rewrConcept`$(A, G_{\mathcal{T}})$:
  $\pi := \mathtt{rpe}(A, G_{\mathcal{T}})$
  foreach $r \in \pi$ **do**
    $\pi := \pi[r\backslash$ `rewriteRole`$(r, \mathcal{T})]$
  return π
function `witnessSets`$(A, G_{\mathcal{T}})$:
  $\mathbf{W_A} := \{\{A\}\}; \mathbf{W'_A} := \emptyset$
  while $W_A \neq W'_A$ **do**
    $\mathbf{W'_A} := \mathbf{W_A}$
    if *B propagates $B_1 \sqcap \cdots \sqcap B_n$ in $G_{\mathcal{T}}$ and $W \in \mathbf{W'}_A$ with $B \in W$* **then**
      $\mathbf{W_A} := \mathbf{W_A} \cup ((W \setminus \{B\}) \cup \{B_1, \cdots B_n\})$
  return $\mathbf{W_A}$

---

replaced by a set of atoms that may match to regular paths, we introduce a fresh variable y_i for each of them. We produce a C2RPQ by forming the conjunction of these atoms. The output of `rewriteAtomQuery` is the union of these C2RPQs.

5 Rewriting Navigational Queries

In this section we provide an algorithm for rewriting navigational queries. Its pseudo-code description is given in Appendix B. Unfortunately, it is in general not possible to rewrite C2RPQs into unions of C2RPQs (UC2RPQs) for any ontology language that allows $\mathcal{T} \models \exists r.\top \sqsubseteq \exists s.\top$ for role names r and s, as we show in Theorem 1. The proofs for this section are in the full version [20].

Theorem 1. *There exist C2RPQs that cannot be rewritten into UC2RPQs w.r.t. TBoxes containing concepts of the form $\exists r.\top$ on both sides of CIs. This holds already for C2RPQs with only one atom.*

This negative result motivates the need for a restricted query language that admits a rewriting into UC2RPQs under $\mathcal{ELH}i^{ql}$ TBoxes. We use a similar language to [14], called *Navigational Conjunctive Queries* (NCQs).

Definition 11 (Navigational Conjunctive Query). *A* Navigational Conjunctive Query *(NCQ) is a C2RPQ, with atoms restricted to the following forms:*

$$\{T\}(x)\ \{T\}(x,y)\ \langle A_1\rangle + \cdots + \langle A_n\rangle(x,x)\ \big(\pi_1 + \cdots + \pi_n\big)(x,y)\ \big(\pi_1 + \cdots + \pi_n\big)^*(x,y)$$

with T a data test, $A_i \in \mathbf{C}$ and π_i a restricted RPE $\pi_i := r \mid r^- \mid r^* \mid (r^-)^*$.

While in NCQs we cannot use concatenation, it can still be simulated outside the scope of Kleene stars as usual.

5.1 Rewriting Algorithm for NCQs

In this section we give the complete algorithm for rewriting NCQs and a given $\mathcal{ELH}i^{ql}$ TBox into UC2RPQs. Let us recall that $\mathcal{ELH}i^{ql}$ does not make assertions about property values, meaning that data tests are only evaluated on individuals in the ABox. For the sake of simplicity, we can therefore omit atoms with data tests during rewriting, as they remain untouched in the query. For the purpose of rewriting NCQs, we make use of the functions for rewriting single roles and concepts given in Sect. 4. We thus introduce a new function, called *clipping*, which makes use of axioms in the TBox to rewrite a given NCQ into a union of NCQs, inspired by [17]. Given an RPE π and a role r, we construct a new expression $\pi|^{\mathcal{T}} r$ in the following way.

Definition 12. *Let π be an expression of the form $\pi_1 + \cdots + \pi_n$, and let r be a role. Then the* r-restriction of π, *written as $\pi|^{\mathcal{T}} r$, is the union of all those π_j s.t. there exists some s with $r \sqsubseteq^*_{\mathcal{T}} s$, where s is in π_j.*

Informally, $\pi|^{\mathcal{T}}r$ matches only those paths of π that contain the role r or a super-role of r. For a set of atoms $\mathbb{A}$ and atom π we use $\pi(x,y) \mathrel{\widehat{\in}} \mathbf{A}$ to mean $\pi(x,y) \in \mathbf{A}$ or $\pi^-(y,x) \in \mathbf{A}$. Note that $\pi^-(x,y)$ is built from $\pi(y,x)$, where $r^- \in \pi^-$ iff $r \in \pi$.

Definition 13 (Clipping Function). *Given a query $q(\vec{x})$, we select a set Y of variables s.t. $Y \cap \vec{x} = \emptyset$, and a CI $A \sqsubseteq \exists r.B \in \mathcal{T}$. Then, do the following:*

(D1) Pick any $y \in Y$, and replace each $y' \in Y$ by y everywhere in $q(\vec{x})$.
(D2) Every atom α where y occurs needs to satisfy one or more of the following:
- *(A) α contains a star, and if it contains a variable different from y, it is an unbound variable.*
- *(B) α contains a concept name C with $\mathcal{T} \models B \sqsubseteq C$, or $\mathcal{T} \models \exists r^-.\top \sqsubseteq C$, and if it contains a variable different from y, it is an unbound variable.*
- *(C) α is of the form $\pi(x,y)$ or $\pi^-(y,x)$ where $x \neq y$ and π contains some s with $r \sqsubseteq^*_{\mathcal{T}} s$.*

(D3) Let
$\mathbf{C}_y = \{\pi(x,y) \mid \pi(x,y)$ satisfies (C)$\} \cup \{\pi^-(x,y) \mid \pi(y,x)$ satisfies (C)$\}$. Define $\mathbf{C}^*_y$ as the set of atoms α in $\mathbf{C}_y$ that contain a role s with $r \sqsubseteq^*_{\mathcal{T}} s$ occurring in α in the scope of a star. Let $\mathbf{C}^1_y = \mathbf{C}_y \setminus \mathbf{C}^*_y$, and let X be the set of all variables different from y that occur in $\mathbf{C}^1_y$.
(D4) Drop from $q(\vec{x})$ every atom satisfying (A) or (B), and every atom $\alpha \mathrel{\widehat{\in}} \mathbf{C}^1_y$.
(D5) Replace each $x \in X$ by y, everywhere in $q(\vec{x})$.
(D6) Replace each atom $\pi(x,y) \mathrel{\widehat{\in}} q$ such that $\pi(x,y) \in \mathbf{C}^*_y$ by $\pi|^{\mathcal{T}}r(x,y)$.
(D7) Add $A(y)$ to $q(\vec{x})$.

The clipping function relies on a well-known property of $\mathcal{ELHI}$: for each $\mathcal{T}$ and $\mathcal{A}$ there is a universal, tree-shaped model $\mathcal{I}^u$ that can be used for answering all C2RPQs. Intuitively, we consider each possible set Y of variables that may be mapped to the same 'anonymous' object d of maximal depth in $\mathcal{I}^u$. Each such d is triggered by one existential axiom $A \sqsubseteq \exists r.B$, i.e., if d was added to $\mathcal{I}^u$, we know that it has a parent d_p that is A, and d was introduced as an r-child of d_p to satisfy $A \sqsubseteq \exists r.B$. Using this, we can modify the query to require that we map a variable to the object d_p that is A, and drop from the query all (parts of) atoms that are already guaranteed by the existence of d. We show in the correctness proofs that each application of the function results in a rewritten query with the same answers, but whose mappings have a strictly lower depth (as at least one variable now is mapped to d_p instead of its child d). By repeated application, we obtain a query with all variables are mapped to ABox individuals.

Example 6. Let us consider the TBox $\mathcal{T} = \{A \sqsubseteq \exists r.B\}$ and the query $q(x_1) := (t^* + r^*)(x_1, x_2) \land s^*(x_2, x_3) \land B(x_3) \land r^-(x_2, x_4) \land C(x_4) \land t^*(x_4, x_5)$. In line 6 of Algorithm 2 we iterate over all subsets Y of $vars(q)$. Suppose that $Y = \{x_2, x_3\}$, then after applying (D1) we get the query $q(x_1) := (t^* + r^*)(x_1, x_2) \land s^*(x_2, x_2) \land B(x_2) \land r^-(x_2, x_4) \land C(x_4) \land t^*(x_4, x_5)$. Each atom of the query fulfills one of the conditions in item (D2). In (D3) we get the sets $\mathbf{C}^*_{x_2} = \{(t^* + r^*)(x_1, x_2)\}$ and $\mathbf{C}^1_{x_2} = \{r^-(x_2, x_4)\}$, note that $\mathbf{C}_{x_2}$ is the union of these two sets. Then, in the next step (D4) we drop from q every atom that satisfies (A), (B), and the set $\mathbf{C}^1_{x_2}$. After replacing the variables in $X = \{x_4\}$ by x_2, we obtain the query $q(x_1) := (t^* + r^*)(x_1, x_2) \wedge C(x_2)$. In step (D6) we replace the atom $(t^* + r^*)(x_1, x_2)$ by $r^*(x_1, x_2)$. Observe that keeping t^* in the query makes the rewriting become not sound, since we either query for a path with exclusive r or t labels. Finally, in (D7) we add $A(x_2)$ to the query and return $q(x_2) := r^*(x_1, x_2) \wedge C(x_2) \wedge A(x_2)$.

In order to reduce the number of redundant queries in the output, we extend Algorithm 2 by a simple containment check that we call *structural subsumption* (see function $\texttt{add}^{\subseteq}$). Intuitively, a query q is structurally subsumed by another query q' w.r.t. some TBox $\mathcal{T}$ (if each atom of q is more specific than an atom of q'); this guarantees that the set of answers of q' always contains all the answers of q. This allows us to drop some queries and obtain a smaller rewriting.

Definition 14. *Let $\mathcal{T}$ be a TBox and q, q' be NCQs, we say that q is* structurally subsumed *by q' w.r.t. $\mathcal{T}$, or short $q \subseteq_{\mathcal{T}} q'$, if for each atom $\beta_1 + \ldots + \beta_j + \ldots + \beta_m(x, y) \in q'$ there exists an atom $\alpha_1 + \ldots + \alpha_i + \ldots + \alpha_n(x, y) \in q$, such that for each α_i there is a β_j such that $\mathcal{T} \models \alpha_i \sqsubseteq \beta_j$ and $\alpha_i, \beta_j \in \mathbf{C} \cup \overline{\mathbf{R}}$.*

Example 7. Consider a TBox $\mathcal{T} = \{r \sqsubseteq s, A_1 \sqsubseteq B_1, A_2 \sqsubseteq B_2\}$ and the queries $q_1(x) := C(x), r(x, y) \wedge A_1 + A_2(y)$ and $q_2(x) := s(x, y) \wedge B_1 + B_2 + B_3(y)$. Then, by Definition 14 q_1 is subsumed by q_2 w.r.t. $\mathcal{T}$. Let's consider the ABox $\mathcal{A}_1 = \{r(a, b), C(a), A_1(b)\}$. Observe that a is an answer to $q_1(x)$ as well as $q_2(x)$ over $(\mathcal{T}, \mathcal{A})$. However, this is not always the case for the opposite direction as we show by the ABox $\mathcal{A}_2 = \{r(a, b), B_2(b)\}$. Here a is indeed an answer to $q_2(x)$ over $(\mathcal{T}, \mathcal{A}_2)$, but not q_1. Hence, q_1 is subsumed by q_2 w.r.t. $\mathcal{T}$, but not vice-versa.

Lemma 4. *Let $\mathcal{T}$ be a TBox, $\mathcal{A}$ be an ABox, and $q_1(\vec{x}), q_2(\vec{x})$ be two NCQs, such that $q_1 \subseteq_{\mathcal{T}} q_2$. Then, $\vec{a}$ is a certain answer to $(\mathcal{T}, \mathcal{A}, q_2(\vec{x}))$ if $\vec{a}$ is a certain answer to $(\mathcal{T}, \mathcal{A}, q_1(\vec{x}))$.*

Algorithm to Rewrite NCQs. We now formally state the definition of the function `rewriteNCQ`. As input we assume an NCQ $q(\vec{x})$ and an $\mathcal{ELHI}^{ql}$ TBox $\mathcal{T}$. The function first iterates over all subsets of variables and axioms in $\mathcal{T}$ of the form (NF3) and (NF6), and applies the clipping function exhaustively, producing a union of NCQs Q. Next the function loops over the queries $q' \in Q$ and for each concept A_i in q', we compute its witnessing sets and between lines 9 to 12, we produce a conjunction using `rewrConcept`, similar to Algorithm 1. For roles r occurring inside q', between lines 13 and 14, we similarly produce a new query, replacing each occurrence of r with the output of `rewriteRole`. In order to reduce the number of redundant queries in Q we check q' against structural query subsumption over Q, and either remove all queries in Q that are contained in q' before adding it, or drop q' if q' itself is already structurally subsumed by some other element of Q. The result Q of this rewriting is a union of C2RPQs.

Theorem 2. *Let $\mathcal{T}$ be an $\mathcal{ELHI}^{ql}$ TBox and let $q(\vec{x})$ be a NCQ. For every ABox $\mathcal{A}$ and tuple $\vec{a}$ of individuals, we have that $\vec{a}$ is a certain answer to $q(\vec{x})$ over $(\mathcal{T}, \mathcal{A})$ if and only if $\vec{a}$ is an answer to* `rewriteNCQ`*$(q(\vec{x}), \mathcal{T})$ over $\mathcal{A}$.*

In the proofs we show that one step of the `clipping` function is sound and complete. In function `rewriteNCQ` we exhaustively apply the clipping function, which means that we have a query that can be evaluated over the plain ABox without the anonymous part. The correctness for the second part of function `rewriteNCQ`, where we substitute unions with conjunctions, is given by Lemma 1 and Lemma 2 and the exhaustive replacement of all reachable conjunctions and rewriting of roles and atoms. By Theorem 2 NCQ answering for $\mathcal{ELH}i^{ql}$ reduces to C2RPQ query evaluation, which is in NL in data complexity (see [6]). This is worst-case optimal.

Theorem 3. *Let $\mathcal{T}$ be an $\mathcal{ELH}i^{ql}$ TBox and q a NCQ. Then, the algorithm* `rewriteNCQ`$(\mathcal{T}, q)$ *terminates.*

6 Implementation and Experiments

We implemented a proof-of-concept prototype that, given an $\mathcal{ELH}i^{ql}$ TBox (in OWL syntax), rewrites NCQs into UC2RPQs and translates them into Cypher, a declarative query language for the Neo4j property graph database. The Cypher query is then evaluated over real-world data stored in Neo4j. The Java source code of the prototype is publicly available [15].

Setup. We execute the experiments on a virtual cluster node running Rocky Linux 8.10 with an AMD EPYC 7513 32-Core CPU clocked at 2.60 GHz and 400 GB RAM; with Neo4j 5.18.1 running on the same machine.

Ontology. As TBox (OWL ontology) we use the Cognitive Task Ontology (COGITO) [21], which integrates concepts of the Cognitive Atlas [25] with the Hierarchical Event Descriptors (HED) [28]. This ontology includes about 4700 concepts and 9200 axioms, all of them expressible in $\mathcal{ELH}i^{ql}$: 122 of the axioms contain conjunction (NF1) and existential quantifiers on the right (NF3). For example, the axiom $\mathsf{ReadingTask} \sqsubseteq (\exists \mathsf{has.Read} \sqcap \exists \mathsf{has.Lang\text{-}item})$ defines a reading task by referring to the HEDs Read and Lang-item (where the conjunction is just a shortcut for two axioms in normal form). For all axioms, COGITO also includes the converse axiom, e.g., $(\exists \mathsf{has.Read} \sqcap \exists \mathsf{has.Lang\text{-}item}) \sqsubseteq \mathsf{ReadingTask}$.

Data. The prototype rewrites an NCQ into a Cypher query assuming that concepts in the ontology correspond to node labels in the database, and roles correspond to relationships (i.e., edge labels). For the experiments we choose a dataset from the domain of cognitive neuroscience [26]. This dataset—stored in our Neo4j database—consists of 396 741 nodes and 2 870 405 relationships. It contains meta-information about fMRI data from OpenNeuro [27].

Queries. One use case of COGITO is to query for fMRI data containing a specific set of HED concepts (e.g. Lang-Item, Read), even if the data has only annotations for cognitive task concepts (e.g. ReadingTask), or vice versa. Our goal is to evaluate the effects of different input queries on our rewriting approach, which is not affected by the presence of data tests. Therefore, we do not include data tests in our queries. We generated a total of 4370 queries without data tests, which can be structurally divided into 5 groups (G1-G5 in Table 1a). The following list shows an example query representative of each query group.

Table 1. Properties of rewritten queries, rewriting and evaluation time in Neo4j.

(a) Queries grouped by type

Group		**Rewritten Queries (Avg.)**		**Runtime [s]**		**#Timeouts**
type	#queries	#answers	#atoms	rewriting	evaluation	(600s)
G1	114	0.35	27.13	0.053 35	2.877 53	0
G2	1041	1.21	5.00	0.035 65	2.389 56	6
G3	2060	0.45	52.46	1.006 72	54.405 65	27
G4	1041	371.65	2.79	0.015 20	0.938 19	0
G5	114	1.04	20.21	0.026 78	2.349 97	0
Total	4370	89.74	27.70	0.486 17	26.435 89	33

(b) Queries grouped by size (number of C2RPQs in union resulting from rewriting)

Group		**Rewritten Queries (Avg.)**		**Runtime [s]**		**#Timeouts**
size	#queries	#answers	#atoms	rewriting	evaluation	(600s)
1-10	3785	101.20	10.60	0.167 22	14.355 77	5
11-20	302	12.23	83.62	1.456 07	80.907 83	0
21-30	129	1.63	134.63	2.730 19	127.833 98	6
30+	154	21.11	289.81	5.309 56	153.257 02	22
Total	4370	89.74	27.70	0.486 17	26.435 89	33

G1 $\mathsf{q(x) := \langle Dataset\rangle(x) \wedge has^*(x,y) \wedge \langle ReadingTask\rangle(y)}$
G2 $\mathsf{q(x) := \langle Dataset\rangle(x) \wedge has^*(x,y) \wedge \langle Lang\text{-}item\rangle(y)}$
G3 $\mathsf{q(x) := \langle Dataset\rangle(x) \wedge has^*(x,y_1) \wedge \langle Read\rangle(y_1), has^*(x,y_2) \wedge \langle Lang\text{-}item\rangle(y_2)}$
G4 $\mathsf{q(x) := has(x,y) \wedge \langle Read\rangle(y)}$
G5 $\mathsf{q(x) := \langle ReadingTask\rangle(x)}$

The queries in the groups G1-G3 request fMRI datasets, with either a specific cognitive task (G1), one specific HED tag (G2), or a combination of two HED tags (G3). Since the depth at which the task or event tags occur varies from an fMRI scan to another, the queries use the Kleene star to navigate to them. In the group G4 and G5 we query for individual HED tags and tasks. While there are early research prototypes that can parse and evaluate GQL queries [22], to the best of our knowledge, there are no publicly available, robust and scalable database systems, which support GQL at the time of our experimental evaluation. Having such systems would allow us to evaluate our queries under the walk-based semantics that coincides with the certain answer semantics, and Cypher only supports the so-called *trail semantics* [19]. The walk-based semantics [5] returns all nodes that match the RPE of a query, while the trail semantics does not visit the same edge twice. Through careful manual inspection, we generated queries for which both semantics coincide. Finding syntactic conditions for the two semantics to match is left for future work.

Results. In Table 1a and Table 1b, we report the results of the experiments grouped by the type of input query and the number of queries in the output union, respectively. In each table we provide the number of queries in that group, the average rewriting and evaluation time, as well as the average number of atoms in the rewritten query. Lastly, we state how often the evaluation timed out at 600 s. We averaged the times for rewriting and evaluation over 10 runs for each of the input queries. Constructing the CDG, on which Algorithm 2 depends, takes around two minutes. In Table 1a we can see that group G3, which has queries with a combination of two HED tags, takes rather long (on average more than 50 s). We attribute that to the number of atoms, which suggest that the output queries are larger compared to the queries in the other groups. In Table 1b we see that queries producing a smaller number of C2RPQs in the output union were evaluated faster (compare average evaluation time for the groups 1-10 to 30+). The runtime also increases with the number of atoms in the query, which in turn grows with the query size. The time it takes to rewrite the queries is on average below 6 s, even for the group with the largest queries. The evaluation time seems to be independent of the number of answers. Additionally, we ran experiments with a version of Algorithm 2 that does not check for structural query subsumption in line 12. However, for our use case this rewriting algorithm often produces a union with more than 2000 C2RPQs. As a consequence the evaluation often times out, so this is no longer practicable.

7 Conclusion and Future Work

We presented an algorithm for rewriting NCQs into UC2RPQs over a lightweight ontology that extends DL-Lite with some of the expressive features of $\mathcal{ELH}$ while keeping the data complexity of reasoning in NL. Our restricted input query language (NCQs) is justified by the fact that we have proven the impossibility of using (U)C2RPQs as both the input and output language of query rewriting. Nested regular path queries seem a promising target language for rewriting ontology-mediated C2RPQs. One of our goals is to find a practicable algorithm, and our prototype implementation, which rewrites the queries into Cypher, suggests that we may be on track. It shows promising results on a real-world dataset from cognitive neuroscience. For future work, we aim to support a richer ontology language, and to target the GQL standard to support full C2RPQs.

Acknowledgments. This research was funded in whole or in part by the Austrian Science Fund (FWF) PIN8884924 and P30873. This work was also partially supported by the Wallenberg AI, Autonomous Systems and Software Program (WASP) funded by the Knut and Alice Wallenberg Foundation. This work was partially supported by the State of Salzburg under grant number 20102-F2101143-FPR (DNI) and the Austrian Federal Ministry of Education, Science and Research (BMBWF) under grant number 2920 (Austrian NeuroCloud). The authors acknowledge the computational resources and services provided by Salzburg Collaborative Computing (SCC), funded by the Federal Ministry of Education, Science and Research (BMBWF) and the State of Salzburg.

Disclosure of Interests. The authors have no competing interests to declare that are relevant to the content of this article.

A Additional Preliminaries

We provide here some further materials, that are not critical to be explicitly given in Sect. 2 since they are standard notions from the cited works given in that section.

Table 2. Semantics of $\mathcal{ELHI}$

Name	Syntax	Semantics
top concept	$\top$	$\Delta^\mathcal{I}$
concept name	A	$A^\mathcal{I} \subseteq \Delta^\mathcal{I}$
negation	$\neg A$	$\Delta^\mathcal{I} \setminus A^\mathcal{I}$
role name	r	$r^\mathcal{I} \subseteq \Delta^\mathcal{I} \times \Delta^\mathcal{I}$
inverse role	r^-	$\{(b, a) \mid (a, b) \in r^\mathcal{I}\}$
exist. restriction	$\exists r.C$	$\{a \mid b \in C^\mathcal{I} : (a, b) \in r^\mathcal{I}\}$
conjunction	$C \sqcap D$	$C^\mathcal{I} \cap D^\mathcal{I}$
concept inclusion	$C \sqsubseteq D$	$C^\mathcal{I} \subseteq D^\mathcal{I}$

Definition 15 (Semantics of C2RPQs). *Consider a C2RPQ $q(\vec{x})$ and a property graph $\mathcal{A} = (N, E, \mathsf{label}, \mathsf{prop})$. We define the evaluation of $q(\vec{x})$ in terms of a function $[\![\cdot]\!]^\mathcal{A}$, given in Table 3, which assigns to every query a set of mappings from $vars(q)$ to nodes in N. A tuple $\vec{a} \subseteq N$ is an* answer *of $q(\vec{x})$ in $\mathcal{A}$ iff there exists a mapping $\mu \in [\![\varphi]\!]^\mathcal{A}$ such that $\mu(\vec{x}) = \vec{a}$.*

Table 3. Walk-based evaluation function for C2RPQs [23] extended with data tests.

$$[\![k \odot v\ (x)]\!]^\mathcal{A} = \{\mu \mid v \text{ and } \mathsf{prop}(\mu(x), k) \text{ in } \mathbf{D}, \mathsf{prop}(\mu(x), k) \odot v \text{ holds}\}$$
$$[\![k \odot v\ (x, y)]\!]^\mathcal{A} = \{\mu \mid v \text{ and } \mathsf{prop}((\mu(x), \mu(y)), k) \text{ in } \mathbf{D}, \mathsf{prop}((\mu(x), \mu(y)), k) \odot v \text{ holds}\}$$
$$[\![T \wedge T'(x)]\!]^\mathcal{A} = [\![T(x)]\!]^\mathcal{A} \cap [\![T'(x)]\!]^\mathcal{A} \qquad [\![T \vee T'(x)]\!]^\mathcal{A} = [\![T(x)]\!]^\mathcal{A} \cup [\![T'(x)]\!]^\mathcal{A}$$
$$[\![T \wedge T'(x, y)]\!]^\mathcal{A} = [\![T(x, y)]\!]^\mathcal{A} \cap [\![T'(x, y)]\!]^\mathcal{A} \quad [\![T \vee T'(x, y)]\!]^\mathcal{A} = [\![T(x, y)]\!]^\mathcal{A} \cup [\![T'(x, y)]\!]^\mathcal{A}$$
$$[\![\neg T(x)]\!]^\mathcal{A} = \{\mu \mid \mu \notin [\![T(x)]\!]^\mathcal{A}\} \qquad [\![\neg T(x, y)]\!]^\mathcal{A} = \{\mu \mid \mu \notin [\![T(x, y)]\!]^\mathcal{A}\}$$
$$[\![\langle A \rangle(x, y)]\!]^\mathcal{A} = \{\mu \mid A \in \mathsf{label}(\mu(x)), \mu(x) = \mu(y)\}$$
$$[\![r(x, y)]\!]^\mathcal{A} = \{\mu \mid r(\mu(x), \mu(y)) \in E\} \qquad [\![r^-(x, y)]\!]^\mathcal{A} = \{\mu \mid r(\mu(y), \mu(x)) \in E\}$$
$$[\![(\pi\pi')(x, z)]\!]^\mathcal{A} = \{\mu \cup \mu' \mid \mu \in [\![\pi(x, y)]\!]^\mathcal{A}, \mu' \in [\![\pi'(y, z)]\!]^\mathcal{A}, \mu(y) = \mu'(y)\}$$
$$[\![(\pi + \pi')(x, y)]\!]^\mathcal{A} = [\![\pi(x, y)]\!]^\mathcal{A} \cup [\![\pi'(x, y)]\!]^\mathcal{A}$$
$$[\![\pi^*(x, y)]\!]^\mathcal{A} = \{\mu \mid \mu(x) \in N, \mu(x) = \mu(y)\} \cup [\![\pi(x, y)]\!]^\mathcal{A} \cup [\![\pi\pi(x, y)]\!]^\mathcal{A} \cup \dots$$

B Algorithm for Rewriting NCQs

We present the rewriting for NCQs in Algorithm 2.

As input we assume an NCQ $q(\vec{x})$ and an $\mathcal{ELHI}^{ql}$ TBox $\mathcal{T}$. We iterate over all variable subsets Y and axioms in $\mathcal{T}$ of the form NF3 and NF6, and apply the clipping function exhaustively. Then we loop over the queries inside $q' \in Q$ and for each atom A_i of a union $A_1 + \dots A_i \dots + A_n(x)$, we compute its witnessing sets and produce a new query by producing a conjunction using `rewrConcept` as shown. For roles r occurring inside q', we similarly produce a new query, replacing each occurrence of r with the output of `rewriteRole`. Note here that we use the function `add`$^{\subseteq}$ to avoid adding any rewritten queries to our output which are structurally contained by another element in the output.

Algorithm 2: Function `rewriteNCQ`

```
Input  : NCQ q = (φ, x⃗), ELHI^ql TBox T
Output: UC2RPQs Q
function rewriteNCQ(q, T):
    Q := {q}, Q' := ∅
    while Q' ≠ Q do
        Q' := Q
        foreach q' ∈ Q' do
            foreach A ⊑ ∃r.B ∈ T and Y ⊆ vars(q'), Y ∩ x⃗ = ∅ do
                Q := Q ∪ clipping(q', A ⊑ ∃r.B ∈ T, Y)
    foreach q' ∈ Q' do
        foreach A_1 + ... A_i ... + A_k(x) occurring in q' do
            foreach {B_1, ..., B_n} ∈ witnessSets(A_i, G_T) do
                q'_rw(x) := rewrConcept(B_1, G_T)(x, y_1) ∧ ··· ∧ rewrConcept(B_n, G_T)(x, y_n)
                Q := add^⊆(Q, q'_rw)
        foreach roles r occurring in q' do
            Q := Q ∪ q'[r\rewriteRole(r, T)]
    return Q
function add^⊆(Q, q):
    foreach q' ∈ Q do
        if q ⊆ q' then
            return Q
        if q' ⊆ q then
            Q := Q\q'
    return Q ∪ q
```

References

1. Mastro. https://obdm.obdasystems.com/mastro/. Accessed 27 Mar 2025
2. Ontopic. https://ontopic.ai. Accessed 27 Mar 2025
3. Stardog. https://www.stardog.com/. Accessed 27 Mar 2025
4. Abiteboul, S., Hull, R., Vianu, V.: Foundations of Databases. Addison-Wesley (1995). http://webdam.inria.fr/Alice/
5. Angles, R., Arenas, M., Barceló, P., Hogan, A., Reutter, J.L., Vrgoc, D.: Foundations of modern query languages for graph databases. ACM Comput. Surv. **50**(5), 68:1–68:40 (2017). https://doi.org/10.1145/3104031

6. Barceló, P.: Querying graph databases. In: Hull, R., Fan, W. (eds.) Proceedings of the 32nd ACM SIGMOD-SIGACT-SIGART Symposium on Principles of Database Systems, PODS 2013, New York, NY, USA - 22–27 June 2013, pp. 175–188. ACM (2013). https://doi.org/10.1145/2463664.2465216
7. Bienvenu, M., Calvanese, D., Ortiz, M., Simkus, M.: Nested regular path queries in description logics. In: Baral, C., Giacomo, G.D., Eiter, T. (eds.) Principles of Knowledge Representation and Reasoning: Proceedings of the Fourteenth International Conference, KR 2014, Vienna, Austria, 20–24 July 2014. AAAI Press (2014). http://www.aaai.org/ocs/index.php/KR/KR14/paper/view/8000
8. Bienvenu, M., Ortiz, M., Simkus, M.: Conjunctive regular path queries in lightweight description logics. In: Rossi, F. (ed.) IJCAI 2013, Proceedings of the 23rd International Joint Conference on Artificial Intelligence, Beijing, China, 3–9 August 2013, pp. 761–767. IJCAI/AAAI (2013). http://www.aaai.org/ocs/index.php/IJCAI/IJCAI13/paper/view/6886
9. Bienvenu, M., Ortiz, M., Simkus, M.: Regular path queries in lightweight description logics: complexity and algorithms. J. Artif. Intell. Res. **53**, 315–374 (2015). https://doi.org/10.1613/jair.4577
10. Calvanese, D., Giacomo, G.D., Lembo, D., Lenzerini, M., Rosati, R.: Tractable reasoning and efficient query answering in description logics: the DL-lite family. J. Autom. Reason. **39**(3), 385–429 (2007). https://doi.org/10.1007/s10817-007-9078-x
11. Calvanese, D., Giacomo, G.D., Lembo, D., Lenzerini, M., Rosati, R.: Data complexity of query answering in description logics. Artif. Intell. **195**, 335–360 (2013). https://doi.org/10.1016/j.artint.2012.10.003
12. Dimartino, M.M.: Integrating and querying linked datasets through ontological rules. Ph.D. thesis, University of London, Birkbeck College, UK (2020). https://ethos.bl.uk/OrderDetails.do?uin=uk.bl.ethos.852364
13. Dimartino, M.M., Calì, A., Poulovassilis, A., Wood, P.T.: Efficient ontological query answering by rewriting into graph queries. In: Cuzzocrea, A., Greco, S., Larsen, H.L., Saccà, D., Andreasen, T., Christiansen, H. (eds.) Flexible Query Answering Systems - 13th International Conference, FQAS 2019, Amantea, Italy, 2–5 July 2019, Proceedings. Lecture Notes in Computer Science, vol. 11529, pp. 75–84. Springer (2019). https://doi.org/10.1007/978-3-030-27629-4_10
14. Dragovic, N.: Querying property graphs with ontologies. Master's thesis, TU Wien (2022)
15. Dragovic, N., Löhnert, B.: Ontology-mediated querying for property graphs (2024). https://gitlab.com/austrian-neurocloud/software/owl2cypher/-/releases/ESWC2025. Accessed 27 Mar 2025
16. Dragovic, N., Okulmus, C., Ortiz, M.: Rewriting ontology-mediated navigational queries into cypher. In: Kutz, O., Lutz, C., Ozaki, A. (eds.) Proceedings of the 36th International Workshop on Description Logics (DL 2023) co-located with the 20th International Conference on Principles of Knowledge Representation and Reasoning and the 21st International Workshop on Non-Monotonic Reasoning (KR 2023 and NMR 2023), Rhodes, Greece, 2–4 September 2023. CEUR Workshop Proceedings, vol. 3515. CEUR-WS.org (2023). https://ceur-ws.org/Vol-3515/paper-9.pdf
17. Eiter, T., Ortiz, M., Simkus, M., Tran, T., Xiao, G.: Query rewriting for Horn-SHIQ plus rules. In: Hoffmann, J., Selman, B. (eds.) Proceedings of the Twenty-Sixth AAAI Conference on Artificial Intelligence, 22–26 July 2012, Toronto, Ontario, Canada, pp. 726–733. AAAI Press (2012). http://www.aaai.org/ocs/index.php/AAAI/AAAI12/paper/view/4931

18. Francis, N., et al.: A researcher's digest of GQL (invited talk). In: Geerts, F., Vandevoort, B. (eds.) 26th International Conference on Database Theory, ICDT 2023, March 28-31, 2023, Ioannina, Greece. LIPIcs, vol. 255, pp. 1:1–1:22. Schloss Dagstuhl - Leibniz-Zentrum für Informatik (2023). https://doi.org/10.4230/LIPIcs.ICDT.2023.1
19. Francis, N., et al.: Cypher: an evolving query language for property graphs. In: Das, G., Jermaine, C.M., Bernstein, P.A. (eds.) Proceedings of the 2018 International Conference on Management of Data, SIGMOD Conference 2018, Houston, TX, USA, 10–15 June 2018, pp. 1433–1445. ACM (2018). https://doi.org/10.1145/3183713.3190657
20. Löhnert, B., Augsten, N., Okulmus, C., Ortiz, M.: Towards practicable algorithms for rewriting graph queries beyond dl-lite. CoRR abs/2405.18181 (2024). https://doi.org/10.48550/arXiv.2405.18181
21. Löhnert, B., Engler, B., Hutzler, F., Augsten, N.: Cognitive task ontology (COGITO) (2025). https://gitlab.com/austrian-neurocloud/ontologies/cogito/-/releases/ESWC2025. Accessed 27 Mar 2025
22. Morra, O.: GQL parser (2021). https://github.com/OlofMorra/GQL-parser. Accessed 19 Dec 2024
23. Pérez, J., Arenas, M., Gutierrez, C.: Semantics and complexity of SPARQL. In: Cruz, I.F., et al. (eds.) The Semantic Web - ISWC 2006, 5th International Semantic Web Conference, ISWC 2006, Athens, GA, USA, 5–9 November 2006, Proceedings. Lecture Notes in Computer Science, vol. 4273, pp. 30–43. Springer (2006). https://doi.org/10.1007/11926078_3
24. Poggi, A., Lembo, D., Calvanese, D., Giacomo, G.D., Lenzerini, M., Rosati, R.: Linking data to ontologies. J. Data Semant. **10**, 133–173 (2008). https://doi.org/10.1007/978-3-540-77688-8_5
25. Poldrack, R.A., et al.: The cognitive atlas: toward a knowledge foundation for cognitive neuroscience. Front. Neuroinform. **5**, 17 (2011). https://doi.org/10.3389/fninf.2011.00017
26. Ravenschlag, A., Denissen, M., Löhnert, B., Pawlik, M., Himmelstoß, N.A., Hutzler, F.: Effective queries for mega-analysis in cognitive neuroscience. In: Fletcher, G., Kantere, V. (eds.) Proceedings of the Workshops of the EDBT/ICDT 2023 Joint Conference, Ioannina, Greece, 28 March 2023. CEUR Workshop Proceedings, vol. 3379. CEUR-WS.org (2023). https://ceur-ws.org/Vol-3379/CoMoNoS_2023_id252_Mateusz_Pawlik.pdf
27. Stanford Center for Reproducible Neuroscience: OpenNeuro MRI platform (2017). https://openneuro.org/search/modality/mri. Accessed 24 May 2024
28. Robbins, K., Truong, D., Appelhoff, S., Delorme, A., Makeig, S.: Capturing the nature of events and event context using hierarchical event descriptors (HED). Neuroimage **245**, 118766 (2021)
29. Sipser, M.: Introduction to the theory of computation. PWS Publishing Company (1997)
30. Xiao, G., et al.: Ontology-based data access: a survey. In: Proceedings of IJCAI 2018, pp. 5511–5519 (2023)
31. Xiao, G., Ding, L., Cogrel, B., Calvanese, D.: Virtual knowledge graphs: an overview of systems and use cases. Data Intell. **1**(3), 201–223 (2019). https://doi.org/10.1162/dint_a_00011

Designing Hierarchies for Optimal Hyperbolic Embedding

Melika Ayoughi(✉), Max van Spengler, Pascal Mettes, and Paul Groth

University of Amsterdam, Amsterdam, The Netherlands
m.ayoughi@uva.nl

Abstract. Hyperbolic geometry has shown to be highly effective for embedding hierarchical data structures. As such, machine learning in hyperbolic space is rapidly gaining traction across a wide range of disciplines, from recommender systems and graph networks to biological systems and computer vision. The performance of hyperbolic learning commonly depends on the hierarchical information used as input or supervision. Given that knowledge graphs and ontologies are common sources of such hierarchies, this paper aims to guide ontology designers in designing hierarchies for use in these learning algorithms. Using widely employed measures of embedding quality with extensive experiments, we find that hierarchies are best suited for hyperbolic embeddings when they are wide, and single inheritance, independent of the hierarchy size and imbalance.

Keywords: Hyperbolic Learning · Ontology Design · Machine Learning

1 Introduction

Knowledge graphs and ontologies provide a rich source of hierarchical information, such as the classification of creative works in Schema.org[1] or the organization of professions in Wikidata[2]. This hierarchical structure is well-suited for machine learning, particularly in hyperbolic learning, which utilizes hyperbolic geometry to embed tree-like structures into low-dimensional spaces [45]. Such embeddings have been shown to enhance performance in tasks like image and video classification [29,36,37], audio understanding [24,52], and recommender systems [34,62,64]. Throughout the literature [45,49], hyperbolic representation learning approaches typically assume that hierarchies are provided *as is*. However, ontology engineers consider multiple factors when designing ontologies beyond representing the domain (e.g. reusing existing ontologies, use concepts for interoperability, end user tasks). This work seeks to offer insights for ontology engineers in crafting hierarchies optimized for hyperbolic hierarchical embeddings. Unlike previous studies, which focus on improving hyperbolic embeddings

[1] https://schema.org/.

[2] https://www.wikidata.org/wiki/Wikidata:Main_Page.

E. Curry et al. (Eds.): ESWC 2025, LNCS 15718, pp. 362–382, 2025.
https://doi.org/10.1007/978-3-031-94575-5_20

for a given hierarchy, we address the reverse question: how can hierarchies be designed to enhance their suitability for embedding in hyperbolic space? Specifically, we conduct controlled experiments to examine how different tree structures affect the quality of embeddings produced by two primary classes of hyperbolic embedding algorithms: gradient-based and construction-based methods. Embedding quality is evaluated using distortion metrics [55], which quantify the discrepancy between distances in the embedding space (calculated via a continuous distance function) and the original graph distances (defined by the edge count between nodes). Our objective is to uncover the key axes of hierarchy design that influence the effectiveness of hyperbolic embeddings. Our results demonstrate that hierarchies optimized for width, rather than height, are best suited for hyperbolic embeddings. Hierarchy imbalance and size are shown to have minimal impact, while multiple inheritance should be avoided. We validate these findings using a real-world scenario, where alternative semantic organizations significantly reduce distortion. These results complement existing approaches to ontology design and evaluation by providing actionable insights for ontology engineers to enhance downstream embedding quality. We hope these recommendations assist ontology designers when downstream hyperbolic embedding performance is a priority. In summary, the contributions of this paper are as follows:

- **In-depth empirical study:** We perform in-depth analyses across four hyperbolic embedding algorithms to examine the impact of hierarchy structure on embedding quality.
- **Practical recommendations:** Our experiments lead to four recommendations for ontology engineers on structuring hierarchical portions of ontologies.
- **Real-world case study:** We validate our recommendations on real-world use cases, highlighting the inherent trade-off between ontological design goals and downstream utility in continuous hyperbolic spaces.

Our recommendations apply to various real-world scenarios requiring hierarchical data in continuous spaces, such as: recommender systems (e.g. product/content hierarchies), drug discovery (e.g., gene Ontology, SNOMED CT), and biological analysis (e.g. protein families). Code is openly available here[3].

2 Related Work

2.1 Hierarchy and Ontology Design

Hierarchies, particularly taxonomic backbones, whether formal or informal, play a critical role in ontology and knowledge graph design [22,28,44]. They organize complex domains [57] into manageable components and enable various forms of reasoning, such as subsumption. Changes in hierarchy structure can have a significant downstream impact on applications [51]. Therefore, ontology design

[3] https://github.com/Melika-Ayoughi/Optimal-Hierarchy.

methodologies provide guidance on crafting hierarchies to reflect domain constraints and ensure proper reasoning outcomes [28,53].

Works focused on the evaluation of ontologies also consider hierarchy [20,43]. These studies assess aspects such as whether a hierarchy correctly partitions instances, whether there are cycles of specialization and generalization, and whether instance assertions are semantically accurate [20]. Other evaluation approaches adopt principled criteria and metrics based on formal notions (e.g. unity) [3,17,18]. Examples of criteria include the complexity of the hierarchy (e.g. number of classes, depth, number of top level classes) as well as conciseness (e.g. cycles, and classes without instances) [50]

In ontology induction and knowledge graph construction, the creation of high-quality hierarchies is a critical consideration [63,69]. Evaluation typically involves expert review, comparison with gold-standard ontologies, or the application of aforementioned established evaluation criteria. Our work complements these existing recommendations, metrics, and evaluation approaches [43] by providing ontology engineers with guidance on designing hierarchical structures to enable machine learning tasks.

2.2 Learning over Knowledge Graphs Using Hierarchies

A significant body of work leverages hierarchical information within knowledge graphs to enhance machine learning tasks [27]. These tasks include link prediction [6,70], question answering [11], and query answering using embedding spaces [25]. Additionally, research has focused on creating embeddings for knowledge graphs with complex semantics [5]. Our work differs by providing guidance to ontology engineers on designing hierarchies, rather than focusing on embedding design for existing knowledge graphs.

2.3 Hyperbolic Representation Learning

We focus on hyperbolic embeddings, because they demonstrate superior performance in representing hierarchical data structures compared to Euclidean methods. In early work, Sarkar [56] introduced Delaunay tree embeddings in hyperbolic space, demonstrating the potential of hyperbolic geometry to achieve tree embeddings with arbitrarily low distortion. However, Sarkar's construction-based algorithm is limited to 2D embeddings, reducing its expressiveness and applicability in deep learning contexts. To address this limitation, Nickel and Kiela [46] proposed a contrastive approach that supports embedding optimization in any dimensionality, significantly outperforming Euclidean embeddings on trees. This line of work has been extended through entailment cones to induce partial hierarchical order [14,66], adapted to the Lorentz model of hyperbolic space [32,47], and improved by incorporating distortion [67] or separation [38] objectives during optimization. Subsequently, Sala et al. [55] expanded Sarkar's construction-based approach to higher-dimensional embeddings. Overall, these algorithms, whether optimized via gradient descent or constructed explicitly, consistently outperform Euclidean embeddings. This superiority stems from the insight that "hyperbolic space can be thought of as a continuous analogue to discrete trees" [47], owing to their shared nature of exponential growth.

In light of the strong performance of hyperbolic representation learning, numerous studies have integrated hyperbolic embeddings into neural networks, enabling deep learning to incorporate hierarchical knowledge. Hyperbolic learning have been shown to improve recognition across various domains, including image and video classification [29,36,37], word embeddings [10,33], recommender systems [34,62,64], audio understanding [24,52], single-cell analysis in biology [31], networks and graphs [60,65,71], and image-text settings [8,26,30,48]. Beyond classification, hyperbolic representations facilitate hierarchical recognition [9,16], learning from limited samples [15,23,41,68], interpretability [19], robustness [35,58], and other tasks. For a comprehensive overview of advances in hyperbolic learning, we refer to recent surveys [45,49]. A common assumption in the literature is that hierarchical information is known and fixed *a priori*. In this work, we flip the perspective to investigate how different hierarchical designs affect hyperbolic embeddings, aiming to potentially enhance integration of hierarchical ontologies in hyperbolic deep learning.

3 Hyperbolic Embedding Algorithms for Hierarchical Data

3.1 Preliminaries

Throughout this work, we are given a tree-like data structure $T = (V, E)$, containing a set of nodes V and a set of edges E, with each edge $e \in E$ connecting two vertices. We strive to obtain a continuous analogue of T by embedding each node $v \in V$ in an embedding space, such that the distance between two nodes corresponds one-to-one to the shortest path between the nodes in the tree, as given by the number of edges between them. Let $\phi : V \mapsto \mathbb{D}^n$ denote the embedding function that takes nodes as input and outputs their embedding in an n-dimensional hyperbolic space $\mathbb{D}^n$.

Following [14,46,55], we will operate in the Poincaré ball model of hyperbolic space for the embeddings. For an n-dimensional space, let $(\mathbb{D}^n, \mathfrak{g}^n)$ denote the Riemannian manifold of the Poincaré ball model, given as:

$$\mathbb{D}^n = \{\mathbf{x} \in \mathbb{R}^n : ||\mathbf{x}||^2 < 1\}, \qquad \mathfrak{g}^n = \lambda_{\mathbf{x}} I_n, \quad \lambda_{\mathbf{x}} = \frac{2}{1 - ||x||^2}. \tag{1}$$

A key operator for hyperbolic embedding algorithms is the distance between two vectors in hyperbolic space. Here, it denotes the distance between the embeddings of two nodes. For nodes $\mathbf{v}_1, \mathbf{v}_2 \in \mathbb{D}^n$, the distance is given as:

$$d_{\mathbb{D}}(\mathbf{v}_1, \mathbf{v}_2) = 2\tanh^{-1}\left(||-\mathbf{v}_1 \oplus \mathbf{v}_2||\right), \tag{2}$$

where $\oplus$ denotes the Möbius addition, defined as:

$$\mathbf{v}_1 \oplus \mathbf{v}_2 = \frac{(1 + 2\langle \mathbf{v}_1, \mathbf{v}_2\rangle + ||\mathbf{v}_2||^2)\mathbf{v}_1 + (1 - ||\mathbf{v}_1||^2)\mathbf{v}_2}{1 + 2\langle \mathbf{v}_1, \mathbf{v}_2\rangle + ||\mathbf{v}_1||^2||\mathbf{v}_2||^2}. \tag{3}$$

Using this manifold and distance function, we outline below how different algorithms generate hyperbolic embeddings for hierarchical data. We focus on two types of algorithms: general-purpose methods that optimize embeddings via gradient descent and hierarchy-specific approaches that constructively embed trees.

3.2 Gradient-Based Hyperbolic Embeddings

Gradient-based hyperbolic embeddings are general-purpose approaches that take any graph structure as input and yield a hyperbolic embedding of each node, where the embedding uses the edges between nodes as objective. In this work, we investigate two canonical approaches: Poincaré Embeddings [46] and Hyperbolic Entailment Cones [14].

Poincaré Embeddings. In the seminal work of Nickel and Kiela [46], the goal is to embed V using contrastive learning with edges E as positive pairs. Let $\Theta = \{\theta_i\}_{i=1}^{|V|}$ denote the embeddings of nodes in hyperbolic space. The estimation of Θ optimized under the following objective:

$$\Theta^* = \arg\min_{\Theta} \mathcal{L}(\Theta), \quad \text{s.t. } \forall \theta_i \in \Theta : ||\theta_i|| < 1. \tag{4}$$

Here, the loss is determined by the edges that connect two nodes. Specifically in the context of tree-like structures, edges denote hypernym-hyponym relations. With D the set of hypernym-hyponym relations the contrastive loss is given as:

$$\mathcal{L}(\Theta) = \sum_{(u,v)\in D} \log \frac{\exp(-d_{\mathbb{D}}(u,v))}{\sum_{v'\in N(u)} \exp(-d_{\mathbb{D}}(u,v'))}, \tag{5}$$

with $N(u)$ denoting the set of nodes not directly connected to u. To optimize Θ, the parameters are initialized as random vectors in a unit ball of dimensionality d and subsequently optimized using gradient descent in hyperbolic space [2,4].

Hyperbolic Entailment Cones. A limitation of the contrastive loss in Poincaré Embeddings is the absence of an explicit objective to preserve hierarchical order. Consequently, nodes deep in the hierarchy may be placed near the origin, reducing the utility of their embeddings. To address this, Ganea et al. [14] reinterpret hierarchical relations as partial orderings defined by cones in hyperbolic space. They extend the contrastive loss to a max-margin variant, aiming for each parent node u to encapsulate its child nodes v. Specifically, each child v must fall within the entailment cone of its parent u. The loss is defined as:

$$\mathcal{L} = \sum_{(u,v)\in D} E(u,v) + \sum_{(u',v')\in A\setminus D} \max(0, \gamma - E(u',v')), \tag{6}$$

with A the set of all node pairs, γ a margin, and the energy loss given as:

$$E(u,v) = \max(0, \Xi(u,v) - \psi(u)), \tag{7}$$

where $\psi(u)$ denotes the aperture of the cone based on its root point u, or equivalently: the size of the entailment cone. The closer u is to the origin, the larger its aperture, reflecting the intuition that points near the origin correspond to higher levels in the tree structure. Lastly, $\Xi(u, v)$ measures the angle between u and v. If v has a higher norm and its angle relative to u is smaller than the aperture of u, the embedding is considered correct, and no loss is incurred. Otherwise, the loss scales with the angular error. Similar to Poincaré Embeddings, this objective can be directly optimized using gradient descent in hyperbolic space. In practice, nodes are initialized with Poincaré Embeddings and refined using the Hyperbolic Entailment Cones objective.

Algorithm 1 Construction-based hyperbolic tree embeddings [55]

1: **Input:** Tree $T = (V, E)$, scaling factor $\tau > 0$ and root node v_1 with $\phi(v_1) = \mathbf{0}$.
2: **for** $v \in V$ **do**
3: Isometrically reflect $\phi(v)$ to the origin and apply the same to its parent.
4: Generate $\mathbf{x}_1, \ldots, \mathbf{x}_{\deg(v)}$ uniformly distributed points on a unit hypersphere.
5: Rotate the points such that $\mathbf{x}_1$ is aligned with the reflected parent embedding.
6: Scale $\mathbf{x}_1, \ldots, \mathbf{x}_{\deg(v)}$ according to τ and the tree distance to v.
7: Reflect rotated and scaled points back.
8: **end for**

3.3 Construction-Based Hyperbolic Embeddings

Gradient-based approaches are general-purpose and operate on a wide range of graphs, including those that are not strictly acyclic or have nodes with multiple inheritance, as in the case of Poincaré Embeddings [46]. However, this versatility often comes at the expense of embedding quality, with the resulting hyperbolic embeddings Θ of the nodes V retaining only partial information from the original graph. In contrast, construction-based methods [55,56] embed trees directly, sacrificing flexibility in the types of graphs they can handle in favor of producing high-quality embeddings that preserve nearly all the original tree structure.

The general approach of construction-based methods is outlined in Algorithm 1. These methods embed a root node at the origin and iteratively traverse the tree, positioning each child node on a sphere centered around its parent. This approach offers strong theoretical guarantees for low distortion and is highly efficient, with linear complexity relative to the number of nodes, avoiding complex optimization problems. However, these methods are limited to tree structures and often require arbitrary-precision arithmetic to achieve low-distortion embeddings.

The core distinction among the construction-based hyperbolic embeddings lies in step 4 of Algorithm 1. The distortion of the resulting embedding depends heavily on the degree of separation between the generated points. However, generating an arbitrary number of uniformly separated points on an n-dimensional hypersphere remains an open problem [54]. Sala et al. [55] propose two approaches for generating hyperspherical points at this step. The first involves placing points at the vertices of a hypercube inscribed within the hypersphere, leveraging coding theory [42]. Specifically, they use the *Hadamard* code, enabling the placement of $2^{\lfloor \log_2 n \rfloor}$ points with a fixed pairwise distance. While this method is computationally efficient and produces predictable results, it suffers from poor separation between points, resulting in higher distortion. Additionally, it imposes a strict requirement on the dimension n, namely:

$$2^{\lfloor \log_2 n \rfloor} \geq \deg_{\max}(V). \quad (8)$$

Their second approach involves *precomputing* 1000 hyperspherical points using the method from [39] and sampling from these as needed. This method often results in lower distortion compared to the first approach and offers greater flexibility regarding the dimension n. However, it has drawbacks, including higher variance in results, with problematic outliers for certain trees, and increased computational cost for smaller trees.

In terms of scalability, the Hadamard method is highly efficient, as an $n \times n$ Hadamard matrix can be constructed in $O(\log_2(n))$ time. The precomputed method incurs minimal initial computation but later only requires tensor sampling. Overall, constructive methods are significantly faster than optimization-based methods.

4 Experimental Setup

4.1 Data

Hierarchies for Controlled Experiments. For our first experiment we generate a variety of tree structures using the NetworkX library [21] with $N = 256, 512, 1024$ nodes to evaluate how different hierarchical structures affect the hyperbolic representation learning. The selected trees encompass diverse structural properties and are defined as follows:

- **Full r-ary trees** *(balanced)* We generate full r-ary trees [59] with r values ranging from 2 to 5. In an r-ary tree, all non-leaf nodes have exactly r children and all levels are full except for some rightmost position of the bottom level (if a leaf at the bottom level is missing, then so are all of the leaves to its right, resulting in trees of varying branching factors and depths. Intuitively, higher values for r result in wider trees lower values for deeper trees r.
- **Binomial tree** *(imbalanced)* A binomial tree is constructed iteratively, with each step having twice the number of nodes as the previous step, forming a hierarchical structure. A binomial tree of order k is defined recursively by linking two binomial trees of order $k-1$, where the root of one is the leftmost child of the root of the other. Thus, the tree grows imbalanced.
- **Barabási–Albert tree** *(long-tailed)* The Barabási–Albert graph [1] of n nodes is generated by attaching new nodes each with m edges that are preferentially attached to existing nodes with high degree, namely preferential attachment. We generate Barabási–Albert graphs with $m = 1$, which is guaranteed to form a tree. This trees' degree distribution follows a power-law distribution ($P(k) = k^{-3}$). The resulting tree captures the scale-free nature

observed in many real-world networks, where a few nodes have a high degree and most other nodes have a small degree.

These trees are chosen to represent a diverse range of topologies, from balanced and uniform (r-ary trees) to skewed (binomial and Barabási–Albert trees) [40], while simultaneously allowing us to have control over the number of nodes to enable direct comparisons between different hierarchical organizations. They are visualized for $n = 16$ in Fig. 1.

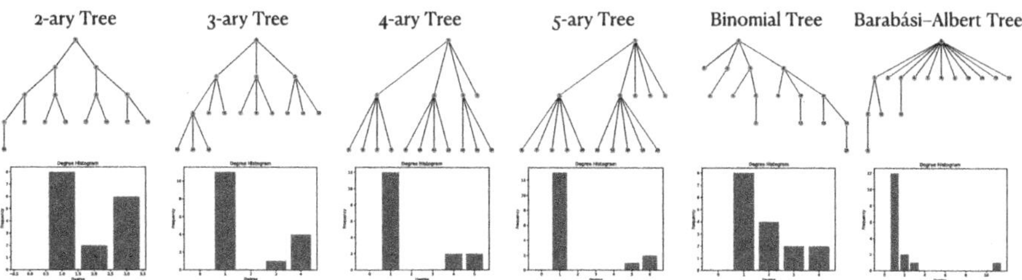

Fig. 1. The different hierarchies and their degree histograms used in our experiments. We investigate various balanced (r-ary trees) and imbalanced trees (binomial and Barabási-Albert trees) to help us understand which dimensions of hierarchy design are most important for their hyperbolic embedding.

Real-world Use Cases. To investigate the potential impact of our findings in real-world scenarios, we use the ImageNet [7][4] (based on WordNet [12]) and Pizza[5] ontologies as case studies, demonstrating the influence of alternative ontology designs on the quality of hyperbolic embeddings. The original ontologies, comprising 1778 and 100 nodes respectively, contain multiple inheritance and are non-tree structures. For each ontology, we construct a single inheritance version by applying the DFS algorithm to extract a spanning tree [61]. To minimize tree height while preserving the number of nodes, we restructure the hierarchy by merging the children of parent nodes into siblings. This reduces the height of ImageNet hierarchy from 13 to 8 and Pizza hierarchy from 7 to 5 in the reorganized ontologies.

4.2 Implementation Details

In our experiments, we analyze the impact of varying embedding dimensions. For the controlled experiment with generated trees, we use embedding dimensions of $d = 10$, $d = 20$, and $d = 130$. Since the Hadamard method encodes a tree with a minimum dimension d determined by Eq. 8, we set the embedding dimensions to $d = 40$ for the Pizza ontology and $d = 70$ for the ImageNet ontology, corresponding to their maximum degrees of 23 and 39, respectively.

[4] https://observablehq.com/@mbostock/imagenet-hierarchy.

[5] https://protege.stanford.edu/ontologies/pizza/pizza.owl.

Each embedding algorithm is configured using its recommended hyperparameter settings from the corresponding papers. For Poincaré embeddings, we adopt the settings used in the WordNet nouns experiment. Specifically, we use a constant learning rate of 1, with an initial burn-in phase of 20 epochs at a reduced learning rate of 0.1. Training continues for a total of 10, 000 epochs, with a batch size of 50. For each positive example, we randomly sample 50 negative examples.

For the entailment cones method, following Ganea et al. [14], we first pretrain using Poincaré embeddings for 100 epochs using a learning rate of 5.0 and a burn-in learning rate of 0.5 for the initial 20 epochs. Subsequently, we train with entailment cones loss for 300 epochs using a learning rate of 1.0. Both pretraining and training use a batch size equal to the number of nodes, meaning each epoch consists of a single step. We observed that increasing the number of steps led to overfitting, which adversely affected some metrics. For the construction-based approaches, following Sala et al. [55], the scaling factor τ is set to

$$\tau = \frac{1}{1.3 * \ell} \log \left(\frac{2 - \frac{\epsilon}{2}}{\frac{\epsilon}{2}} \right), \tag{9}$$

where ϵ is the machine precision of the applied floating point format and ℓ is the maximum path length of the tree, to avoid numerical problems while still obtaining near optimal results.

4.3 Embedding Evaluation Metrics

Following the conventions in hyperbolic embedding literature [46,55,56], we focus on three metrics to evaluate the quality of tree embeddings. The first metric is average relative distortion, which measures the average relative embedding error between all pairs of nodes in V, given as follows for $N = |V|$ nodes:

$$D_{avg}(\phi) = \frac{1}{N(N-1)} \sum_{u \neq v} \frac{|d_{\mathbb{D}}(\phi(u), \phi(v)) - d_T(u, v)|}{d_T(u, v)}. \tag{10}$$

This metric measures how much the hyperbolic distance on the embeddings differs from the tree distance between all node pairs. The second metric is worst-case distortion, which specifically measures the ratio between the largest stretching and shrinking factor of pairwise distances:

$$D_{wc}(\phi) = \max_{u \neq v} \frac{d_{\mathbb{D}}(\phi(u), \phi(v))}{d_T(u, v)} \left(\min_{u \neq v} \frac{d_{\mathbb{D}}(\phi(u), \phi(v))}{d_T(u, v)} \right)^{-1}. \tag{11}$$

where the average distortion measures the global distortion, the worst-case distortion captures large local distortions. The third metric is the mean average precision (MAP), given here as:

$$\mathrm{MAP}(\phi) = \frac{1}{N} \sum_{u \in V} \frac{1}{\deg(u)} \sum_{v \in \mathcal{N}_V(u)} \frac{\left| \mathcal{N}_V(u) \cap \phi^{-1}\Big(B_{\mathbb{D}}(u, v)\Big) \right|}{\left| \phi^{-1}\Big(B_{\mathbb{D}}(u, v)\Big) \right|}, \tag{12}$$

with deg(u) the degree of node u, $\mathcal{N}_V(u)$ the neighboring nodes of u, and $B_{\mathbb{D}}(u, v) \subset \mathbb{D}^n$ a closed ball centered at the embedding $\phi(u)$ of u with hyperbolic radius $d_{\mathbb{D}}(\phi(u), \phi(v))$. Intuitively, the MAP is a reconstruction measure, which identifies how well we can find back neighboring nodes in the area surrounding each embedded node.

5 Experiments

For our experiments, we focus on four research questions to explore key aspects of embedding hierarchies: (i) width versus depth, (ii) balanced versus imbalanced structures, (iii) few versus many nodes, and (iv) few versus many embedding dimensions. We address each question sequentially. Finally, we apply the insights gained to a case study, where we revisit the ImageNet and Pizza hierarchies and propose an alternative organization that improves hyperbolic embeddings.

5.1 Is It Better to Design Deep or Wide Hierarchies?

In the first experiment, we address a fundamental question: given the same number of nodes, should hierarchies be designed to be deep or wide? r-ary trees, as visualized in Fig. 2, provide an ideal structure for this investigation. Depth versus width inherently carries semantic implications for hierarchical organization. Our objective is to quantify its impact on the resulting hyperbolic embeddings.

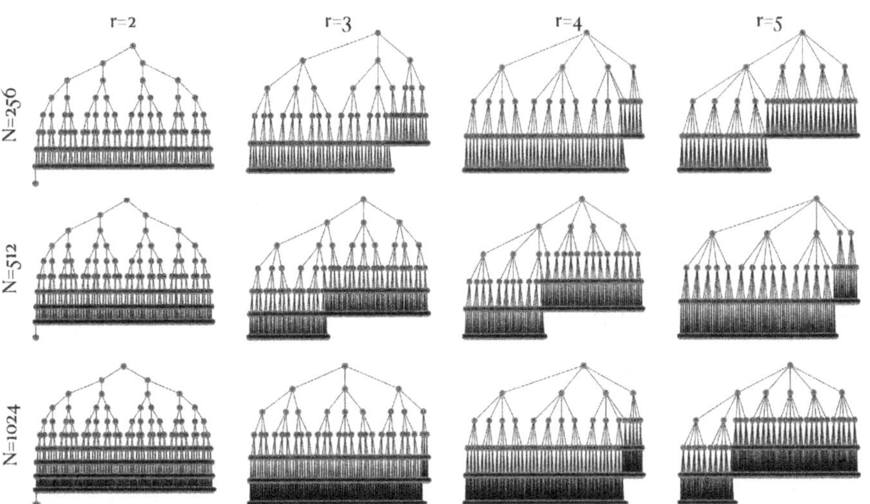

Fig. 2. Visualizing depth versus width in r-ary trees. We show hierarchies for 256, 512, and 1024 nodes for four branching factors, ranging from 2 to 5. The higher the branching factor, the wider the tree and the fewer hierarchical layers that are required to reach the same number of nodes in the hierarchy.

Figure 3 presents the average distortion, worst-case distortion, and MAP as functions of r for all r-ary trees. As the branching factor r increases, the hierarchy becomes wider, requiring fewer layers to reach the same number of nodes. For Poincaré embeddings, wider hierarchies result in higher distortions because the algorithm relies solely on contrastive learning, ignoring the partial order between hierarchical layers. Thus, it performs the worst among the approaches. Poincaré embeddings with $r = 2$ achieve lower distortion and higher MAP than the entailment method at high branching factors, effectively capturing deeper hierarchies.

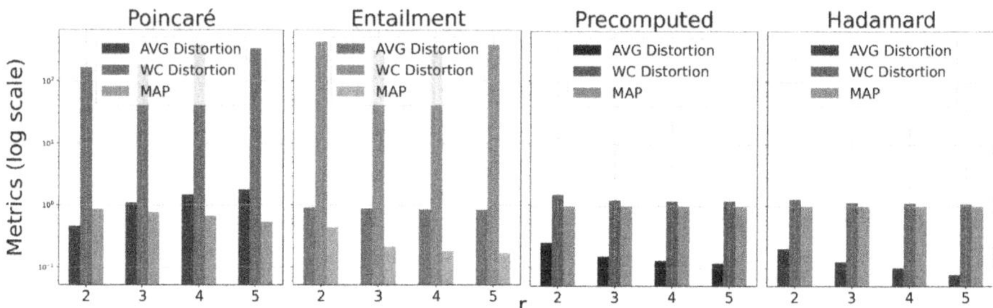

Fig. 3. Investigating depth versus width across four hyperbolic embedding algorithms, using r-ary trees with r ranging from 2 to 5, with hierarchies of 512 nodes and 20 embedding dimensions. For all methods except Poincaré embeddings, we find that wide and shallow hierarchies lead to lower distortion than thin and deep hierarchies.

In contrast to Poincaré, all other hyperbolic embedding methods exhibit the opposite trend: wider hierarchies improve embedding quality, particularly reducing average distortion. Construction-based methods outperform Poincaré with $r = 2$ in both metrics, highlighting a key trade-off. These methods achieve the best overall scores. Thus, we recommend using construction-based algorithms paired with wide hierarchies to achieve optimal hyperbolic embeddings.

5.2 What is the Impact of Hierarchy Imbalance?

In the second experiment, we examine the impact of hierarchy imbalance on embedding performance. Real-world hierarchies naturally tend to follow long-tailed distributions and power laws [13]. However, current embedding algorithms are agnostic to imbalance, and its effect on distortion remains largely unexplored. To address this, we compare two imbalanced hierarchies—based on binomial distributions and the Barabási–Albert model with four balanced r-ary trees. The results for all algorithms and evaluation metrics are presented in Table 1.

Table 1. Balanced versus imbalanced hyperbolic embeddings. We show the results across 4 embedding methods, 6 hierarchies, and 3 evaluation metrics with 512 nodes and 20 embedding dimensions. BA denotes the hierarchy from the Barabási–Albert construction, which can't be performed for the Hadamard construction due to having a maximum degree of 86 (Eq. 8), hence the dash. For each algorithm, we highlight the **best** and **worst** score over the hierarchies. Overall, we find that balance is not critical for hyperbolic embedding. It is better to have a wide imbalanced hierarchy than a deep balanced hierarchy.

	Gradient-based						Construction-based					
	Poincaré			Entailment			Precomputed			Hadamard		
	D_{avg}	D_{wc}	MAP	D_{avg}	D_{wc}	MAP	D_{avg}	D_{wc}	MAP	D_{avg}	D_{wc}	MAP
Balanced												
2-ary	0.459	164.777	0.866	0.914	434.177	0.439	0.259	1.539	1	0.207	1.297	1
3-ary	1.085	183.974	0.770	0.878	316.338	0.217	0.156	1.252	1	0.127	1.155	1
4-ary	1.471	390.397	0.671	0.855	323.967	0.183	0.133	1.201	1	0.103	1.121	1
5-ary	1.770	336.711	0.534	0.837	383.626	0.169	0.120	1.201	1	0.080	1.092	1
Imbalanced												
Binomial	1.439	69.530	0.171	0.863	224.731	0.304	0.249	1.542	1	0.186	1.257	1
BA	2.791	3607.95	0.020	0.802	731.914	0.231	0.140	1.329	1	-	-	-

Interestingly, imbalanced hierarchies do not necessarily yield the highest distortion. Focusing on the two construction-based methods, which outperform the others overall, we observe that the binomial and Barabási-Albert trees perform competitively with the r-ary trees. Their distortion is consistently better than the 2-ary trees but worse than the 5-ary trees. In summary, a wide imbalanced hierarchy is preferable to a deep balanced hierarchy, indicating that enforcing hierarchical balance is not a strict requirement. However, the best results are achieved when hierarchies are both wide and balanced, as imbalance increases depth.

5.3 What is the Impact of More Nodes on Embedding Quality?

The larger the hierarchy, the deeper the knowledge it represents, but this also increases the complexity of the corresponding embedding. In the third experiment, we examine how the average distortion of various hierarchies changes as a function of the number of nodes. Table 2 presents results for hierarchies with 256, 512, and 1024 nodes. Table 5 reports the MAPs of the same experiments. Notably, for Poincaré embeddings, larger hierarchies reduce distortion. This is due to the contrastive learning objective, which benefits from more node pairs, improving optimization. For all other methods, results remain largely stable, with a slight positive correlation between the number of nodes and distortion. This outcome relates to the findings of the first experiment: larger hierarchies tend to be deeper, and embedding algorithms that incorporate partial order perform better on shallower hierarchies. We conclude that for most embedding

Table 2. The effect of the number of nodes on average distortion across all hyperbolic embedding algorithms. BA represents the hierarchy generated by the Barabási–Albert model. For Poincaré embeddings, increasing the number of nodes reduces distortion, as more node pairs are available for contrastive learning. In contrast, for other methods, adding more nodes slightly increases distortion due to the added complexity from greater depth.

	Gradient-based						**Construction-based**					
	Poincaré			**Entailment**			**Precomputed**			**Hadamard**		
	256	512	1024	256	512	1024	256	512	1024	256	512	1024
Balanced												
2-ary	0.880	0.459	0.229	0.816	0.914	0.960	0.220	0.259	0.300	0.176	0.207	0.240
3-ary	1.439	1.085	0.752	0.742	0.878	0.940	0.124	0.156	0.160	0.102	0.127	0.130
4-ary	2.129	1.471	1.092	0.695	0.855	0.928	0.102	0.133	0.137	0.079	0.103	0.105
5-ary	2.472	1.770	1.385	0.657	0.837	0.919	0.115	0.120	0.156	0.078	0.080	0.103
Imbalanced												
Binomial	1.736	1.439	0.988	0.717	0.863	0.932	0.207	0.249	0.298	0.161	0.186	0.211
BA	3.444	2.791	2.206	0.595	0.802	0.903	0.108	0.140	0.178	-	-	-

algorithms, enriching hierarchies with more nodes only slightly increases distortion, highlighting that a strong increase in semantic complexity has minimal impact on embedding quality.

5.4 How Many Embedding Dimensions are Sufficient?

Hyperbolic geometry enables representation learning in compact spaces [45]. In the fourth experiment, we analyze the effect of embedding dimensionality on average distortion across different algorithms and hierarchies. The results, shown in Table 3, indicate that all approaches perform better with fewer embedding dimensions. Notably, hyperbolic entailment cones and the Hadamard construction are largely agnostic to dimensionality, exhibiting consistent performance across all dimensions. These findings align with existing literature on the efficiency of hyperbolic geometry in lower-dimensional settings.

Table 3. The effect of embedding dimensionality d across all four embedding algorithms for average distortion, with all hierarchies using 512 nodes. Hyperbolic geometry allows for embedding in low-dimensional spaces. Across all algorithms, using fewer dimensions does not hamper performance, and can even lead to better scores for Poincaré embeddings and the precomputed construction-based approach.

	Gradient-based						**Construction-based**					
	Poincaré			**Entailment**			**Precomputed**			**Hadamard**		
	10	20	130	10	20	130	10	20	130	10	20	130
Balanced												
2-ary	0.365	0.459	0.461	0.915	0.914	0.914	0.151	0.259	0.842	0.207	0.207	0.207
3-ary	1.130	1.085	1.105	0.880	0.878	0.878	0.095	0.156	0.467	0.127	0.127	0.127
4-ary	1.487	1.471	1.458	0.857	0.855	0.854	0.087	0.133	0.374	0.103	0.103	0.103
5-ary	1.838	1.770	1.834	0.840	0.837	0.836	0.088	0.120	0.325	0.080	0.080	0.080
Imbalanced												
Binomial	1.435	1.439	1.435	0.865	0.863	0.861	0.118	0.249	0.682	-	0.186	0.186
BA	2.801	2.791	2.784	0.805	0.802	0.802	0.109	0.140	0.371	-	-	0.114

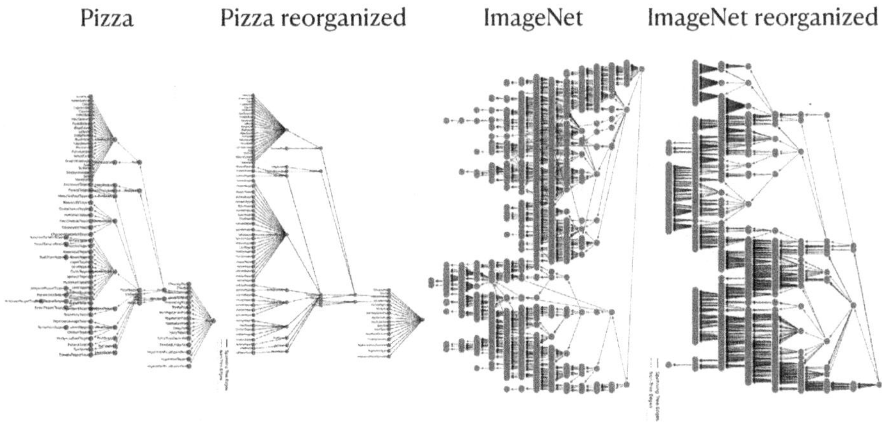

Fig. 4. Original and reorganized real-world ontologies Red edges indicate non-tree edges. Reorganization removes multiple inheritance and reduces tree height. (Color figure online)

5.5 Case Study: The Pizza and ImageNet Ontologies

Lastly, we analyze the practical impact of our recommendation—favoring width over height—on real-world ontologies and evaluate the effects of multiple inheritance. The Pizza and ImageNet ontologies provide suitable case studies, as shown in Fig. 4. The figure illustrates the original ontologies alongside their single inheritance and reorganized versions, which have been adjusted to reduce height while maintaining the same number of nodes.

Table 4 summarizes the results for the original, single inheritance, and reorganized versions of the ontologies.

Table 4. The effect of hierarchy re-organization on their hyperbolic embedding. To showcase our own recommendations, we take the existing ImageNet and Pizza hierarchies. Both have edges that create multiple inheritance, rendering them unusable for three out of four methods. Removing multiple inheritance allows for the use of more effective hyperbolic embedders. By including our other lessons as well, we arrive at a re-organization that leads to vastly better distortion and MAP scores.

	Gradient-based						Construction-based					
	Poincaré			Entailment			Precomputed			Hadamard		
	D_{avg}	D_{wc}	MAP	D_{avg}	D_{wc}	MAP	D_{avg}	D_{wc}	MAP	D_{avg}	D_{wc}	MAP
Pizza												
Original	3.321	7066.671	0.059	-	-	-	-	-	-	-	-	-
+ single inheritance	3.387	10509.346	0.051	0.499	511.594	0.195	0.234	1.538	1	0.126	1.180	1
+ reorganized	3.422	9343.566	0.045	0.452	1454.972	0.164	0.167	1.329	1	**0.089**	**1.118**	**1**
ImageNet												
Original	0.809	3983.563	0.087	-	-	-	-	-	-	-	-	-
+ single inheritance	0.722	2745.952	0.220	0.961	2364.827	0.293	0.725	885.622	0.725	0.297	1.647	1
+ reorganized	1.008	12715.625	0.156	0.955	4096.000	0.164	0.507	2.698	1	**0.171**	**1.232**	**1**

Notably, Poincaré embeddings are the only method applicable to the original multiple inheritance graphs, as other, more effective methods are incompatible with such structures. The results clearly demonstrate that reorganizing the ontologies significantly improves the average distortion across all methods except for Poincaré. Interestingly, for Poincaré embeddings, deeper hierarchies perform better, whereas for all other methods, wider hierarchies yield superior results.

However, there is a trade-off between expressivity and the structural adjustments made to reduce hierarchy depth. Enforcing a wider and less deep hierarchy can result in the loss of certain semantics, which may diminish the ontology's expressivity. For example, in the Pizza ontology, *Cheese Vegetable Topping* is both a child of *Cheese Topping* and *Vegetable Topping*, but one of these relationships is removed during the single inheritance process. Similarly, in the reorganized hierarchy, *Cheese Burger* which is originally connected through *Hamburger* to *Sandwich*, are flattened to become direct children of *Sandwich*. These adjustments improve downstream performance, particularly for construction-based methods. If preserving semantic expressivity is more critical, deeper hierarchies and Poincaré embeddings may be preferable. These findings highlight a nuanced trade-off: wider hierarchies generally optimize hyperbolic embeddings for performance, they may not be ideal for tasks requiring more semantics.

6 Recommendations

We offer the following recommendations for ontology engineers when designing ontologies or knowledge graph schemas for use with hyperbolic embeddings:

➠ **Design hierarchies for width:** The most effective embedding algorithms leverage the hierarchical order between nodes to generate embeddings. Consequently, these algorithms perform best with wide hierarchies that have high branching factors, rather than deep narrow trees with slower branching.

- **Do not worry about balance:** Current algorithms are largely agnostic to the balance between subtrees. Interestingly, our findings indicate that when balance is not prioritized or feasible, embedding performance is not significantly impacted. It is better to have a wide imbalanced hierarchy than a deep balanced hierarchy. Achieving both high width and balance leads to the best performance.
- **Hyperbolic embeddings can handle additional node complexity:** We find that a significant increase in the number of nodes only moderately impacts distortion. While more complex data structures lead to more challenging embedding optimization, strong enforcement of node sparsity is not required to maintain effective embeddings.
- **Avoid multiple inheritance:** While Poincaré embeddings can handle hierarchies with multiple inheritance, high-performance embedding algorithms do not support them. Therefore, to minimize distortion, it is best to have single inheritance. This approach is also recommended in many current ontology evaluation methodologies.

These recommendations should be seen as augmenting the main aim of ontology design which is to reflect the domain for the identified task. Hence, these recommendations serve to help ontology engineers balance the need to reflect the domain and the resulting ontologies effectiveness for use in downstream tasks.

7 Conclusion

In this work, we shed light on the relationship between hierarchy design and hyperbolic embeddings. Current hyperbolic literature assumes that hierarchies are fixed prior knowledge and focuses on minimizing embedding distortion. Here, we take the opposite approach by empirically investigating how different design choices can help improve hyperbolic embeddings. In the future, we plan to move beyond structural features of hierarchies to incorporate semantic aspects, including the ontology languages used and the information embedded in labels or literals. This could be done by introducing additional loss functions to balance distortion with semantic constraints. If ontology details (e.g. labels) are available, we could employ them via, e.g., embeddings. We hope that this study encourages future work in how knowledge graphs and data can be designed from the outset to improve down-stream machine learning performance.

Acknowledgments. This work was partially supported by the EU's Horizon Europe research and innovation programme within the ENEXA project (grant Agreement no. 101070305).

A Appendix

A.1 What is the Impact of More Nodes on Embedding Reconstruction?

In Sect. 5.3, we investigate the effect of the number of nodes on average distortion. Table 5 presents the MAP values for all hyperbolic embedding algorithms applied to trees with 256, 512, and 1024 nodes. The results show that construction-based methods achieve the highest MAP values, with the exception of the Barabási–Albert construction, which cannot be applied to the Hadamard method due to the maximum tree degree constraint (Eq. 8). For gradient-based methods, increasing the number of nodes generally leads to a slight reduction in MAP values. We conclude that for gradient-based algorithms, enriching hierarchies with additional nodes only marginally increases distortion, while construction-based methods remain unaffected. This highlights that even a significant increase in semantic complexity has minimal impact on embedding quality.

Table 5. The effect of the number of nodes on MAPs across all hyperbolic embedding algorithms. BA represents the hierarchy generated by the Barabási–Albert model. For gradient-based methods, increasing the number of nodes in most cases reduces MAP, whereas for construction-based methods, all experiments result in the maximum MAP of 1.

	Gradient-based						Construction-based					
	Poincaré			Entailment			Precomputed			Hadamard		
	256	512	1024	256	512	1024	256	512	1024	256	512	1024
Balanced												
2-ary	0.949	0.866	0.791	0.404	0.439	0.397	1	1	1	1	1	1
3-ary	0.862	0.770	0.620	0.254	0.217	0.206	1	1	1	1	1	1
4-ary	0.798	0.671	0.566	0.219	0.183	0.158	1	1	1	1	1	1
5-ary	0.650	0.534	0.509	0.205	0.169	0.136	1	1	1	1	1	1
Imbalanced												
Binomial	0.169	0.171	0.154	0.342	0.304	0.251	1	1	1	1	1	1
BA	0.028	0.020	0.0136	0.249	0.231	0.192	1	1	1	-	-	-

References

1. Barabási, A.L., Albert, R.: Emergence of scaling in random networks. Science **286**(5439), 509–512 (1999)
2. Becigneul, G., Ganea, O.E.: Riemannian adaptive optimization methods. In: International Conference on Learning Representations (2018)
3. Beydoun, G., Lopez-Lorca, A.A., García-Sánchez, F., Martínez-Béjar, R.: How do we measure and improve the quality of a hierarchical ontology? J. Syst. Softw. **84**(12), 2363–2373 (2011). https://doi.org/10.1016/j.jss.2011.07.010, https://linkinghub.elsevier.com/retrieve/pii/S0164121211001853

4. Bonnabel, S.: Stochastic gradient descent on Riemannian manifolds. IEEE Trans. Autom. Control **58**(9), 2217–2229 (2013)
5. Bourgaux, C., Guimarães, R., Koudijs, R., Lacerda, V., Ozaki, A.: Knowledge base embeddings: semantics and theoretical properties. In: Proceedings of the 21st International Conference on Principles of Knowledge Representation and Reasoning, pp. 823–833 (August 2024). https://doi.org/10.24963/kr.2024/77
6. Cao, Z., Xu, Q., Yang, Z., Cao, X., Huang, Q.: Geometry interaction knowledge graph embeddings. In: Proceedings of the AAAI Conference on Artificial Intelligence, vol. 36, no. 5, pp. 5521–5529 (2022). https://doi.org/10.1609/aaai.v36i5.20491
7. Deng, J., Dong, W., Socher, R., Li, L.J., Li, K., Fei-Fei, L.: Imagenet: a large-scale hierarchical image database. In: 2009 IEEE Conference on Computer Vision and Pattern Recognition, pp. 248–255. IEEE (2009)
8. Desai, K., Nickel, M., Rajpurohit, T., Johnson, J., Vedantam, S.R.: Hyperbolic image-text representations. In: International Conference on Machine Learning, pp. 7694–7731. PMLR (2023)
9. Dhall, A., Makarova, A., Ganea, O., Pavllo, D., Greeff, M., Krause, A.: Hierarchical image classification using entailment cone embeddings. In: Proceedings of the IEEE/CVF Conference on Computer Vision and Pattern Recognition Workshops, pp. 836–837 (2020)
10. Dhingra, B., Shallue, C., Norouzi, M., Dai, A., Dahl, G.: Embedding text in hyperbolic spaces. In: Proceedings of the Twelfth Workshop on Graph-Based Methods for Natural Language Processing, pp. 59–69 (2018)
11. Dong, J., Zhang, Q., Huang, X., Duan, K., Tan, Q., Jiang, Z.: hierarchy-aware multi-hop question answering over knowledge graphs. In: Proceedings of the ACM Web Conference 2023, pp. 2519–2527. ACM, Austin TX USA (April 2023). https://doi.org/10.1145/3543507.3583376
12. Fellbaum, C.: Wordnet. In: Theory and applications of ontology: computer applications, pp. 231–243. Springer (2010)
13. Fix, B.: Hierarchy and the power-law income distribution tail. J. Comput. Soc. Sci. **1**(2), 471–491 (2018). https://doi.org/10.1007/s42001-018-0019-8
14. Ganea, O., Bécigneul, G., Hofmann, T.: Hyperbolic entailment cones for learning hierarchical embeddings. In: International Conference on Machine Learning, pp. 1646–1655. PMLR (2018)
15. Gao, Z., Wu, Y., Jia, Y., Harandi, M.: Curvature generation in curved spaces for few-shot learning. In: International Conference on Computer Vision (2021)
16. Ghadimi Atigh, M., Schoep, J., Acar, E., Van Noord, N., Mettes, P.: Hyperbolic image segmentation. In: Proceedings of the IEEE/CVF Conference on Computer Vision and Pattern Recognition, pp. 4453–4462 (2022)
17. Guarino, N., Welty, C.: Evaluating ontological decisions with OntoClean. Commun. ACM **45**(2), 61–65 (2002). https://doi.org/10.1145/503124.503150
18. Guizzardi, G., Fonseca, C.M., Almeida, J.P.A., Sales, T.P., Benevides, A.B., Porello, D.: Types and taxonomic structures in conceptual modeling: a novel ontological theory and engineering support. Data Knowl. Eng. **134**, 101891 (2021). https://doi.org/10.1016/j.datak.2021.101891, https://linkinghub.elsevier.com/retrieve/pii/S0169023X21000185
19. Gulshad, S., Long, T., van Noord, N.: Hierarchical explanations for video action recognition. In: Proceedings of the IEEE/CVF Conference on Computer Vision and Pattern Recognition, pp. 3703–3708 (2023)
20. Gmez-Prez, A.: Evaluation of ontologies. Int. J. Intell. Syst. **16**(3), 391–409 (2001)

21. Hagberg, A., Swart, P.J., Schult, D.A.: Exploring network structure, dynamics, and function using networkx. Tech. rep, Los Alamos National Laboratory (LANL), Los Alamos, NM (United States) (2008)
22. Hofer, M., Obraczka, D., Saeedi, A., Köpcke, H., Rahm, E.: Construction of knowledge graphs: current state and challenges. Information **15**(8), 509 (2024). https://doi.org/10.3390/info15080509
23. Hong, J., Fang, P., Li, W., Han, J., Petersson, L., Harandi, M.: Curved geometric networks for visual anomaly recognition. IEEE Trans. Neural Netw. Learn. Syst. (2023)
24. Hong, J., Hayder, Z., Han, J., Fang, P., Harandi, M., Petersson, L.: Hyperbolic audio-visual zero-shot learning. In: Proceedings of the IEEE/CVF International Conference on Computer Vision, pp. 7873–7883 (2023)
25. Huang, Z., Chiang, M.F., Lee, W.C.: LinE: logical query reasoning over hierarchical knowledge graphs. In: Proceedings of the 28th ACM SIGKDD Conference on Knowledge Discovery and Data Mining, pp. 615–625. ACM, Washington DC USA (August 2022). https://doi.org/10.1145/3534678.3539338
26. Ibrahimi, S., Atigh, M.G., Van Noord, N., Mettes, P., Worring, M.: Intriguing properties of hyperbolic embeddings in vision-language models. Trans. Mach. Learn. Res. (2024)
27. Ji, S., Pan, S., Cambria, E., Marttinen, P., Yu, P.S.: A survey on knowledge graphs: representation, acquisition, and applications. IEEE Trans.. Neural Netw. Learn. Syst. **33**(2), 494–514 (2022). https://doi.org/10.1109/TNNLS.2021.3070843, https://ieeexplore.ieee.org/document/9416312/
28. Kendall, E.F., McGuinness, D.L.: Ontology Engineering. Synthesis Lectures on Data, Semantics, and Knowledge, Springer International Publishing, Cham (2019). https://doi.org/10.1007/978-3-031-79486-5
29. Kim, S., Jeong, B., Kwak, S.: Hier: metric learning beyond class labels via hierarchical regularization. In: Proceedings of the IEEE/CVF Conference on Computer Vision and Pattern Recognition, pp. 19903–19912 (2023)
30. Kim, W., Chun, S., Kim, T., Han, D., Yun, S.: Hype: hyperbolic entailment filtering for underspecified images and texts. In: European Conference on Computer Vision, pp. 247–265. Springer (2025)
31. Klimovskaia, A., Lopez-Paz, D., Bottou, L., Nickel, M.: Poincaré maps for analyzing complex hierarchies in single-cell data. Nat. Commun. **11**(1), 2966 (2020)
32. Law, M., Liao, R., Snell, J., Zemel, R.: Lorentzian distance learning for hyperbolic representations. In: International Conference on Machine Learning, pp. 3672–3681. PMLR (2019)
33. Le, M., Roller, S., Papaxanthos, L., Kiela, D., Nickel, M.: Inferring concept hierarchies from text corpora via hyperbolic embeddings. arXiv preprint arXiv:1902.00913 (2019)
34. Li, A., Yang, B., Huo, H., Chen, H., Xu, G., Wang, Z.: Hyperbolic neural collaborative recommender. IEEE Trans. Knowl. Data Eng. **35**(9), 9114–9127 (2022)
35. Li, H., Chen, Z., Xu, Y., Hu, J.: Hyperbolic anomaly detection. In: Proceedings of the IEEE/CVF Conference on Computer Vision and Pattern Recognition, pp. 17511–17520 (2024)
36. Li, Y.L., et al.: From isolated islands to pangea: unifying semantic space for human action understanding. In: Proceedings of the IEEE/CVF Conference on Computer Vision and Pattern Recognition, pp. 16582–16592 (2024)
37. Liu, S., Chen, J., Pan, L., Ngo, C.W., Chua, T.S., Jiang, Y.G.: Hyperbolic visual embedding learning for zero-shot recognition. In: Proceedings of the IEEE/CVF Conference on Computer Vision and Pattern Recognition, pp. 9273–9281 (2020)

38. Long, T., Mettes, P., Shen, H.T., Snoek, C.G.: Searching for actions on the hyperbole. In: Proceedings of the IEEE/CVF Conference on Computer Vision and Pattern Recognition, pp. 1141–1150 (2020)
39. Lovisolo, L., Da Silva, E.: Uniform distribution of points on a hyper-sphere with applications to vector bit-plane encoding. IEE Proc.-Vis. Image Signal Process. **148**(3), 187–193 (2001)
40. Łuczak, T., Magner, A., Szpankowski, W.: Asymmetry and structural information in preferential attachment graphs. Random Struct. Algorithms **55**(3), 696–718 (2019)
41. Ma, R., Fang, P., Drummond, T., Harandi, M.: Adaptive poincaré point to set distance for few-shot classification. In: AAAI Conference on Artificial Intelligence (2022)
42. MacWilliams, S.: The theory of error-correcting codes. Elsevier Sci. Publishers BV Google Schola **2**, 39–47 (1977)
43. McDaniel, M., Storey, V.C.: Evaluating domain ontologies: clarification, classification, and challenges. ACM Comput. Surv. **52**(4), 1–44 (2020). https://doi.org/10.1145/3329124
44. McGuinness, D.L.: Ontologies come of age. In: Fensel, D., Hendler, J.A., Lieberman, H., Wahlster, W. (eds.) Spinning the Semantic Web, pp. 171–194. The MIT Press (January 2003). https://doi.org/10.7551/mitpress/6412.003.0008, https://direct.mit.edu/books/book/3817/chapter/125261/Ontologies-Come-of-Age
45. Mettes, P., Ghadimi Atigh, M., Keller-Ressel, M., Gu, J., Yeung, S.: Hyperbolic deep learning in computer vision: a survey. Int. J. Comput. Vis. 1–25 (2024)
46. Nickel, M., Kiela, D.: Poincaré embeddings for learning hierarchical representations. Adv. Neural Inf. Process. Syst. **30** (2017)
47. Nickel, M., Kiela, D.: Learning continuous hierarchies in the lorentz model of hyperbolic geometry. In: International Conference on Machine Learning, pp. 3779–3788. PMLR (2018)
48. Pal, A., van Spengler, M., di Melendugno, G.M.D., Flaborea, A., Galasso, F., Mettes, P.: Compositional entailment learning for hyperbolic vision-language models. arXiv preprint arXiv:2410.06912 (2024)
49. Peng, W., Varanka, T., Mostafa, A., Shi, H., Zhao, G.: Hyperbolic deep neural networks: a survey. IEEE Trans. Pattern Anal. Mach. Intell. **44**(12), 10023–10044 (2021)
50. Peng, Y., Bonald, T., Alam, M.: Refining wikidata taxonomy using large language models. In: Proceedings of the 33rd ACM International Conference on Information and Knowledge Management, pp. 5395–5399. CIKM '24, Association for Computing Machinery, New York, NY, USA (2024). https://doi.org/10.1145/3627673.3679156
51. Pernisch, R., Dell'Aglio, D., Bernstein, A.: Beware of the hierarchy — an analysis of ontology evolution and the materialisation impact for biomedical ontologies. J. Web Semant. **70**, 100658 (2021). https://doi.org/10.1016/j.websem.2021.100658, https://linkinghub.elsevier.com/retrieve/pii/S1570826821000330
52. Petermann, D., Wichern, G., Subramanian, A., Le Roux, J.: Hyperbolic audio source separation. In: ICASSP 2023-2023 IEEE International Conference on Acoustics, Speech and Signal Processing (ICASSP), pp. 1–5. IEEE (2023)
53. Poveda-Villalón, M., Fernández-Izquierdo, A., Fernández-López, M.,García-Castro, R.: LOT: An industrial oriented ontology engineering framework. Eng. Appl. Artif. Intell. **111**, 104755 (2022). https://doi.org/10.1016/j.engappai.2022.104755, https://linkinghub.elsevier.com/retrieve/pii/S0952197622000525

54. Saff, E.B., Kuijlaars, A.B.: Distributing many points on a sphere. Math. Intell. **19**, 5–11 (1997)
55. Sala, F., De Sa, C., Gu, A., Ré, C.: Representation tradeoffs for hyperbolic embeddings. In: International Conference on Machine Learning, pp. 4460–4469. PMLR (2018)
56. Sarkar, R.: Low distortion delaunay embedding of trees in hyperbolic plane. In: International Symposium on Graph Drawing, pp. 355–366. Springer (2011)
57. Shimizu, C., Hammar, K., Hitzler, P.: Modular ontology modeling. Semantic Web **14**(3), 459–489 (2023). https://doi.org/10.3233/SW-222886
58. van Spengler, M., Zahálka, J., Mettes, P.: Adversarial attacks on hyperbolic networks. In: European Conference on Computer Vision Workshops (2024)
59. Storer, J.A.: An introduction to data structures and algorithms. Springer Science & Business Media (2012)
60. Sun, L., et al.: Perfect: a hyperbolic embedding for joint social network alignment. In: IEEE International Conference on Data Mining (2020)
61. Tarjan, R.: Depth-first search and linear graph algorithms. SIAM J. Comput. **1**(2), 146–160 (1972)
62. Tu, L., Meng, S., Qi, L., Xu, X., Zhang, X.: Hyperbolic contrastive learning with second order sampling for collaborative filtering. In: 2024 IEEE International Conference on Web Services (ICWS), pp. 281–290. IEEE (2024)
63. Weikum, G., Dong, X.L., Razniewski, S., Suchanek, F.: Machine knowledge: creation and curation of comprehensive knowledge bases. Found. Trends® Databases **10**(2-4), 108–490 (Jul 2021). https://doi.org/10.1561/1900000064
64. Yang, M., Li, Z., Zhou, M., Liu, J., King, I.: Hicf: hyperbolic informative collaborative filtering. In: Proceedings of the 28th ACM SIGKDD Conference on Knowledge Discovery and Data Mining, pp. 2212–2221 (2022)
65. Yang, Y., et al.: Hyperbolic graph learning for social recommendation. IEEE Trans. Knowl. Data Eng. (2023)
66. Yu, T., Liu, T.J., Tseng, A., De Sa, C.: Shadow cones: unveiling partial orders in hyperbolic space. arXiv preprint arXiv:2305.15215 (2023)
67. Yu, Z., et al.: Skin lesion recognition with class-hierarchy regularized hyperbolic embeddings. In: International Conference on Medical Image Computing and Computer-assisted Intervention, pp. 594–603. Springer (2022)
68. Zhang, B., Jiang, H., Feng, S., Li, X., Ye, Y., Ye, R.: Hyperbolic knowledge transfer with class hierarchy for few-shot learning. In: IJCAI, pp. 3723–3729 (2022)
69. Zhang, Y., Pietrasik, M., Xu, W., Reformat, M.: Hierarchical topic modelling for knowledge graphs. In: Groth, P., et al. (eds.) The Semantic Web, LNCS, vol. 13261, pp. 270–286. Springer International Publishing, Cham (2022). https://doi.org/10.1007/978-3-031-06981-9_16
70. Zhang, Z., Cai, J., Zhang, Y., Wang, J.: Learning hierarchy-aware knowledge graph embeddings for link prediction. In: Proceedings of the AAAI Conference on Artificial Intelligence, vol. 34, no. 03, pp. 3065–3072 (2020). https://doi.org/10.1609/aaai.v34i03.5701
71. Zhou, M., Yang, M., Xiong, B., Xiong, H., King, I.: Hyperbolic graph neural networks: a tutorial on methods and applications. In: Proceedings of the 29th ACM SIGKDD Conference on Knowledge Discovery and Data Mining, pp. 5843–5844 (2023)

RDF-Based Semantics for Selective Disclosure and Zero-Knowledge Proofs on Verifiable Credentials

Christoph H.-J. Braun(✉) and Tobias Käfer

Karlsruhe Institute of Technology, Karlsruhe, Germany
{braun,tobias.kaefer}@kit.edu

Abstract. Our work connects the W3C Verifiable Credentials (VC) data model and zero-knowledge proofs (ZKPs) to allow for minimised information disclosure.More generally, as VCs are Resource Description Framework (RDF) datasets, our work enables the application of the following ZKPs on such an RDF dataset:selective disclosure, proof of numeric bounds, and proof of set non-membership – all on the level of RDF terms. To pre-process RDF datasets for such ZKP applications, we introduce schema-free and schema-based approaches, and highlight their differences. We then present a data model for credential presentation and its semantics, where we show that selective disclosure is equivalent to RDF's simple entailment, and that verifications of the presented proofs are validity checks on the processed RDF dataset.

Keywords: RDF · Verifiable Credentials · Zero-knowledge Proofs

1 Introduction

Governments [19] and industry [6] are pushing for the introduction of digital credentials to digitise physical ID cards [23]. In the near future, using digital credentials in the physical world and on the Web will become part of daily life, e.g. for receiving discounts (Fig. 1): Alice is CEO of `aCompany`, which issues a credential attesting that Alice is both member and head of `aCompany`. For lunch, Alice visits a restaurant offering a company discount. To receive the discount, Alice presents her credential to the waiter for verification using her digital wallet. In the process, only the fact that Alice is indeed `aCompany`'s employee needs to be disclosed. No unnecessary information should be revealed, not even the fact that an employee ID card was used. Still, the waiter must be able to verify the credential's integrity, i.e. that `aCompany` indeed signed that Alice is an employee.

To model such credentials, W3C Verifiable Credentials (VCs) [36] provide a standard data model based on the Resource Description Framework (RDF) [17]. Thus, VCs are small personal knowledge graphs that are cryptographically signed by an issuer. Preserving a person's privacy as in our example can be achieved by using cryptographic methods such as zero-knowledge proofs (ZKPs) [22].

E. Curry et al. (Eds.): ESWC 2025, LNCS 15718, pp. 383–402, 2025.
https://doi.org/10.1007/978-3-031-94575-5_21

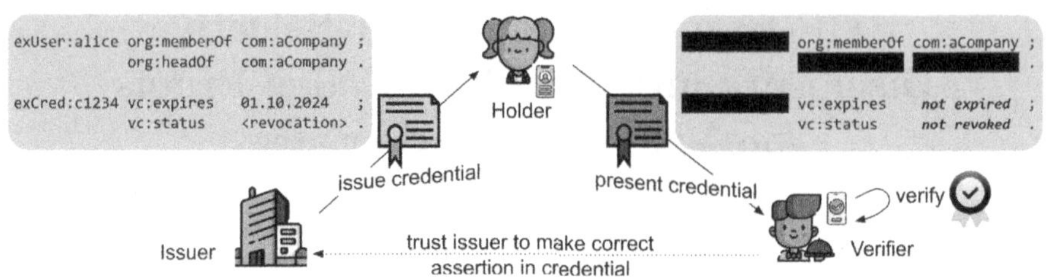

Fig. 1. A high-level illustration of our running example.

This has been strongly advocated by Europe's leading cryptographers [3], which sparked a major discussion[1] about the cryptographic mechanisms in the EU Digital Identity Wallet [18]. In this paper, we therefore investigate the connection between VCs as RDF datasets and cryptographic methods for ZKPs.

Corresponding cryptographic methods are being implemented, e.g., in AnonCreds (cf. Sect. 6) which only focus on a JSON- and schema-based credential format. They do not consider an RDF-based perspective and its formal logical foundations as relevant. While there has been related work [38] on RDF-based credentials and ZKPs, their approach only allows selective disclosure of complete triples of an RDF graph. Upon presentation, the full triple is either hidden or revealed. To prove that e.g. the expiration date has not yet passed without revealing it is impossible, as the RDF literal is inseparable from its triple.

We thus present a more granular approach that enables ZKPs on RDF term level. Our approach allows to selectively disclose RDF terms within a quad of an RDF dataset, which allows, e.g. to cryptographically prove the equality of RDF terms across quads, or to conceal this equality. We also support proofs of numeric bounds (e.g. expiration date not yet passed) or set (non-)membership on RDF terms (e.g. credential id not revoked), while also proving that the values in both cases were signed by the issuer. Such proofs work on any term of a dataset's quads, and without a schema. Proof composition is not new in cryptography [12], yet to our knowledge, it has not been investigated for RDF, let alone at term granularity. Moreover for our approach, we prove using RDF's formal underpinnings [25] that a presented credential graph logically follows from its original credential graph. Thus, it is logically sound in our approach to apply querying and reasoning on VCs and their presentations, even when they use the advanced cryptography of ZKPs. We provide the following contributions:

- Two approaches to transform RDF datasets for ZKP application (Sect. 3): one schema-free and one schema-based,
- A data model and its semantics for credential presentation (Sect. 4): for selective disclosure of RDF terms, and for associated ZKP verification.
- We show in our evaluation (Sect. 5):

[1] https://github.com/eu-digital-identity-wallet/eudi-doc-architecture-and-reference-framework/discussions/211.

· selective disclosure is equivalent to RDF's simple entailment
· proof verification poses validity checks on the processed RDF dataset

We introduce VCs and ZKPs in Sect. 2. After presenting our approach in Sects. 3–4 and evaluation in Sect. 5, we cover related work in Sect. 6, and close by highlighting aspects of our work in Sect. 7. A proof-of-concept implementation of our approach and our vocabulary are available online[2].

2 Preliminaires

In this section, we cover W3C Verifiable Credentials [36] and provide a non-technical introduction to Selective Disclosure using ZKPs [22].

2.1 W3C Verifiable Credentials Are RDF Datasets

The Verifiable Credentials (VCs) data model [36] is a W3C Recommendation for modeling digitally signed information using JSON-LD, e.g. digital credentials such as an employee ID card. Such digital credential – and thus a VC as one way to model it – consists of at least two fundamental parts: the information which is claimed to be "true", and the digital signature by the credential's issuer which indicates that the issuer asserts the claims to be indeed "true".

This two-part structure is reflected in the VC data model [36]. A VC is an RDF dataset[3] comprised of two graphs: the *credential graph* and the *proof graph*. The credential graph contains claims and credential metadata. The credential graph on its own is unknown to be "true", i.e. asserted [15]. A claim that a person is a company's employee is meaningless without additional assurance. To decide if the credential graph is asserted/"true", the digital signature of the issuer, e.g. the company, is attached to the credential in the proof graph. Only if the digital signature is valid, and the verifier trusts the issuer to accurately attest the credential's data, the credential graph may be considered asserted/"true".

When showing a credential in the physical world, e.g. the employee ID card to get the lunch discount, the verifier/waiter also physically verifies that the person *presenting the credential* is the same person that the employee ID card was issued to. In the digital world, one way of achieving this is authenticating the credential holder by attaching their digital signature to the VC in presentation. Such additional attachment for presentation form two additional graphs: the *presentation graph* that may include e.g. a usage policy for the presented data and the *presentation proof graph* that indicates that the holder asserts the presented information, e.g. using a signature or cryptographic proof. These two graphs are linked to the VC, thereby forming a *Verifiable Presentation (VP)* [36].
We clarify a technical detail about the interconnections of the different graphs in a VC: The W3C VC data model mandates linking the proof graph (not the proof itself) from a node within the default graph. This implies that the proof

[2] https://github.com/uvdsl/rdf-zkp/.
[3] For a common formalisation of RDF datasets, see Appendix A.

specified in the proof graph was calculated over all triples in the (default) graph it was linked from. If two VCs (RDF datasets) would simply merge their default graphs, the proofs of neither would be verifiable anymore. Therefore, we make our choice of semantics [40] explicit: Blank nodes are scoped across graphs of an RDF dataset. When merging two RDF datasets, blank nodes between the two RDF datasets must be re-labelled to avoid co-references, and new graph names must be minted for the default graphs of the respective RDF datasets. We express the opinion that the W3C VC data model should mandate asserting claims in a named graph, and have the proof link to the graph it covers.

2.2 Selective Disclosure Using Zero-Knowledge Proofs

Selective disclosure refers to the idea of proving properties of credential data while hiding (some of) the actual data [7]. When showing a credential in the physical world, e.g. the employee ID card to get the lunch discount, the verifier/waiter sees all the information on the card. They actually only need to see that the presenting person is indeed an employee of an eligible company. Superfluous information should be kept secret such that (a) privacy of the presenting person is preserved and (b) the verifier only receives data that they are prepared and expected to handle. A restaurant would only be expected to handle information on how many customers belong to a company, e.g. receiving a linear payment to offer this benefit to company employees. Handling arbitrary personal information of customers should not be the restaurant's concern.

Considering VCs, a first step towards selective disclosure is to hide or reveal exact information in a VP, e.g. to reveal the employment relation while hiding all other data. A second step is adding proofs on hidden pieces, e.g. prove that the hidden expiration date has not yet passed. These kinds of proofs are enabled by cryptographic protocols that are already mature today: zero-knowledge proofs. The concept of a *zero-knowledge proof (ZKP)* [22] is that a prover is able to convince a verifier that a statement is true without the verifier learning any additional information. Considering VCs for example, a prover needs to prove that their employee ID card is not expired. The verifier will gain zero knowledge beyond the fact that the prover's credential is still valid.

Technically, this non-expiration proof is a composition of multiple distinct yet connected proofs: As expected, there is a *proof of numeric bounds*. This proof specifically includes proving that the prover knows (a) a certain date and (b) that this date is in the future. But because the prover could just pick a suitable expiration date, there must also be a proof that the picked date is the same as in the employee ID card, signed by the issuing company. Proving the attestation of the expiration date is done using a *proof of knowledge of signature* of the VC (and its content) without revealing the date itself. The fact that both proofs are about the same secret value is ensured by cryptographically tying them together. One way of doing that is to use the *Schnorr Protocol*[4] [34]. The Schnorr Protocol

[4] For a mathematical description of the Schnorr Protocol, see Appendix B.

Fig. 2. A non-technical illustration of how two proofs (each represented as a seal on an envelope) are cryptographically tied together (blue color pattern in the seals' right-hand bottom quarter) by the Schnorr protocol (mechanism to ensure matching color pattern on seals) when the proofs are about the same value. (Color figure online)

allows us to prove knowledge of a secret value, and subsequently, that the same secret value must have been used in both proofs.

Figure 2 provides a non-technical illustration: Each proof is represented by a seal on an envelope. The values that the proof is about are the content of the envelope. The envelope is indestructible and cannot be opened such that the contents of the envelopes remain secret. The contents of the envelope induce a certain color pattern on the seal, e.g. the date `01.01.2030` induces a blue section in the bottom right-hand corner of the seal. A proof is validated by checking the integrity of the colored seal. When two seals share a specific colored section, the blue bottom-right hand corner, then they share a specific content value, the date `01.01.2030`. The mechanism that ensures that matching color pattern is the Schnorr Protocol. The blue section of the seal represents the so-called *Schnorr response value* that corresponds to a particular secret value. In this way, the Schnorr Protocol allows us to prove knowledge of a secret value – the date `01.01.2030` – and with that, that the same value must have been used in the two proofs – by using its Schnorr response value, the blue section in both seals. Note that a Schnorr response value is not only dependent on the secret value, called *witness*, but also an an additional random value called *blinding factor*, meaning that a single witness can have multiple Schnorr responses using different blinding factors. This allows us to prove or hide the fact that a specific RDF term occurs in multiple distinct RDF triples.

Using this technique, arbitrary combinations of proofs can be arranged. The above is just one example. To present multiple VCs together, proving e.g. that they are about the same person or share a secret value, two proofs of knowledge of signature can be combined. Or, maybe in a non-VC context, a proof of set-membership and a proof of numeric bounds may be combined on the same value. Circling back to the two levels of selective disclosure: VCs provide information that is attested by the issuer's signature. Initially, using a proof of knowledge of that signature [2], single pieces of the attested information can be disclosed or remain hidden in a VP. Then, additional proofs such as proof of numeric bounds [12] or proof of set (non-)membership [37] can be performed on the hidden yet attested pieces. The glue that keeps it all together is the Schnorr Protocol [34]. On top: A verifier learns no additional information from the ZKPs.

There is, however, a step zero: Creating a VC signature that allows for such proofs to be performed. A simple digital signature covers only a single value (called *message*), e.g. the string value of the VC's credential graph as a whole. To allow for selective disclosure of specific parts of the VC, the digital signature must be a *multi-message* signature that allows for deriving a proof of knowledge of the signature value and of all the messages that were used in its creation. How our work dissects a VC into these multiple messages is presented in Sect. 3.

3 Preparing W3C Verifiable Credentials for ZKPs

To create a multi-message signature on a VC and then later be able to perform ZKPs on credential data, two conceptual steps are required: First the RDF dataset at hand must be transformed into a list of RDF terms (Sect. 3.1), i.e., a list of terms that we aim to create signature and proofs on. Second, the RDF terms must be transformed to numeric representations such that the cryptographic algorithms of the ZKPs can be applied on these terms (Sect. 3.2).

Recall our running example from Fig. 1 and consider Listing 1.1 as the issued credential, which adheres to the W3C VC data model [36] and includes an expiration date and revocation status (Sect. 3.3). Note that Listing 1.1 already includes the signature, a BBS+ multi-message signature [10] (lines 14–18).

```
1  # claims
2  <http://example.org/users#alice>
3    org:headOf <http://example.org/organisations#aCompany>;
4    org:memberOf <http://example.org/organisations#aCompany>.
5  # credential metadata
6  <http://example.org/credentials#cred123>
7    cred:credentialSubject <http://example.org/users#alice>;
8    cred:issuer <http://example.org/organisations#aCompany>;
9    cred:validUntil "2023-11-07T13:20:00Z"^^xsd:dateTimeStamp;
10   cred:credentialStatus <http://example.org/revoc#accumulatorId123>;
11   cred:proof _:signatureGraph.
12 # proof graph
13 GRAPH _:signatureGraph {
14   [] a bbsp16:BbbsPlus16GLDS;
15      cred:verificationMethod <http://example.org/keys#aCompanyKey>;
16      cred:proofValue [  bbsp16:A "g1Em1..."^^xsd:base64Binary ;
17                         bbsp16:e "gq1/f..."^^xsd:base64Binary ;
18                         bbsp16:s "FpwmJ..."^^xsd:base64Binary ] .
19 }
```

Listing 1.1: Our example VC (prefixes omitted for brevity) in TriG notation: Graphs are explicit; JSON-LD notation would hide them in the VC's `@context`.

3.1 Transforming RDF Datasets into a List of RDF Terms

To dissect an RDF dataset into a list of RDF terms, we present two foundational transformation approaches, each stemming from a different perspective on data modeling: VCs may be built with or without using credential schemas [36]. The first foundational approach is schema-free and reflects an RDF purist's perspective who cherishes RDF's flexibility. The second foundational approach is schema-based and reflects the perspective of many JSON-oriented credential system developers. A schema-based transformation is theoretically more efficient as it results in a smaller list of messages to sign; reducing computational expense. However, this approach requires a schema to be linked from the VC and thus

Table 1. Example schema-free (left) and schema-based (right) lists of RDF terms.

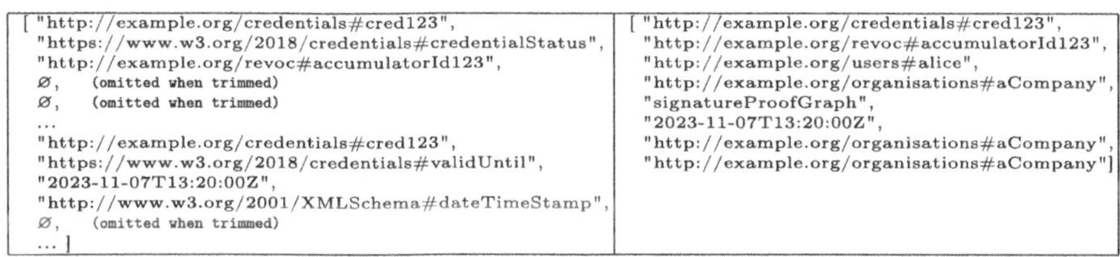

["http://example.org/credentials#cred123",	["http://example.org/credentials#cred123",
"https://www.w3.org/2018/credentials#credentialStatus",	"http://example.org/revoc#accumulatorId123",
"http://example.org/revoc#accumulatorId123",	"http://example.org/users#alice",
∅, (omitted when trimmed)	"http://example.org/organisations#aCompany",
∅, (omitted when trimmed)	"signatureProofGraph",
...	"2023-11-07T13:20:00Z",
"http://example.org/credentials#cred123",	"http://example.org/organisations#aCompany",
"https://www.w3.org/2018/credentials#validUntil",	"http://example.org/organisations#aCompany"]
"2023-11-07T13:20:00Z",	
"http://www.w3.org/2001/XMLSchema#dateTimeStamp",	
∅, (omitted when trimmed)	
...]	

may require a remote lookup by a verifier, to determine how to interpret the credential data. With that public schema, certain information, e.g., predicates or property names, cannot be hidden at all. This may lead to inferring information about a credential holder: e.g. from a driver's license schema, when only used for age verification, the verifier may presume that the holder is allowed to drive.

Schema-Free Transformation. Our schema-free transformation is based on RDF canonicalisation and dissecting the resulting dataset into atomic elements. First, we canonicalise the dataset following the W3C Recommendation [30]. This includes ordering quads lexicographically in N-Quads serialisation. Then, we dissect the quads into atomic components. We consider the individual RDF terms' values: URIs and blank nodes each consist of a simple bitstring value: The URI without the enclosing brackets and the blank node's label without _:. Literals have a lexical form, optionally a language tag [1], and a possibly implicit datatype. Literals consist therefore of two parts [25], a value and a suffix, which is either a data type or a language tag. Thus, the suffix always specifies the literal's data type. In summary, a quad consists of at most 5 components: subject, predicate, object value, object suffix, and graph name. Canonical ordering of quads and terms is preserved in the list of messages.

We emphasize the distinction between the literal value and the literal suffix because range proofs require numeric comparison. Inclusion of the literal suffix is required because agents need to be informed and assured on how to correctly interpret the literal's value *as intended by the issuer*. Thus, it would be unclear if e.g. value `0x41` should be interpreted as integer `65` or as string `"a"`.

A technical detail provides an optional optimisation: trimming zero elements from the message list to reduce its size. If a term in object position is not a literal, the object suffix component is the zero element (∅). If there is no graph name, i.e. for the default graph, the graph name component is the zero element. The resulting list is thus reduced by (a) the number of triples in the default graph, and (b) the number of URIs and blank nodes in object position.

External Schema-Based Transformation. Public schemas are commonly used to indicate which information to expect to be asserted in a VC [36]. Taking inspiration from [9] in adhering to W3C recommended specifications, we assume usage of SHACL shapes [29] to be used as RDF schemas. We outline a rough proposal on how to convert a SHACL-based credential schema with an associated claim schema to a value list: By convention in our work, the first element of the resulting value list is the identifier of the credential itself. For credential data, in lexicographical ordering of the predicate URIs the corresponding object values are appended, i.e., URIs and the values of RDF literals without their suffix. Object suffix information is provided by the schema. Similarly then for claim data, the corresponding object values are appended in lexicographical ordering of predicate URIs. For our example, a resulting list of RDF terms is in Table 1.

In this way, only the credential identifier and all the object values need to be signed, proven and verified. There is no need to re-iterate the subject for each triple. In addition, there is no need to include the predicate of each triple and object suffix if present, as these are provided by the public schema anyways. This reduces the number of messages to be signed significantly in trade-off for usage of a public external schema, i.e., the fixed structure of the credential and its predicates are public knowledge. These are then necessary to be obtained for proof verification, which may require a remote call.

3.2 Interpreting RDF Terms for Signatures and ZKPs

After transforming the RDF dataset (Sect. 3.1), we need to consider the values of individual RDF terms, either in lexical space or in value space. Only then we can apply the cryptographic algorithms for creation of signature and proof.

URIs and blank node labels are considered in lexical space. Similarly as in a SPARQL `FILTER` expression [24], literals are considered according to the XML Schema Datatypes (XSD) definition [33]. All literals that exhibit the *numeric* fundamental facet [33] are considered in the numeric value space. We note that in an implementation, non-integer-based datatypes require specialised treatment, which does not limit the generality of our approach. Dates also need specialised treatment, see Appendix C. All other literals are considered in the lexical space.

More specifically, we map RDF term representations in the lexical space to sequences of bits (i.e. bitstrings of dynamic length) using UTF-8 encoding [39]. For the numeric value space, we map value representations of RDF Literals to unsigned 64-bit integers (i.e. bitstrings of 64 bits). These bitstrings are then further mapped to field representations of the elliptic curves used for signatures and proofs, as usual in elliptic curve cryptography.

While blank nodes are allowed in VCs, their label may pose a privacy risk, specifically on the security property of unlinkability [32]: A sufficiently unique blank node label may allow a verifier to track that the same credential is used in multiple presentations, and then to identify the presenting holder.

Re-labeling blank nodes for a presentation does not pose a solution: The original blank node labels are used in calculation of the signature value. When simply re-labeling blank nodes in presentation, the original blank node label is not available to a verifier (nor any cryptographic stand-in) and they cannot verify the proofs. There is just no connection between the new labels and the proof. Instead, we cryptographically hide all blank node labels in a VC when creating a presentation. In this case, there exists a Schnorr response value for each blank node which allows for verifying the proof of knowledge of signature (and of the values it was calculated on). Moreover, this enables to cryptographically prove and verify that two blank nodes in the presentation graph refer to the same node in the credential graph, or to cryptographically hide exactly that fact.

3.3 Expiration and Revocation

Once the VC has been signed and handed over from the issuer to the holder, expiration and revocation mechanisms ensure the issuer's control over a VC's validity. The expiration date is specified using a `cred:validUntil` relation (Listing 1.1 line 9) and should be hidden in presentation. If a holder proves this date has not yet passed, a verifier may assume the VC to not be expired.

For revocation, an issuer may manage a Universal Accumulator [37], where the identifiers of revoked VCs are accumulated. Using a corresponding witness, a holder is able to create a proof that the VC's id is not part of the accumulator, and thus that the VC is not yet revoked.

4 Data Model and Semantics in Credential Presentation

Alice presents her credential to the waiter to receive the discount using her digital wallet. The waiter only needs to know that she is indeed an eligible company's employee, and that the credential is not expired and not revoked. All other information should not be revealed; and thus remain hidden. To this end, Alice's wallet derives a presentation of the credential graph (Sect. 4.1) and corresponding proofs (Sect. 4.2) from the VC.

4.1 The Presentation's Credential Graph

Instead of presenting the VC in plain, to be able to selective disclose information of the credential graph, a named graph is derived from the VC's default graph (cf. Listing 1.1 lines 1–11). The terms that are to be hidden are replaced by blank nodes. The remaining terms are revealed. Listing 1.2 lines 2–10 illustrate this selective disclosure of RDF terms.

Recall that a presented credential is an RDF dataset [36]. To hide terms, we cannot simply omit them: If we omitted single terms, the quads would be incomplete. If we omitted entire quads if all their elements are hidden, we could not assert properties about the individual terms in the quad. Thus, when applying selective disclosure and hiding terms, we need a stand-in for the hidden terms.

We use blank nodes for this purpose: Blank nodes are treated as indicating the existence[5] of a thing [25]. Two blank nodes with the same label refer to the

[5] In a graph not deemed "true", the fact that a blank node indicates existence is meaningless. Recall Sect. 2.1: Whether or not a (presented) credential graph may be assumed to be asserted/"true", and with that the existence of a thing denoted by a blank node, depends on whether or not the verifier trusts the issuer to accurately attest information and on the validity of their digital proof/signature.

same thing. We note that two blank nodes with different labels may or may not refer to the same thing. This ambiguity aids the desired features in selective disclosure of RDF terms: Hidden terms, i.e., hidden messages, are represented by blank nodes in the presented credential graph. Preserving the semantics of blank nodes, we consider all blank nodes to indicate hidden messages. URIs and Literals are revealed messages. To allow for selective disclosure of predicates, we allow blank nodes in predicate position similar to SPARQL [24].

```
# the presented credential graph
GRAPH _:4 {
  _:0 cred:credentialStatus <http://example.org/revoc#accumulatorId123>.
  _:0 cred:credentialSubject _:7 .
  _:0 cred:issuer <http://example.org/organisations#aCompany>.
  _:0 _:16 _:17 .
  _:0 cred:validUntil _:22 .
  _:25 _:26 _:27 .
  _:7 org:memberOf <http://example.org/organisations#aCompany>.
}
# the presentation proof graph
GRAPH <#presentationProofGraph> {
  # Schnorr response values for hidden RDF terms
  _:0 spok:hasSchnorrResponse "YT+CO..."^^xsd:base64Binary.
  _:4 spok:hasSchnorrResponse "zuN2D..."^^xsd:base64Binary .
  _:7 spok:hasSchnorrResponse "QW+Hb..."^^xsd:base64Binary.
  _:7 spok:hasSchnorrResponseForSuffix "m761h..."^^xsd:base64Binary.
  ... # _:16, _:17, _:22, _:25, _:26 omitted for brevity
  _:27 spok:hasSchnorrResponse "hU3N1..."^^xsd:base64Binary .
  _:27 spok:hasSchnorrResponseForSuffix "CGUG+..."^^xsd:base64Binary .
  # composite proof
  _:cproof a zkp:CompositeProof ;
      zkp:comprisedOf <#poks> , <#rp> , <#snmp> .
  # proof of knowledge of signature over original credential graph
  <#poks> a bbsp16:PoKS;
      bbsp16:hasVerificationKey <http://example.org/keys#aCompanyKey>;
      bbsp16:isProofOfKnowledgeOfSignatureOverGraph _:4 ;
      bbsp16:A_prime "og4GMd...e"^^xsd:base64Binary ;
      bbsp16:A_bar   "oh2fn..."^^xsd:base64Binary ;
      bbsp16:d       "ihCgp..."^^xsd:base64Binary ;
      bbsp16:pi      _:spk1, _:spk2 .
    _:spk1 a bbsp16:SPK1
      spok:hasCommitmentToRandomness "gi01T..."^^xsd:base64Binary ;
      bbsp16:hasResponseValueFor_e   "U2Rte..."^^xsd:base64Binary ;
      bbsp16:hasResponseValueFor_r2  "VrS/w..."^^xsd:base64Binary .
    _:spk2 a bbsp16:SPK2
      spok:hasCommitmentToRandomness     "jWnoZ..."^^xsd:base64Binary ;
      bbsp16:hasResponseValueFor_r3      "6fh2i..."^^xsd:base64Binary ;
      bbsp16:hasResponseValueFor_s_prime "vUfqW..."^^xsd:base64Binary .
  # range proof of expiration date
  <#rp> a lg16:PoRM ;
      lg16:hasVerificationKey <http://example.org/keys#verifierLg16VerificationKey> ;
      lg16:hasWitness _:n22 ;
      lg16:hasLowerBound "1383830400"^^xsd:nonNegativeInteger ;          # transformed dates
      lg16:hasUpperBound "18446744073709551615"^^xsd:nonNegativeInteger; # see Appendix C
      lg16:hasProofValue [ ... ] ; # details omitted for brevity
      lg16:pok [ ... ] .           # details omitted for brevity
  # set non-membership proof of credential id (URI) in revocation accumulator
  <#snmp> a uacc:PoSNMP ;
      uacc:hasAccumulator <http://example.org/revocation#accumulatorId123> ;
      uacc:hasWitness _:0 ;
      uacc:hasProvingKey <http://example.org/keys#proverUaccProvingKey> ;
      uacc:hasRandomizedWitness [ ... ]; # details omitted for brevity
      uacc:hasCommitments [ ... ];       # details omitted for brevity
      uacc:hasResponses [ ... ].         # details omitted for brevity
}
```

Listing 1.2: The example credential presentation (in schema-free approach).

4.2 Modeling Zero-Knowledge Proofs on W3C Verifiable Credentials

So far, in the *presented credential graph*, RDF terms are revealed or hidden. To be able to verify the validity of the presented information and to perform additional proofs on properties of hidden terms, the corresponding proofs are modelled in the *presentation proof graph*. Our data model[6] thus aims to provide verifier with all the information required for proof verification. In particular, we focus on proof composition: The proof of a credential presentation is a composition of distinct sub-proofs (Listing 1.2 line 22–23) that are interlinked by proving statements on the same secret witness values, and by shared parameters.

Center of proving that a hidden RDF term is witness across multiple proofs are *Schnorr response values* (see Sect. 2.2). For each hidden RDF term, there exists a Schnorr response value[7] (Listing 1.2 line 14–20).

For proof of knowledge of signature, Listing 1.2 line 25–39 provides a BBS+ proof tuple according to its definition in [10]: $(A', \bar{A}, d, \pi)$. The proof links to its hidden and revealed messages, i.e. the presented credential graph. For proof verification, parameters of the proof itself, verification key, the revealed RDF terms, and the hidden RDF terms' Schnorr response values are required.

To prove properties of the hidden RDF terms, proofs of numeric bounds or set (non-)membership, link to their particular witnesses in the presented credential graph: The proof of numeric bounds, a LegoGroth16 proof (Appendix H of [12]), links to the witness for which the specified upper and lower bounds are proven. The witness is the hidden object value of the `cred:validUntil` relation (Listing 1.2 line 7), and is thus part of the graph that the BBS+ proof of knowledge of signature covers. The proof of set non-membership, using a Universal Accumulator as defined in [37], applies the same principle and is thus linked to the BBS+ proof via the witness' blank node in the presented credential graph.

5 Evaluation

To evaluate our approach, we first look at the semantics for selective disclosure of RDF terms (Sect. 5.1). Then, we examine proof verification as validity checks on the RDF dataset (Sect. 5.2). Last, we compare the two transformations (Sect. 5.3).

5.1 Semantics for Selective Disclosure of RDF Terms

A verifier, who receives a presentation with hidden terms, will process the dataset using querying or reasoning techniques. Thus, a statement about the truthfulness in the logical sense of such query or reasoning results is an open question. We thus aspire a logical connection between the original graph and the selectively disclosing graph: If the truth of the selectively disclosing graph follows from the

[6] https://github.com/uvdsl/rdf-zkp/tree/main/vocab/.

[7] There may also exist a Schnorr response value for a quad's object suffix (Sect. 3.1).

truth of the original graph, the said querying and reasoning results are true (in the logical sense). We prove[8]:

Theorem 1. *Every selectively disclosing graph can be entailed under simple entailment from the underlying original graph that it selectively discloses.*

Proof. Let $g2m : \mathcal{D} \rightarrow (t_1, t_2, \ldots, t_n) | t_i \in \mathcal{U} \cup \mathcal{B} \cup \mathcal{L}$, a transformation function as outlined in Sect. 3.1 from an RDF dataset to an array of RDF terms (the list of messages). Let its inverse function be $m2g$. Let $\mathtt{I_D}$ be the set of indices of revealed (Disclosed) messages from the message list.

From graph G, create a presentation P such that P selectively discloses G:

1. Let $m := g2m(G)$, i. e., graph G to RDF term array (the list of messages).
2. Let $\forall i \notin \mathtt{I_D} : \exists r_i$ (blinding factor) s. t. $\exists i, j \notin \mathtt{I_D} : r_i = r_j \Rightarrow m_i = m_j$. We note that $\exists i, j \notin \mathtt{I_D} : m_i = m_j \not\Rightarrow r_i = r_j$ which means that an RDF term that occurs at different indicies may have two distinct blinding factors, e. g., to hide the fact that the same RDF term occurs in two quads.
3. Let $\forall i \notin \mathtt{I_D} : z_i = r_i + c\, m_i$ (the Schnorr response value), with c the overall proof challenge which is the same for all messages.
4. It follows immediately: $\exists i, j \notin \mathtt{I_D} : z_i = z_j \Rightarrow m_i = m_j$. We again note that $\exists i, j \notin \mathtt{I_D} : m_i = m_j \not\Rightarrow z_i = z_j$.
5. Let B be a set of fresh blank nodes (that are not in G): $\forall i \notin \mathtt{I_D} : \exists b \in B$.
6. If two messages have in the same Schnorr Response value z, then they map to the same blank node in B: $t2b : \{i \notin \mathtt{I_D} : m_i\} \rightarrow \mathcal{B} \mid \forall (m_i, m_j) : z_i = z_j \Rightarrow t2b(m_i) = t2b(m_j)$.
7. Then replace the hidden terms with their stand-in blank node: $m' := m$ and $\forall i \notin \mathtt{I_D} : m'_i := t2b(m_i)$.
8. Finally, convert the array back[9] to an RDF graph: $P := m2g(m')$. Thus, P is G with some terms replaced by blank nodes, where multiple blank node may refer to the same RDF term.

Recall the definition of a proper instance I of an RDF graph H, where in H two distinct blank nodes may refer to one term in I [25].

It follows: original graph G is proper instance of the presented graph P.

It follows: original graph G is instance of the presented graph P.

Recall that "a graph is entailed under simple entailment by any of its instances" [25].

It follows: original graph G entails under simple entailment presented graph P. □

In our example, the original credential graph G (Listing 1.1 lines 1–11) is a proper instance of the presented credential graph P (Listing 1.2 lines 2–10). Notice how blank nodes `_:7` and `_:25` are both mapped to the same node http://example.org/users#alice. The presented credential graph P can be entailed under simple entailment from the original credential graph G.

[8] See Appendix A and B as a refresher on the usual formalisation of RDF and Schnorr.

[9] In this step, there exists a matching from object suffix to the blank node of the corresponding object value: The blank node stands-in for the object as a whole.

5.2 Proof Verification as Validity Check

Our vocabulary and model aims to supply a verifier with all the information to execute a validity check on the RDF dataset. That is, for an RDF dataset to be a valid representation of a presented credential, the modeled values must satisfy the proofs' mathematical verification equations. We thus ask:

Does Our Data Model Support Validity Checks using Proofs' Verification Equations?

In this evaluation, we only cover the proof of knowledge of BBS+ signature; the other proofs trivially follow the same line of argument. The BBS+ proof a tuple $(A', \bar{A}, d, \pi)$, and the prover proves as defined in [10]:

$$\pi \in SPK\{(\{m_i\}_{i \in \mathtt{I_D}}, e, r_2, r_3, s'): \\ \bar{A}/d = A'^{-e} * h_0^{r_2} \ \wedge \ g_1 \prod_{i \in \mathtt{I_D}} h_i^{m_i} = d^{r_3} * h^{-s'} \prod_{i \notin \mathtt{I_D}} h_i^{-m_i}\}$$

where SPK is short for signature proof of knowledge. For the SPKs, we use the Schnorr protocol (see Appendix B) to prove knowledge of the discrete logs within the equations[10]. The resulting *verification equations* (among other checks) are

SPK1	$A'^{\mathcal{S}(-e)} * h_0^{\mathcal{S}(r_2)} / (\bar{A}/d)^c = t_{\mathtt{SPK1}}$	
SPK2	$d^{\mathcal{S}(-r_3)} * h_0^{\mathcal{S}(s')} * \prod_{i \notin \mathtt{I_D}} h_i^{\mathcal{S}(m_i)} * (g_1 \prod_{i \in \mathtt{I_D}} h_i^{m_i})^c = t_{\mathtt{SPK2}}$	

where $\mathcal{S}(\cdot)$ is a Schnorr response, t the commitment to randomness and c the proof challenge. g_1, h_0 and h_i are verification key parameters [10]. $\mathtt{I_D}$ is the set of indices of revealed messages: $m_i | i \in \mathtt{I_D}$ is revealed; $m_i | i \notin \mathtt{I_D}$ is hidden.

For the RDF dataset to be a valid representation of the presented credential, every proof's verification equations must hold (including the above). We thus connect our proof model to the corresponding verification equations by creating a mapping from variables to values, e.g. trivially implemented using SPARQL (cf. values from Listing 1.2, e.g., line 34 for $\mathcal{S}(-e)$):

$$\mu = \{\mathcal{S}(-e) \mapsto \texttt{"U2Rte..."\^{}\^{}xsd:base64Binary}, \mathcal{S}(r_2) \mapsto ...\}$$

For all variables from the verification equations, there exists a corresponding value in our model. We then apply the mapping to the verification equations. If verification equations hold for all proofs, then the composite proof is successfully verified: The RDF dataset is a valid representation of the presented credential and its proofs. Our vocabulary thus provides a data model that supports validity checks on the RDF dataset using the proofs' mathematical verification equations.

---

[10] After inverting the equation for SPK2 to $g_1^{-1} \prod_{i \in \mathtt{I_D}} h_i^{-m_i} = d^{-r_3} * h^{s'} \prod_{i \notin \mathtt{I_D}} h_i^{m_i}$.

5.3 Evaluating Transformation Approaches

We presented two foundational approaches to transform RDF datasets to a list of messages usable for ZKP application in Sect. 3.1. We look at potential trade-offs:

Efficiency. We compare the transformation approaches regarding the number of resulting messages to sign, prove and verify. Schema-free transformation always results in a message list whose length is 5 times the number of quads in the RDF dataset. Each quad is dissected into subject, predicate, object value, object suffix and graph name. The optional trimming removes any zero-elements from the message list resulting in a equal or smaller length list. Schema-based transformation results in a message list whose length depends on the schema. In our version, the resulting list is the credential id followed by all object values of the VC. This results in a list length of one fifth of the number of quads, plus one. However, our performance comparison indicates that the theoretical efficiency advantage does not translate to practical performance gains (see Appendix D). Therefore, we conclude that the schema-free transformation is not worse (and not impractical at all) compared to the schema-based transformation.

(Proof) Generality. We compare the transformation approaches regarding their support of different kinds of proofs on an RDF datasets' terms. Schema-free transformation, in a generic fashion, enable proofs on any RDF term of an RDF dataset: Full selective disclosure on RDF term level; on subject, predicate, objects and their suffix. The fact that the same term occurred in two RDF triples can also be hidden by assigning two different blank nodes. Proofs of numeric bounds on numeric RDF literals and set (non-)membership proofs on RDF terms are also supported. Schema-based transformation only allows proofs on focal terms of the dataset depending on the schema. Selective disclosure is therefore partial as e.g. predicates are known from the schema, and existence relations between nodes can be inferred from the schema. Numeric bounds and set proofs are supported on the focal terms. Therefore, we conclude that the schema-free transformation offers a more general and thus flexible approach to performing proofs on credential data compared to using a pre-defined schema.

Conformance. We compare the transformation approaches regarding their conformance with the W3C VC data model [36]. For the credential itself, all approaches are compliant. For the presentation, we discuss: Schema-free transformation require that the triples of the presented credential graph are provided in the correct order as the ordering of RDF terms need to be the same for verification as for proof creation. Additionally for proofs on terms in predicate position, the data model must allow blank nodes as predicates, e.g. as SPARQL or Generalised RDF. This is not necessarily an issue: The mandatory serialisation of the VC data model [36] is JSON-LD, which supports Generalised RDF by default[11]. Schema-based transformation does not impose additional requirements. Using the schema (known a priori or looked up), the list of RDF terms

[11] https://www.w3.org/TR/json-ld11/#relationship-to-rdf.

can be re-constructed. Predicates are public knowledge; they cannot be hidden. Therefore, we conclude that the schema-free transformation imposes minor syntactical requirements on presented credential graphs in general, while the schema-based transformation requires a specific schema on specific credential instances.

6 Related Work

We do not claim originality of proof composition: LegoSNARK [12] is a framework to combine multiple proofs on the same witness using a commit-and-prove scheme [13,28]. Our work applies such composition of ZKPs to W3C VCs. Therefore, we now look at related work combining W3C VCs [36] and ZKPs.

First practical work on selective disclosure of RDF triples was introduced by MATTR, resulting in a community draft specification [31]. [38] presents a general formalisation of this approach to ultimately construct RDF-based verifiable digital credentials. As an application of this, [8] aspires privacy-preserving authentication and attribute-based authorization on Solid Pods [14]. Recently, a candidate recommendation [5] to achieve selective disclosure based on the Data Integrity specification [35] was published by the W3C VC Working Group. These works are limited to selective disclosure on RDF triple level. They do not consider proofs on an RDF term level, other proof types like numeric bounds, or proof composition in general. Our work, on the other hand, enables these features.

Hyperledger AnonCreds [16] use CL signatures [11] to create "anonymous credentials". Hyperledger AnonCreds can be constructed such that they can be transformed to a representation compliant with the W3C VC data model [36]. Proof-wise, AnonCreds allow for selective disclosure of credential values (the schema reveals the credential properties), predicate proofs like CL numeric bounds proofs, and revocation in zero-knowledge using CL accumulators. Recent work on the next version AnonCreds 2.0 aims to expand the feature set to support different signature schemes, statement proof protocols and verifiable encryption.

AnonCreds 2.0 is similar to the presented work in that it investigates flexible composition of ZKPs for digital credentials. It is different in that it focuses on a credential data format which is both JSON- and schema-based. Similar to AnonCreds 1.0, a transformation to the VC data model may be possible. But AnonCreds 2.0 does not recognise the RDF-based W3C VC Recommendation [36] as foundational. As such, there is no considerations of the logical connection between the original credential data and the selectively disclosed data – a connection that our work provides. Investigating selective disclosure in credential presentation from an RDF perspective allows our work to prove that applying querying and reasoning techniques on a presented credential is logically sound. On top and at the very least, our work provides an alternative perspective to AnonCreds: Our work shows that VCs can also be built without using schemas.

7 Conclusion

With our work, we seek to bridge the gap between the W3C recommended VC data model [36] and European cryptographers' recommendation for ZKPs [3]. In particular, we aim to contribute to the W3C's ongoing efforts in improving the securing mechanisms of VCs [5,35] by expanding their feature set while providing formal logical underpinnings: We highlighted the semantics for selective disclosure of RDF terms and proved that selective disclosure of terms from the credential graph is equivalent to simple entailment. Our work provides a foundational understanding of how compositions of ZKPs may be applied on terms in an RDF dataset – with or without using a schema.

Our work provides a foundation to build advanced semantic credential systems, e.g.: VC-based anonymous credentials that allow for user anonymity or pseudonymity while offering semantic querying and reasoning capabilities. But even using regular VCs, research on reasoning-enabled access control, i.e. based on entailed relations, would push the frontier of authorization mechanisms.

Acknowledgements. This work is supported in part by the German federal ministry of education and research (BMBF) in MANDAT (FKZ 16DTM107B). We thank Ruben Verborgh for shepherding the paper and Sebastian Rudolph for his help with logics terminology.Icons (used in figures) created by Freepik.

A Common Formalisation of RDF Graphs and Datasets

We re-use the common formalisation: Let $\mathcal{U}$ denote the set of all HTTP URIs [4, 21], $\mathcal{B}$ the set of all blank nodes, and $\mathcal{L}$ the set of all literals. Let $\mathcal{G}$ denotes the set of all RDF graphs. An RDF graph $G \in \mathcal{G}$ is defined as a set of triples. A triple t is defined as $t \in (\mathcal{U} \cup \mathcal{B}) \times \mathcal{U} \times (\mathcal{U} \cup \mathcal{B} \cup \mathcal{L})$. Let $\mathcal{D}$ denote the set of all RDF datasets. An RDF dataset $D \in \mathcal{D}$ is a set of named graphs and an unnamed *default graph*. A named graph [15] is a couple (n, G_n) where $n \in (\mathcal{U} \cup \mathcal{B})$ and $G_n \in \mathcal{G}$. The default graph does not have a graph name, $(\_, G)$ with $G \in \mathcal{G}$. A triple of a graph G_n within an RDF dataset is also referred to as a quad q with $q \in (\mathcal{U} \cup \mathcal{B}) \times \mathcal{U} \times (\mathcal{U} \cup \mathcal{B} \cup \mathcal{L}) \times \{n\}$, with n being the graph name.

B Schnorr Protocol Proof of Knowledge of Discrete Log

A prover wants to convince a verifier that they know a secret $x \in \mathbb{Z}_q$, the group of integers modulo q. To this end, the prover calculates $h = g^x$ and makes h public. In general terms, we call $h = g^x$ a *statement* and x a *witness* to that statement. In this proof, h is used as a *verification key* for the proof.

Prover and verifier know that $\mathbb{G}_q$ is a cyclic group of prime order q, g is a generator of $\mathbb{G}_q$, and $h \in \mathbb{G}_q$. The non-interactive protocol under the Fiat-Shamir transformation [20] makes use of a hash function that serves as a *random oracle*. The protocol is as follows:

Prover		**Verifier**
$r \xleftarrow{\$} \mathbb{Z}_q$		
$u = g^r$		
$c = H(g, q, h, u)$		
$z = r + x \cdot c$		
	$\xrightarrow{\pi=(u,c,z)}$	
		$u, h \overset{?}{\neq} 0$ && $u, h \overset{?}{\in} \mathbb{G}_q$
		$z \overset{?}{\neq} 0 \mod q$
		$c \overset{?}{=} H(g, q, h, u)$
		$g^z \overset{?}{=} u \cdot h^c$

To prove knowledge of a secret $x \in \mathbb{Z}_q$, the prover has *verification key* $h = g^x$. The prover creates: A randomly sampled $r \leftarrow \mathbb{Z}_q$ referred to as *randomness* or *blinding factor*. Statement $u = g^r$ is called *commitment to randomness*. A *Schnorr response value* is $z = r + c \cdot x$ where c is called the proof *challenge*. The challenge is computed by $c = H(g, q, h, u)$. Its argument list (g, q, h, u) is the proof's *challenge contribution*. The Schnorr proof of knowledge itself is $\pi = (u, c, z)$ which is provided by the prover to a verifier with the verification key h. The verifier knows (g, h, π) and validates their inputs. Then, the verifier checks the *verification equations* $g^z = u \cdot h^c$ and $c = H(g, q, h, u)$. If true, the verifier is convinced that the prover knows the secret x.

C Treatment of Dates and Time in Signature Creation

The W3C VC data model recommends using `xsd:dateTimeStamp` for indicating creation and expiration dates. `xsd:dateTimeStamp` does not exhibit the numeric fundamental facet and thus would be interpreted in the lexical space. With that, range proofs of "later/earlier than" will not work out-of-the-box. Consider `2004-04-12T13:20:00Z`, which is 1:20 pm on April 12, 2004, Coordinated Universal Time (UTC), and `2004-04-12T13:20:00-05:00`, which is the same time, albeit in different time zones. To compare the two, a common "reference point in time" is required. By convention, the time zone offsets could be resolved and all date time stamps could be transformed to UTC. Then, we could compare the lexical values of the adjusted dateTimeStamps as these values are now totally ordered. JSON Web Tokens [27] already solved this issue for their `iat` (issued at) and `exp` (expiration time) claims. Their values must be a `NumericDate` value. `NumericDate` is defined as the number of seconds from `1970-01-01T00:00:00Z UTC` until the specified UTC time ignoring leap seconds [26]. Similar to JWTs, we internally transform an `xsd:dateTimeStamp` value to a `NumericDate` value. In our example, `"2023-11-07T13:20:00Z"` is transformed to `1699363200`.

D Example Implementation

We implemented[12] our example with credential issuance, presentation and verification to evaluate the transformation approaches (Sect. 3.1). In theory, the distinguishing factor should be the message list length as the proof of knowledge of signature becomes more expensive with more messages. Table 2 indicates however that performance difference is negligible. The first number is the time in milliseconds of the overall process and the second number is only the part of creating/verifying a proof in the process. We presume the credentials are too small scale such that any algorithmic difference is not relevant given the hardware; a consumer laptop with an AMD Ryzen 7 PRO 5850U.

Table 2. Performance benchmark using `criterion.rs`.

Transformation	Schema-free (no trim)	Schema-based
Number of messages	35	8
Prove (incl. RDF transform)/proof creation time	∼158 ms/∼59 ms	∼157 ms/∼60 ms
Verify (incl. RDF transform)/verification time	∼114 ms/∼18 ms	∼116 ms/∼18 ms

References

1. Alvestrand, H.: Tags for the identification of languages. Best current practice, IETF (2001). https://www.ietf.org/rfc/rfc3066.txt
2. Au, M.H., Susilo, W., Mu, Y.: Constant-size dynamic k-taa. In: Proceedings of the 5th SCN (2006)
3. Baum, C., et al.: Cryptographers' Feedback on the EU Digital Identity's ARF (June 2024). https://github.com/user-attachments/files/15904122/cryptographers-feedback.pdf, see also https://github.com/eu-digital-identity-wallet/eudi-doc-architecture-and-reference-framework/issues/200
4. Berners-Lee, T., Fielding, R., Masinter, L.: Uniform resource identifier (uri): generic syntax. Internet standards track document, IETF (January 2005). https://www.ietf.org/rfc/rfc3986.txt
5. Bernstein, G., Sporny, M.: Data integrity bbs cryptosuites v1.0. W3c candidate recommendation draft, W3C Verifiable Credentials Working Group (2024). https://www.w3.org/TR/vc-di-bbs/
6. Bitkom Arbeitskreis Digitale Identitäten: EU Digital Identity Wallet (2025). Bitkom e.V. https://www.bitkom.org/sites/main/files/2025-01/bitkom-whitepaper-organisationsidentitaeten.pdf. Accessed 11 Mar 2025
7. Brands, S.: A technical overview of digital credentials (2002). https://api.semanticscholar.org/CorpusID:18284690
8. Braun, C., Käfer, T.: Attribute-based access control on solid pods using privacy-friendly credentials. In: Proceedings of the Posters & Demos at the 18th SEMANTiCS. CEUR Workshop Proceedings, vol. 3235. CEUR-WS.org (2022)

[12] https://github.com/uvdsl/rdf-zkp/.

9. Braun, C.H.J., Papanchev, V., Käfer, T.: SISSI: an architecture for semantic interoperable self-sovereign identity-based access control on the Web. In: Proceedings of the 32nd Web Conference (WWW), pp. 3011–3021. ACM, New York, NY, USA (2023). https://doi.org/10.1145/3543507.3583409
10. Camenisch, J., Drijvers, M., Lehmann, A.: Anonymous attestation using the strong diffie hellman assumption revisited. IACR Cryptol. ePrint Arch, p. 663 (2016). http://eprint.iacr.org/2016/663
11. Camenisch, J., Lysyanskaya, A.: A signature scheme with efficient protocols. In: Revised Papers of the 3rd SCN. LNCS, vol. 2576, pp. 268–289. Springer (2002)
12. Campanelli, M., Fiore, D., Querol, A.: Legosnark: modular design and composition of succinct zero-knowledge proofs. IACR Cryptol. ePrint Arch. p. 142 (2019)
13. Canetti, R., Lindell, Y., Ostrovsky, R., Sahai, A.: Universally composable two-party and multi-party secure computation. IACR Cryptol. ePrint Arch. p. 140 (2002). http://eprint.iacr.org/2002/140
14. Capadisli, S., Berners-Lee, T., Verborgh, R., Kjernsmo, K.: Solid protocol. Version 0.9.0, W3C Solid Community Group (December 2021). https://solidproject.org/TR/protocol
15. Carroll, J.J., Bizer, C., Hayes, P.J., Stickler, P.: Named graphs, provenance and trust. In: Ellis, A., Hagino, T. (eds.) Proceedings of the 14th international conference on World Wide Web, WWW 2005, Chiba, Japan, 10–14 May 2005, pp. 613–622. ACM (2005). https://doi.org/10.1145/1060745.1060835
16. Curran, S., et al.: Anoncreds specification. v1.0 draft, Hyperledger AnonCreds Working Group (2024). https://hyperledger.github.io/anoncreds-spec/
17. Cyganiak, R., Wood, D., Lanthaler, M.: Rdf 1.1 concepts and abstract syntax. W3C Recommendation, W3C (2014). https://www.w3.org/TR/rdf11-concepts/
18. European Commission: EU Digital Identity Wallet (2024). https://ec.europa.eu/digital-building-blocks/sites/display/EUDIGITALIDENTITYWALLET/. Accessed 17 Sep 2024
19. European Commission: Regulation (eu) 2024/1183 of the european parliament and of the council amending regulation (eu) no 910/2014 as regards establishing the european digital identity framework (2024). https://eur-lex.europa.eu/legal-content/EN/TXT/?uri=OJ:L_202401183. Accessed 17 Sep 2024
20. Fiat, A., Shamir, A.: How to prove yourself: practical solutions to identification and signature problems. In: Odlyzko, A.M. (ed.) CRYPTO 1986. LNCS, vol. 263, pp. 186–194. Springer, Heidelberg (1987). https://doi.org/10.1007/3-540-47721-7_12
21. Fielding, R., Reschke, J.: Hypertext transfer protocol (http/1.1): Message syntax and routing. Internet standards track document, IETF (June 2014). https://www.ietf.org/rfc/rfc7230.txt
22. Goldwasser, S., Micali, S., Rackoff, C.: The knowledge complexity of interactive proof-systems. In: Proceedings of the 17th ACM STOC, pp. 291–304. ACM (1985)
23. Hanssens, S., Vonner, F., Lisoir, X.: Digital identity (2021). PricewaterhouseCoopers, Société coopérative. https://www.pwc.lu/en/smart-identity/docs/pwc-digital-identity.pdf. Accessed 11 Mar 2025
24. Harris, S., Seaborne, A.: Sparql 1.1 query language. W3C Recommendation, W3C (2013). https://www.w3.org/TR/sparql11-query/
25. Hayes, P., Patel-Schneider, P.F.: Rdf 1.1 semantics. W3C Recommendation, W3C (2014). https://www.w3.org/TR/rdf11-mt/
26. IEEE, The Open Group: The open group base specifications issue 7, 2018 edition. IEEE Std 1003.1-2017, IEEE and The Open Group (2017). https://pubs.opengroup.org/onlinepubs/9699919799/basedefs/V1_chap04.html#tag_04_15

27. Jones, M., Bradley, J., Sakimura, N.: Json web token (jwt). Internet standards track document, IETF (May 2015). https://www.ietf.org/rfc/rfc7519.txt
28. Kilian, J.: Uses of Randomness in Algorithms and Protocols. MIT Press, Cambridge (1990)
29. Knublauch, H., Kontokostas, D.: Shapes constraint language (shacl). W3C Recommendation, W3C (July 2017). https://www.w3.org/TR/shacl/
30. Longley, D., Kellogg, G., Yamamoto, D.: RDF dataset canonicalization a standard RDF dataset canonicalization algorithm. W3C recommendation, W3C (2024). https://www.w3.org/TR/rdf-canon/
31. Looker, T., Steele, O.: Bbs+ signatures 2020. Draft cg report, W3C Credentials Community Group (2023). https://w3c-ccg.github.io/ldp-bbs2020/#the-bbs-signature-suite-2020
32. National Security Agency: Common Criteria for information technology security evaluation (CCMB-2017-04-002) (2017). https://www.commoncriteriaportal.org/files/ccfiles/CCPART2V3.1R5.pdf
33. Peterson, D., Gao, S., Malhotra, A., Sperberg-McQueen, C.M., Thompson, H.S.: W3C XML schema definition language (XSD) 1.1 part 2: Datatypes. W3C recommendation, W3C (2012). https://www.w3.org/TR/xmlschema11-2/
34. Schnorr, C.P.: Efficient identification and signatures for smart cards. In: Brassard, G. (ed.) CRYPTO 1989. LNCS, vol. 435, pp. 239–252. Springer, New York (1990). https://doi.org/10.1007/0-387-34805-0_22
35. Sporny, M., Longley, D., Bernstein, G., Zagidulin, D., Crane, S.: Verifiable credential data integrity 1.0. W3c candidate recommendation draft, W3C Verifiable Credentials Working Group (2024). https://www.w3.org/TR/vc-data-integrity/
36. Sporny, M., Noble, G., Longley, D., Burnett, D.C., Zundel, B., Hartog, K.D.: Verifiable credentials data model v1.1. W3C recommendation, W3C (2022). https://www.w3.org/TR/vc-data-model/
37. Vitto, G., Biryukov, A.: Dynamic universal accumulator with batch update over bilinear groups. IACR Cryptol. ePrint Arch. p. 777 (2020). https://eprint.iacr.org/2020/777
38. Yamamoto, D., Suga, Y., Sako, K.: Formalising linked-data based verifiable credentials for selective disclosure. In: IEEE European Symposium on Security and Privacy, EuroS&P 2022 - Workshops, Genoa, Italy, 6–10 June 2022. pp. 52–65. IEEE (2022). https://doi.org/10.1109/EUROSPW55150.2022.00013
39. Yergeau, F.: Utf-8, a transformation format of iso 10646. Internet standards track document, IETF (November 2003). https://www.ietf.org/rfc/rfc3629.txt
40. Zimmermann, A.: RDF 1.1: On semantics of RDF datasets. W3C working group note, W3C (February 2014). https://www.w3.org/TR/rdf11-datasets/

Taxonomy Inference for Tabular Data Using Large Language Models

Zhenyu Wu(✉), Jiaoyan Chen, and Norman W. Paton

The University of Manchester, Mancheste, UK
{zhenyu.wu,jiaoyan.chen,norman.paton}@manchester.ac.uk

Abstract. Taxonomy inference for tabular data is a critical task of schema inference, aiming at discovering entity types (i.e., concepts) of the tables and building their hierarchy. It can play an important role in data management, data exploration, ontology learning, and many data-centric applications. Existing schema inference systems focus more on XML, JSON or RDF data, and often rely on lexical formats and structures of the data for calculating similarities, with limited exploitation of the semantics of the text across a table. Motivated by recent works on taxonomy completion and construction using Large Language Models (LLMs), this paper presents two LLM-based methods for **t**axonomy inference for **t**ables: (i) EmTT which **em**beds columns by fine-tuning with contrastive learning encoder-alone LLMs like BERT and utilises clustering for hierarchy construction, and (ii) GeTT which **ge**nerates table entity types and their hierarchy by iterative prompting using a decoder-alone LLM like GPT-4. Extensive evaluation on three real-world datasets with six metrics covering different aspects of the output taxonomies has demonstrated that EmTT and GeTT can both produce taxonomies with strong consistency relative to the Ground Truth.

Keywords: Taxonomy Inference · Tabular Data · Large Language Models · Contrastive Learning · Prompt Learning · Schema Inference

1 Introduction

Schema inference, which is to identify the structure, meta information and semantics of a dataset such as relationships between data fields and data types, plays a critical role in data management, ontology learning, and data-centric applications [2,3,46]. In particular, inferring the entity types (i.e., semantic types or concepts like *School* and *Hotel*) of tables in a given dataset as well as their hierarchies is one of the fundamental tasks in schema inference. It not only provides necessary semantics for other tasks like mining constraints on the relationships, but also directly supports data exploitation in quite a few scenarios such as Knowledge Graph population, table retrieval and table question answering [14,29].

E. Curry et al. (Eds.): ESWC 2025, LNCS 15718, pp. 403–422, 2025.
https://doi.org/10.1007/978-3-031-94575-5_22

However, most of the current schema inference methods consider XML or JSON documents, or graph data composed of RDF triples [23,38], while inferring type hierarchies for sets of heterogeneous tables is paid little attention. Furthermore, early proposals mostly rely on lexical formats and the structure of the data for calculating similarities, without fully exploiting the semantics of the unstructured or semi-structured text [9,24]. Recently, there are some works that attempt to train neural networks, especially Transformer-based architectures, or fine-tune their pre-trained versions, for embedding tabular data for measuring similarities and conducting prediction tasks such as column type annotation and joinable table discovery (e.g., TURL [11] and DeepJoin [13]), but there is a shortage of exploration into complex tasks including hierarchy inference, which has as input a set of heterogeneous tables and a structured output, and may rely on multiple steps.

Meanwhile, quite a few recent works explore Large Language Models (LLMs) of different architectures including encoder-alone, encoder-decoder and decoder-alone for taxonomy construction and curation, utilising text. Most of these works focus on taxonomy completion, such as the insertion of new concepts or new subsumption relationships, by transforming the problem into machine learning classification based on the encoding of LLMs (e.g., [7,28,42]). There are also several works that develop complex prompts, which are often iterative, for generative LLMs like the Llama [44] and GPT series for constructing taxonomies from scratch or from a given set of concepts (e.g., [18,53]). However, none of these LLM-based works have explored taxonomy construction for a given set of heterogeneous tabular data, which involves not only the manipulation of the concepts but also learning from the raw data.

The contributions of this work are as follows:

1. A proposal for an *Embedding-based Method*, that clusters column embeddings to identify top-level concepts, their attributes and concept hierarchies, in sequence, and which can be used with pretrained or fine-tuned language models.
2. A proposal for a *Generative Method* that prompts a pre-trained generative LLM such as GPT-4, DeepSeek-R1 [10] or Qwen2 [48] to infer table semantic types, and then applies an iterative prompting method named Chain-of-Layer [53] for taxonomy construction.
3. An evaluation of (1) and (2), each building on several language models, on three real-world table sets with annotated ground truth hierarchies, using six metrics that consider quality of both top-level types and the overall taxonomy, with positive results.

For conciseness, we refer to these two methods as EmTT and GeTT, respectively, where TT is short for **t**axonomy inference for **t**abular data. We study two methods that use different language model techniques to thoroughly explore the potential of language models for solving the problem of table hierarchy inference.

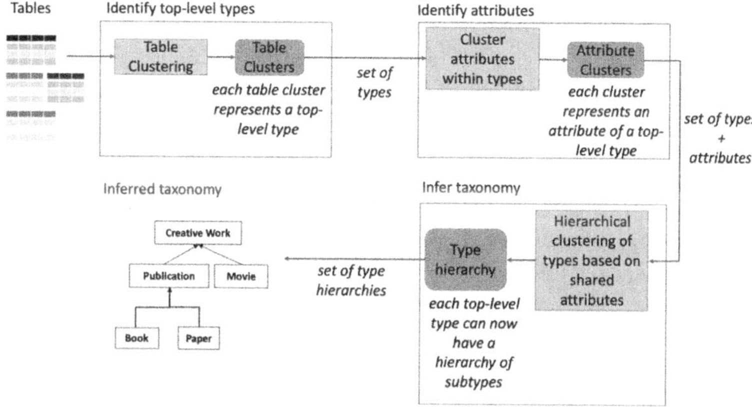

Fig. 1. The Framework of the Embedding-based Method EmTT

2 Problem Statement

Given a set of tables $\mathcal{D}$, this study aims to: *(i)* for each table $d \in \mathcal{D}$, infer entity types that the corresponding entities of the table rows all belong to; *(ii)* with the entity types of all the tables, denoted as $\mathcal{T}$, build an *entity type taxonomy* (taxonomy in short), denoted as $\mathcal{H} = (\mathcal{T}, \mathcal{E})$, which is a directed acyclic graph (DAG) for representing the hierarchy of $\mathcal{T}$, with $\mathcal{E}$ being a set of directed edges representing "is-a" (i.e., subsumption) relationships between the entity types. This formulation assumes that each table is associated with to a single entity type; non-entity tables can be converted into entity tables by a recent method [27].

3 Embedding-Based Method

In this section, we introduce the embedding-based method EmTT. As illustrated in Fig. 1, it includes the following three steps: *Identify top-level types*, which clusters the tables according to their column embeddings, with each cluster representing one top-level type; *Identify attributes*, which clusters columns of the tables associated with each top-level type in turn, with each cluster representing an attribute of the top-level type; and *Infer taxonomy*, which utilises these type attributes to group the tables of each top-level type into hierarchical sub-types based on their shared attributes. We now introduce each step.

3.1 Identify Top-Level Types

Each table is composed of rows and columns (a.k.a. table attributes), with each row representing an entity. Such flat entity tables are common in real-world data such as web pages and government data[1]. It is assumed that each table contains a column representing the *entity type* (i.e., semantic type or concept) that the table is about, which is known as the *subject column* or *subject attribute*. For example, in a table of *companies*, the subject column could be the *company name*, while the other columns provide additional properties of the company such as its head office address or turnover. Subject columns have been used to support a variety of tasks, such as web table extraction [45] and table annotation [47], and several methods have been proposed for their inference (e.g., [45,55]).

To identify top-level types, a set of clusters $\mathcal{C}$ is created, where each $c \in \mathcal{C}$ groups tables from $\mathcal{D}$ that share a common high-level type like *Organization* or *Person*. The approach involves two steps: (i) identifying the subject column of each table using an existing technique proposed in [55]; and (ii) clustering the tables based on the semantic similarity of the embeddings of their subject columns, where the number of clusters is chosen with the Silhouette Coefficient [41].

In experiments, we use Agglomerative Clustering with Euclidean distance as the metric, as this approach was found to perform well in comparison with other clustering algorithms for this task. As the subject columns capture the identifying property of each table, such as the *name* of a *company* or the *title* of a *movie*, clustering by subject column should bring together all the tables representing companies (or perhaps organizations) in one cluster and all the tables representing movies (or perhaps creative works) to provide candidate top-level types. We note that the choice of top-level types is subjective; should *movie* or *creative work* be the top level type? We depend on clustering based on embeddings to make decisions on the granularity of the top-level types, and infer taxonomies to identify finer-grained types. The experiments compare the inferred top-level types of tables to manually annotated tables. The inferred top-level types can be considered to be conceptual types, in that they aim to reflect the concepts represented in the tables.

3.2 Identify Attributes

After identifying the (conceptual) top-level types, we derive the (conceptual) attributes of each type based on the attributes (columns) of the tables in the type's cluster. These table attributes are grouped into clusters, where those within the same cluster are expected to share similar semantics or belong to the same semantic domain. For example, the *location* attribute of an *Organization* table and the *place* attribute of another *Organization* table are likely to represent similar properties of organizations and thus could be consolidated into a single attribute of the top-level type.

Therefore, we apply a clustering algorithm to the embeddings of all the table attributes (columns) of tables of each top-level type, and use the resultant clusters to define its conceptual attributes. Note that after this process, each table

[1] Studies indicate that 30–50% of spreadsheets contain non-entity tables [27]. However, recent techniques have been proposed to automate the conversion of non-entity tables into entity tables [27], which broadens the applicability of methods relying on subject attributes.

Algorithm 1: Dendrogram Pruning

```
Data: dendrogram; maxSilhouette;
Result: Type Hierarchy
Initialize hierarchy;
current_y = max(y values in dendrogram);
while current_y >= 0 do
    clusters = getClustersAtHeight(dendrogram, current_y);
    if getSilhouette(clusters) > (maxSilhouette − Δ) then
        for cluster in clusters do
            if cluster is not a single table then
                tables = getTable(cluster);
                parentCluster = findParentCluster(hierarchy, cluster);
                if parentCluster exists then
                    AddEdge(hierarchy, parentCluster, cluster);
                else
                    addRoot(hierarchy, cluster)
    current_y− = δ;
return hierarchy;
```

attribute is mapped to a single conceptual (top-level type) attribute. As in the identification of top-level types, Agglomerative Clustering and Euclidean distance are used, having been shown to provide good performance in experiments.

3.3 Infer Taxonomy

Within each top-level type, tables may reflect different perspectives. For example, a cluster of tables representing *Organization* might encompass specific subtypes such as *University* and *Company*. In this method, we assume that the subtypes' distinctions are evident through their conceptual attributes, and that a sub-taxonomy can be developed for each top-level type by applying hierarchical clustering to the conceptual attributes that its tables have. In particular, for each top-level type t, the approach includes:

1. **Hierarchical Clustering**: a hierarchical clustering algorithm is applied to the tables of t to construct a dendrogram, where the distance between two tables is the Jaccard similarity of their sets of conceptual attributes.
2. **Dendrogram Pruning**: The resulting Dendrogram is sliced (Algorithm 1) at various y-axis levels to generate clusters representing potential subtypes. This approach identifies groups of attributes that co-occur across multiple tables in the cluster. To ensure taxonomy quality, we retain only slices where cluster silhouette scores fall within the range $[maxSilhouette - \Delta, maxSilhouette]$, where $maxSilhouetteScore$ is the highest silhouette score that can be obtained by slicing the dendrogram and Δ helps determine the minimum

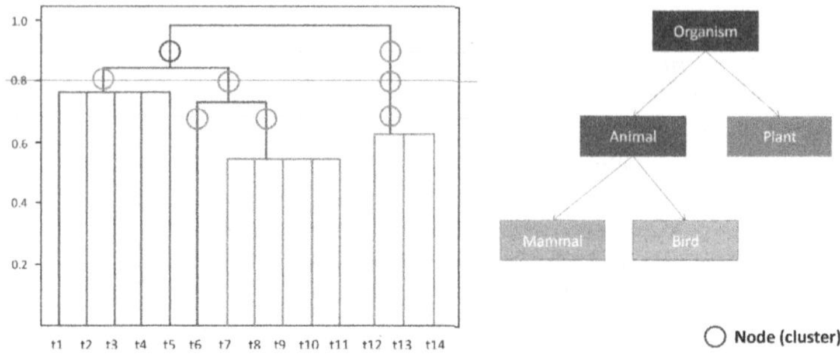

Fig. 2. An Example of Dendrogram Prunning from WDC.

silhouette score for the resulting clusters when slicing the dendrogram. Each retained cluster is designated as a new subtype in the taxonomy, and the dendrogram's hierarchy is reflected in the taxonomy's structure.

Figure 2 demonstrates the taxonomy inference process for the top-level type *Organism* in WDC. The dendrogram, generated from clustering 14 tables (*t1* to *t14*), identifies three slices with silhouette scores within Δ of the highest score. Each slice is color-coded to match the corresponding types in the GT hierarchy. The slicing algorithm (Algorithm 1) begins at the top of the dendrogram and moves downward, including clusters in the resulting *hierarchy* if their average silhouette scores fall within the predefined range.

In the experiments, Δ is set to 0.15—a value that has been shown empirically to produce hierarchies that are both at intuitive levels of detail and have good levels of consistency. Broadly speaking, a higher Δ leads to a deeper hierarchy.

4 Generative Method

The framework of GeTT is shown in Fig. 3. It includes two modules: entity type generation for tables and type hierarchy construction. We will next introduce these two modules with details. Note that complete prompts used in both modules can be found in GeTT/prompts.txt in the code and data repository.

4.1 Table Entity Type Generation

GeTT uses a generative LLM to generate potential entity types for each given table, with the structure of the prompt demonstrated in Fig. 3. This prompt includes the following parts: (a) a description of the task, (b) a specification of the input including a simple but effective and widely adopted table serialization which uses commas to separate the values [35], (c) a rule for specifying the output—solely the names of the entity types. Current LLMs support a limited context window, which typically encompasses a few thousand tokens. Therefore, a sampling operation needs to be applied to the original table before it is used

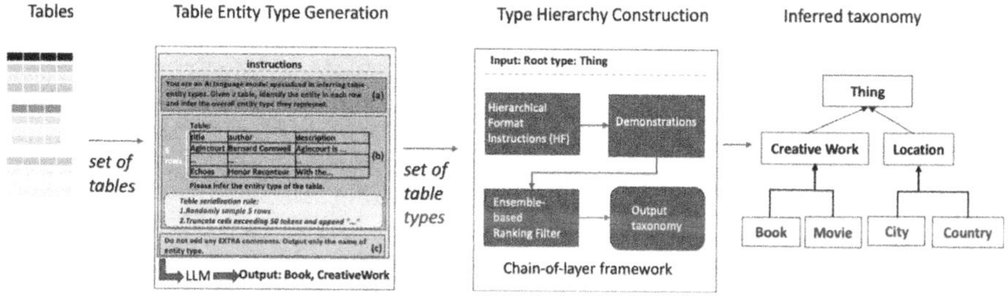

Fig. 3. The Framework of the Generative Method GeTT

to construct the prompt. Following [22], we randomly sampled 5 rows from each table. For some columns like "descriptions", an individual cell may contain so many tokens that a prompt with one or two rows exceeds the maximum context window size. We thus truncate cells with more than 50 tokens to 50 tokens and append "..." to indicate the truncation. The output of the Table Entity Type Generation step in Fig. 3 is a set of entity type names.

4.2 Type Hierarchy Construction

GeTT first transforms the generated entity types of all the tables into a flat list (denoted as $\mathcal{V}$), where types with the same name are regarded as one type with their associated tables merged, and then feeds $\mathcal{V}$ into an LLM together with a root type (denoted v_0) for constructing a coherent hierarchy. Instead of developing prompts from scratch, we use a state-of-the-art LLM prompt for hierarchy construction named Chain-of-Layer (CoL) [53]. There are other methods that construct taxonomies with an LLM (e.g., [18,32]), but their settings are relatively different from ours which has types given and requires no LLM training. CoL builds the taxonomy from the top down, starting from the initial layer $\mathcal{T}^0$ (at a top level type resulting from Sect. 4.1) that is composed of v_0 alone, and iteratively adding new layers of types. At the k-th iteration, given the current layer $\mathcal{T}^k$, CoL selects the appropriate child types from $\mathcal{V}$, forming the next layer $\mathcal{T}^{k+1}$, and removes them from $\mathcal{V}$. This process continues until $\mathcal{V}$ becomes empty. For the technical details of CoL, please refer to [53]. Here is a brief introduction to its three main components:

- **Hierarchical Format Instructions:** Each iteration is guided by an instruction that directs the LLM to generate plausible child types of the types in the current layer from the given list $\mathcal{V}$.
- **Demonstrations:** CoL provides example taxonomies to the LLM. They can be taxonomies either annotated by experts or generated by the LLM. Our method GeTT adopts the latter, which is called the zero-shot setting of CoL. For each demonstration, the taxonomy is decomposed in a hierarchical order and simulated from top to bottom. After each induction step, the LLM checks if all target entities are included. If not, the taxonomy is further expanded until it encompasses the entire set of entities.

Table 1. Statistics of the Datasets

Dataset	# Tables	# Attributes	# Top-level Types	# Lowest-level Types	# Entity Types	Depth
WDC	602	4200	7	43	71	4
GDS	660	15195	6	53	66	3
OpenData	10361	313822	6	49	62	3

- **Ensemble-based Ranking Filter:** At each iteration, to mitigate hallucinations, i.e., incorrect or semantically inconsistent parent-child relationships introduced by the LLM, CoL filters out the generated low-quality parent-child relationships by transforming them into sentences with multiple templates and feeding them into a pre-trained mask language model for scoring and ranking.

5 Evaluation

5.1 Experiment Settings

Datasets. We adopt the following three table sets: WDC which includes 602 tables from Web Data Commons [34] and a Web table set named T2DV2 [40], GDS which includes 660 tables from Google Dataset Search [4] and OpenData, which includes 10361 tables collected from various Open Data portals worldwide, covering a diverse range of sources beyond those from the UK, US, and Australia [19–21]. Each table set is annotated with a ground truth (GT) taxonomy composed of entity types from Schema.org, and each table is also annotated with entity types from Schema.org, with its most specific entity type and top-level type specified. The statistics of the three datasets are shown in Table 1. Note that lowest-level types are the leaf types in the taxonomy.

Metrics Applied to the Top-Level Types. Top-level types, as the most fundamental output of schema inference, include important meta information of the table set, and also influence the quality of the constructed taxonomy. We regard the top-level type inference as a problem of table clustering, and accordingly calculate the widely used metric Rand Index (RI) $= \frac{\text{TP+TN}}{\text{TP+FP+FN+TN}}$, where TP (resp. TN) is the number of pairs of tables that belong to the same (resp. different) top-level type in both the output taxonomy $\mathcal{H}_o$ and the GT taxonomy $\mathcal{H}_{gt}$, and FP (resp. FN) is the number of pairs of tables that belong to the same (resp. different) top-level type in $\mathcal{H}_o$ but different (resp. same) top-level types in $\mathcal{H}_{gt}$. We also calculate Purity of each top-level type t in $\mathcal{H}_o$ as the ratio of the tables whose top-level type in $\mathcal{H}_{gt}$ is $m(t)$ (i.e., matched with t), among all the tables associated to t. With the Purities of all the top-level types in $\mathcal{H}_o$, we average them as the final Purity of top-level types of $\mathcal{H}_o$.

Metrics Applied to the Whole Taxonomy. We first use some taxonomy statistics, including the number of types (T#) and the maximum depth of leaf types, i.e., the number of levels (L#), for assessment. Richer taxonomies have higher T# and/or L#. There have been metrics to assess the correctness of taxonomies but they usually independently assess each edge (i.e., the subsumption of two concepts) ignoring the taxonomy structure, such as the pre-trained language model-based RaTE [26]. Therefore, we propose a new metric named Tree Consistency Score (TCS) to measure the overall structure consistency between $\mathcal{H}_o$ and $\mathcal{H}_{gt}$, with the basic idea of comparing all the ancestors of two matched concepts in their taxonomies. A higher TCS indicates a higher-quality output taxonomy. It is calculated as follows:

1. For each type t from $\mathcal{H}_o$, we match it with a type that is from $\mathcal{H}_{gt}$ and is the most frequent entity type annotation of the associated tables of t. This matched type is denoted as $m(t)$.
2. We calculate the consistency of each type t from $\mathcal{H}_o$ as

$$C_{type}(t) = \frac{|\{a \in A(t, \mathcal{H}_o) | m(a) \in A(m(t), \mathcal{H}_{gt})\}|}{|A(t, \mathcal{H}_o)|} \tag{1}$$

   where the function $A(\cdot, \cdot)$ calculates the set of ancestors of a given type in a given taxonomy, and $|\cdot|$ denotes the set cardinality.
3. The TCS score of $\mathcal{H}_o$ is computed as

$$C_{taxo}(\mathcal{H}_o) = \frac{\sum_{t \in \mathcal{T}_o} C_{type}(t)}{|\mathcal{T}_o|} \tag{2}$$

   where $\mathcal{T}_o$ is all the types of $\mathcal{H}_o$.

Embeddings. For EmTT, we employ the following approaches to create column embeddings, the similarity of which is used to identify top-level types and their attributes, as described in Sect. 3:

- SBERT [39] is a pretrained language model developed for encoding sentences, that has not been fine-tuned for the specific task of column embedding. This is in contrast with the other embedding models used. To represent a column as a sentence, we provide a structured format: using $< s >$ as the start token, enclosing the column header in $<$ *header* $>$ ... $<$ */header* $>$, and appending unique cell values concatenated with spaces.
- *Starmie* was designed primarily for the Table Union Search problem [36], which identifies similar table attributes to determine table unionability [15]. It applies the SimCLR [8] contrastive learning framework to table attributes, treating table attributes as data items and their subsets as positive variants. Embeddings are generated by fine-tuning RoBERTa [31] through self-supervised learning to maximize similarity between positive examples and minimize it for negatives.

- *DeepJoin* was developed for discovering joinable attributes, supporting both exact and semantic joins [13]. Positive training examples are derived from attribute pairs with SBERT cosine similarity above 0.9, while unmatched attributes act as negatives. DeepJoin optimizes embeddings using a contrastive loss function to differentiate between positive and negative examples.
- *Unicorn* [16] trains an architecture that first encodes data element pairs of multiple matching tasks by a pre-trained language model and an additional Mixture-of-Experts layer, and then predicts the matchings by attaching a classifier. In our application, Unicorn matches attributes between tables to infer top-level types and conceptual attributes. Matched pairs are represented as edges in a graph, where nodes correspond to attributes. Attributes are grouped into clusters as connected subgraphs, with each cluster treated as a top-level type's attribute.
- *SwAV* [5], is a contrastive learning technique originally developed for image embedding and clustering [5]; here we apply it for the first time to tabular data. Unlike traditional contrastive learning approaches such as SimCLR [8], which rely heavily on both positive and negative samples and require pairwise feature comparisons within large batches, SwAV reduces the need for explicit negative pairs and large batch sizes by learning consistent cluster assignments across multiple augmentations (views) of the same data item. Here, we use SwAV to fine tune SBERT embeddings for column comparison.

Other Features of Experiment Setup. EmTT and GeTT are implemented using PyTorch, the Hugging Face Transformers, and the Ollama library. Details of the fine-tuning in EmTT with Starmie, DeepJoin and SwAV are presented in the Appendix A. We evaluate GeTT using two closed-source LLMs (GPT-3.5 and GPT-4) and three open-source LLMs, namely deepseek-R1 and two different-sized variants of Qwen2.5[2], as they are widely adopted and often achieve promising results.

We run each configuration 5 times, reporting the averages as well as the standard deviations. To run EmTT and GeTT with Qwen2.5 (14B/32B), which rely on local GPU, we use an NVIDIA A100 (80GB) with 2 x 24-core AMD Epyc 7413 2.65 GHz processors and 512GB RAM. For GeTT with GPT-3.5/4, we use a workstation with a 14-core Intel Xeon E5-2680 v4 processor with 128 GB RAM.

Availability. All the datasets and code can be accessed from: https://github.com/PierreWoL/TwoMethods.

[2] Llama 3 was also tested but inconsistent output representations made it difficult to incorporate in the Chain-of-Layer workflow.

Table 2. Results of the baselines, EmTT and GeTT over all the metrics, with the best results in bold and the second best underlined.

Method	WDC					GDS					OpenData				
	TCS	**T#**	**L#**	**RI**	**Purity**	**TCS**	**T#**	**L#**	**RI**	**Purity**	**TCS**	**T#**	**L#**	**RI**	**Purity**
EmTT (SBERT)	**1.000**	158	5	0.770	0.814	**0.929**	231	5	0.792	0.674	**0.845**	936	7	0.889	0.884
EmTT (Starmie)	**1.000**	117	2	0.771	0.754	0.895	134	3	0.813	0.726	0.813	743	5	0.841	0.772
EmTT (DeepJoin)	0.836	162	4	0.845	0.842	0.848	214	5	**0.896**	0.857	0.821	897	7	0.879	0.878
EmTT (Unicorn)	-	-	-	0.734	0.890	-	-	-	0.709	**0.903**	-	-	-	-	-
EmTT (SwAV)	0.856	213	**6**	0.865	0.870	0.800	276	**7**	0.885	0.856	0.783	1228	**10**	0.881	0.851
GeTT (GPT-3.5)	0.490	354	4	0.516	0.747	0.437	293	5	0.852	0.526	0.457	1034	5	0.879	0.914
GeTT (GPT-4)	0.816	**358**	5	0.976	**0.961**	0.425	296	5	0.760	0.900	0.516	1162	6	**0.912**	0.906
GeTT (DeepSeek-R1)	0.813	348	5	**0.980**	0.957	0.596	**302**	5	0.819	0.787	0.532	1216	6	0.907	0.898
GeTT (Qwen2.5-14b)	0.602	352	4	0.727	0.738	0.562	294	4	0.784	0.610	0.453	**1344**	7	0.836	0.797
GeTT (Qwen2.5-32b)	0.681	356	5	0.814	0.930	0.493	295	6	0.751	0.673	0.545	1258	6	0.809	0.782

5.2 Result Analysis

The experimental results are reported in Tables 2, 3 and 4. We will next analyse the quality of the top-level types and the overall taxonomies, and compare EmTT and GeTT w.r.t. efficiency and stability.

Quality of the Top-Level Types. In EmTT, among the embedding methods, it can be seen from Table 2 that DeepJoin and SwAV provide the most consistently strong performance for RI and Purity. SwAV fine tunes SBERT for column similarity, so its advantage w.r.t. SBERT stems from its contrastive learning mechanism, which pulls attributes with mutual information closer together in the embedding space, resulting in small, dense and high-purity clusters. DeepJoin benefits from the use of (automatically identified) positive attribute examples during training, which helps the model embed similar attributes closer together. In contrast, Starmie, which is based on SimCLR, struggles to separate attributes that share overlapping values but have different meanings. For instance, *music album* names such as *The Mozart Album* and *Mozart Momentum: 1785* may be confused with *event* names like *Mozart Gala* or *Mozart's Violin Concerto No. 5.* This issue arises because Starmie's loss function maximizes the distance between negative samples within each batch but does not account for negatives across batches. As a result, Starmie sometimes fails to distinguish semantically distinct but plausible attributes, leading to fewer, larger clusters that mix attribute types and yield lower Rand Index scores compared to EmTT.

GeTT's performance varies significantly on different datasets, depending on the adopted LLMs. On the WDC dataset, GeTT (DeepSeek-R1) achieves a RI of 0.980 and Purity of 0.957, while GeTT (GPT-4) attains 0.976 and 0.961; these values surpass EmTT (SwAV) by approximately 10%. By contrast, GeTT with Qwen2.5-14b produces more modest RI and Purity scores of 0.727 and 0.738, which fall below most EmTT embedding models. On the GDS dataset, GeTT (DeepSeek-R1) trails EmTT (DeepJoin) by roughly 8% in RI, whereas GeTT (GPT-4) achieves a Purity of 0.900, comparable to EmTT (Unicorn). Nevertheless, GeTT with Qwen2.5-14b/32b remains slightly below EmTT's low-performing variants on both RI and Purity. On OpenData, GeTT (GPT-4) and

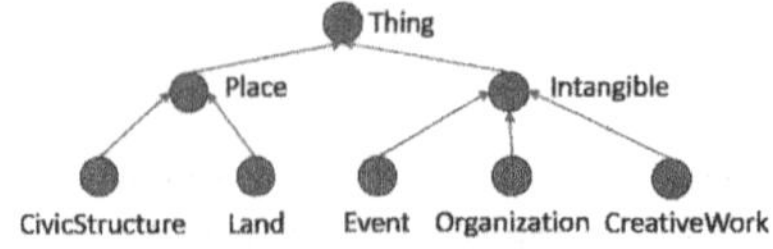

(a) Taxonomy with Abstract Types

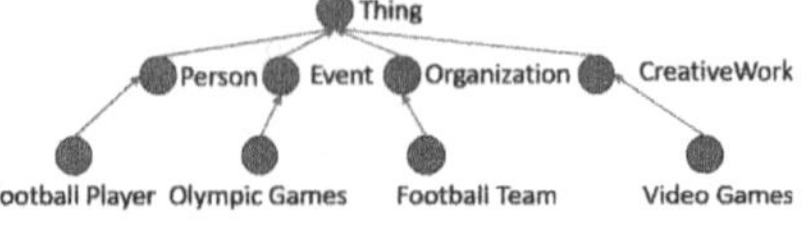

(b) Taxonomy with Specific Types

Fig. 4. Examples of the top-3 levels of the inferred taxonomies by two separated runs, using GeTT (GPT-4) on the GDS dataset.

GeTT (DeepSeek-R1) achieve the highest Rand Index (0.912 and 0.907), whereas GeTT (GPT-3.5) and GeTT (GPT-4) yield the top Purity (0.914 and 0.906).

Overall, GeTT produces the best Rand Index and Purity with DeepSeek-R1 and GPT-family models for WDC and OpenData, but EmTT (DeepJoin) and EmTT (SwAV) are competitive in GDS.

Quality of the Taxonomies. In Table 2 all of TCS, T# and L# report features of the generated taxonomies.

In EmTT, it is noteworthy that different methods produce different numbers of types (T#) and depths of hierarchy (L#). For example, on all three datasets, EmTT (SwAV) generates more types and layers than with the other embeddings. This means that EmTT (SwAV) often produces richer hierarchies than Starmie or DeepJoin. Unicorn fails to complete the type taxonomy construction due to scalability limitations.

Increased taxonomy complexity from EmTT (SwAV) comes at a cost in terms of the tree consistency score. EmTT (SwAV) creates more small column clusters, where each cluster corresponds to a specific top-level type attribute. This leads to a more complex hierarchy driven by shared attributes. This richer hierarchy, however, provides more opportunities for mismatches with the GT taxonomy, leading to a lower tree consistency score. In GDS, for instance, *School* is sometimes misclassified as a subtype of *LocalBusiness* because both types share similar address-related attributes. Notably, when no hierarchy is constructed under a top-level type, the tree consistency score defaults to *1*, indicating perfect consistency.

Compared to EmTT with all embedding methods, GeTT infers more types but with reduced overall tree consistency. GeTT generates entity types directly from tables, resulting in more types than EmTT in most cases. Specifically, with four different LLMs, GeTT in average generates 353.6 types on WDC, 296 on GDS and 1202.8 on OpenData—66.7%, 7.8% and 2% more than EmTT (SwAV), respectively. For example, while EmTT (SwAV) clusters three tables into a single type category representing *Healthcare Facility*, GeTT infers more granular types like *Healthcare Facility*, *Hospital*, and *Medical Clinic*. The taxonomy depth of GeTT with all language models is comparable to that of EmTT (SBERT), aligning more closely with the ground truth layers, though slightly shallower. The increased semantic granularity of inferred types can lead to inconsistencies

Table 3. Results of standard deviation (STD) of EmTT and GeTT

Method	WDC			GDS			OpenData		
	RI-STD	Purity-STD	TCS-STD	RI-STD	Purity-STD	TCS-STD	RI-STD	Purity-STD	TCS-STD
EmTT (SBERT)	0	0	0	0	0	0	0	0	0
EmTT (SwAV)	0.013	0.012	0.021	0.011	0.013	0.017	0.009	0.014	0.021
GeTT (GPT-3.5)	0.091	0.090	0.111	0.134	0.129	0.138	0.112	0.067	0.146
GeTT (GPT-4)	0.015	0.040	0.243	0.245	0.163	0.102	0.058	0.073	0.103
GeTT (DeepSeek-R1)	0.087	0.033	0.089	0.104	0.094	0.126	0.067	0.093	0.117
GeTT (Qwen2.5-14b)	0.164	0.265	0.206	0.178	0.349	0.288	0.091	0.104	0.121
GeTT (Qwen2.5-32b)	0.089	0.056	0.177	0.162	0.174	0.233	0.059	0.060	0.183

Table 4. Overall Running Times for EmTT and GeTT.

Method	Total Time		
	WDC	**GDS**	**OpenData**
EmTT (SBERT)	4.59 min	7.61 min	594.44 mins
EmTT (Unicorn)	45.21 min	54.84 min	-
EmTT (Starmie)	202.86 mins (train: 200 mins)	207.74 mins (train: 201 mins)	1532.75 min (train: 927 min)
EmTT (DeepJoin)	339.41 mins (train: 334 mins)	415.79 mins (train: 408 mins)	2241.26 min (train:1647 min)
EmTT (SwAV)	239.52 mins (train: 234 mins)	277.81 mins (train: 270 mins)	2156.75 min (train: 1591.65 min)
GeTT (GPT-3.5)	11.08 mins	13.18 mins	164.72 min
GeTT (GPT-4)	17.26 mins	20.62 mins	135.13 min
GeTT (DeepSeek-R1)	65.28 mins	70.83 mins	343.97 min
GeTT (Qwen2.5-14b)	12.58 mins	14.69 mins	95.64 min
GeTT (Qwen2.5-32b)	21.27 mins	26.51 mins	109.84 min

when the types become too specific or overly abstract, resulting in instability over taxonomy consistency. As a result, GeTT underperforms in consistency compared to EmTT with different embedding models.

Figure 4 illustrates that performance instability arises when the LLM infers overly abstract types. In Fig. 4(a), the LLM infers an abstract type *Intangible*, which serves as a top-level type subsuming multiple semantically distinct subtypes. This abstract type groups plausible but unrelated types together, causing difficulties in distinguishing between levels in the taxonomy. In contrast, Fig. 4(b) demonstrates an example where the LLM infers more specific top-level types that better capture the semantics of the underlying table. As a result, the generated taxonomy is clearer and more aligned with the expected GT taxonomy. This highlights a limitation of the GeTT method: it relies heavily on organizing types inferred by the LLM without adequately considering the overall semantic content of the tables. When the types are overly abstract, the taxonomy is likely to lack support from actual table content. Consequently, the quality of the generated taxonomy depends on the accuracy and granularity of the inferred table types.

Efficiency and Stability. We compare the stability of GeTT and EmTT across metrics relating to top-level types (RI and Purity) and the taxonomy (TCS) in Table 3. For TCS, GeTT exhibits significant variability, with standard deviations ranging from 0.089 to 0.288 across three datasets, far exceeding those of EmTT.

Similarly, for top-level type inference, GeTT again shows higher variability with the standard deviation of the RI ranging from 0.104 to 0.349 on GDS, 0.058 to 0.112 on OpenData. This indicates that GeTT is short of robustness in the generation of taxonomies. Such instability without repeatable performance in LLM prompting-based methods is reported in other studies [1,43]. Meanwhile, as the inference steps in EmTT are deterministic, EmTT (SBERT) has standard deviations of 0. Even when EmTT involves fine-tuning large language models multiple times and then encoding, as in EmTT (SwAV), the resulting performance fluctuations remain very small, with standard deviations only around 0.01–0.02.

As for the efficiency, the overall running times of GeTT are much less than those of EmTT, where fine-tuning is required, as shown in Table 4. The EmTT variations, except those using SBERT and Unicorn, require a fine-tuning phase that costs about 3.5–4.5 h on smaller datasets like WDC and GDS, and 16–25 h on the large-scale OpenData. For example, when running on the same computational resource, GeTT (Qwen2.5-14b) takes only around 5% of the overall time of EmTT across the three datasets. Even when a larger Qwen2.5 model is applied locally or an online closed-source LLM like GPT-4 is used, the running time of GeTT remains significantly lower than that of EmTT where fine-tuning is required. Furthermore, for large datasets with a high number of attributes—such as OpenData, which contains 310K attributes—EmTT still exhibits noticeably longer clustering times, even when employing a pre-trained LM like SBERT, requiring 1.5–5 times more time than GeTT variations. Meanwhile, the Unicorn variant of EmTT incurs a high computational cost due to its reliance on pairwise attribute matching, making it unable to complete all tests on OpenData.

6 Related Work

Schema Inference. A relevant line of work to taxonomy inference for tabular data is schema inference. Schema inference techniques have primarily focused on semi-structured data formats such as XML, JSON, and RDF. These techniques produce output schemas that range from high-level integration schemas [25] and concise summaries [49,51] to full disjunctions capturing structural and semantic patterns [25]. For XML and JSON, schema inference often relies on element names and structural patterns identified through graph-based partitioning [3]. In contrast, RDF schemas leverage type annotations to handle heterogeneous and inconsistent data sources [9,24], while some approaches employ similarity metrics to merge distinct schemas [17]. Our work addresses a gap in schema inference: inferring a conceptual taxonomy specifically tailored to heterogeneous tabular datasets. Unlike existing methods that rely heavily on consistent naming or type annotations, our approach EmTT uses LLMs to embed column-level semantics, which accommodates inconsistent terminologies and encodes nuanced contextual clues, thereby enabling similarity-based type inference and the construction of a taxonomy, while our another approach GeTT directly generates table semantics of entity types in an end-to-end way using generative LLMs.

Column Semantics. Language model-based encoding is increasingly leveraged to capture column semantics across various data integration tasks beyond schema inference. For instance, Unicorn [16] learns embedding-based classifiers for schema matching and column type annotation; DeepJoin [12] fine-tunes a pretrained language model to identify joinable columns; and Starmie [15] leverages column-level semantics for data search and table compatibility. We have included these methods, along with the first use of SwAV [5] with tabular data, in our experiments, allowing comparison of their effectiveness on a new problem.

LLM-Based Taxonomy Construction. Recently, methods like Chain-of-Layer (CoL) [53] and TaxonomyGPT [6] have exploited the in-context learning capabilities of LLMs to construct taxonomies. CoL, in particular, employs an iterative prompting strategy and a filter module to reduce LLM's hallucination, improving the reliability of the inferred taxonomy. Our approach GeTT uses an LLM to infer each table's entity types and leverages CoL to assemble these types into a hierarchical structure. It is worth mentioning that while some prior methods infer entity types from textual contexts [30,50], including zero-shot approaches [33,37,52,54], they are not designed for tabular data, which include semantics of not only text and data values but also structures with rows representing entities and columns representing their attributes. Among existing methods, Chorus [22] is the closest to ours. However, it uses predefined ontology classes, while our approaches stand out as open-domain solutions that directly infer table taxonomies using LLMs, without relying on predefined semantics, enabling more generalizable taxonomy construction across diverse structured data sources.

7 Conclusion and Future Work

In this study, we propose two LLM-based methods for taxonomy inference with a given set of tables: EmTT which is based on similarities and hierarchical clustering over column embeddings achieved by fine-tuning encoding LLMs; and GeTT which generates entity types for the tables and organizes the types by prompting decoding LLMs. Our empirical evaluations show that both EmTT and GeTT can infer appropriate top-level types and taxonomies that show good consistency with manually annotated tables. However, there are interesting differences within and between the methods. In EmTT, the embedding method used has a significant impact on the richness of the taxonomy. In GeTT, both the language model used and the dataset have a significant bearing on result quality. Overall, while GeTT-based proposals using GPT-4 and DeepSeek-R1 have tended to give rise to the highest scores for top-level type inference, EmTT (SwAV) and EmTT (DeepJoin) have provided more dependable performance for taxonomy inference. GeTT relies on no training with higher efficiency than fine-tuned EmTT methods, but it is relatively unstable. Furthermore, fine-tuned EmTT solutions can be used for different tasks and datasets. In the future, we will explore more robust frameworks with an ensemble of effective prompts and multiple LLMs, and consider instruction tuning decoder-alone LLMs for not only higher-quality output taxonomies but also higher stability. We will also explore other related tasks in schema inference including learning type level relationships.

A Fine-Tuning Configurations in EmTT

A.1 SwAV and Starmie Configurations

To investigate the optimal parameter configurations of the column encoder of EmTT, we used the WDC dataset and evaluated the performance of the column encoder model based on the top-level type inference results.

SwAV and Starmie are both based on constrastive learning, which are self-supervised, allowing for the use of the same datasets for both training and testing, and this is the approach taken in the experiments. For training, the following parameters are used with the AdamW optimization algorithm: Batch Size: 64; Epochs: 100 (SwAV)/64 (Starmie); Contrastive loss: decay to $1e^{-6}$; Learning rate: $5e^{-5}$;

For data augmentation in Starmie and SwAV, we set the number of views (t) to 2 and sampled cells in proportion to their TFIDF value as the data augmentation operator (OP). In SwAV, we adopted a $Serial$ strategy that includes both the header and the cells, serializing each column into a string. The choices for OP and $Serial$ strategy were determined by evaluating each parameter individually while keeping all other parameters fixed.

To determine the optimal sampling fractions (p_1, p_2) for generating the two views of a column during fine-tuning, we started with $(0.5, 0.5)$ to align with Starmie's initial setting. We then empirically tested additional pairs, including $(0.5, 0.3)$, $(1.0, 0.5)$, $(0.8, 0.4)$, $(0.6, 0.3)$, $(0.5, 0.25)$ and $(0.3, 0.3)$, and found that $(p_1, p_2) = (0.5, 0.3)$ achieved the best performance. Detailed experimental results are provided in "README.md" in our data and code repository.

Regarding the parameters for online clustering and the loss function in SwAV, we set the number of prototypes (K) to 500 for training only subject attributes and 3000 for all attributes on WDC and GDS. For OpenData, we use 8000 for subject attributes and 30000 for all attributes. The recommended practice from original SwAV [5] framework is to initialise K around an order of magnitude larger than the expected number of clusters. The dimension of the projection head in the model (d) is set to 768, implemented as a sequence of fully connected layers. We use a temperature coefficient (τ) of 0.07, consistent with Starmie's settings [15]. The parameter ϵ in the Sinkhorn algorithm, which controls the smoothness of the assignment process during online clustering, was set to 0.03, as suggested by the SwAV framework.

A.2 DeepJoin Configurations

For DeepJoin, we follow the original configuration from its paper, using a batch size of 32, a learning rate of $2e^{-5}$, and a weight decay of 0.01. The model is trained for 25 epochs with SBERT (all-mpnet-base-v2) as the backbone. Additionally, we set the shuffle rate to 0.3 and adopt the "colname-stat-col" format for column-to-text processing.

References

1. Atil, B., Chittams, A., Fu, L., Ture, F., Xu, L., Baldwin, B.: LLM stability: a detailed analysis with some surprises. *CoRR*, abs/2408.04667, 2024
2. Baazizi, M.A., Lahmar, H.B., Colazzo, D., Ghelli, G., Sartiani, C.: Schema inference for massive JSON datasets. In: Proceedings of the 20th International Conference on Extending Database Technology, EDBT 2017, Venice, Italy, 21–24 March 2017, pp. 222–233. OpenProceedings.org, 2017
3. Barret, N., Manolescu, I., Upadhyay, P.: Computing generic abstractions from application datasets. In: Tanca, L., et al., (eds.), Proceedings 27th International Conference on Extending Database Technology, EDBT 2024, pp. 94–107. OpenProceedings.org, 2024
4. Brickley, D., Burgess, M., Noy, N.: Google dataset search: building a search engine for datasets in an open web ecosystem. In: The World Wide Web Conference, WWW '19, pp. 1365–1375, New York, NY, USA. Association for Computing Machinery, 2019
5. Caron, M., Misra, I., Mairal, J., Goyal, P., Bojanowski, P., Joulin, A.: Unsupervised learning of visual features by contrasting cluster assignments. In: Advances in Neural Information Processing Systems 33: Annual Conference on Neural Information Processing Systems 2020, NeurIPS 2020, 6–12 December 2020, virtual, 2020
6. Chen, B., Yi, F., Varro, D.: Prompting or fine-tuning? A comparative study of large language models for taxonomy construction. In: 2023 ACM/IEEE International Conference on Model Driven Engineering Languages and Systems Companion (MODELS-C), pp. 588–596, Los Alamitos, CA, USA. IEEE Computer Society, October 2023
7. Chen, J., He, Y., Geng, Y., Jiménez-Ruiz, E., Dong, H., Horrocks, I.: Contextual semantic embeddings for ontology subsumption prediction. World Wide Web (WWW) **26**(5), 2569–2591 (2023)
8. Chen, T., Kornblith, S., Norouzi, M., Hinton, G.: A simple framework for contrastive learning of visual representations. In: Proceedings of the 37th International Conference on Machine Learning, ICML 2020, 13–18 July 2020, Virtual Event, pp. 1597–1607, 2020
9. Christodoulou, K., Paton, N.W., Fernandes, A.A.: Structure inference for linked data sources using clustering. Trans. Large Scale Data Knowl. Centered Syst. **19**, 1–25 (2015)
10. DeepSeek-AI. Deepseek-r1: Incentivizing reasoning capability in llms via reinforcement learning, 2025
11. Deng, X., Sun, H., Lees, A., You, W., Cong, Yu.: TURL: table understanding through representation learning. Proc. VLDB Endow. **14**(3), 307–319 (2020)
12. Dong, Y., Takeoka, K., Xiao, C., Oyamada, M.: Efficient joinable table discovery in data lakes: a high-dimensional similarity-based approach. In: 37th IEEE International Conference on Data Engineering, ICDE 2021, Chania, Greece, 19–22 April 2021, pp. 456–467, 2021
13. Dong, Y., Xiao, C., Nozawa, T., Enomoto, M., Oyamada, M.: Deepjoin: joinable table discovery with pre-trained language models. Proc. VLDB Endow. **16**(10), 2458–2470 (2023)
14. Edge, D., et al.: From local to global: a graph RAG approach to query-focused summarization. *CoRR*, abs/2404.16130, 2024
15. Fan, G., Wang, J., Li, Y., Zhang, D., Miller, R.J.: Semantics-aware dataset discovery from data lakes with contextualized column-based representation learning. Proc. VLDB Endow. **16**(7), 1726–1739 (2023)

16. Fan, J.: Unicorn: a unified multi-tasking matching model. SIGMOD Rec. **53**(1), 44–53 (2024)
17. Flores, J., et al.: Incremental schema integration for data wrangling via knowledge graphs. Semantic Web **15**(3), 793–830 (2024)
18. Funk, M., Hosemann, S., Jung, J.C., Lutz, C.: Towards ontology construction with language models. In: Joint proceedings of the 1st workshop on Knowledge Base Construction from Pre-Trained Language Models (KBC-LM) and the 2nd challenge on Language Models for Knowledge Base Construction (LM-KBC) co-located with the 22nd International Semantic Web Conference (ISWC 2023), Athens, Greece, 6 November 2023, volume 3577 of *CEUR Workshop Proceedings*. CEUR-WS.org, 2023
19. Australian Government. data.gov.au, n.d. Accessed 21 Aug 2025
20. UK Government. data.gov.uk, n.d. Accessed 21 Aug 2025
21. U.S. Government. data.gov, n.d. Accessed 21 Aug 2025
22. Kayali, M., Lykov, A., Fountalis, I., Vasiloglou, N., Olteanu, D., Suciu, D.: Chorus: foundation models for unified data discovery and exploration. Proc. VLDB Endow. **17**(8), 2104–2114 (2024)
23. Kellou-Menouer, K., Kardoulakis, N., Troullinou, G., Kedad, Z., Plexousakis, D., Kondylakis, H.: A survey on semantic schema discovery. VLDB J. **31**(4), 675–710 (2022)
24. Kellou-Menouer, K., Kedad, Z.: Schema discovery in RDF data sources. In: Johannesson, P., Lee, M.L., Liddle, S.W., Opdahl, A.L., López, Ó.P. (eds.) ER 2015. LNCS, vol. 9381, pp. 481–495. Springer, Cham (2015). https://doi.org/10.1007/978-3-319-25264-3_36
25. Khatiwada, A., Shraga, R., Gatterbauer, W., Miller, R.J.: Integrating data lake tables. Proc. VLDB Endow. **16**(4), 932–945 (2022)
26. Langlais, P., Gao, T.L.: Rate: a reproducible automatic taxonomy evaluation by filling the gap. In: Proceedings of the 15th International Conference on Computational Semantics, pp. 173–182, 2023
27. Li, P., He, Y., Yan, C., Wang, Y., Chaudhuri, S.: Auto-tables: relationalize tables without using examples. SIGMOD Rec. **53**(1), 76–85 (2024)
28. Liu, H., Perl, Y., Geller, J.: Concept placement using BERT trained by transforming and summarizing biomedical ontology structure. J. Biomed. Inform. **112**, 103607 (2020)
29. Liu, J., Chabot, Y., Troncy, R., Huynh, V.-P., Labbé, T., Monnin, P.: From tabular data to knowledge graphs: a survey of semantic table interpretation tasks and methods. J. Web Semant. **76**, 100761 (2023)
30. Liu, Q., Lin, H., Xiao, X., Han, X., Sun, L., Wu, H.: Fine-grained entity typing via label reasoning. In: Moens, M.F., Huang, X.J., Specia, L., Yih, W.T. (eds.), Proceedings of the 2021 Conference on Empirical Methods in Natural Language Processing, pp. 4611–4622, Online and Punta Cana, Dominican Republic, November 2021. Association for Computational Linguistics, November 2021
31. Liu, Y., et al.: Roberta: a robustly optimized BERT pretraining approach. *CoRR*, abs/1907.11692, 2019
32. Lo, A., Jiang, A.Q., Li, W., Jamnik, M.: End-to-end ontology learning with large language models. In: Globersons, A. et al. (eds.), Advances in Neural Information Processing Systems, vol. 38: Annual Conference on Neural Information Processing Systems 2024, NeurIPS 2024, Vancouver, BC, Canada, 10–15 December 2024
33. Ma, Y., Cambria, E., Gao, S.: Label embedding for zero-shot fine-grained named entity typing. In: Matsumoto, Y., Prasad, R. (eds.), Proceedings of COLING 2016,

the 26th International Conference on Computational Linguistics: Technical Papers, pp. 171–180, Osaka, Japan. The COLING 2016 Organizing Committee, December 2016

34. Meusel, R., Petrovski, P., Bizer, C.: The WebDataCommons microdata, RDFa and microformat dataset series. In: Mika, P., et al. (eds.) ISWC 2014. LNCS, vol. 8796, pp. 277–292. Springer, Cham (2014). https://doi.org/10.1007/978-3-319-11964-9_18
35. Min, D., et al.: Exploring the impact of table-to-text methods on augmenting llm-based question answering with domain hybrid data. In: Yang, Y., Davani, A., Sil, A., Kumar, A. (eds.), Proceedings of the 2024 Conference of the North American Chapter of the Association for Computational Linguistics: Human Language Technologies: Industry Track, NAACL 2024, Mexico City, Mexico, 16–21 June 2024, pp. 464–482. Association for Computational Linguistics, 2024
36. Nargesian, F., Zhu, E., Pu, K.Q., Miller, R.J.: Table union search on open data. Proc. VLDB Endow. **11**(7), 813–825 (2018)
37. Obeidat, R., Fern, X., Shahbazi, H., Tadepalli, P.: Description-based zero-shot fine-grained entity typing. In: Burstein, J., Doran, C., Solorio, T. (eds.), Proceedings of the 2019 Conference of the North American Chapter of the Association for Computational Linguistics: Human Language Technologies, Volume 1 (Long and Short Papers), pp. 807–814, Minneapolis, Minnesota, June 2019. Association for Computational Linguistics
38. Paton, N.W., Chen, J., Wu, Z.: Dataset discovery and exploration: a survey. ACM Comput. Surv. **56**(4), 102:1–102:37 (2024)
39. Reimers, N., Gurevych, I.: Sentence-bert: Sentence embeddings using siamese bert-networks. In: Proceedings of the 2019 Conference on Empirical Methods in Natural Language Processing and the 9th International Joint Conference on Natural Language Processing, EMNLP-IJCNLP 2019, Hong Kong, China, 3–7 November 2019, pp. 3980–3990. Association for Computational Linguistics, 2019
40. Ritze, D., Bizer, C.: Matching web tables to dbpedia - a feature utility study. In: International Conference on Extending Database Technology, 2017
41. Rousseeuw, P.J.: Silhouettes: a graphical aid to the interpretation and validation of cluster analysis. J. Comput. Appl. Math. **20**, 53–65 (1987)
42. Shi, J., et al.: Subsumption prediction for e-commerce taxonomies. In: The Semantic Web - 20th International Conference, ESWC 2023, Hersonissos, Crete, Greece, May 28 - 1 June 2023, Proceedings, vol.13870 of LNCS, pp. 244–261. Springer, 2023
43. Stewart, I., Horawalavithana, S., Kennedy, B., Munikoti, S., Pazdernik, K.: Surprisingly fragile: Assessing and addressing prompt instability in multimodal foundation models, 2024. *CoRR*, abs/2408.14595
44. Touvron, H., et al.: Llama: Open and efficient foundation language models. *CoRR*, abs/2302.13971, 2023
45. Venetis, P., et al.: Recovering semantics of tables on the web. Proc. VLDB Endow. **4**(9), 528–538 (2011)
46. Völker, J., Niepert, M.: Statistical schema induction. In: Antoniou, G., et al. (eds.) ESWC 2011. LNCS, vol. 6643, pp. 124–138. Springer, Heidelberg (2011). https://doi.org/10.1007/978-3-642-21034-1_9
47. Wang, J., Wang, H., Wang, Z., Zhu, K.Q.: Understanding tables on the web. In: Atzeni, P., Cheung, D., Ram, S. (eds.) ER 2012. LNCS, vol. 7532, pp. 141–155. Springer, Heidelberg (2012). https://doi.org/10.1007/978-3-642-34002-4_11
48. Yang, A., et al.: Qwen2 technical report. *CoRR*, abs/2407.10671, 2024

49. Yang, X., Procopiuc, C.M., Srivastava, D.: Summarizing relational databases. Proc. VLDB Endow. **2**(1), 634–645 (2009)
50. Yosef, M.A., Bauer, S., Hoffart, J., Spaniol, M., Weikum, G.: Hyena: hierarchical type classification for entity names. In: International Conference on Computational Linguistics, 2012
51. Yu, C., Jagadish, H.V.: Schema summarization. In: Proceedings of the 32nd International Conference on Very Large Data Bases, Seoul, Korea, 12–15 September 2006, pp. 319–330, 2006
52. Yuan, Z., Downey, D.: Otyper: a neural architecture for open named entity typing. AAAI'18/IAAI'18/EAAI'18. AAAI Press, 2018
53. Zeng, Q., et al.: Chain-of-layer: iteratively prompting large language models for taxonomy induction from limited examples. In: Proceedings of the 33rd ACM International Conference on Information and Knowledge Management, CIKM 2024, Boise, ID, USA, 21–25 October 2024, pp. 3093–3102. ACM, 2024
54. Zhang, T., Xia, C., Lu, C.T., Yu, P.: MZET: memory augmented zero-shot fine-grained named entity typing. In: Scott, D., Bel, N., Zong, C. (eds.), Proceedings of the 28th International Conference on Computational Linguistics, pp. 77–87, Barcelona, Spain (Online). International Committee on Computational Linguistics, December 2020
55. Zhang, Z.: Effective and efficient semantic table interpretation using tableminer$^+$. Semantic Web **8**(6), 921–957 (2017)

GeoRDF2Vec–Learning Location–Aware Entity Representations in Knowledge Graphs

Martin Böckling[1(✉)], Heiko Paulheim[1], and Sarah Detzler[2]

[1] Data and Web Science Group, University of Mannheim, 68160 Mannheim, Germany
martin.boeckling@uni-mannheim.de
[2] Corporate State University of Mannheim, Mannheim, Germany

Abstract. Many knowledge graphs contain a substantial number of spatial entities, such as cities, buildings, and natural landmarks. For many of these entities, exact geometries are stored within the knowledge graphs. However, most existing approaches for learning entity representations do not take these geometries into account. In this paper, we introduce a variant of RDF2Vec that incorporates geometric information to learn location-aware embeddings of entities. Our approach expands different nodes by flooding the graph from geographic nodes, ensuring that each reachable node is considered. Based on the resulting flooded graph, we apply a modified version of RDF2Vec that biases graph walks using spatial weights. Through evaluations on multiple benchmark datasets, we demonstrate that our approach outperforms both non-location-aware RDF2Vec and GeoTransE.

Keywords: Entity Representation · RDF2Vec · Embedding · Geographic Knowledge Graph

1 Introduction

Knowledge Graph Embeddings are a family of methods designed to learn numeric, low-dimensional representations of entities in knowledge graphs. These representations can be used in downstream tasks to make predictions about entities, under the assumption that similar entities are mapped to similar embedding vectors [12,26]. Many large knowledge graphs not only encode relational information between entities but also capture the geographic geometries of some or all of these entities. Prominent knowledge graphs such as DBpedia [17], YAGO [24], and Wikidata [32] contain geographic information on entities like places and buildings. Additionally, dedicated geographic knowledge graphs, such as KnowWhereGraph [14], WorldKG [8], and OSMh3KG [3], explicitly model geographic relationships.

When geographic relationships are explicitly represented such as by assigning entities to grid cells and modeling spatial relations between these cells in

E. Curry et al. (Eds.): ESWC 2025, LNCS 15718, pp. 423–440, 2025.
https://doi.org/10.1007/978-3-031-94575-5_23

the knowledge graph they can be captured by embedding methods as well [2]. However, most knowledge graphs store geographic information as literals using the WKT format, which is not well captured by most embedding methods [10].

In this paper, we propose an approach that integrates geographic data into the well-known RDF2Vec method for knowledge graph embedding [28]. Our method consists of two steps. First, a geographic geometry is assigned to all entities in the graph, particularly those that either lack a predefined geometry or are not inherently geographic entities. Second, based on these geometries, the RDF2Vec graph walk mechanism is modified so that edges connecting geographically closer entities are assigned higher transition probabilities. We demonstrate that by incorporating geometric information, our approach achieves superior results compared to standard RDF2Vec while maintaining the same computational complexity as the original method.

2 Related Work

Within the domain of Knowledge Graphs (KGs), previous research has explored the use of spatial data in various tasks. Below, we provide an overview of different studies that have incorporated spatial data using various strategies for respective downstream tasks.

For geographical question answering, several models have been developed to generate spatially explicit KG embeddings. Unlike standard QA benchmark datasets, QA datasets that involve spatial data must account for additional constraints. One model that employs a specialized embedding approach for geographic data is TransGeo, an adaptation of TransE [19]. A similar approach is proposed by Qiu et al. [27], who also present an adaptation of TransE. Like TransGeo, their method incorporates a penalization term using a spatial weight function. However, while Mai et al. use geodesic distance to compute spatial weights, Qiu et al. rely on Euclidean distance [27]. Since spherical distance functions provide more accurate distance calculations between geographic instances [13], we will focus on the TransGeo implementation proposed by Mai et al.

In TransGeo, the spatial weight is modeled using a stepwise function with a predefined distance threshold. If the distance exceeds this threshold, a constant value is assigned to the corresponding weight node instance. Geographic entities are identified by filtering DBpedia using the *geo:geometry* property. Once the spatial weights for each edge are computed, PageRank is applied to the graph, using these edge weights to determine the PageRank value for each entity. This process captures the patterns of incoming and outgoing edges from specific entities. Finally, the computed entity weights are incorporated into an adapted objective function of TransGeo. Experimental results demonstrate that TransGeo outperforms various baselines in link prediction tasks as well as in QA datasets [19].

Another approach, proposed by Mai et al. [18], relies on an encoder-decoder architecture that captures various aspects of spatial information. By directly integrating spatial features such as point coordinates and bounding boxes of geographic entities into the embedding space, this method enables effective spatial reasoning.

The model consists of multiple components. An entity encoder separately captures both the semantics of the KG entity through type relations and its spatial embedding. If an entity has a point geometry, its coordinates are used as input for the spatial embedding. For non-point geometries, a point is randomly sampled from a uniform distribution within the bounding box. If an entity is not geographic, a random spatial embedding is assigned. These embeddings are then concatenated to form a unified entity embedding. Additionally, a projection operator maps entity-relation pairs to specific embeddings, enabling link prediction tasks. This allows the model to predict tail entities while also performing semantic reasoning to determine the most probable linked entity [18].

Mann et al. [20] address the challenge of predicting spatial links in geographic KGs, which are often sparsely interlinked. Traditional link prediction methods typically rely on existing entity relations, which may be missing in such KGs. To overcome this limitation, the authors propose two approaches:

Supervised Spatial Link Prediction (SSLP): This method leverages spatial and semantic embeddings derived from the literal values of geographic entities to predict links. Unsupervised Inductive Spatial Link Prediction (USLP): Unlike SSLP, this approach does not require labeled training data. Instead, it uses the haversine distance between geohash grid cell centroids to infer spatial links. The authors evaluate these methods using WorldKG, demonstrating that both SSLP and USLP outperform existing state-of-the-art link prediction techniques. Their results highlight the effectiveness of incorporating spatial and semantic embeddings to enhance the completeness of geographic knowledge graphs [20].

In this paper, we propose a location-aware variant of RDF2Vec [22]. Our method modifies random walks by assigning different weights and consequently, different transition probabilities to edges. While prior studies have explored enhancing RDF2Vec with edge weights, using internal metrics such as PageRank [7] or externally sourced relevance metrics [31], our approach differs by leveraging geographic proximity as edge weights.

3 Theoretical Background

This section defines key concepts used throughout the paper and discusses various approaches for calculating weights between different spatial geometries.

3.1 Definitions

In many public knowledge graphs, whether they model only spatial data or general information, spatial geometries are represented as literals within a KG. Our approach aims to leverage these geometries to model spatial relationships between individual nodes. To ensure consistency in terminology, we define key concepts in this section. A knowledge graph KG is defined as a directed labeled graph with a set of vertices V and edges E. Therefore, it is defined as $KG =$

(V, E). The vertices v are a finite set of individual nodes. Edges E are defined as a finite set of individual edges connecting two nodes and having a relation type as a label, i.e., $E \subseteq V \times R \times V$, where R is the set of different types of relations.

In a KG, each node can has a *neighborhood*. The neighborhood of a node $v \in V$ is defined as $H(v) := \{(s,p,o) : \exists p, o | (v,p,o) \in E\} \cup \{(s,p,o) : \exists s, p | (s,p,v) \in E\}$. The set of neighbors $N(v)$ of a node $v \in V$ is defined as $N(v) := \{v' : \exists r | (v,r,v') \in H(v) \vee (v',r,v) \in H(v)\}$.

Geometries can be points, lines, polygons, multipoints, multilines, multipolygons and geometry collections. We define a *spatial knowledge graph* SKG as a knowledge graph where a part of the nodes has a geometry attached, i.e., $SKG = (V, E, S)$, where $S \subseteq V \times G$ is the set of nodes that have a geometry attached, and G is the (theoretically infinite) set of all possible geometries. The subset $GV \subseteq V$ of geographic nodes is the set of nodes which have at least one geometry, i.e., $GV := \{v \in V | \exists g : (v,g) \in S\}$.

We further distinguish *fully spatial KGs* as those KGs where *every* node has a geometry attached (i.e., $V = GV$), and *partially spatial KGs* where some, but not all nodes have a geometry attached ($GV \subset V$). The majority of common knowledge graphs, such as DBpedia or Wikidata, are partially geographic KGs.

Another important concept in our work are *weighted knowledge graphs*. They are augmented with a weighting function $w : E \to \mathbb{R}^+$, which assigns a weight to each edge.

GeoSPARQL has been introduced as a standard for geographical representation in KGs. It defines six classes and 36 object properties. The superclass of GeoSPARQL is named *SpatialObject*, which represents spatial entities and includes two primary subclasses: *Feature* and *Geometry*. The *Geometry* subclass supports various geometry types and provides the ability to query them accordingly. Spatial geometries are represented as geographic literals, often using the WKT (Well-Known Text) format, which allows geometries to be encoded within the KG [4]. In our approach, we adhere to the WKT standard for modeling geometries.

3.2 From Spatial Distances to Edge Weights

To calculate distances between two spatial entities i and j, we use the *geodesic distance*, which accounts for the ellipsoidal curvature of the Earth. Compared to other distance measurements, the geodesic distance provides a more accurate representation of the Earth's shape. In contrast, the *great-circle distance* assumes the Earth is a perfect sphere. The difference between the great-circle distance and the geodesic distance becomes more pronounced over long distances, where the former introduces greater inaccuracies.

Therefore, we adopt the geodesic distance, as defined in Eq. 1. The distance d_{ij} between spatial entities i and j is computed using the Earth's radius R and the central angle $\Delta\sigma$, as defined in Eq. 2 [15]. When using the WGS84 ellipsoid projection, the Earth's radius R is taken as a constant value of 6,378,137 m.

$$d_{ij} = R \cdot \Delta\sigma \tag{1}$$

To calculate the central angle between two points, the longitude λ and latitude ϕ must be expressed in radians. The central angle $\Delta\sigma$ represents the angular separation between two points on an ellipsoid, as observed from the ellipsoid's center. It is derived by combining latitude and longitude differences using trigonometric relationships.

The calculation of $\Delta\sigma$ incorporates the difference in longitude $\Delta\lambda$ and includes the sum of the product of the sine of each latitude ϕ, the cosine of each latitude, and the cosine of the difference between the longitudes $\Delta\lambda$.

$$\Delta\sigma = \arccos(\sin\phi_i \cdot \sin\phi_j + \cos\phi_i \cdot \cos\phi_j \cdot \cos\Delta\lambda) \tag{2}$$

Using these distances, we define a spatial weight matrix W, where each entry w_{ij} represents the spatial weight between a spatial entity i and another spatial entity j. One commonly used approach for spatial weighting is *distance-based spatial weighting*. For a given distance function d_{ij} between spatial entities i and j, the weight can be computed using various weighting functions. A widely used method is the *threshold-based weight function*. Given a threshold δ, the pairwise function described in Eq. 3 is applied [11].

$$w_{ij} = \begin{cases} 1 & \text{if } d_{ij} \leq \delta \\ 0 & \text{if } d_{ij} > \delta \end{cases} \tag{3}$$

Similarly to contiguity-based spatial weights, the threshold-based weight function assigns weights from the set $\{0, 1\}$. However, unlike these discrete spatial weighting methods, the *inverse distance weighting* in Eq. 4 and the *exponential kernel function* in Eq. 5 calculate continuous weights. Inverse distance weighting employs a power parameter $\alpha \in \mathbb{N}$, where a higher value of α increases the influence of nearer points relative to more distant points. The value range of inverse distance weighting is within $(0, \infty)$, where $w_{ij} \in \mathbb{R}^+$ [29].

$$w_{ij} = (d_{ij})^{-\alpha} \tag{4}$$

In contrast to inverse distance weighting, the calculation of the spatial weight w_{ij} using an exponential function results in weights w_{ij} within the range $(0, 1]$. By using the inverse of the distance d_{ij}, which falls within the value range $(-\infty, 0]$, the exponential function assigns a value close to 0 for large distances and a value close to 1 for small distances.

$$w_{ij} = \exp(-d_{ij}) \tag{5}$$

When comparing different spatial weighting approaches, it becomes evident that each method offers specific advantages and disadvantages depending on the type of influence being measured. Inverse distance weighting has been shown to perform particularly well for global-scale measurements. In contrast, the threshold-based distance function (Eq. 3) and contiguity-based weighting are limited to capturing only local dependencies, disregarding more distant influences. The exponential-based weighting, as described in Eq. 5, provides a balance by allowing for *quasi-globality*, where nearer spatial entities are assigned higher weights

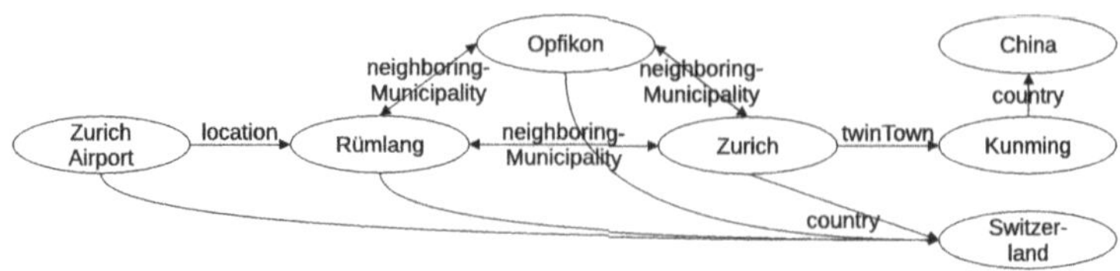

Fig. 1. Example graphexcerpt from DBpedia

compared to those further away. Since our research focuses on interdependencies that are predominantly local due to the structure of the utilized KGs, we adopt the exponential function outlined in Eq. 5 to calculate spatial weights [6].

4 The GeoRDF2Vec Approach

The GeoRDF2Vec approach consists of two major steps. First, to account for partially geographic knowledge graphs, we conduct a *geographic information flooding* process to assign geometries to non-geographic entities. Based on these assigned geometries, we compute weights for all edges in the knowledge graph. Finally, we apply a weight-aware variant of RDF2Vec, where walks are extracted from the graph using the assigned weights to define transition probabilities.

The core idea behind GeoRDF2Vec is to assign higher weights to relations that connect geographically close entities. Considering the graph illustrated in Fig. 1, classic RDF2Vec would extract the following walks with equal likelihood:

```
Zurich Airport -> location -> Rümlang -> neighb.Mun. -> Opfikon -> neighb.Mun. -> Zurich
Zurich Airport -> location -> Rümlang -> neighb.Mun. -> Zurich -> twinTown -> Kunming
```

However, the second walk is less useful for learning representations because Kunming is not a relevant entity for Zurich Airport, and vice versa. The main idea of GeoRDF2Vec is to reduce the likelihood of walks that connect geographically distant entities, as illustrated in the second example.

To visualize the concept of our approach, we use a generated Erdős-Rényi graph with ten vertices, where two nodes are labeled as geographic nodes and are assigned random coordinates from Romania. In the following subsections, we provide an overview of how our approach is applied to the example graph.

4.1 Geographic Information Flooding in Knowledge Graphs

Since most KGs are only partially spatial knowledge graphs, we first perform a preparatory step to assign geometries to non-geographic nodes. In this approach, each non-geographic node inherits the geometries of its geographic neighbors. Formally, the set S is extended as follows:

$$S = S \cup \{(v, g) : v \notin GV \wedge v' \in n(v) \cap GV \wedge (v', g) \in S\}.$$

Algorithm 1. Geographic information flooding

Require: Partially geographic Knowledge Graph (G, V, S)
  $GV \leftarrow \{v \in V | \exists g : (v, g) \in S\}$
  repeat
    $NodesToVisit \leftarrow \{v \in V \setminus GV | \exists v' \in N(v) : v' \in GV\}$
    for $v \in NodesToVisit$ **do**
      for $v' \in N(v)$ **do**
        for $(v', g) \in S$ **do**
          $S = S \cup \{(v, g)\}$
        end for
      end for
    end for
  until $NodesToVisit = \emptyset$
  return (G, V, S)

Since some non-geographic nodes may still remain (i.e., those without a neighbor in GV), the process is iteratively repeated until no further geometries can be assigned.

We begin with the set GV, which includes all nodes in the knowledge graph that already have a geometry. The direct neighbors of these nodes, which lack a geometry, are considered for processing in the next iteration. In each iteration, the set GV is expanded by assigning the geometries of neighboring nodes in GV to those without geometries. This process continues as long as there are non-geographic nodes adjacent to geographic nodes that have not yet been assigned a geometry.[1] To propagate geographic information to all nodes in the network, we implement a flooding mechanism, as outlined in Algorithm 1. For efficient execution, the set S is stored in a hash table, using the node ID as the hash key for the corresponding values. As shown in Table 1, a single node can inherit multiple geometries if it has multiple geographic neighbors.

To visualize the results, we start with our example where nodes (3, 9) are marked as geographic nodes. Applying the geographic information flooding algorithm to our generated sample graph, we observe that the algorithm terminates after two iterations, as all nodes have been assigned a geographic geometry. As shown in Table 1, once the flooding process is complete, three nodes (IDs: 4, 1, and 8) inherit both geometries. Nodes 5, 6, and 7 are assigned the geometry of node 3, while nodes 0 and 2 inherit the geometry of node 9.

4.2 Spatial Weighting for Graph Walks

To assign weights to the edges, we apply the geodesic distance as defined in Eq. 1, in conjunction with the exponential weight function in Eq. 5. As shown in Table 1, in certain cases, a node may be associated with multiple geometries. To account for this, we combine nodes with multiple geometries into a *geometry collection*.

[1] Some nodes may lack a geometry at the end of the algorithm if their subgraph contains only non-geographic nodes, with no geographic neighbors to inherit from.

Table 1. Iterations of the geographic flooding algorithm with the assignment of the geometries based on their node indices

Iteration	Node IDs	Geometry Set
1	5,6,7	{3}
1	0,2	{9}
1	4	{9,3}
2	1,8	{9,3}

This approach enables the integration of different geometry types into a unified set of geometries for each vertex v_i in a KG G. Since the geodesic distance is defined between two points, we use the centroid of individual geometries or the centroid of the combined geometry set.

Due to the properties of the exponential function, large distances result in weights close to zero. For example, while a distance of 1 km corresponds to a weight of 0.3678, a distance of 5 km results in a weight of only 0.0067. In graphs with large spatial extents, many edge weights approach zero, limiting effective exploration. To mitigate this issue and enable graph traversal over larger distances, we normalize the distances. For an individual node v_i, the distances to all its neighbors are normalized using a *min-max scaler*. Compared to other scaling methods, the min-max scaler preserves the distribution of vertex-neighbor distances. The min-max normalization of the distance $d'(v_i, v_j)$ is defined in Eq. 6.

$$d'(v_i, v_j) = \frac{d(v_i, v_j) - \min_{(s,p,o)\in H(v_i)} d(s, o)}{\max_{(s,p,o)\in H(v_i)} d(s, o) - \min_{(s,p,o)\in H(v_i)} d(s, o)} \tag{6}$$

The normalized distance d' is then used to compute the spatial weights. For each edge e_i, we calculate the spatial weight w_{ij} and assign it to the corresponding edge e_i. To visualize the effects of different weighting strategies on our constructed graph, we compute the spatial weight for each edge. In Fig. 2, the calculated weights are compared based on plain distance and normalized distance. Even after the flooding process, some nodes may remain without geographic coordinates. For edges connecting such nodes, a uniform weight of 1 is assigned.[2]

4.3 RDF2Vec on Weighted Knowledge Graph

Based on the weighted graph, we apply RDF2Vec to generate an embedding for each individual vertex v_i. RDF2Vec consists of two main steps: (i) the generation of graph walks and (ii) the computation of embedding vectors using word2vec. For the graph walks, two key parameters are considered. The depth d, which

[2] This can only happen between two non-geographic nodes, not between a geographic and a non-geographic node, due to the flooding algorithm.

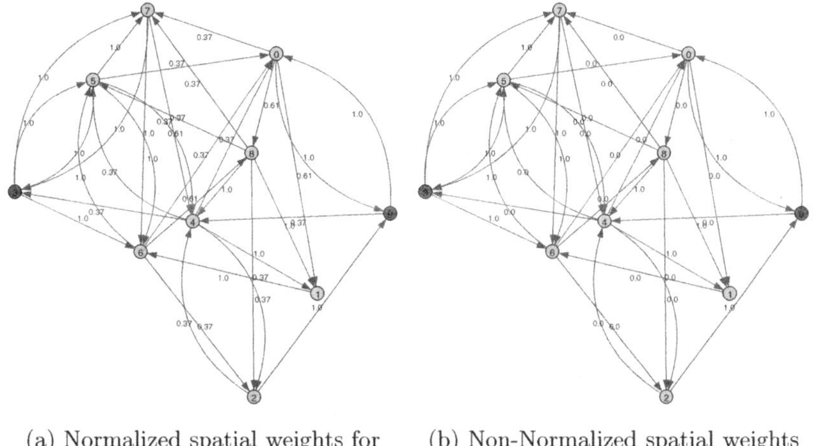

(a) Normalized spatial weights for graph edges rounded to two decimal places

(b) Non-Normalized spatial weights for graph edges rounded to two decimal places

Fig. 2. Different calculation of distances and resulting spatial weights for edges using the generated Erdős-Rényi graph

defines the (maximum) length of each extracted walk and The number of walks per vertex w. For the random walks, we use a transition probability matrix P, where each entry p_{ij} represents the probability of transitioning from v_i to v_j. It is defined in Eq. 7:

$$p_{ij} = \begin{cases} \frac{w_{ij}}{\deg(v_i)} & \text{if}(v_i, v_j) \in E, \\ 0 & \text{if}(v_i, v_j) \notin E. \end{cases} \tag{7}$$

The weighted degree of a vertex v_i is defined as $\deg(v_i) = \sum_{j=i}^{n} w_{ij}$. A single walk P_v is a sequence of nodes, where each subsequent node v_j is selected based on the transition probabilities between the connected nodes. The selection of neighboring nodes is repeated until either the random walk reaches a terminal node or the maximum walk distance d is reached. The complete corpus of extracted walks is the union of the maximum walks per entity and all extracted walks $\bigcup_{i=1}^{V} \bigcup_{i=1}^{w} P_{t_i}$. One single walk is represented by the set of combined triples t_i [28]. The constructed walk corpus serves as input for the word2vec model. The word2vec model embeds each individual word in the corpus based on the generated walking sequences. Generally, word2vec differentiates between two training approaches: Skip-Gram and Continuous Bag of Words (CBOW). CBOW predicts a missing word from its surrounding words, while Skip-Gram predicts surrounding words given a specific word [28].

5 Evaluation

We evaluate our approach using two different benchmark frameworks. Additionally, we assess the effectiveness of the information flooding algorithm in isolation, as well as the impact of different hyperparameters.[3]

5.1 Evaluation Frameworks

We utilize two evaluation frameworks: KGBench and GEval. KGBench provides a collection of curated KGs and evaluation tasks designed to assess relational or multimodal information encoding. For our study, we employ the dmg777k KG and its corresponding node classification task, as the DMG datasets in KGBench are the only ones containing spatial information. The dmg777k KG represents the locations of Dutch monuments using WKT geometries, which capture both the position and structure of each monument. The spatial extent of the KG is confined to the national boundaries of the Netherlands [34]. The node classification benchmarks for dmg777k include a total of five different node classes [1]. To evaluate the classification performance on the dmg777k benchmark dataset, we use Accuracy, Macro F1 score, and the Matthews Correlation Coefficient (MCC) score as evaluation metrics. As the classification algorithm for node classification, we employ XGBoost, optimizing its hyperparameters using Bayesian Search Optimization based on the recommended hyperparameters.[4]

GEval is an evaluation framework designed to assess the performance of embedding approaches across various downstream datasets. It supports a range of learning tasks, including classification and unsupervised clustering. For each task, the framework specifies the evaluation metrics against which the results should be measured. GEval is built on a subset of DBpedia and assumes that the generated embeddings for entities are associated with DBpedia identifiers. The framework evaluates different embedding approaches using multiple algorithms, facilitating comparability across various methods [23]. From the GEval evaluation framework, we utilize the three largest tasks: node classification, node regression, and node clustering.

For our evaluation of the TransGeo method, we implemented our own version based on the original paper, as the source code was not publicly available. The implemented code for TransGeo is provided within our GitHub repository. Following the original work by Mai et al., we heavily reused code from Wang et al., which implements a context-preserving variant of TransE [33]. The primary adaptation in our implementation was in the data preparation process, where we calculated entity-specific weights for context selection. For the experiments with TransGeo, we used the same set of hyperparameters as described in the original paper [19].

---

[3] The code and data for this paper are available here.

[4] The selection of hyperparameters follows XGBoost recommendations.

Table 2. Evaluation metrics for kgbench dmg777k node classification

Model	Accuracy	F1	MCC
GeoRDF2Vec	**0.6672** ± 0.0203	**0.6060** ± 0.023	**0.5126** ± 0.0296
RDF2Vec	0.5882 ± 0.0217	0.536 ± 0.024	0.4439 ± 0.0282
TransGeo	0.5132 ± 0.0212	0.4901 ± 0.022	0.4219 ± 0.0280

5.2 Knowledge Graph Characteristics

In the dmg777k KG two different geographic projections are used: the WGS 84 projection (EPSG:4326) and the Amersfoort/RD New projection (EPSG:28992). Among more than 20,000 geometries, 371 utilize the Amersfoort projection. Since these projections differ, we first convert the Amersfoort projection to the WGS 84 projection as a preprocessing step. Geographic entities using the Amersfoort projection are modeled as objects within a triple set with the specific predicate: http://data.pdok.nl/def/pdok#asWKT-RD. The dmg777k KG consists of 331,194 nodes and 776,920 edges. It is not a fully connected graph and contains multiple edges between certain vertices. The average degree of the graph is 4.6775. After performing geographic information flooding, the average geographic distance of an edge is 46.3675 km, and the average weight assigned to each individual edge is 0.8330 [1].

For our experiments on GEval, we use the 2016 version of DBpedia, as GEval is based on this version. In DBpedia, geographic entities are implicitly modeled within the graph itself. In 2016, DBpedia used the GeoRSS standard for representing geographic data, which differs from the currently established GeoSPARQL standard that models geometries in the WKT format. To ensure compatibility, we transform all geometries present within the relevant subsets of DBpedia into the WKT format, allowing us to accurately identify geographic nodes within the KG. The DBpedia extract used in our experiments contains a total of 10,021,706 nodes and 84,087,224 triples. Similar to dmg777k, DBpedia is not fully connected and does not represent a simple graph. For our experiments with DBpedia, we utilize the following dataset components: Article categories, Instance types, Transitive instance types, Mapping-based object properties, SKOS categories.

5.3 Evaluation Results

For the dmg777k node classification task, we evaluate the multi-class classification performance using Accuracy, the Macro F1 score, and the MCC score. For GEval, we use the evaluation metrics assigned to each specific evaluation task.

The results of our evaluation comparing different embedding methodologies on dmg777k are presented in Table 2. Across all evaluation metrics, GeoRDF2Vec outperforms both plain RDF2Vec and TransGeo, the latter serving as a spatially weighted alternative to our approach. Similar to other evaluation datasets, the

Table 3. Evaluation metrics for DBpedia GEval benchmark datasets

Task	Dataset	RDF2Vec	GeoRDF2Vec
Classification (accuracy)	AAUP	0.676 ± 0.0043	$\mathbf{0.775} \pm 0.0036$
	Cities	0.810 ± 0.0128	0.809 ± 0.0134
	Forbes	0.610 ± 0.0093	$\mathbf{0.744} \pm 0.0152$
	Albums	0.774 ± 0.0053	$\mathbf{0.812} \pm 0.0049$
	Movies	0.739 ± 0.0037	$\mathbf{0.746} \pm 0.0029$
Clustering (accuracy)	Cities and countries (2k)	$\mathbf{0.758} \pm 0.0038$	0.712 ± 0.0029
	Cities and countries	0.696 ± 0.0267	$\mathbf{0.812} \pm 0.0142$
	Cities, albums, movies, AAUP, Forbes	$\mathbf{0.926} \pm 0.0434$	0.821 ± 0.0371
	Teams	0.917 ± 0.0174	$\mathbf{0.928} \pm 0.0073$
Regression (RMSE)	AAUP	68.745 ± 1.4563	$\mathbf{62.123} \pm 1.3591$
	Cities	15.601 ± 1.2734	$\mathbf{14.102} \pm 0.9571$
	Forbes	36.459 ± 1.0231	$\mathbf{32.241} \pm 1.2467$
	Albums	$\mathbf{11.930} \pm 0.0915$	12.131 ± 0.0743
	Movies	19.648 ± 0.3428	$\mathbf{11.254} \pm 0.5310$

translation-based method performs worse compared to RDF2Vec. While TransGeo implicitly models spatial weights associated with its nodes, its performance is inferior to the plain RDF2Vec implementation. This aligns with previous research, which indicates that non-simple KGs containing multiple predicates between the same set of entities tend to perform worse compared to RDF2Vec [5]. The spatially weighted variant of GeoRDF2Vec demonstrates significant performance improvements over plain RDF2Vec. Compared to TransGeo, our approach achieves a 26% improvement in accuracy and a 21% improvement in the F1 score.

For the GEval benchmark, we evaluated all three approaches: GeoRDF2Vec, RDF2Vec, and TransGeo. Due to scalability limitations, TransGeo did not successfully scale to the full DBpedia KG. In its current implementation, the negative sampling process based on the entity context requires approximately 13 h per sampling iteration on our hardware. A more scalable approach, such as using a subset of DBpedia or optimizing the negative sampling implementation, could improve feasibility for large KGs like DBpedia. However, such a reimplementation was beyond the scope of our research. As shown in Table 3, GeoRDF2Vec outperforms plain RDF2Vec in the majority of evaluated tasks.

5.4 Analysis of the Information Flooding Approach

For a deeper analysis of our information flooding approach, we tested five different queries on entities in DBpedia that do not have an associated geometry. These entities are linked to a country, for example, based on the location of an event or the nationality of a person. To ensure a manageable dataset size, we randomly limit the query results to 10,000 entities from each of the following queries:

```
select ?x ?c ?g where {?x a dbo:Event . ?x dbo:country ?c . ?c geo:geometry ?g}
select ?x ?c ?g where {?x a dbo:Person . ?x dbo:nationality ?c . ?c geo:geometry ?g}
select ?x ?c ?g where {?x a dbo:RecordLabel . ?x dbo:location ?l . ?l dbo:country ?c .
?c geo:geometry ?g}
select ?x ?c ?g where {?x a dbo:SportsTeam . ?x dbo:city ?y . ?y dbo:country ?c .
?c geo:geometry ?g}
select ?x ?c ?g where {?x a dbo:Currency . ?x dbo:usingCountry ?c . ?c geo:geometry ?g}
```

We determine whether the centroid of the geometry assigned to the entity `x`, computed as described in Sect. 4, is in the geographic boundaries of the country `c`. The country boundaries are retrieved from OpenDataSoft [21]. If the centroid is within those boundaries, we count the example as a true prediction, otherwise as a false prediction. We can see that in the majority of the cases, the information flooding assigns a sensible geometry to a non-geographic entity. The accuracy of the information flooding approach varies by query type, achieving 94.70% for Event, 93.30% for Person, 94.09% for RecordLabel, 98.74% for SportsTeam, and 84.86% for Currency.

While the majority of cases outlined in above paragraph achieve an accuracy above 90%, we observe lower accuracy for the currency subset. Incorrect geometries are assigned to currencies such as the New Zealand Pound, which has been used both in New Zealand and remote territories like Tokelau. Since DBpedia only contains point geometries, an unweighted mean is computed, distorting the centroid calculation. In contrast, for geometry collections that include polygons, the centroid would be weighted based on the area covered by each polygon.[5] To further evaluate the accuracy of our flooding approach, we conducted an additional experiment on the dmg777k dataset. We split the geographic nodes into an 80% training set and a 20% test set. Using the prior evaluation method, our flooding approach achieves an accuracy of 87.3590% on the dmg777k dataset.

5.5 Influence of GeoRDF2Vec Hyperparameters

On the dmg777k node classification dataset, we conducted an ablation study to assess the influence of walk length and the number of walks in GeoRDF2Vec compared to the plain RDF2Vec implementation. To analyze the effect of both variables, we separate the grid search results for the weighted and unweighted graphs in the dmg777k evaluation dataset. We evaluate the grid search results by examining the maximum F1 score achieved for each individual parameter during the hyperparameter search. Figure 3 illustrates the walk distance corresponding to the maximum F1 score for both the plain RDF2Vec variant and GeoRDF2Vec. For each parameter value, the confidence interval is visualized in the blue-shaded area of the line chart.

For the plain RDF2Vec variant shown in Fig. 3a, the F1 score increases from 0.48 to 0.58, reaching peak performance at a walk distance of 6. Beyond this distance, performance declines, with the F1 score decreasing to 0.54 at a walk

[5] The exact formula for centroid calculation can be found in the documentation.

distance of 10. In contrast, GeoRDF2Vec exhibits a similar initial performance increase, from 0.5 at a walk distance of 2 to 0.56 at a walk distance of 3. As depicted in the upper section of Fig. 3b, the F1 score remains within the confidence interval until a walk distance of 7. However, at a walk distance of 10, the F1 score increases to 0.61. These results indicate that introducing spatial weights during walk generation stabilizes performance and supports longer walk distances in graph embeddings. This allows for capturing more contextual information around individual entities, leading to more stable prediction results in downstream tasks. The likely reason for this improvement is that entities a few hops away in the graph can still be relevant if they are geographically close, whereas geographically distant entities are less relevant in the absence of spatial weighting.

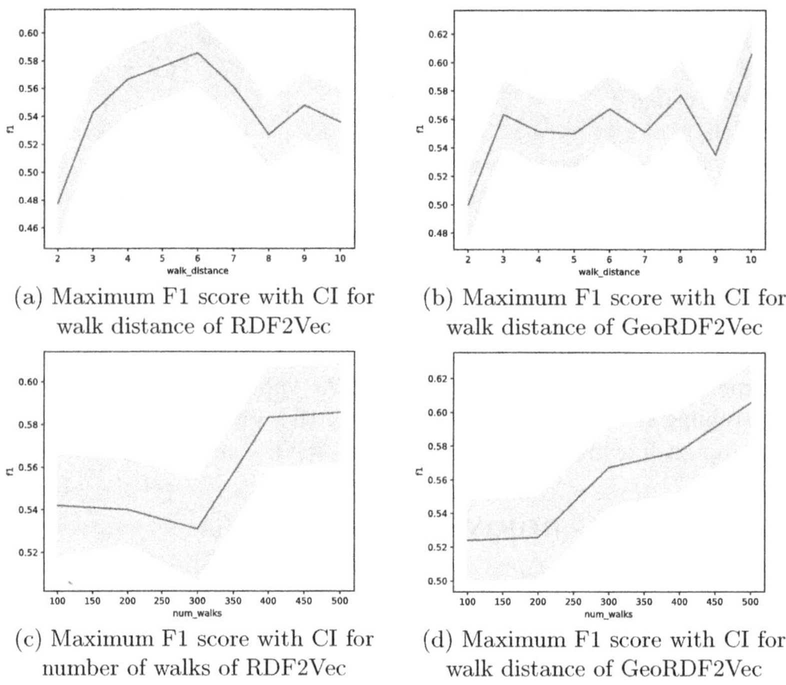

(a) Maximum F1 score with CI for walk distance of RDF2Vec

(b) Maximum F1 score with CI for walk distance of GeoRDF2Vec

(c) Maximum F1 score with CI for number of walks of RDF2Vec

(d) Maximum F1 score with CI for walk distance of GeoRDF2Vec

Fig. 3. Influence of walk distance and number of walks of RDF2Vec variants

The impact of the number of walks on RDF2Vec and GeoRDF2Vec is shown in Fig. 3. In both variants, an increase in the number of walks leads to an improvement in the F1 score. However, the improvement is more substantial and continuous for GeoRDF2Vec compared to plain RDF2Vec. In standard RDF2Vec, all neighboring nodes are sampled with equal likelihood, leading to greater diversity in walk sequences as the number of walks increases. In contrast, when using geographically weighted walks, the diversity of walks is reduced since not all edges are sampled at the same rate. This less diverse sampling strategy, which prioritizes geographically proximal entities, evidently results in better performance.

6 Conclusion and Outlook

In this paper, we have demonstrated that incorporating geographic proximity when computing entity representations in knowledge graphs can significantly improve entity embeddings. By using geographic proximity as a proxy for transition probabilities in the walk generation mechanism, we have developed an effective, overhead-free method to integrate geographic distances into RDF2Vec. Additionally, through the geographic information flooding mechanism, we have introduced a way to leverage geographic information in partially geographic KGs. Previous research has proposed various walk generation mechanisms for RDF2Vec [30]. One such approach, *community hops*, presents an interesting candidate for incorporating geographic information, as it could define hops based on geographic distance. While GeoRDF2Vec, as proposed here, only considers geographic proximity when entities share a direct connection in the knowledge graph, such an approach could also exploit geographic proximity for otherwise unconnected nodes.

Our approach incorporates spatial edge weights into the walk distance, filtering out entities that are geographically distant during walk generation. Future research could explore adapting established embedding methods such as TransE or ComplEx to incorporate geographic penalization in their loss functions. Specifically, introducing a penalization term for the geodesic distance between two entities could refine training processes by penalizing geographically distant entity pairs. Incorporating geographic penalization in loss functions has been shown to enhance learning outcomes when geographic data is integrated into a dataset [25]. In addition to geographic information, many knowledge graphs also contain temporal facts, such as timestamps and timespans for events. In principle, our approach can be extended to temporal knowledge graphs as well. Research has shown that, particularly in link prediction tasks, a deterministic approach using weights can outperform state-of-the-art models [9,16]. Ultimately, both geographic and temporal factors can be integrated into a unified weighting function to learn embeddings that incorporate both spatial and temporal aspects.

Resources

The coding used for this paper together with datasets is provided via the following GitHub repository accessible under the following link.

References

1. Bloem, P., Wilcke, X., van Berkel, L., de Boer, V.: `kgbench`: A Collection of Knowledge Graph Datasets for Evaluating Relational and Multimodal Machine Learning. In: Verborgh, R., Hose, K., Paulheim, H., Champin, P.-A., Maleshkova, M., Corcho, O., Ristoski, P., Alam, M. (eds.) ESWC 2021. LNCS, vol. 12731, pp. 614–630. Springer, Cham (2021). https://doi.org/10.1007/978-3-030-77385-4_37
2. Böckling, M., Paulheim, H., Detzler, S.: Comparing spatial-temporal knowledge graph on spatial downstream tasks. In: Proceedings of the 32nd ACM International Conference on Advances in Geographic Information Systems, pp. 581–584 (2024)
3. Böckling, M., Paulheim, H., Detzler, S.: A planet scale spatial-temporal knowledge graph based on OpenStreetMap and H3 grid. In: Geospatial Linked Data Workshop. arXiv (2024)
4. Car, N.J., Homburg, T.: Geosparql 1.1: Motivations, details and applications of the decadal update to the most important geospatial LOD standard. ISPRS Int. J. Geo-Inf. **11**(2), 117 (2022). https://doi.org/10.3390/ijgi11020117
5. Celebi, R., Uyar, H., Yasar, E., Gumus, O., Dikenelli, O., Dumontier, M.: Evaluation of knowledge graph embedding approaches for drug-drug interaction prediction in realistic settings. BMC Bioinform. **20**(1) (2019). https://doi.org/10.1186/s12859-019-3284-5
6. Chen, Y.: On the four types of weight functions for spatial contiguity matrix. Lett. Spat. Resour. Sci. **5**(2), 65–72 (2012). https://doi.org/10.1007/s12076-011-0076-6
7. Cochez, M., Ristoski, P., Ponzetto, S.P., Paulheim, H.: Biased graph walks for RDF graph embeddings. In: Proceedings of the 7th International Conference on Web Intelligence, Mining and Semantics, pp. 1–12 (2017)
8. Dsouza, A., Tempelmeier, N., Yu, R., Gottschalk, S., Demidova, E.: WorldKG: a world-scale geographic knowledge graph. In: Proceedings of the 30th ACM International Conference on Information & Knowledge Management, pp. 4475–4484. ACM, Virtual Event Queensland Australia (2021). https://doi.org/10.1145/3459637.3482023
9. Gastinger, J., Meilicke, C., Errica, F., Sztyler, T., Schuelke, A., Stuckenschmidt, H.: History repeats itself: a baseline for temporal knowledge graph forecasting (2024). https://doi.org/10.48550/ARXIV.2404.16726
10. Gesese, G.A., Biswas, R., Alam, M., Sack, H.: A survey on knowledge graph embeddings with literals: which model links better literal-ly? Semant. Web **12**(4), 617–647 (2021)
11. Getis, A.: Spatial weights matrices. Geogr. Anal. **41**(4), 404–410 (2009). https://doi.org/10.1111/j.1538-4632.2009.00768.x
12. Hubert, N., Paulheim, H., Brun, A., Monticolo, D.: Do similar entities have similar embeddings? In: European Semantic Web Conference, pp. 3–21. Springer, Cham (2024)
13. Ivis, F.: Calculating geographic distance: concepts and methods. In: Proceedings of the 19th Conference of Northeast SAS User Group, pp. 17–20. sn (2006)

14. Janowicz, K., Hitzler, P., Li, W., et al.: Know, know where, knowwheregraph: a densely connected, cross-domain knowledge graph and geo-enrichment service stack for applications in environmental intelligence. AI Mag. **43**(1), 30–39 (2022). https://doi.org/10.1002/aaai.12043
15. Karney, C.: Algorithms for geodesics. J. Geodesy **87**(1), 43–55 (2012). https://doi.org/10.1007/s00190-012-0578-z
16. Krause, F., Weller, T., Paulheim, H.: On a generalized framework for time-aware knowledge graphs. In: Towards a Knowledge-Aware AI, pp. 69–74. IOS Press (2022)
17. Lehmann, J., et al.: DBpedia-a large-scale, multilingual knowledge base extracted from Wikipedia. Semant. Web **6**(2), 167–195 (2015)
18. Mai, G., et al.: SE-KGE: a location-aware knowledge graph embedding model for geographic question answering and spatial semantic lifting. Trans. GIS **24**(3), 623–655 (2020). https://doi.org/10.1111/tgis.12629
19. Mai, G., Yan, B., Janowicz, K., Zhu, R.: Relaxing unanswerable geographic questions using a spatially explicit knowledge graph embedding model. In: Kyriakidis, P., Hadjimitsis, D., Skarlatos, D., Mansourian, A. (eds.) AGILE 2019. LNGC, pp. 21–39. Springer, Cham (2020). https://doi.org/10.1007/978-3-030-14745-7_2
20. Mann, G., Dsouza, A., Yu, R., Demidova, E.: Spatial link prediction with spatial and semantic embeddings, pp. 179–196. Springer, Cham (2023). https://doi.org/10.1007/978-3-031-47240-4_10
21. OpenDataSoft: World Administrative Boundaries - Countries and Territories — public.opendatasoft.com. https://public.opendatasoft.com/explore/dataset/world-administrative-boundaries/export/. Accessed 20 Dec 2024
22. Paulheim, H., Ristoski, P., Portisch, J.: Embedding Knowledge Graphs with RDF2vec. Springer, Cham (2023)
23. Pellegrino, M.A., Cochez, M., Garofalo, M., Ristoski, P.: A configurable evaluation framework for node embedding techniques. In: Hitzler, P., et al. (eds.) ESWC 2019. LNCS, vol. 11762, pp. 156–160. Springer, Cham (2019). https://doi.org/10.1007/978-3-030-32327-1_31
24. Pellissier Tanon, T., Weikum, G., Suchanek, F.: YAGO 4: a reason-able knowledge base. In: Harth, A., et al. (eds.) ESWC 2020. LNCS, vol. 12123, pp. 583–596. Springer, Cham (2020). https://doi.org/10.1007/978-3-030-49461-2_34
25. Peng, J., Huang, Y., Sun, W., Chen, N., Ning, Y., Du, Q.: Domain adaptation in remote sensing image classification: a survey. IEEE J. Sel. Top. Appl. Earth Obs. Remote Sens. **15**, 9842–9859 (2022)
26. Portisch, J., Heist, N., Paulheim, H.: Knowledge graph embedding for data mining vs. knowledge graph embedding for link prediction–two sides of the same coin? Semant. Web **13**(3), 399–422 (2022)
27. Qiu, P., Gao, J., Yu, L., Lu, F.: Knowledge embedding with geospatial distance restriction for geographic knowledge graph completion. ISPRS Int. J. Geo Inf. **8**(6), 254 (2019). https://doi.org/10.3390/ijgi8060254
28. Ristoski, P., Paulheim, H.: RDF2Vec: RDF graph embeddings for data mining. In: Groth, P., et al. (eds.) ISWC 2016. LNCS, vol. 9981, pp. 498–514. Springer, Cham (2016). https://doi.org/10.1007/978-3-319-46523-4_30

29. Shepard, D.: A two-dimensional interpolation function for irregularly-spaced data. In: Proceedings of the 1968 23rd ACM National Conference. ACM Press (1968). https://doi.org/10.1145/800186.810616
30. Steenwinckel, B., et al.: Walk extraction strategies for node embeddings with rdf2vec in knowledge graphs. In: Database and Expert Systems Applications-DEXA 2021 Workshops: BIOKDD, IWCFS, MLKgraphs, AI-CARES, ProTime, AISys 2021, Virtual Event, September 27–30, 2021, Proceedings 32, pp. 70–80. Springer, Cham (2021)
31. Taweel, A.A., Paulheim, H.: Towards exploiting implicit human feedback for improving rdf2vec embeddings. arXiv preprint arXiv:2004.04423 (2020)
32. Vrandečić, D., Krötzsch, M.: Wikidata: a free collaborative knowledgebase. Commun. ACM **57**(10), 78–85 (2014)
33. Wang, M., Wang, R., Liu, J., Chen, Y., Zhang, L., Qi, G.: Towards empty answers in SPARQL: approximating querying with RDF embedding. In: Vrandečić, D., et al. (eds.) ISWC 2018. LNCS, vol. 11136, pp. 513–529. Springer, Cham (2018). https://doi.org/10.1007/978-3-030-00671-6_30
34. Wilcke, X., Bloem, P., Berkel, L.V., Boer, V.D.: kgbench: dmgfull and dmg832k (2020). https://doi.org/10.5281/ZENODO.4361779

RelCheck: Improving Relation Extraction with Ontology-Guided and LLM-Based Validation

Mounir Ourekouch[1,2(✉)], Mohammed-Amine Koulali[1,4], and Mohammed Erradi[1,2,3]

[1] College of Computing, Mohammed VI Polytechnic University, Ben Guerir, Morocco
{mounir.ourekouch,amine.koulali,mohammed.erradi}@um6p.ma
[2] CID Developpement, Rabat, Morocco
[3] ENSIAS, Mohammed V University, Rabat, Morocco
[4] Laboratoire de Modélisation et Calcul Scientifique, ENSAO, Mohammed I University, Oujda, Morocco

Abstract. Relation extraction (RE) is a key task in natural language processing (NLP) and a core component of information extraction. It focuses on identifying semantic relations between entities in text. Pre-trained language models (PLMs), such as transformer-based models like BERT, XLNet and RoBERTa, have made notable progress in RE. Predictions of relations from these models are provided with varying confidence levels. While high-confidence predictions of relations are generally accurate, low-confidence predictions tend to be less precise and often lead to inaccuracies. The current research question is how to re-evaluate the low confidence predictions to ensure the overall confidence of a PLM. To solve this problem we propose a framework using automatically generated ontology schemas and LLMs. We first propose an algorithm that constructs ontology schemas from the RE datasets (TACRED and ReTACRED). Then we use LLMs to validate these low-confidence predictions through prompting to further improve the precision of final predictions. Experimental results on transformer-based models, GCN and LSTM-based models across two large-scale RE datasets (TACRED and ReTACRED) show significant improvements in precision and overall performance.

Keywords: Relation Extraction · Large Language Models (LLMs) · Ontology Schema · Knowledge Graph Construction

1 Introduction

Relation extraction (RE) is the task of identifying and classifying relationships between named entities in text. It serves as a key component in applications such as information extraction and knowledge graph population. Earlier neural RE methods relied on Graph Neural Networks (GNNs) and Recurrent Neural Networks (RNNs) [10,20,24] to model relationships, but they struggled with long-distance dependencies and complex sentence structures due to vanishing gradients and limited global context representation. The introduction of Pre-trained

E. Curry et al. (Eds.): ESWC 2025, LNCS 15718, pp. 441–459, 2025.
https://doi.org/10.1007/978-3-031-94575-5_24

Language Models (PLMs), such as BERT [4] and SpanBERT [8], marked a significant breakthrough in RE [12,18,25]. These models leverage large-scale language data and fine-tuning techniques, resulting in enhanced contextual understanding and overall performance. PLMs balance strong performance with efficient resource utilization, making them well-suited for diverse applications.

Large Language Models (LLMs) have demonstrated exceptional capabilities across various NLP tasks, including RE. Through their pre-training knowledge, LLMs can infer relationships directly from natural language instructions, giving rise to Generative Relation Extraction (GRE) [7,9,16,21]. GRE extracts relations without task-specific fine-tuning, allowing models to infer relationships from context while avoiding reliance on labeled data. Their scale and pre-training knowledge enhance reasoning, enabling them to handle complex relations.

PLMs achieve strong performance at a lower computational cost, making them a widely adopted choice for RE. However, their low-confidence predictions often lack reliability, which requires consideration before using them for downstream tasks such as knowledge graph construction. Due to their efficiency, improving these uncertain predictions rather than discarding them is important. LLMs, with their advanced reasoning capabilities, present a potential solution, though their high computational requirements make complete dependence impractical. This creates a significant challenge: developing a systematic approach to re-evaluate low-confidence predictions from PLMs while selectively utilizing LLMs' reasoning strengths to ensure overall prediction confidence.

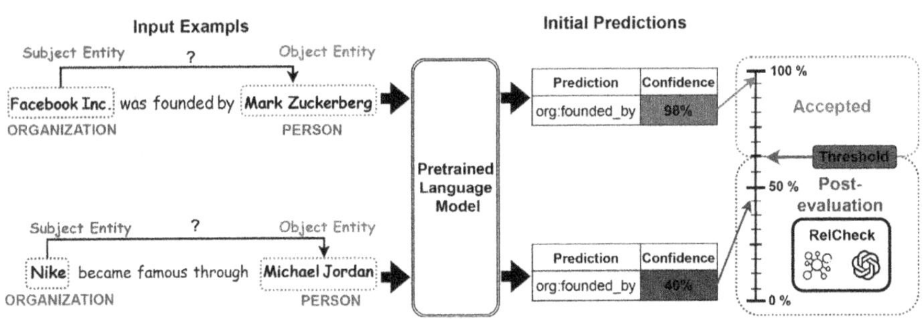

Fig. 1. Overview of RE with PLMs and re-evaluation using RelCheck. Given two example sentences with highlighted entities, the PLM predicts relations along with confidence scores. High-confidence predictions are accepted, while low-confidence ones undergo re-evaluation with RelCheck.

To address this challenge, we propose RelCheck, a post-evaluation framework that refines low-confidence predictions from PLMs while maintaining efficiency. As shown in Fig. 1, a PLM generates relation predictions with their respective confidence scores. High-confidence predictions are accepted, while low-confidence ones require validation. RelCheck refines these cases using ontology-guided validation and selective LLM reasoning. Ontology schemas check alignment with relational constraints, while LLMs reassess borderline cases via prompting. This hybrid approach improves precision without the cost of full-scale LLM inference. Experiments on multiple RE models across TACRED and ReTACRED show significant precision gains. Our code is publicly available at Github[1]

Our contributions include the following:

1. **Confidence-Based Validation**: Using confidence scores to identify predictions requiring further refinement.
2. **Prediction Improvements**: We introduce RelCheck, a post-evaluation framework that enhances prediction precision through ontology schemas and LLM prompting.
3. **Empirical Evaluation**: Validating RelCheck's effectiveness through extensive experiments on RE datasets.

RelCheck improves RE reliability, leading to three broader impacts. First, by validating predictions for schema compliance and accuracy, it enhances prediction quality and supports the construction of more reliable knowledge graphs. This benefits downstream tasks such as information retrieval and question answering, which rely on accurate structured data. Second, its selective use of LLMs for uncertain cases optimizes efficiency, enabling scalable deployment in resource-constrained settings. Finally, the RelCheck mechanism could be extended beyond RE, providing a flexible framework for uncertainty-aware refinement in classification-based NLP tasks, including named entity recognition and sentiment analysis.

2 Related Work

Relation Extraction. RE has significantly evolved across multiple paradigms. Early supervised methods, such as CNNs and RNNs [10,20], captured sequential dependencies but struggled with long-distance relationships. Enhanced approaches like PaLSTM [24] introduced positional embeddings, improving attention to relevant sentence regions. Graph-based models [5,23] leveraged syntactic structures to handle complex, non-sequential relationships effectively. Transformer-based models such as BERT [4] and SpanBERT [8] have transformed RE by introducing the fine-tuning paradigm with deep bidirectional contexts and span-level predictions, enabling richer semantic representations. Zhou and Chen [25] further refined these models with effective marking strategies, demonstrating substantial performance improvements. Recent LLM-based generative methods have shifted RE from fine-tuning to instruction-based learning. VanillaRE [16] explored generative RE with LLMs, achieving results close to the state of the art. SuRE [9] reframed RE as a summarization task, utilizing indirect supervision to improve performance. QA4RE [21] further improved generative RE by aligning models with a question-answering approach, enhancing

[1] Code repository: https://github.com/ourekouch/Rel_Check.

robustness. Despite recent progress, transformer-based models like BERT and SpanBERT [8,25] remain widely used for RE, offering a strong trade-off between performance and efficiency. In contrast, LLMs leverage instruction tuning for greater flexibility, excelling in zero-shot and few-shot RE across domains. These complementary strengths motivate hybrid approaches that integrate transformer efficiency with LLM reasoning.

Large Language Models (LLMs). The evolution of LLMs has significantly reshaped natural language processing, enhancing both reasoning and generalization capabilities. The introduction of BERT by Google in 2018 [4] marked a turning point, leveraging bidirectional contextual representations to improve NLP tasks through fine-tuning. Progress accelerated with OpenAI's GPT-3 in 2020 [2], which, with 175 billion parameters, set new benchmarks in generative modeling, enabling effective zero-shot and few-shot learning. Building on this, Google's PaLM [3] demonstrated strong multitask learning, while Meta's OPT [22] showcased efficient zero-shot performance. The release of LLaMA [14] expanded open-access LLM research by training solely on public datasets. More recently, GPT-4 [1] introduced larger architectures, further improving reasoning and adaptability. Despite these advances, large-scale deployment remains resource-intensive and costly, requiring specialized hardware. This has driven research into more efficient strategies, such as selective LLM application. Our framework, RelCheck, follows this approach, using LLMs to validate low-confidence predictions while optimizing resource costs.

3 RelCheck

In this section, we introduce RelCheck, a framework for re-evaluating uncertain predictions in RE. As shown in Fig. 2, RelCheck improves reliability through a multi-step process. Step 1 filters low-confidence predictions based on confidence scores. Step 2 validates them using an automatically constructed ontology schema. Step 3 applies LLM-based validation using binary-choice prompting to verify correctness. Finally, Step 4 updates predictions based on validation results, ensuring improved precision and reliability.

3.1 Filtering Low-Confidence Predictions

In deep learning architectures, including PLMs, confidence scores quantify a model's certainty in its predictions. In RE, these scores indicate the likelihood that a predicted relationship is correct among all possible relationships. They are typically derived from the model's output probabilities, obtained by applying a normalization function such as the Softmax to the logits.

RE models, as described by [17] and illustrated in Fig. 3, follow a standard framework. The model processes input through representation stages, extracts features, and then passes them to a classification layer. This final layer outputs raw scores, or logits, for each relationship type. A logit is an unnormalized value representing the model's association strength for a specific relationship.

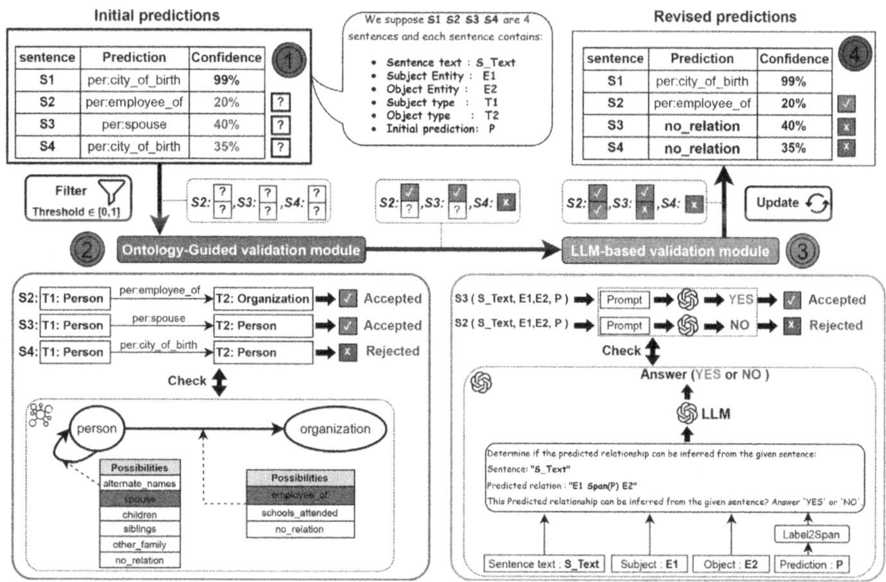

Fig. 2. Overall design of RelCheck framework. Sentences with low-confidence predictions (S2, S3, S4) go through RelCheck validation steps. Initial predictions for S3 and S4 are discarded while S2 prediction is validated.

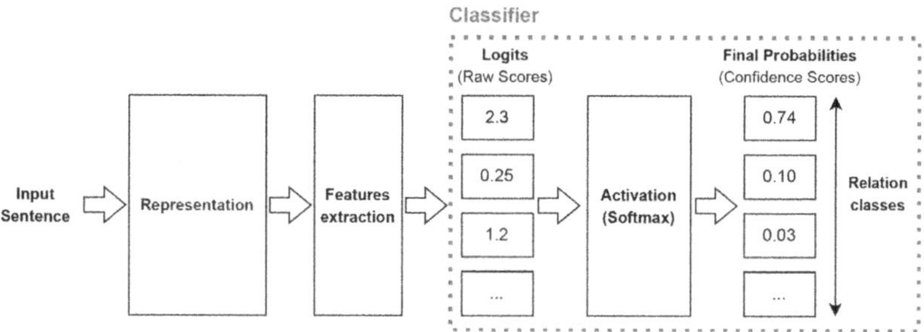

Fig. 3. Framework for RE Models: From Input Sentence to Final Logits and Probabilities.

The Softmax function transforms these logits into probabilities, normalizing them to a range between 0 and 1. The confidence score for a specific relationship type is computed as follows:

$$\sigma(z_i) = \frac{\exp(z_i)}{\sum_{j=1}^{K} \exp(z_j)}, \tag{1}$$

where z_i is the logit output for a specific relationship i ($1 \leq i \leq K$), and K is the total number of relationship types.

The resulting probability vector consists of values between 0 and 1 that sum to 1, representing the model's confidence in each relationship type. For example, if a relationship i has a probability of 0.85, the model estimates an 85% likelihood that it is correct.

In RE, high confidence scores indicate strong certainty, while low scores reflect uncertainty. These scores help filter predictions requiring further validation. Predictions below a set threshold are flagged for reassessment, ensuring that only high-confidence predictions are accepted. This improves precision by carefully validating uncertain predictions.

3.2 Ontology-Guided Validation Module

Ontology has been defined in various ways in the literature. [15] described it as a worldview of a domain, encompassing concepts (e.g., entities, attributes, processes), their definitions, and their interrelationships.

In the context of RE, we define our ontology schema as follows: concepts represent types of named entities, while relations encompass all possible connections between these entity types. Algorithm 1 outlines schema construction steps by iterating through dataset samples to cover all cases. Figure 4 illustrates the extraction process and presents the possible combinations between subject and object entity types. The resulting table shows possible combinations between subject and object entity types. For each subject type paired with an object type, the schema lists all possible relations.

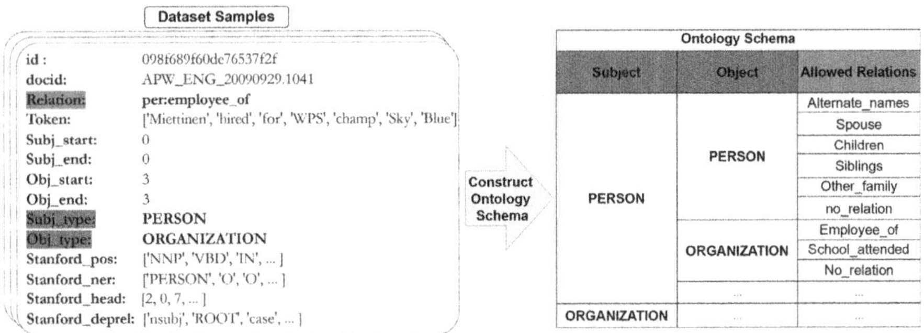

Fig. 4. Construct RE ontology schema from dataset samples attributes (Relation, Subject Type, Object Type).

As an example, in Fig. 2, for a subject entity of type *PERSON* and an object entity of type *ORGANISATION*, the possible relations could be either ***employee_of*** or ***schools_attended***, or there might be **no_relation**. This means that the predicted relation should be within the scope of those possibilities, and any other prediction will be automatically considered false.

Algorithm 1. Construct Ontology Schema from RE Dataset

Input: Dataset with samples containing {subject_type , object_type, relation} .
Output: Ontology schema in JSON format.

```
Initialize an empty set S to store unique triplets {subject_type, object_type, relation}.
for each sample in Dataset do
    Extract subject_type, object_type, and relation from the sample.
    Add {subject_type, object_type, relation} to S if not already present.
end for
Initialize an empty dictionary ontology.
for each {subject_type, object_type, relation} in S do
    if subject_type not in ontology then
        ontology[subject_type] ← {}
    end if
    if relation not in ontology[subject_type] then
        ontology[subject_type][object_type] ← []
    end if
    if relation not in ontology[subject_type][object_type] then
        Append relation to ontology[subject_type][object_type]
    end if
end for
Convert ontology to JSON format.
return ontology in JSON format.
```

3.3 LLM-Based Validation Module

The LLM-Based Validation Module in RelCheck builds on the evaluation of various LLM-based templates for RE as explored in Sect. 5.2. It adapts the QA4RE prompt model to a binary-choice prompting style, as shown in Fig. 5. Instead of requiring the model to select the correct relationship from multiple options, RelCheck provides an initial prediction and asks the model to validate its correctness.

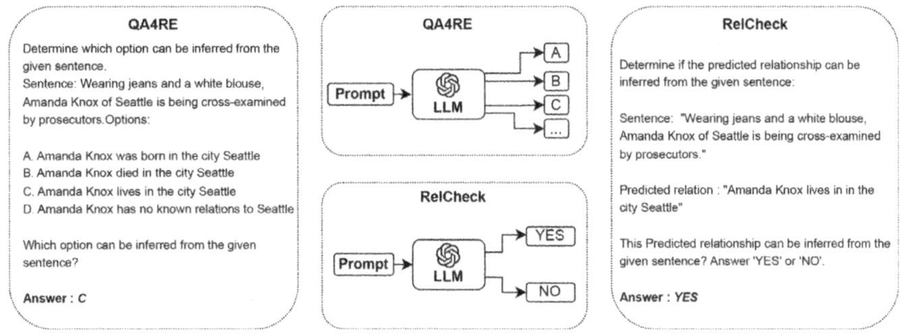

Fig. 5. Adaptation of multiple-choice setting (QA4RE) to validation setting.

This binary-choice format simplifies the task by presenting a validation question that requires a "YES" or "NO" response, improving the efficiency of the process. Relation labels were transformed into verbalized textual spans, following prior studies using the same format [11,21]. This ensures compatibility and interpretability in prompts.

Multiple templates were evaluated, as discussed in Subsect. 5.2, and the best-performing binary-choice design was selected for RelCheck. By focusing on validating predictions rather than generating them, this adaptation enhances the precision and reliability of RE tasks.

4 Results

4.1 RE Models

To evaluate the effectiveness of RelCheck as a post-validation method, we first generate initial predictions using various RE models. We experiment with transformer-based models, including $\text{BERT}_{\text{BASE}}$, $\text{BERT}_{\text{LARGE}}$, $\text{RoBERTa}_{\text{BASE}}$, $\text{RoBERTa}_{\text{LARGE}}$, and $\text{XLNet}_{\text{LARGE}}$ [19], following configurations described by Zhou and Chen [25]. Additionally, we evaluate a graph convolutional network (GCN)-based model, C-GCN [23], and an RNN-based model, PaLSTM [24], for broader comparison. All models are trained on the large-scale open-domain datasets TACRED [24] and its revised version, ReTACRED [13].

4.2 Experimental Setup

In this setup, we obtain initial predictions from the models described in Subsect. 4.1. The ontology schema is automatically generated from the evaluation RE datasets (Subsect. 3.2) and defines valid entity-relation combinations. For prompt-based LLMs, we use GPT-4 [1] via the OpenAI API.

Since this is a post-validation step, the experiments require only API access and baseline model predictions, eliminating the need for retraining. Each baseline model follows its official training guidelines, with confidence scores extracted from the last layer. We use fixed thresholds: 0.6 for TACRED and 0.8 for ReTACRED. These values balance precision gains with slight F1 score improvements. Section 5.5 provides further analysis.

4.3 Experimental Analysis

Our main experimental results on the TACRED and ReTACRED datasets highlight the impact of RelCheck's post-validation process across different models. In general, precision is the most affected metric. Since RelCheck focuses on validating uncertain positive predictions, it consistently improves precision across all models. As shown in Table 1, precision with RelCheck is always higher than that of the baseline model.

However, recall is negatively affected, as the framework only removes incorrect predictions without reclassifying them. Failed predictions are assigned the label `no_relation`, which can introduce false negatives and lower recall. This is evident in Table 1, where baseline models maintain higher recall compared to those after RelCheck post-validation. Furthermore, RelCheck has a stronger impact on older models such as C-GCN and PaLSTM. These lower-performing models show the most significant precision gains, demonstrating RelCheck's effectiveness in refining weaker predictions and enhancing model performance.

Table 1. Performance scores (%) of baseline models with and without the RelCheck module on the TACRED and ReTACRED test sets. Metrics include precision (P), recall (R), and F1 score (F1). The best scores for each metric are highlighted in bold.

Model	TACRED			ReTACRED		
	P	R	F1	P	R	F1
PaLSTM	67.120	**63.789**	65.412	79.885	**78.754**	79.315
PaLSTM + RelCheck	**70.995**	62.647	**66.560**	**85.325**	77.001	**80.949**
CGCN	69.857	**63.218**	66.372	80.063	**80.984**	80.521
CGCN + RelCheck	**73.962**	61.594	**67.214**	**87.166**	78.647	**82.688**
BERT BASE	73.102	**70.947**	72.009	87.847	**88.049**	87.948
BERT BASE + RelCheck	**75.876**	69.714	**72.665**	**89.969**	87.181	**88.553**
BERT LARGE	72.858	**74.917**	73.873	89.477	**88.669**	89.071
BERT LARGE + RelCheck	**74.939**	73.474	**74.199**	**91.046**	88.031	**89.513**
RoBERTa BASE	71.315	**72.752**	72.026	88.975	**89.023**	88.999
RoBERTa BASE + RelCheck	**74.459**	70.406	**72.376**	**91.915**	87.960	**89.894**
XLnet LARGE	74.110	**72.662**	73.379	89.964	**91.732**	90.839
XLnet LARGE + RelCheck	**76.749**	71.278	**73.912**	**91.199**	91.183	**91.191**
RoBERTa LARGE	76.123	**72.872**	74.462	91.467	**90.528**	90.995
RoBERTa LARGE + RelCheck	**77.959**	71.910	**74.812**	**92.870**	89.713	**91.264**

5 Discussion and Analysis

5.1 Ablation Study

We analyze the contribution of each component in RelCheck by evaluating its impact on TACRED and ReTACRED. Table 2 presents the results. Ontology-guided validation provides a modest precision gain (+0.10%) without affecting recall, while LLM-based validation offers a larger precision improvement (>1.5% gain), effectively refining low-confidence predictions. However, applying ontology-guided and LLM-based validation (full RelCheck) improves precision but slightly reduces recall (−1.20% in TACRED, −0.74% in ReTACRED). This suggests that RelCheck favors precision over recall, filtering out uncertain predictions at the expense of potentially discarding some correct predictions.

Ontology validation acts as a structural safeguard, ensuring schema consistency by enforcing subject-object type constraints. While its role in this setting is

Table 2. Ablation results on TACRED and ReTACRED. Values in parentheses indicate variation from the baseline.

Model	TACRED			ReTACRED		
	P	R	F1	P	R	F1
Baseline Model	74.92	**74.29**	74.60	90.68	**91.25**	90.96
+ Relcheck (Ontology only)	75.05 (+0.13)	**74.29** (+0.00)	74.67 (+0.07)	90.80 (+0.12)	**91.25** (+0.00)	91.03 (+0.06)
+ Relcheck (LLM only)	76.87 (+1.95)	73.05 (−1.23)	74.91 (+0.31)	92.17 (+1.50)	90.49 (−0.76)	91.33 (+0.36)
+ Relcheck (Ontology + LLM)	**77.00** (+2.08)	73.08 (−1.20)	**74.99** (+0.39)	**92.29** (+1.62)	90.51 (−0.74)	**91.39** (+0.43)

minimal, a more complex schema with additional rules could further improve performance. Figure 6 shows the number of ontology violations per model. Although relatively few, these violations indicate schema misalignment, with some models exhibiting higher error rates. Detecting and addressing these violations is crucial, as they reflect inconsistencies in model predictions that could impact downstream applications like knowledge graph construction. Strengthening schema adherence through enriched ontologies and human-defined rules could mitigate these errors and further enhance performance.

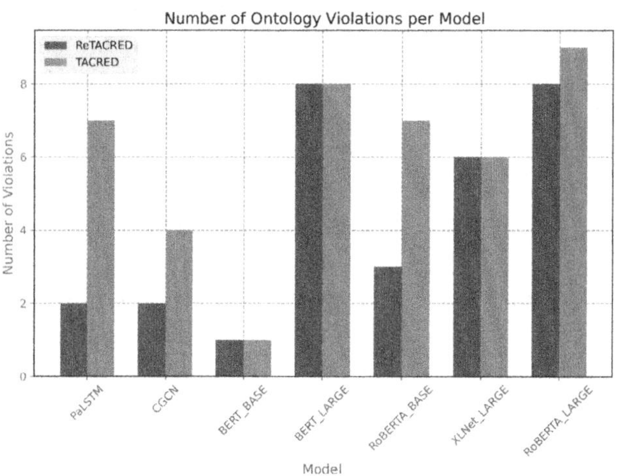

Fig. 6. Number of Ontology Violations per Model in TACRED and ReTACRED.

5.2 Prompt Design for RelCheck Validation

In previous sections, we presented results using the best-performing prompt for RelCheck validation. Here, we analyze how prompt structure influences validation performance, focusing on the RoBERTa model.

LLMs are highly sensitive to prompt structure, as different templates can significantly alter their outputs. Choosing an optimal prompt is essential for maximizing performance. For RelCheck, we adapted RE templates, including SURE [9], NLI [6], QA4RE [21], and Vanilla [7], for binary YES/NO validation. We also examined templates with and without entity type annotations. Below, we outline each adapted template and its refinements for RelCheck:

- **Template 1: Summarization-Based**: Inspired by [9], this method reframes RE as a summarization task. The prompt presents the sentence and entities, asking whether the verbalized relation accurately summarizes the contextual relationship.
- **Template 2: Entailment-Based**: Based on the natural language inference (NLI) framework used in medical RE [6], this approach treats the sentence as the premise and the predicted relation as the hypothesis. The model evaluates whether the relation logically follows from the sentence, leveraging its strength in entailment assessment.
- **Template 3: Question-Answering Based**: Derived from QA4RE [21], this template employs a question-answering format, explicitly asking whether the relation can be inferred from the sentence. The binary YES/NO format simplifies and clarifies the validation process.
- **Template 4: Direct Relation Validation**: Following the vanilla prompting approach used in [7], this template provides direct instructions for validation. The model assesses whether the specified relation can be inferred from the sentence, maintaining a straightforward structure.

Table 3. Performance of Baseline RoBERTa (LARGE) and RoBERTa + RelCheck Using Different Templates (With and Without Entity Types in Prompts).

Model	Entity Types	TACRED			ReTACRED		
		P	R	F1	P	R	F1
Baseline RoBERTa LARGE	-	76.123	72.872	74.462	91.467	90.528	90.995
+ RelCheck (T1)	No	**78.949**	70.496	74.484	**93.542**	88.739	91.078
	Yes	78.898	70.617	74.528	93.501	88.651	91.012
+ RelCheck (T2)	No	78.197	71.729	74.824	93.077	89.501	91.254
	Yes	78.171	71.729	74.812	92.949	89.625	91.257
+ RelCheck (T3)	No	78.086	**71.910**	**74.871**	92.938	**89.713**	**91.297**
	Yes	78.003	71.880	74.816	92.965	89.607	91.255
+ RelCheck (T4)	No	77.967	71.729	74.718	92.936	89.678	91.278
	Yes	78.065	71.820	74.812	92.985	89.412	91.163

The adapted QA4RE template, designed for binary-choice prompting without entity types, proved effective across both datasets. Although performance differences between templates were minimal, its structured question-answering format leveraged LLM capabilities efficiently. Consequently, it achieved the highest F1 score, as shown in Table 3. These findings confirm the adapted QA4RE template as the most suitable choice for binary-choice prompting in RelCheck.

5.3 Testing RelCheck with Different GPT Models

In this section, we evaluate the performance of RelCheck using various GPT-based models. We test models including **GPT-3.5-turbo**, **GPT-3.5-turbo-instruct**, **ChatGPT-4.0-latest**, **GPT-4-turbo**, **GPT-4o-mini**, and

GPT-4o to assess and compare their performances regarding relation validation. The models are evaluated on the TACRED and ReTACRED datasets. Precision, recall, and F1 score are measured to determine which model offers the best performance when integrated with RelCheck.

As shown in Table 4, all models demonstrate efficacy and align with the framework's purpose, with slight performance differences. The latest GPT-4o achieved the best overall F1 score, as it is one of the most recently enhanced models in the GPT series.

Table 4. Performance of Baseline RoBERTa and RoBERTa + RelCheck with Different GPT Models on TACRED and ReTACRED.

Model	TACRED			ReTACRED		
	P	R	F1	P	R	F1
Baseline RoBERTa LARGE	76.123	72.872	74.462	91.467	90.528	90.995
+ RelCheck (gpt-3.5-turbo)	78.293	71.158	74.555	93.136	89.129	91.088
+ RelCheck (gpt-3.5-turbo-instruct)	**79.385**	68.331	73.444	**93.947**	87.943	90.846
+ RelCheck (gpt-4-turbo)	78.579	70.827	74.502	93.558	88.970	91.206
+ RelCheck (gpt-4o-mini)	78.285	71.669	74.831	93.202	89.324	91.221
+ RelCheck (gpt-4o)	78.086	**71.910**	**74.871**	92.938	**89.713**	**91.297**

5.4 Confidence Scores and RelCheck Impact

In this section, we analyze the confidence scores for each model on the TACRED and ReTACRED datasets. As shown in Fig. 7, higher confidence scores strongly correlate with correct predictions. In contrast, lower confidence scores are less reliable for precision. This observation supports our hypothesis that predictions with lower confidence levels represent the weakest aspect of any model output and necessitate validation.

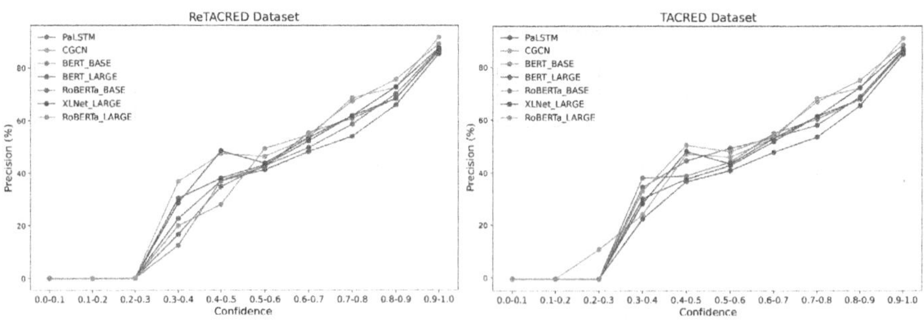

Fig. 7. Correlation between confidence and precision for baseline models on ReTACRED and TACRED datasets.

We further analyze the impact of RelCheck on the precision of low-confidence predictions. Specifically, RelCheck is applied to predictions with confidence scores below 0.6 for the TACRED dataset and below 0.8 for the ReTACRED dataset. These thresholds were selected based on an analysis presented in Sect. 5.5, balancing precision and recall to achieve the highest overall F1-score. As illustrated in Fig. 8, applying RelCheck results in clear improvements in precision within these low-confidence intervals. This highlights RelCheck's effectiveness in enhancing prediction reliability in cases where model confidence is weak.

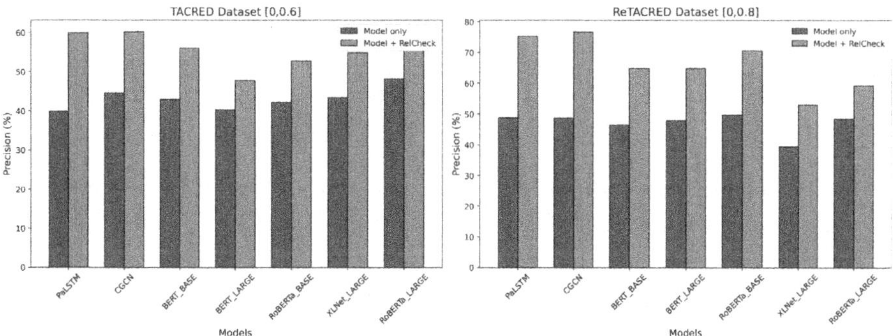

Fig. 8. Impact of RelCheck on the precision of low-confidence predictions for different models.

5.5 Threshold Choice

The choice of threshold significantly affects the results obtained. As shown in Fig. 9, we evaluated the effect of RelCheck for different threshold values using the RoBERTa model on the TACRED and ReTACRED datasets. The results show that increasing the threshold improves precision. However, this comes at the cost of reduced recall. This behavior aligns with the principle of RelCheck, which enhances precision by removing false positives (FP) but reduces recall by introducing additional false negatives (Δ_{FN}) due to reclassification errors. Specifically, not all incorrect FP are true negatives (TN); some may belong to another class, as this is a multi-class classification task. This relationship can be formalized as:

$$P' = \frac{\text{TP}}{\text{TP} + \text{FP} - \Delta_P}, \quad R' = \frac{\text{TP}}{\text{TP} + \text{FN} + \Delta_{FN}}. \tag{2}$$

where:

- $\Delta_P > 0$: False positives corrected, improving precision ($P' > P$).
- $\Delta_{FN} \geq 0$: False negatives added due to misclassifications, reducing recall ($R' < R$).

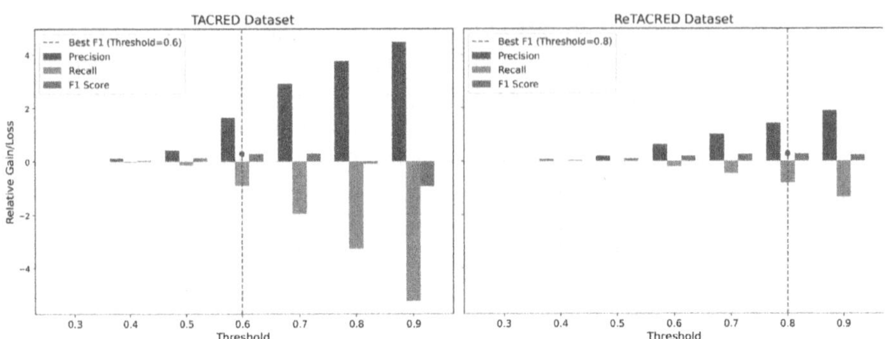

Fig. 9. Impact of the validation threshold on evaluation results (RoBERTa model + RelCheck results in TACRED and ReTACRED datasets).

While RelCheck consistently improves precision ($P' > P$) by reducing FP, its negative impact on recall ($R' < R$) arises from the introduction of Δ_{FN}, where some false positives are misclassified as *no_relation* but actually belong to another class. This limitation underscores the trade-off inherent in RelCheck. The selection of an appropriate threshold is essential to balance precision and recall, as it directly impacts the overall F1 score. In this work, we empirically determined thresholds of 0.6 for TACRED and 0.8 for ReTACRED, optimizing the precision-recall trade-off on the RoBERTa model before generalizing these values to other models. Unlike confidence scores, which are sample-specific and vary across instances, threshold selection should be grounded in global dataset performance metrics such as F1 score, ensuring a uniform and fair decision rule that applies consistently across all samples. The difference in threshold values reflects variations in data quality between TACRED and ReTACRED, yet both thresholds lead to similar validation ratios as shown in Table 5. For real-world applications, automating threshold selection is highly desirable for practical deployment. Techniques such as percentile-based thresholding (e.g., selecting the lowest 2–5% confidence predictions) can help standardize decision rules across datasets. However, Confidence scores need to be calibrated to mitigate inconsistencies caused by differences in data quality, model architecture, and training distributions.

5.6 Cost Comparison

RelCheck significantly reduces costs by acting as a selective post-validation framework. It validates only a subset of predictions below a confidence threshold. This subset covers fewer than 4% of total predictions, as shown in Table 5.

To quantify costs, we estimated the average prompt length at 75 tokens, including prompt templates and sentence context. For the TACRED test set (15,509 examples), validation using RelCheck costs approximately \$0.05 per model (assuming GPT-4o pricing at \$0.0000025/token). By comparison, fully prompting the entire dataset would cost about $3.60. Thus, RelCheck reduces the cost to nearly 1% of fully prompting methods.

Despite processing less than 4% of predictions and using just around 1% of fully LLM-based approaches cost, RelCheck achieves notable performance improvements, including an average precision gain of around 3% and an increase in F1 scores ranging from 0.3% to 2%.

Table 5. RelCheck Results: Verified Sentences, Ratio, Gains in Precision and F1 Score, and Cost for TACRED and ReTACRED Datasets.

Dataset	Model	Verified Sentences (#)	Ratio	Cost ($)	Gain in Precision (%)	Gain in F1 Score (%)
TACRED (15,509)	PaLSTM	486	0.0313	0.0911	+3.875	+1.148
	CGCN	599	0.0386	0.1123	+4.105	+0.842
	BERT $_{\text{BASE}}$	207	0.0133	0.0388	+2.774	+0.656
	BERT $_{\text{LARGE}}$	159	0.0103	0.0298	+2.081	+0.326
	RoBERTa $_{\text{BASE}}$	293	0.0189	0.0549	+3.144	+0.350
	XLNet $_{\text{LARGE}}$	86	0.0055	0.0161	+2.639	+0.533
	RoBERTa $_{\text{LARGE}}$	169	0.0109	0.0317	+1.836	+0.350
	Average	285.57	0.0184	0.0535	+2.922	+0.600
ReTACRED (13,418)	PaLSTM	257	0.0192	0.0482	+5.440	+1.634
	CGCN	336	0.0250	0.0630	+7.103	+2.167
	BERT $_{\text{BASE}}$	248	0.0185	0.0465	+2.122	+0.585
	BERT $_{\text{LARGE}}$	195	0.0145	0.0366	+1.569	+0.442
	RoBERTa $_{\text{BASE}}$	245	0.0183	0.0459	+2.940	+0.895
	XLNet $_{\text{LARGE}}$	239	0.0178	0.0448	+1.235	+0.352
	RoBERTa $_{\text{LARGE}}$	205	0.0153	0.0384	+1.403	+0.264
	Average	246.43	0.0184	0.0462	+3.116	+0.906

6 Conclusion and Future Work

In this work, we introduced RelCheck, a post-validation framework designed to address low-confidence predictions in RE models. By combining Ontology-guided, which relies on automatically constructed schemas, and LLM-based validation through binary-choice prompting, RelCheck improves prediction precision while maintaining computational efficiency.

Experimental results demonstrate that RelCheck significantly improves precision, making it valuable for downstream tasks such as knowledge graph construction. Its cost-effectiveness enables deployment in resource-constrained environments. However, as it focuses on validating positive predictions without reclassifying errors, it slightly reduces recall. Careful threshold selection is therefore required to maintain a balance between precision and recall.

Future work will explore integrating error reclassification mechanisms to improve recall. Additionally, we will extend RelCheck to other NLP classification tasks, such as document classification and sentiment analysis. This research provides a balanced, cost-effective approach to combining LLMs with traditional models, leveraging their strengths while minimizing computational overhead.

A Appendix: Prompts Templates

Appendix: Template Descriptions

Template 1: Summarization-Based

Summarization-Based Template

With Entity Types: `The subject is [Entity1]. The object is [Entity2]. The type of Entity 1 is [Entity1Type]. The type of Entity 2 is [Entity2Type]. [Sentence]. Does the summary '[Entity1] [RelationshipSpan] [Entity2]' accurately and comprehensively capture the relationship between the subject and object based on the context provided? Answer YES or NO.` **Without Entity Types:** `The subject is [Entity1]. The object is [Entity2]. [Sentence]. Does the summary '[Entity1] [RelationshipSpan] [Entity2]' accurately and comprehensively capture the relationship between the subject and object based on the context provided? Answer YES or NO.`

Template 2: Entailment-Based

Entailment-Based Template

With Entity Types: `You are given a sentence and a hypothesis below. If the sentence entails the relationship described in the hypothesis, return 'YES'. Otherwise, return 'NO'. Sentence: [Sentence] Hypothesis: [Entity1] (Type: [Entity1Type]) [RelationshipSpan] [Entity2] (Type: [Entity2Type]).` **Without Entity Types:** `You are given a sentence and a hypothesis below. If the sentence entails the relationship described in the hypothesis, return 'YES'. Otherwise, return 'NO'. Sentence: [Sentence] Hypothesis: [Entity1] [RelationshipSpan] [Entity2].`

Template 3: Question-Answering Based

Question-Answering Based Template

With Entity Types: `Determine if the predicted relationship can be inferred from the given sentence. Sentence: [Sentence] Predicted relationship: [Entity1] (Type: [Entity1Type]) [RelationshipSpan] [Entity2] (Type: [Entity2Type]). This predicted relationship can be inferred from the given sentence? Answer YES or NO.` **Without Entity Types:** `Determine if the predicted relationship can be inferred from the given sentence. Sentence: [Sentence] Predicted relationship: [Entity1] [RelationshipSpan] [Entity2]. This predicted relationship can be inferred from the given sentence? Answer YES or NO.`

Template 4: Direct Relation Validation

Direct Relation Validation Template

With Entity Types: `Given a sentence and two entities within the sentence, confirm if the given predicted relationship between the entities is correct. Sentence: [Sentence] Entity 1: [Entity1] (Type: [Entity1Type]) Entity 2: [Entity2] (Type: [Entity2Type]). Predicted Relationship: [Relationship]. Answer YES or NO.` **Without Entity Types:** `Given a sentence and two entities within the sentence, confirm if the given predicted relationship between the entities is correct. Sentence: [Sentence] Entity 1: [Entity1] Entity 2: [Entity2]. Predicted Relationship: [Relationship]. Answer YES or NO.`

References

1. Achiam, J., et al.: Gpt-4 technical report. arXiv preprint arXiv:2303.08774 (2023)
2. Brown, T.B., et al.: Language models are few-shot learners. Adv. Neural. Inf. Process. Syst. **33**, 1877–1901 (2020)
3. Chowdhery, A., et al.: Palm: scaling language modeling with pathways. J. Mach. Learn. Res. **24**(240), 1–113 (2023)
4. Devlin, J., Chang, M.W., Lee, K., Toutanova, K.: BERT: pre-training of deep bidirectional transformers for language understanding. In: Burstein, J., Doran, C., Solorio, T. (eds.) Proceedings of the 2019 Conference of the North American Chapter of the Association for Computational Linguistics: Human Language

Technologies, vol. 1 (Long and Short Papers), pp. 4171–4186. Association for Computational Linguistics, Minneapolis, Minnesota (2019). https://doi.org/10.18653/v1/N19-1423
5. Guo, Z., Zhang, Y., Lu, W.: Attention guided graph convolutional networks for relation extraction. In: Korhonen, A., Traum, D., Màrquez, L. (eds.) Proceedings of the 57th Annual Meeting of the Association for Computational Linguistics, pp. 241–251. Association for Computational Linguistics, Florence, Italy (2019). https://doi.org/10.18653/v1/P19-1024
6. Hogan, W., Shang, J.: Entangled relations: leveraging NLI and meta-analysis to enhance biomedical relation extraction. arXiv preprint arXiv:2406.00226 (2024)
7. Jimenez Gutierrez, B., et al.: Thinking about GPT-3 in-context learning for biomedical IE? think again. In: Goldberg, Y., Kozareva, Z., Zhang, Y. (eds.) Findings of the Association for Computational Linguistics: EMNLP 2022, pp. 4497–4512. Association for Computational Linguistics, Abu Dhabi, United Arab Emirates (2022). https://doi.org/10.18653/v1/2022.findings-emnlp.329
8. Joshi, M., Chen, D., Liu, Y., Weld, D.S., Zettlemoyer, L., Levy, O.: SpanBERT: improving pre-training by representing and predicting spans. Trans. Assoc. Comput. Linguist. **8**, 64–77 (2020)
9. Lu, K., Hsu, I.H., Zhou, W., Ma, M.D., Chen, M.: Summarization as indirect supervision for relation extraction. In: Goldberg, Y., Kozareva, Z., Zhang, Y. (eds.) Findings of the Association for Computational Linguistics: EMNLP 2022. pp. 6575–6594. Association for Computational Linguistics, Abu Dhabi, United Arab Emirates (2022). https://doi.org/10.18653/v1/2022.findings-emnlp.490
10. Miwa, M., Bansal, M.: End-to-end relation extraction using LSTMs on sequences and tree structures. In: Erk, K., Smith, N.A. (eds.) Proceedings of the 54th Annual Meeting of the Association for Computational Linguistics (Volume 1: Long Papers), pp. 1105–1116. Association for Computational Linguistics, Berlin, Germany (2016). https://doi.org/10.18653/v1/P16-1105
11. Sainz, O., Lopez de Lacalle, O., Labaka, G., Barrena, A., Agirre, E.: Label verbalization and entailment for effective zero and few-shot relation extraction. In: Moens, M.F., Huang, X., Specia, L., Yih, S.W.t. (eds.) Proceedings of the 2021 Conference on Empirical Methods in Natural Language Processing, pp. 1199–1212. Association for Computational Linguistics, Online and Punta Cana, Dominican Republic (2021). https://doi.org/10.18653/v1/2021.emnlp-main.92
12. Soares, L.B., FitzGerald, N., Ling, J., Kwiatkowski, T.: Matching the blanks: distributional similarity for relation learning. In: Proceedings of the 57th Annual Meeting of the Association for Computational Linguistics, pp. 2895–2905 (2019)
13. Stoica, G., Platanios, E.A., Póczos, B.: Re-TACRED: addressing shortcomings of the tacred dataset. In: Proceedings of the AAAI Conference on Artificial Intelligence, vol. 35, pp. 13843–13850 (2021)
14. Touvron, H., et al.: Llama: open and efficient foundation language models. ArXiv **abs/2302.13971** (2023). https://api.semanticscholar.org/CorpusID:257219404
15. Uschold, M., Gruninger, M.: Ontologies: principles, methods and applications. Knowl. Eng. Rev. **11**(2), 93–136 (1996)
16. Wadhwa, S., Amir, S., Wallace, B.: Revisiting relation extraction in the era of large language models. In: Rogers, A., Boyd-Graber, J., Okazaki, N. (eds.) Proceedings of the 61st Annual Meeting of the Association for Computational Linguistics, vol. 1: Long Papers, pp. 15566–15589. Association for Computational Linguistics, Toronto, Canada (2023). https://doi.org/10.18653/v1/2023.acl-long.868
17. Wang, H., Qin, K., Zakari, R.Y., Lu, G., Yin, J.: Deep neural network-based relation extraction: an overview. Neural Comput. Appl. 1–21 (2022)

18. Wu, S., He, Y., Wang, K., Liu, K., Zhao, J.: Enriching pre-trained language model with entity information for relation classification. Proceedings of the 57th Annual Meeting of the Association for Computational Linguistics, pp. 2363–2368 (2019)
19. Yang, Z.: Xlnet: generalized autoregressive pretraining for language understanding. arXiv preprint arXiv:1906.08237 (2019)
20. Zeng, D., Liu, K., Chen, Y., Zhao, J.: Distant supervision for relation extraction via piecewise convolutional neural networks. In: Màrquez, L., Callison-Burch, C., Su, J. (eds.) Proceedings of the 2015 Conference on Empirical Methods in Natural Language Processing, pp. 1753–1762. Association for Computational Linguistics, Lisbon, Portugal (2015). https://doi.org/10.18653/v1/D15-1203
21. Zhang, K., Gutiérrez, B.J., Su, Y.: Aligning instruction tasks unlocks large language models as zero-shot relation extractors. arXiv preprint arXiv:2305.11159 (2023)
22. Zhang, S., et al.: OPT: open pre-trained transformer language models (2022)
23. Zhang, Y., Qi, P., Manning, C.D.: Graph convolution over pruned dependency trees improves relation extraction. arXiv preprint arXiv:1809.10185 (2018)
24. Zhang, Y., Zhong, V., Chen, D., Angeli, G., Manning, C.D.: Position-aware attention and supervised data improve slot filling. In: Conference on Empirical Methods in Natural Language Processing (2017)
25. Zhou, W., Chen, M.: An improved baseline for sentence-level relation extraction. In: He, Y., Ji, H., Li, S., Liu, Y., Chang, C.H. (eds.) Proceedings of the 2nd Conference of the Asia-Pacific Chapter of the Association for Computational Linguistics and the 12th International Joint Conference on Natural Language Processingm, vol. 2: Short Papers, pp. 161–168. Association for Computational Linguistics (2022). https://aclanthology.org/2022.aacl-short.21

Analyzing the Influence of Knowledge Graph Information on Relation Extraction

Cedric Möller[1(✉)] and Ricardo Usbeck[2]

[1] Department of Informatics, Semantic Systems, Universität Hamburg, Hamburg, Germany
cedric.moeller@uni-hamburg.de

[2] Institute for Information Systems, Artificial Intelligence and Explainability, Leuphana Universität Lüneburg, Lüneburg, Germany
ricardo.usbeck@leuphana.de

Abstract. We examine the impact of incorporating knowledge graph information on the performance of relation extraction models across a range of datasets. Our hypothesis is that the positions of entities within a knowledge graph provide important insights for relation extraction tasks. We conduct experiments on multiple datasets, each varying in the number of relations, training examples, and underlying knowledge graphs. Our results demonstrate that integrating knowledge graph information significantly enhances performance, especially when dealing with an imbalance in the number of training examples for each relation. We evaluate the contribution of knowledge graph-based features by combining established relation extraction methods with graph-aware Neural Bellman-Ford networks. These features are tested in both supervised and zero-shot settings, demonstrating consistent performance improvements across various datasets.

1 Introduction

Populating an existing knowledge graph (KG) with new information is an essential challenge. An integral subtask for this is relation extraction (RE). RE is the task of identifying the expressed relation between two entities. The focus of this subtask usually resides on a relation expressed in a sentence, document, or between multiple documents. In contrast to that, link prediction [35] infers potential relations based on the structure of a knowledge graph [11].

In this paper, our goal is to combine both ways of tackling the task, based on the text and based on the graph, in a single framework. For that, we incorporate a Neural Bellman Ford (NBF) network [48] into the RE process, allowing us to include graph-based information while being generalizable to new entities in the knowledge graph. This leads to a model that jointly considers information in the KG and the document.

We study the impact of the KG data on several established datasets, focusing on supervised and zero-shot scenarios. To tackle the supervised and zero-shot

E. Curry et al. (Eds.): ESWC 2025, LNCS 15718, pp. 460–480, 2025.
https://doi.org/10.1007/978-3-031-94575-5_25

scenarios, we use two different versions of the NBF network, one that assumes that all encountered relations in the graph are known before, and another that creates the relation representations on the fly. As we focus solely on RE, we assume that the actual entity mentions in the text are pre-annotated. Therefore, the task is to determine whether a relationship exists between a pair of entities and, if so, identify which specific relationship holds.

Unlike existing methods [1,9], our approach does not rely on learned entity embeddings, allowing it to generalize to new entities. Furthermore, our developed post-prediction mechanism shows further improvements in document-level RE. Additionally, we introduce a method that is also suitable for zero-shot settings.

Our contributions are an analysis of the impact of knowledge graph information in the following relation extraction settings:

1. **Supervised Setting:** We investigate how incorporating knowledge graph information enhances model performance. For that, we modify the NBF network to be usable in the relation extraction problem. This improves accuracy by providing richer contextual information and better resolving ambiguities.
2. **Zero-shot Setting:** We explore the role of knowledge graphs in zero-shot relation extraction, demonstrating their potential to enable models to generalize to unseen categories by providing semantic background and auxiliary data for informed predictions.

Code: Our code is available at
https://github.com/semantic-systems/kg-based-re.

2 Method

2.1 Problem Definition

We target the RE problem. Given a text and a pair of annotated entities within the text, the goal is to identify the relation expressed between the two entities out of the set of all relations R. Depending on the input text, multiple entities might be mentioned in the text. For each pair of entities multiple relations are potentially expressed. We refer to this as document-level RE. If there are only two entities between which the relation needs to be extracted, we refer to it to sentence-level RE.

In a **supervised setting**, the set of potential relations is known during training. It is possible that no relation is expressed between two entities, which can also be interpreted as an additional `no_relation` relation. When one considers the **zero-shot setting**, the assumption is that the set of relations seen during training differs from the set of relations encountered during evaluation. To solve this task, the method has to generalize to unseen relations.

In addition to textual information, we assume the availability of a knowledge graph. A knowledge graph is a graph consisting of nodes and edges. Each node corresponds to a specific entity, either depicting some individual or concept. The edges between nodes are equipped with a relation that denotes some relationship between two nodes. Two nodes linked by a relation are commonly also denoted as a triple. Formally, we define the graph as $G = (V, E)$, where V is the set of nodes and $E \subseteq V \times V \times R_G$ is the set of edges with R_G being the edges existing in the graph. R_G does not need to overlap with R. For example, an edge equipped with the `married_to` relation between the subject `Barack Obama` and the object `Michelle Obama` expresses that both those persons are married to each other. Multiple edges may hold between two nodes. In this work, we assume that the corresponding nodes of the entities marked in the text are known and can be used to incorporate KG-internal information.

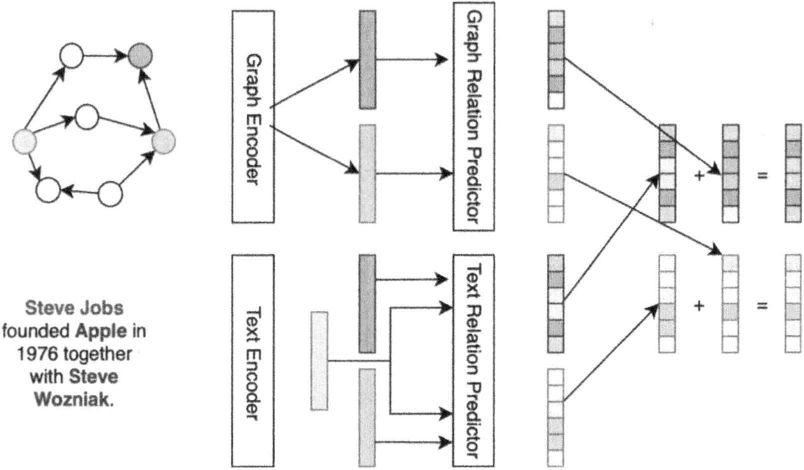

Fig. 1. Model architecture: The figure illustrates relation prediction between a subject (blue) and two objects (red and green). Text and graph are encoded to predict relations involving Steve Jobs. The graph predictor identifies likely graph-based relations, while the text predictor identifies text-expressed relations, leading to two vectors per entity. Both vectors are combined, giving the final predictions. Identically colored mentions and nodes represent the same entity, and the predictors output a relation distribution, operating in either supervised or zero-shot modes. (Color figure online)

Our method consists of two main components (see Fig. 1): the textual module and the graph module. The textual module purely works on the textual input, while the graph module considers the potential relation between two mentioned entities as expressed in the graph.

2.2 Textual Module

Supervised. The input text consists of the marked entities where each entity mention is surrounded by $-signs. The textual encoder follows a well-established architecture as proposed by Zhou et al. [46]. Initially, the textual input is encoded using an encoder-only model

$$H = \text{Encoder}(X), \quad H = [h_1, h_2, \ldots, h_N], \quad h_i \in \mathbb{R}^d.$$

Then, for each of the marked entity mentions m_j, the encoded token of its left-side \$-sign is taken, denoted as $h_{m_j} = h_{p_j}$ where p_j is the position of the sign. Additionally, the attention scores for the \$ to all other tokens throughout all layers are averaged over all layers and normalized

$$\bar{a}_{p_j} = \frac{1}{L} \sum_{\ell=1}^{L} a_{p_j}^{(\ell)}$$

with $a_{p_j}^{(\ell)} = A^{(\ell)}[p_j, :]$. As multiple entity mentions M_{entity} might exist in a document for a single *entity*, the attention scores are averaged over all entity-specific mentions

$$\tilde{a}_{\text{entity}} = \frac{1}{|M_{\text{entity}}|} \sum_{j=1}^{|M|} \bar{a}_{p_j}$$

and the \$-sign encodings are pooled by applying the `logsumexp` operation to get the entity-encoding

$$\tilde{h}_{\text{entity}} = \texttt{logsumexp}(\{h_{m_1}, h_{m_2}, \ldots, h_{m_{|M_{\text{entity}}|}}\}).$$

This gives us two key representations per entity: the attention $\tilde{a}_{\text{entity}}$ to all other tokens and the entity's \$-encoding $\tilde{h}_{\text{entity}}$, which are used in the next steps. For each pair of entities, we compute a combined representation by first point-wise multiplying the attention scores of both entities $\tilde{a}_k \odot \tilde{a}_l$ and normalizing them, giving $\tilde{a}_{k,l}$. Then, we compute an attention-based pair representation by performing a weighted sum over all encoded tokens, resulting in

$$c = \sum_{i=1}^{N} \tilde{a}_{k,l}[i] \cdot h_i.$$

Finally, the subject encoding is calculated by

$$s_k = W_s \text{concat}\left(\tilde{h}_k, c\right) + b_s$$

where W_s is a weight matrix and b_s a bias. The same is done to get an object encoding o_l. Both are used to predict the scores for all relations by applying a bilinear mapping

$$p_t = s_k^\top W o_l \in \mathbb{R}^R$$

where $W \in \mathbb{R}^{d \times R \times d}$ and R is the number of relations.

Zero-Shot. For zero-shot learning, we employ an RE approach via multiple-choice classification [13]. For each document s, we concatenate it with the label l_r and description d_r of a relation r

$$x_r = \text{concat}(s, l_r, d_r).$$

The concatenated input x_r is then fed into an encoder-only model to obtain the token encoding

$$H_r = \text{Encoder}(x_r), \quad H_r = [h_1^{(r)}, h_2^{(r)}, \ldots, h_N^{(r)}].$$

We extract the encoding of the first token $h_1^{(r)}$ and project it using a two-layer mapping to a score

$$\text{score}(r) = W_2\,\sigma(W_1 h_1^{(r)} + b_1) + b_2$$

where W_1 and W_2 are weight matrices, b_1 and b_2 are biases, and σ is the tanh function.

This process is repeated for all relations $r \in R$ to compute their respective predictions, giving $p_t \in \mathbb{R}^R$.

2.3 Graph Module

The graph module uses the background KG to predict likely relations. The exception is when the Post-Prediction step is also applied (see Sect. 2.5), where both the background KG and text-based triples are used.

Supervised. For the graph-encoder, we follow a modified version of the Neural Bellmann-Ford (NBF) graph neural network [48]. Originally designed for link prediction, this method takes a subject entity and a relation as input. Given our dataset's large number of relations, running this model for each relation introduces significant computational overhead, which was not feasible with our available resources.

To address this, we eliminate the need to specify an input relation. Instead, we predict all possible output relations based on the final representation of each node.

Initially, each node in the graph is initialized with a zero vector, except for a designated start node, which is assigned a specific vector g_{start}:

$$g_v^{(0)} = \begin{cases} g_{\text{start}}, & \text{if } v = v_{\text{start}}, \\ \mathbf{0}, & \text{otherwise.} \end{cases}$$

The graph undergoes T iterations of message passing, where each edge (u, v) is represented by its direction and associated relation r. The message passing consists of three main steps for each node v:

1. Propagation: Information is propagated along the edges using the DistMult operation [38]

$$m_{u \to v}^{(t)} = g_u^{(t-1)} \odot r_{(u,v)}$$

where $m_{u \to v}^{(t)}$ is the message from node u to node v, $r_{(u,v)}$ is a relation-specific representation and $\odot$ denotes element-wise multiplication.

2. Aggregation: Incoming messages are aggregated for each node v using Principal Neighbourhood Aggregation (PNA) [5]

$$m_v^{(t)} = \text{Aggregate}\left(\{m_{u \to v}^{(t)} \mid u \in \mathcal{N}(v)\}\right)$$

where $\mathcal{N}(v)$ denotes the neighbors of v.

3. Update: The node's representation is updated via a linear projection and a non-linear activation with

$$g_v^{(t)} = \mu(W^{(t)} m_v^{(t)} + b^{(t)})$$

where $W^{(t)}$ and $b^{(t)}$ are the weights and biases of the linear transformation at iteration t and μ is the ReLU activation function.

After T iterations, the message passing is stopped, and the representation of each node is retrieved. This representation captures the relation of any node with respect to the start node.

We initialize the start node as the node corresponding to the subject entity and retrieve the representations of each object entity node g_l. In contrast to the original method, this representation g_l is then passed through a two layer network to predict the scores for all relations instead of a single relation

$$p_g = W_4\, \mu(W_3 g_l + b_3) + b_4 \in \mathbb{R}^R$$

where W_3, W_4, b_3, and b_4 are weights and biases.

Zero-Shot. For zero-shot learning, we rely on the zero-shot variant of the NBF network, denoted as ULTRA [7]. ULTRA consists of two components: a relation graph encoder and an entity graph encoder, both implemented as NBF networks.

The relation graph is constructed with nodes representing relations and edges connecting them if the subject/object of one relation is the subject/object of another. A designated relation is set as the start, and the relation graph encoder produces a representation for each relation node h_r

$$h_r = \text{RelationGraphEncoder}(r_{\text{start}})$$

where $r \in R_G$.

These relation representations h_r are used in the entity graph encoder, which outputs entity representations h_l conditioned on a start node and the relation. These are then fed into a multi-layer projection to compute a single score

$$\text{score}(r) = W_6\, \mu(W_5 h_{\text{entity}} + b_5) + b_6$$

where W_5, W_6, b_5, and b_6 are weights and biases, and μ is the ReLU activation function.

To classify all relations, ULTRA is run for each relation in the set R, and the predictions are concatenated:

$$p_g = [\text{score}(r_1), \text{score}(r_2), \ldots, \text{score}(r_R)] \in \mathbb{R}^R$$

2.4 Final Prediction

To compute the final prediction, we integrate the logits from both sources by accumulating them. Let α and β represent the respective weights for each source. The final logits are computed as

$$p = p_t + p_g \in \mathbb{R}^R$$

where $p_t \in \mathbb{R}^R$ and $p_g \in \mathbb{R}^R$ are the logits from the two predictive models (either for the supervised or zero-shot setting), and R represents the number of relations.

2.5 Post-prediction

In document-level RE, rule-learning techniques are frequently employed to enhance the predictive capability of models [6,20,24]. These approaches leverage an initial set of relations generated by a text-only model and perform reasoning using learned rules to infer additional relations.

We adopt a similar methodology for document-level relation extraction tasks. Specifically, starting with the initial predictions produced by the textual component, we enrich the underlying graph. This enrichment involves adding new edges to the graph between each pair of nodes (v_i, v_j) whenever a relation is identified between the corresponding entities in the input text.

R_G denotes the set of predefined relation types in the initial graph. To distinguish between a priori known relations and those inferred through reasoning, we define an additional set of relation types R' corresponding exclusively to the predicted relations. The graph $G = (V, E)$ is extended by introducing new edges:

$$E' = \{(v_i, v_j, r') \mid r' \in R', v_i, v_j \in V, \text{ and } r' \text{ is predicted for } (v_i, v_j)\}.$$

The extended graph is then represented as $G' = (V, E \cup E')$.

By explicitly encoding these predicted relations as a separate set R', the model can effectively distinguish between the original relations R and the newly inferred ones, thus enabling more nuanced reasoning and relation classification. Post-Prediction is only used for document-level RE. See Fig. 2 for an overview.

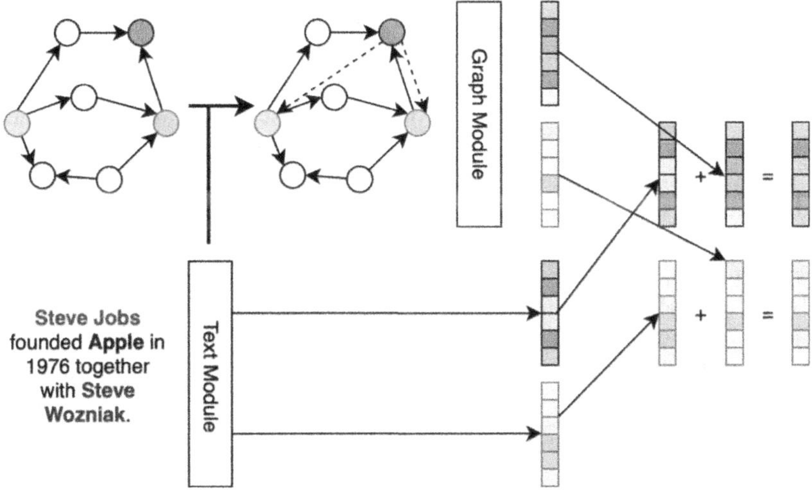

Fig. 2. The figure illustrates the post-prediction mechanism. The input text is initially processed through the textual module to identify relations. These identified relations are then incorporated into the input graph for the graph module. Using the updated graph and the initial textual predictions, the final predictions are generated.

2.6 Losses

For datasets with more than two entities per example, we use the HingeABL loss [34], which is designed to more-gracefully handle the problem of imbalance in multi-class classification optimization in contrast to a regular binary cross-entropy loss. For sentence-level RE datasets, we use the cross-entropy loss to optimize the models.

3 Evaluation

3.1 Setup

We evaluate the influence of the graph information on several datasets. As the encoder models, we used RoBERTa-large for Re-DocRED, BERT-base for DWIE, and BioBERT for BioRel. For Wiki-ZSL and FewRel, we relied on BERT-base. We chose those to be comparable to other existing methods.

Supervised Datasets. As supervised datasets, we use the Re-DocRED [28], DWIE [40] and BioRel [36] dataset. We chose those datasets as they are available with annotated entity mentions and have been evaluated by past methods that utilised KG information. An overview of them can be found in Table 1.

For Re-DocRED and DWIE, we rely on Wikidata as the corresponding knowledge graph. For BioRel, we rely on two available ontologies: Medline [39] and NCIt [12].

We gathered each ontology and linked the entity mentions to the corresponding nodes. On Re-DocRED, 83% of mentions matched a Wikidata node, while on DWIE, only 53% did. The rest were literals or unlinked during dataset creation. On BioRel, 82% of mentions were covered; the rest were unidentifiable in the available ontologies. DWIE and Re-DocRED are document-level RE datasets, with DWIE containing longer documents. BioRel is a sentence-level RE dataset.

Our main metric is F1. For Re-DocRED and DWIE, we use micro F1, where true positives are correctly classified relations, false negatives are relations missed between entities, and false positives are incorrect relation predictions. For Re-DocRED, we also report Ign-F1, which excludes triples seen during training.

For BioRel, we use Macro-F1, averaging F1 over all relation classes.

Each method is trained three times, and we report the averaged metrics.

Table 1. Supervised RE datasets.

Dataset Name	# Documents	# Relations	# Mentions	# Triples
DWIE	802	65	43,373	317,204
Re-DocRED	4053	97	132,375	120,539
BioRel	533,560	125	1,067,120	533,560

Zero-Shot Datasets. For the zero-shot evaluations we rely on the popular FewRel [4,8] and Wiki-ZSL [2] datasets (see Table 2 for an overview). These are sentence-level RE datasets. Both contain annotated entity mentions linked to Wikidata [33]. The datasets are prepared in three versions, with 5, 10, or 15 test relations. Furthermore, for each of the versions, the full dataset is resampled five times. That means, the model is evaluated on five different runs for each version to compensate for the high variance due to the small number of test relations. All mentioned entities are linked to Wikidata. Macro F1-measure is calculated to evaluate the performance of methods.

Graph. For each entity in a document, we gather its two-hop neighborhood in the knowledge graph, limiting edges to 100 per hop, resulting in up to 10,000 entities per neighborhood. This enables exploration of four-hop connections. We remove nodes appearing in only one triple unless they belong to the original entity set, helping manage datasets with up to 200,000 nodes per subgraph.

Table 2. Zero-shot RE datasets.

Dataset Name	# Documents	# Relations	# Mentions	# Triples
FewRel	70,000	100	140,000	70,000
Wiki-ZSL	94,383	113	132,375	188,766

For sentence-level RE, we remove direct triples between subjects and objects to prevent trivial relations, particularly in distantly-supervised datasets.[1] There, 80–90% of triples overlap between graph and text, whereas in document-level RE, less than 45% do. In the zero-shot setting, we sample 1000 triples per relation, extract two-hop neighborhoods, and construct the relation graph, filtering noise by keeping only relation-relation edges found in at least 10% of sampled neighborhoods.

The GNN runs four layers of message passing to utilize the max-hop distance between entities. More hops did not improve performance.

Methods. In our analysis on DWIE, we evaluate a range of document-level RE methods, referencing works such as [31,32,37]. We specifically focus on comparing the performance of the top models utilizing rule-learning techniques, as highlighted in [6,20,24]. Additionally, we incorporate two methods that integrate knowledge graph information, as explored by [1,9].

On ReDocRED, we compare against several BERT-based and RoBERTa-based methods [15,27,37,43,46] again including two that incorporate knowledge graph information [1,9] as well.

On BioREL we compare against the best-performing non-biomedical [1,21, 26,47] and biomedical RE methods [10] as reported by Jain et al. [10].

On Wiki-ZSL and Fewrel, we compare against several zero-shot methods, ranging from entailment-based methods [23], encoding-based methods [2,29,30, 44], generative methods [4], and discriminative prompting methods [13,14]

In our supervised experiments, we refer to our method as ATLOP-KG when incorporating the knowledge graph component into the ATLOP architecture, and as ATLOP-KG-PP when including both the KG component and the post-prediction module. Additionally, we present our baseline results using the ATLOP architecture with the HingeABL loss, labeled as ATLOP-Hinge. Furthermore, we denote a method solely using the graph component as NBF.

For the zero-shot experiments, we designate our approach as MC-BERT-KG, and we also present results for MC-BERT with added descriptions, referred to as MC-BERT w/ descriptions.

Hyperparameters. All training and inference runs were performed on a single NVIDIA A6000 GPU. We used a batch size of 8 for each supervised training run and a batch size of 16 for the unsupervised runs. We use learning rates of $3e-5$ for the text encoders and of $1e-4$ for the graph encoders. We trained each method by relying on early stopping based on the validation F1-measure.

3.2 Results

Supervised. Our method outperforms recent state-of-the-art methods on the DWIE dataset by 1.5 F1-measure points, as illustrated in Table 3. Notably, even

[1] Distant supervision annotates documents using the KG, so keeping these triples could let the model simply verify existing relations.

without graph information, our text-based RE architecture performs comparably to existing state-of-the-art models. The inclusion of graph information further enhances performance, which is remarkable given our reliance on the same architectural framework as the state-of-the-art. While including KG information already leads to a larger improvement, using the post-prediction step also improves it even more. We hypothesize that our adaptations for handling long documents, typical in DWIE, contribute significantly to this improvement. This adaptation involves splitting input documents based on the encoder's maximum token length while introducing a stride to maintain contextual coherence. This approach seems to compensate for adjustments potentially overlooked in previous state-of-the-art methods. Without that, our performance diminishes to the one as reported in the paper by Qi et al. [20].

Table 3. F1 Scores on DWIE.

Model	F1
DRN*GloVe [37]	56.04
RESIDE [31]	66.78
RECON [1]	66.94
KB-Graph [32]	66.89
ATLOP [46]	75.13
DocRE-CLiP [9]	67.10
LogicRE-ATLOP [24]	75.67
MILR-ATLOP [6]	76.51
JMLR-ATLOP [20]	77.85
ATLOP-Hinge (Ours)	77.35
NBF (Ours)	44.27
ATLOP-KG (Ours)	<u>78.50</u>
ATLOP-KG-PP (Ours)	**79.46**

Conversely, our approach does not achieve state-of-the-art performance on the Re-DocRED dataset (Table 4). Although incorporating KG information enhances performance over the text-only model, the post-prediction step contributes minimally. This outcome may stem from the typically smaller document sizes in Re-DocRED, reducing the advantage offered by the KG enhancements. Additionally, our lack of pre-training and evidence fusion, which distinguish the leading approaches [15], might explain this discrepancy. Nevertheless, we chose to forego these complex stages to minimize computational overhead, focusing instead on exploring the fundamental effects of graph integration. The unexpectedly high performance of DocRE-CLiP [9] is notable, particularly given their claims regarding the significant impact of integrating graph information. However, upon reviewing their paper and accompanying code, we were unable to

Table 4. F1 Scores on ReDocRED.

Model	Ign-F1	F1
ATLOP [46]	76.82	77.56
DRN* [37]	74.3	75.6
KG-DocRE [27]	80.32	81.04
DocuNet [43]	78.52	79.64
DREEAM [15]	80.39	81.44
DocRE-CLiP [9]	**80.57**	**81.55**
ATLOP-Hinge (Ours)	76.97	78.07
NBF (Ours)	46.94	49.58
ATLOP-KG (Ours)	77.70	78.70
ATLOP-KG-PP (Ours)	77.80	78.83

Table 5. Accuracy and F1 Scores on BioREL.

Model	Accuracy	Macro F1 Score
GPGNN [48]	85	84.00
ContextAware [26]	89	87.00
T5 [21]	88	86.00
RECON [1]	89.6	86.00
DocRE-CLiP [10]	92	90.00
NBF	75.35	74.74
ATLOP-Hinge (Ours)	96.69	95.71
ATLOP-KG (Ours)	**97.93**	**96.90**

ascertain the specific methods they employed to incorporate this information. Additionally, there is an indication that their training process encompasses entities from not only the training set but also the test and validation sets.

While further inspecting the performance of our method on Re-DocRED, we also saw that the biggest difficulty on the document-level RE datasets was to identify whether any relation is expressed between two entities, not which. Only six percent of errors on Re-DocRED stem from the problem of disambiguating between different relations. We assume that graph information is less helpful to decide whether any relation is expressed as this is more a problem related to the textual RE component. An additional cause might be that the number of relations for both datasets is rather small being fewer than 100. Therefore, the ambiguity between different relations might be rather low.

On BioRel (see Table 5), our method performs the best. The textual RE component has again a large impact; however, the inclusion of graph information leads to a performance far beyond the previous SOTA. We suspect that the impact is higher on this dataset as there are more relations to disambiguate in comparison to DWIE and Re-DocRED.

Zero-Shot. In zero-shot RE, the influence of graph information exhibits variation across tasks. As shown in Table 6, graph information significantly enhances performance in the most challenging scenario ($m = 15$) for the Wiki-ZSL dataset. While its impact on FewRel is comparatively modest, our method still surpasses the previous state-of-the-art by two F1-measure points. This indicates that graph information is particularly beneficial when data presents higher complexity or ambiguity, whereas its advantage diminishes in simpler scenarios such as $m = 5$ or $m = 10$. In almost every setting, except for FewRel with $m = 5$, incorporating graph information results in a substantial performance enhancement compared to not utilizing it, as evident when comparing to MC-BERT w/ descriptions. It is important to note that integrating graph information is independent of other model improvements, meaning that the current state-of-the-art methods across several datasets could potentially be further enhanced by incorporating this additional information as well.

3.3 Ablation Studies

In addition to the ablation studies already included in the previous section, we further investigate our method on the $m = 15$ split of the Wiki-ZSL dataset. Specifically, we analyse the influence of the number of hops and keeping the links between subject and object entities.

Analyzing the ablation results for using only the graph model (Table 7), we observe competitive performance with other state-of-the-art approaches, indicating that the KG context provides substantial cues to discern the correct relation, especially in scenarios with a clear single correct relation. However, augmenting this with textual information results in a substantial performance increase of approximately 12 F1-measure points. This underscores the complementary nature of text and graph information, highlighting the importance of leveraging both to maximize RE efficacy. One advantage of the Wiki-ZSL dataset is that a relation is consistently expressed between any two entities. If this were not the case, the model would have to rely more heavily on textual context, making the graph information less dependable. This is evident in the significantly poorer performance of KG-only methods on document-level RE datasets, as reflected in Tables 3 and 4.

If we use fewer than 4 hops, we see a diminishing performance. Surprisingly, only using 2-hops outperforms using at maximum 3-hops of path lengths. As the 1-hop distance is effectively not using any knowledge graph information due to us filtering the single hops out, its performance reduces to the text-only model. Interestingly, including direct triples during training actually reduces performance. We attribute this to the model's tendency to rely on straightforward single-hop information rather than considering the broader context of the surrounding neighborhood, which is not always the optimal approach.

Table 6. Results on FewRel and Wiki-ZSL.

m	Model	Wiki-ZSL			FewRel		
		P	R	F1	P	R	F1
5	CIM [23]	49.63	48.81	49.22	58.05	61.92	59.92
	ZS-BERT [2]	71.54	72.39	71.96	76.96	78.86	77.90
	Tran et al. (2022) [30]	87.48	77.50	82.19	87.11	86.29	86.69
	RelationPrompt NG [4]	51.78	46.76	48.93	72.36	58.61	64.57
	RelationPrompt [4]	70.66	83.75	76.63	90.15	88.50	89.30
	RE-Matching [44]	78.19	78.41	78.30	92.82	92.34	92.58
	DSP-ZRSC [14]	<u>94.1</u>	77.1	84.8	93.4	92.5	92.9
	Tran et al. (2023) [29]	**94.50**	**96.48**	**95.46**	**96.36**	**96.68**	**96.51**
	MC-BERT [13]	80.28	84.03	82.11	90.82	90.13	90.47
	MC-BERT w/ descriptions (Ours)	85.00	84.41	84.68	93.33	92.50	92.91
	MC-BERT-KG (Ours)	88.89	89.46	88.92	88.39	91.37	92.02
10	CIM	46.54	47.90	45.57	47.39	49.11	48.23
	ZS-BERT	60.51	60.98	60.74	56.92	57.59	57.25
	Tran et al. (2022)	71.59	64.69	67.94	64.41	62.61	63.50
	RelationPrompt NG	54.87	36.52	43.80	66.47	48.28	55.61
	RelationPrompt	68.51	74.76	71.50	80.33	79.62	79.96
	RE-Matching	74.39	73.54	73.96	83.21	82.64	82.93
	DSP-ZRSC	80.0	74.0	76.9	80.7	**88.0**	84.2
	Tran et al. (2023)	**85.43**	**88.14**	**86.74**	81.13	82.24	81.68
	MC-BERT	72.81	73.96	73.38	86.57	<u>85.27</u>	85.92
	MC-BERT w/ descriptions (Ours)	74.89	76.05	75.46	85.16	83.36	84.24
	MC-BERT-KG (Ours)	<u>81.72</u>	<u>80.52</u>	<u>81.10</u>	**88.63**	84.80	**86.63**
15	CIM	29.17	30.58	29.86	31.83	33.06	32.43
	ZS-BERT	34.12	34.38	34.25	35.54	38.19	36.82
	Tran et al. (2022)	38.37	36.05	37.17	43.96	39.11	41.36
	RelationPrompt NG	54.45	29.43	37.45	66.49	40.05	49.38
	RelationPrompt	63.69	<u>67.93</u>	65.74	74.33	72.51	73.40
	RE-Matching	67.31	67.33	67.32	73.80	73.52	73.66
	DSP-ZRSC	<u>77.5</u>	64.4	<u>70.4</u>	<u>82.9</u>	78.1	<u>80.4</u>
	Tran et al. (2023)	64.68	65.01	65.30	66.44	69.29	67.82
	MC-BERT	65.71	67.11	66.40	80.71	<u>79.84</u>	80.27
	MC-BERT w/ desc (Ours)	68.53	69.81	69.16	79.22	78.19	78.69
	MC-BERT-KG (Ours)	**79.28**	**76.95**	**78.07**	**84.46**	**80.90**	**82.64**

Table 7. Ablation on Wiki-ZSL $m = 15$.

Model	P	R	F1
MC-BERT-KG - graph only	69.63	66.16	67.78
MC-BERT-KG - 1 - hop	70.00	71.06	70.05
MC-BERT-KG - 2 - hop	76.74	75.95	76.25
MC-BERT-KG - 3 - hop	75.44	75.46	75.44
MC-BERT-KG - with direct triples	76.95	75.57	76.25
MC-BERT-KG	**79.28**	**76.95**	**78.07**

4 Related Work

Regular RE is usually tackled as a classification problem. The input text is encoded, and a classification head is attached. To encode text, CNNs [41], RNNs [16] or transformers [45] are usually employed. Recently, pre-trained models have been extensively used which are fine-tuned on the RE task [43,46].

Document-level RE is tackled mostly in two different ways: either by improving the capabilities of pre-trained language models (PLMs) to identify expressed long-range relations [43,46] or by representing the text information in a more structured way by either modeling the document as a graph [37,42] or by learning additional reasoning rules [6,20,24]. Our method is connected to both as we combine the SOTA-performing models relying on PLMs with the graph representation using graph neural networks.

Regarding zero-shot RE, representation-learning-based methods [2,29,30,44] usually try to embed textual and relational information in the same vector space. The relational information, such as labels or descriptions, is transformed into a representation of the relation. Representations are learned such that the representation of the true relation resides close to a representation of the text in the vector space, while the false relation representations are pushed further away. Recently, generative language models have been increasingly utilized for the task [3,4,14,18]. Here, the model is prompted with the input text and information on the potential relations. The model is then fine-tuned to either generate the relation as expressed in the input text or a full triple consisting of the two entities and the relation. Some methods model the problem as a textual entailment problem [19,22,25]. The method by Lan et al. [13] models relation classification as a multiple-choice problem where the text is encoded with relation information and a score is calculated. This is done for all relations, and the relation with the highest score is taken. As this method can be equipped with knowledge graph information, we extend this method in our zero-shot RE experiments.

RE under the use of knowledge graph information is an underexplored area. Recently, there have been only a few methods investigating this problem. Most methods either incorporate only one-hop information of entities or rely on trained static representations of entities making the generalization to unseen entities difficult [1,9,17,31,32]. Others linearize the underlying graph information while not considering the structural information [10]. In contrast, our method can generalize to new entities and even relations while considering multi-hop information.

5 Conclusion and Future Work

We showed that incorporating graph information consistently improves RE, particularly when textual components are undertrained due to data scarcity. However, its benefit diminishes in datasets where the main challenge is determining whether a relation exists, such as document-level RE datasets. Thus, this inclusion is especially valuable in zero-shot settings.

A post-prediction step that integrates relations predicted via text also enhances document-level RE, particularly for very long documents.

Future work will focus on improving efficiency. The current subgraph sampling introduces many irrelevant nodes; a more refined strategy would mitigate this. Scalability is another challenge in zero-shot settings, as GNNs must run separately for each relation, causing high computational overhead. Addressing this is crucial for both RE and link prediction. Lastly, the method assumes an existing path between entities, which is reasonable for link prediction. However, leveraging graph information beyond direct paths could be valuable when textual evidence is available.

We omitted LLMs for comparability with state-of-the-art methods. Our approach remains orthogonal to improvements from fine-tuned LLMs, which could also integrate graph components.

Acknowledgments. This project was supported by the Hub of Computing and Data Science (HCDS) of Hamburg University within the Cross-Disciplinary Lab program. Additionally, support was provided by the Ministry of Research and Education within the SifoLIFE project "RESCUE-MATE: Dynamische Lageerstellung und Unterstützung für Rettungskräfte in komplexen Krisensituationen mittels Datenfusion und intelligenten Drohnenschwärmen" (FKZ 13N16836). ChatGPT was utilized to improve the readability of certain sections of this text.

References

1. Bastos, A., et al.: RECON: relation extraction using knowledge graph context in a graph neural network. In: Leskovec, J., Grobelnik, M., Najork, M., Tang, J., Zia, L., (eds.) WWW 202121: The Web Conference 2021, Virtual Event / Ljubljana, Slovenia, April 19-23, 2021, p[p. 1673–1685. ACM/IW3C2 (2021)
2. Chen, C-Y., Li, C-T.: ZS-BERT: towards zero-shot relation extraction with attribute representation learning. In: Toutanova, K., (eds.) Proceedings of the 2021 Conference of the North American Chapter of the Association for Computational Linguistics: Human Language Technologies, NAACL-HLT 2021, Online, June 6-11, 2021, pp. 3470–3479. Association for Computational Linguistics (2021)
3. Chen, X., et al.: Knowprompt: knowledge-aware prompt-tuning with synergistic optimization for relation extraction. In: Laforest, F., et al (eds.) WWW 2022: The ACM Web Conference 2022, Virtual Event, Lyon, France, April 25–29, 2022, pp. 2778–2788. ACM (2022)
4. Chia, Y K., Bing, L., Poria, S., Si, L.: Relationprompt: leveraging prompts to generate synthetic data for zero-shot relation triplet extraction. In: Muresan, S., Nakov, P., Villavicencio, A., (eds.), Findings of the Association for Computational Linguistics: ACL 2022, Dublin, Ireland, May 22–27, 2022, pp. 45–57. Association for Computational Linguistics (2022)
5. Corso, G., Cavalleri, L., Beaini, D., Pietro Liò, Velickovic, P.: Principal neighbourhood aggregation for graph nets. In: Larochelle, H., Ranzato, M., Hadsell, R., Balcan, M-F., Lin, H-T., (eds.) Advances in Neural Information Processing Systems 33: Annual Conference on Neural Information Processing Systems 2020, NeurIPS 2020, December 6-12, 2020, virtual (2020)
6. Fan, S., Mo, S., Niu, J.: Boosting document-level relation extraction by mining and injecting logical rules. In: Goldberg, Y., Kozareva, Z., Zhang, Y., (eds.) Proceedings of the 2022 Conference on Empirical Methods in Natural Language Processing, EMNLP 2022, Abu Dhabi, United Arab Emirates, December 7-11, 2022, pp. 10311–10323. Association for Computational Linguistics (2022)
7. Galkin, M., Yuan, X., Mostafa, H., Tang, J., Zhu, Z.: Towards foundation models for knowledge graph reasoning. In: The Twelfth International Conference on Learning Representations, ICLR 2024, Vienna, Austria, May 7–11, 2024. OpenReview.net (2024)
8. Han, X., et al.: Fewrel: a large-scale supervised few-shot relation classification dataset with state-of-the-art evaluation. In: Riloff, E., Chiang, D., Hockenmaier, J., Tsujii, J., (eds.) Proceedings of the 2018 Conference on Empirical Methods in Natural Language Processing, Brussels, Belgium, October 31 - November 4, 2018, pp. 4803–4809. Association for Computational Linguistics (2018)
9. Jain, M., Mutharaju, R., Kavuluru, R., Singh, K.: Revisiting document-level relation extraction with context-guided link prediction. In: Wooldridge, M.J., Dy, J.G., Natarajan, S., (eds.) Thirty-Eighth AAAI Conference on Artificial Intelligence, AAAI 2024, Thirty-Sixth Conference on Innovative Applications of Artificial Intelligence, IAAI 2024, Fourteenth Symposium on Educational Advances in Artificial Intelligence, EAAI 2014, February 20-27, 2024, Vancouver, Canada, pp. 18327–18335. AAAI Press (2024)

10. Jain, M., Singh, K., Mutharaju, R.: Reonto: a neuro-symbolic approach for biomedical relation extraction. In: Koutra, D., Plant, C., Rodriguez, M.G., Baralis, E., Bonchi, F., (eds.) Machine Learning and Knowledge Discovery in Databases: Research Track - European Conference, ECML PKDD 2023, Turin, Italy, September 18-22, 2023, Proceedings, Part IV. LNCS, vol. 14172, pp. 230–247. Springer (2023)
11. Ji, S., Pan, S., Cambria, E., Marttinen, P., Philip, S.Y.: A survey on knowledge graphs: representation, acquisition, and applications. IEEE Trans. Neural Netw. Learn. Syst. **33**(2), 494–514 (2022)
12. Kumar, A., Smith, B.: Oncology ontology in the NCI thesaurus. In: Miksch, S., Hunter, J., Keravnou, E.T. (eds.) AIME 2005. LNCS (LNAI), vol. 3581, pp. 213–220. Springer, Heidelberg (2005). https://doi.org/10.1007/11527770_30
13. Lan, Y., Li, D., Zhang, Y., Zhao, H., Zhao, G.: Modeling zero-shot relation classification as a multiple-choice problem. In: International Joint Conference on Neural Networks, IJCNN 2023, Gold Coast, Australia, June 18-23, 2023, pp. 1–8. IEEE (2023)
14. Lv, B., et al.: DSP: discriminative soft prompts for zero-shot entity and relation extraction. In: Rogers, A., Boyd-Graber, J.L., Okazaki, N., (eds.) Findings of the Association for Computational Linguistics: ACL 2023, Toronto, Canada, July 9-14, 2023, pp. 5491–5505. Association for Computational Linguistics (2023)
15. Ma, Y., Wang, A., Okazaki, N.: DREEAM: guiding attention with evidence for improving document-level relation extraction. In: Vlachos, A., Augenstein, I., (eds.) Proceedings of the 17th Conference of the European Chapter of the Association for Computational Linguistics, EACL 2023, Dubrovnik, Croatia, May 2-6, 2023, pp. 1963–1975. Association for Computational Linguistics (2023)
16. Miwa, M., Bansal, M.: End-to-end relation extraction using LSTMs on sequences and tree structures. In: Proceedings of the 54th Annual Meeting of the Association for Computational Linguistics, ACL 2016, August 7-12, 2016, Berlin, Germany, vol. 1: Long Papers. The Association for Computer Linguistics (2016)
17. Möller, C., Usbeck, R.: Incorporating type information into zero-shot relation extraction. In: Proceedings of the Third International Workshop on Knowledge Graph Generation from Text (TEXT2KG 2024), pp. 26–30, Hersonissos, Greece, May 2024. Co-located with the Extended Semantic Web Conference (ESWC)
18. Ni, J., Rossiello, G., Gliozzo, A., Florian, R.: A generative model for relation extraction and classification. *CoRR*, abs/2202.13229 (2022)
19. Obamuyide, A., Vlachos, A.: Zero-shot relation classification as textual entailment. In: Proceedings of the First Workshop on Fact Extraction and VERification (FEVER), pp. 72–78 (2018)
20. Qi, K., Du, J., Wan, H.: End-to-end learning of logical rules for enhancing document-level relation extraction. In: Ku, L-W., Martins, A., Srikumar, V., (eds.) Proceedings of the 62nd Annual Meeting of the Association for Computational Linguistics, vol. 1: Long Papers, ACL 2024, Bangkok, Thailand, August 11-16, 2024, pp. 7247–7263. Association for Computational Linguistics (2024)
21. Raffel, C., et al.: Exploring the limits of transfer learning with a unified text-to-text transformer. J. Mach. Learn. Res. **21**, 140:1–140:67 (2020)
22. Rahimi, M., Surdeanu, M.: Improving zero-shot relation classification via automatically-acquired entailment templates. In: Can, B., et al. (eds.) Proceedings of the 8th Workshop on Representation Learning for NLP, RepL4NLP@ACL 2023, Toronto, Canada, July 13, 2023, pp. 187–195. Association for Computational Linguistics (2023)

23. Rocktäschel, T., Grefenstette, E., Hermann, K.M., Kociský, T., Blunsom, P.: Reasoning about entailment with neural attention. In: Bengio, Y., LeCun, Y (eds.) 4th International Conference on Learning Representations, ICLR 2016, San Juan, Puerto Rico, May 2-4, 2016, Conference Track Proceedings (2016)
24. Ru, D., et al.: Learning logic rules for document-level relation extraction. In: Moens, M-F., Huang, X., Specia, L., Yih, S.W.T., (eds.) Proceedings of the 2021 Conference on Empirical Methods in Natural Language Processing, EMNLP 2021, Virtual Event / Punta Cana, Dominican Republic, 7-11 November, 2021, pp. 1239–1250. Association for Computational Linguistics (2021)
25. Sainz, O., Lacalle, O.L., Labaka, G., Barrena, A., Agirre, E.: Label verbalization and entailment for effective zero and few-shot relation extraction. In: Moens, M-F., Huang, X., Specia, L., Yih, S.W., (eds.) Proceedings of the 2021 Conference on Empirical Methods in Natural Language Processing, EMNLP 2021, Virtual Event / Punta Cana, Dominican Republic, 7-11 November, 2021, pp. 1199–1212. Association for Computational Linguistics (2021)
26. Sorokin, D., Gurevych, I.: Context-aware representations for knowledge base relation extraction. In: Palmer, M., Hwa, R., Riedel, S., (eds.) Proceedings of the 2017 Conference on Empirical Methods in Natural Language Processing, EMNLP 2017, Copenhagen, Denmark, September 9-11, 2017, pp. 1784–1789 Association for Computational Linguistics (2017)
27. Tan, Q., He, R., Bing, L., Ng, H.T.: Document-level relation extraction with adaptive focal loss and knowledge distillation. In: Muresan, S., Nakov, P., Villavicencio, A., (eds.) Findings of the Association for Computational Linguistics: ACL 2022, Dublin, Ireland, May 22-27, 2022, pp. 1672–1681. Association for Computational Linguistics (2022)
28. Tan, Q., Xu, L., Bing, L., Ng, H.T., Aljunied, S.M.: Revisiting docred - addressing the false negative problem in relation extraction. In: Goldberg, Y., Kozareva, Z., Zhang, Y., (eds.) Proceedings of the 2022 Conference on Empirical Methods in Natural Language Processing, EMNLP 2022, Abu Dhabi, United Arab Emirates, December 7-11, 2022, pp. 8472–8487. Association for Computational Linguistics (2022)
29. Tran, V.-H., Ouchi, H., Shindo, H., Matsumoto, Y., Watanabe, T.: Enhancing semantic correlation between instances and relations for zero-shot relation extraction. J. Nat. Lang. Process. **30**(2), 304–329 (2023)
30. Tran, V.H., Ouchi, H., Watanabe, T., Matsumoto, Y.: Improving discriminative learning for zero-shot relation extraction. In: Proceedings of the 1st Workshop on Semiparametric Methods in NLP: Decoupling Logic from Knowledge, pp. 1–6 (2022)
31. Vashishth, S., Joshi, R., Prayaga, S.S., Bhattacharyya, C., Talukdar, P.P.: RESIDE: improving distantly-supervised neural relation extraction using side information. In: Riloff, E., Chiang, D., Hockenmaier, J., Tsujii, J., (eds.) Proceedings of the 2018 Conference on Empirical Methods in Natural Language Processing, Brussels, Belgium, October 31 - November 4, 2018, pp. 1257–1266. Association for Computational Linguistics (2018)
32. Verlinden, S., Zaporojets, K., Deleu, J., Demeester, T., Develder, C.: Injecting knowledge base information into end-to-end joint entity and relation extraction and coreference resolution. In: Zong, C., Xia, F., Li, W., Navigli, R., (eds.) Findings of the Association for Computational Linguistics: ACL/IJCNLP 2021, Online Event, August 1-6, 2021, volume ACL/IJCNLP 2021 of *Findings of ACL*, pp. 1952–1957. Association for Computational Linguistics (2021)

33. Vrandečić, D., Krötzsch, M.: Wikidata: a free collaborative knowledgebase. Commun. ACM **57**(10), 78–85 (2014)
34. Wang, J., Le, X., Peng, X., Chen, C.: Adaptive hinge balance loss for document-level relation extraction. In: Bouamor, H., Pino, J., Bali, K., (eds.) Findings of the Association for Computational Linguistics: EMNLP 2023, Singapore, December 6-10, 2023, pp. 3872–3878. Association for Computational Linguistics (2023)
35. Wang, M., Qiu, L., Wang, X.: A survey on knowledge graph embeddings for link prediction. Symmetry **13**(3), 485 (2021)
36. Xing, R., Luo, J., Song, T.: Biorel: towards large-scale biomedical relation extraction. BMC Bioinform. **21-S**(16), 543 (2020)
37. Xu, W., Chen, K., Zhao, T.: Discriminative reasoning for document-level relation extraction. In: Zong, C., Xia, F., Li, W., Navigli, R., (eds.) Findings of the Association for Computational Linguistics: ACL/IJCNLP 2021, Online Event, August 1-6, 2021, volume ACL/IJCNLP 2021 of Findings of ACL, pp. 1653–1663. Association for Computational Linguistics (2021)
38. Yang, B., Yih, W., He, X., Gao, J., Deng, L.: Embedding entities and relations for learning and inference in knowledge bases. In: Bengio, Y., LeCun, Y., (eds.) 3rd International Conference on Learning Representations, ICLR 2015, San Diego, CA, USA, May 7-9, 2015, Conference Track Proceedings (2015)
39. Yang, J.-J.: An ontology-based intelligent agent system for semantic search in medicine. In: Lee, J., Barley, M. (eds.) PRIMA 2003. LNCS (LNAI), vol. 2891, pp. 182–193. Springer, Heidelberg (2003). https://doi.org/10.1007/978-3-540-39896-7_16
40. Zaporojets, K., Deleu, J., Develder, C., Demeester, T.: DWIE: an entity-centric dataset for multi-task document-level information extraction. Inf. Process. Manag. **58**(4), 102563 (2021)
41. Zeng, D., Liu, K., Lai, S., Zhou, G., Zhao, J.: Relation classification via convolutional deep neural network. In: Hajic, J., Tsujii, J., (eds.) COLING 2014, 25th International Conference on Computational Linguistics, Proceedings of the Conference: Technical Papers, August 23-29, 2014, Dublin, Ireland, pp. 2335–2344. ACL (2014)
42. Zeng, S., Xu, R., Chang, B., Li, L.: Double graph based reasoning for document-level relation extraction. In: Webber, B., Cohn, T., He, Y., Liu, Y., (eds.) Proceedings of the 2020 Conference on Empirical Methods in Natural Language Processing, EMNLP 2020, Online, November 16-20, 2020, pp. 1630–1640. Association for Computational Linguistics (2020)
43. Zhang, N., et al.: Document-level relation extraction as semantic segmentation. In: Zhou, Z-H., (ed.) Proceedings of the Thirtieth International Joint Conference on Artificial Intelligence, IJCAI 2021, Virtual Event / Montreal, Canada, 19-27 August 2021, pp. 3999–4006. ijcai.org (2021)
44. Zhao, J., et al.: Re-matching: a fine-grained semantic matching method for zero-shot relation extraction. In: Rogers, A., Boyd-Graber, J.L., Okazaki, N., (eds.) Proceedings of the 61st Annual Meeting of the Association for Computational Linguistics (Volume 1: Long Papers), ACL 2023, Toronto, Canada, July 9-14, 2023, pp. 6680–6691. Association for Computational Linguistics (2023)
45. Zhong, Z., Chen, D.: A frustratingly easy approach for entity and relation extraction. In: Toutanova, K., et al (eds.) Proceedings of the 2021 Conference of the North American Chapter of the Association for Computational Linguistics: Human Language Technologies, NAACL-HLT 2021, Online, June 6-11, 2021, pp. 50–61. Association for Computational Linguistics (2021)

46. Zhou, W., Huang, K., Ma, T., Huang, J.: Document-level relation extraction with adaptive thresholding and localized context pooling. In: Thirty-Fifth AAAI Conference on Artificial Intelligence, AAAI 2021, Thirty-Third Conference on Innovative Applications of Artificial Intelligence, IAAI 2021, The Eleventh Symposium on Educational Advances in Artificial Intelligence, EAAI 2021, Virtual Event, February 2-9, 2021, pp. 14612–14620. AAAI Press (2021)
47. Zhu, H., Lin, Y., Liu, Z., Fu, J., Chua, T-S., Sun, M.: Graph neural networks with generated parameters for relation extraction. In: Korhonen, A., Traum, D.R., Màrquez, L., (eds.) Proceedings of the 57th Conference of the Association for Computational Linguistics, ACL 2019, Florence, Italy, July 28- August 2, 2019, Volume 1: Long Papers, pp. 1331–1339. Association for Computational Linguistics (2019)
48. Zhu, Z., Zhang, Z., Xhonneux, L-P., Tang, J.: Neural bellman-ford networks: a general graph neural network framework for link prediction. In: Ranzato, M., Beygelzimer, A., Dauphin, Y.N., Liang, P., Vaughan, J.W., (eds.) Advances in Neural Information Processing Systems 34: Annual Conference on Neural Information Processing Systems 2021, NeurIPS 2021, December 6-14, 2021, virtual, pp. 29476–29490 (2021)

Shape Expressions with Inheritance

Iovka Boneva[1(✉)], Jose Emilio Labra Gayo[2], Eric Prud'hommeaux[3], Katherine Thornton[4], and Andra Waagmeester[5]

[1] Univ. Lille, CNRS, Centrale Lille, UMR 9189 CRIStAL, 59000 Lille, France
iovka.boneva@univ-lille.fr

[2] WESO (WEb Semantics Oviedo) Research Group, University of Oviedo, Oviedo, Spain
labra@uniovi.es

[3] Janeiro Digital, Boston, USA
eric@uu3.org

[4] Yale University Library, New Haven, CT, USA
katherine.thornton@yale.edu

[5] Micelio BV, Ekeren, Belgium
andra@micelio.be

Abstract. We formally introduce an inheritance mechanism for the Shape Expressions language (ShEx). It is inspired by inheritance in object-oriented programming languages, and provides similar advantages such as reuse, modularity, and more flexible data modelling. Using an example, we explain the main features of the inheritance mechanism. We present its syntax and formal semantics. The semantics is an extension of the semantics of ShEx 2.1. It also directly yields a validation algorithm as an extension of the previous ShEx validation algorithms, while maintaining the same algorithmic complexity.

Keywords: RDF · schemas for graphs · formal semantics of languages

1 Introduction

The Shape Expressions language (ShEx) was proposed in 2014 as a concise, high-level language to describe and validate RDF data [12]. It allows us to define ShEx schemas which are collections of shapes. A shape describes the required properties of an RDF node and constrains their values. ShEx 2.1 was published as a W3C final community report in 2019[1], and it introduced logical operators on top of shapes, yielding *shape expressions*. They allow describing alternatives with disjunction, or further constraining the shape of the data and data values with conjunction. ShEx has been increasingly adopted in different domains such as HL7 FHIR[2] and Wikidata [17], where it is used to describe data models and to validate entity data.

[1] https://shex.io/shex-semantics-20191008/.

[2] https://www.hl7.org/fhir/.

E. Curry et al. (Eds.): ESWC 2025, LNCS 15718, pp. 481–499, 2025.
https://doi.org/10.1007/978-3-031-94575-5_26

The adoption of ShEx for describing large data models has led to the appearance of collaborative shape schema ecosystems like that of Wikidata[3]. An important requirement for such ecosystems is to support reusability. The simplest feature for reuse is to import a shape expression that has been defined elsewhere. ShEx 2.1 does offer this possibility and the imported shape expression can be reused as is. Data modelling and reuse often require adapting the existing data and its model to one's particular needs. Two common adaptation patterns are to *restrict* the allowed values of existing properties, and to *extend* the data model and the data with properties that are specific to a new application. Additionally, a commonly-used feature is to define frequently occurring patterns which serve as building blocks of shape expressions.

In response to submitted use cases, the ShEx Community Group[4] has proposed an inheritance mechanism for ShEx that provides the afore-mentioned features. Inspired by inheritance in programming languages, it allows for the definition of a hierarchy of shape expressions. Child shapes can extend their parents with new required properties for RDF nodes, and can impose additional constraints on the properties of their parents. Additionally, as in object-oriented programming languages, a node with a child shape can be used where a parent shape is expected. Multiple inheritance is also allowed. Finally, shape expressionss can be made *abstract* indicating that they cannot be used on their own, but only as building blocks of other expressions. An early version of the inheritance mechanism has been used in [13] to translate in ShEx the user-facing documentation of data structures in FHIR.

In this paper, we introduce the formal semantics for ShEx with inheritance, built as an extension of the formal semantics of ShEx 2.1 [4]. The formalization posed some challenges, notably for handling multiple inheritance. In particular, we introduce a syntactic restriction on the combined use of inheritance with disjunction. The restriction avoids unnecessary complexity in the language, while still providing the functionality required by the use cases. The ShEx formalization presented here will be submitted for the next version of the ShEx language, currently under standardization by IEEE.[5]

The paper is organized as follows. In Sect. 2 we give an example illustrating the main features of the inheritance mechanism, and define some preliminary notions. Section 3 gives the syntax, and Sect. 4 the semantics of the language. Section 5 presents a validation algorithm. We close with a discussion, an overview of related work and a conclusion in Sect. 6. The proofs omitted in the paper are presented in appendix. A companion webpage [1] lists the current implementations and provides source code and demonstrations for the examples.

---

[3] https://www.wikidata.org/wiki/Wikidata:Database_reports/EntitySchema_directory.

[4] https://www.w3.org/community/shex/.

[5] https://standards.ieee.org/ieee/3330/11119/.

2 Motivating Example and Background

Assume the following constraints on nodes in an RDF graph: T_{str} is satisfied by string literals, T_{float} by float literals, T_{any} by all RDF nodes, $T_{\text{“colour”}}$ is satisfied by the literal “colour”, and $T_{\text{“radius”}}$ by the literal “radius”. A ShEx schema using inheritance is presented in Fig. 1, we will describe it section by section. The source code for that schema is available on the companion webpage [1]. The shape expression named *Coord* describes nodes representing coordinates, i.e. having properties x and y whose values are floats. The shape expression *Attribute* requires a property *name* which is a string, and a property *value* that can be anything. The extends _ indicates that the shape expression is being extended. A *Colour* is a specific kind of attribute (extends on *Attribute*) that additionally to the *name* and *value* properties requires a *scope* property whose value is a string. This is captured by the first part of the definition, preceding the and keyword. The second part states that the *name* of a *Colour* attribute is “colour”. Thus, inheritance provides a mechanism for requiring additional properties (before the and) and restricting existing properties (after the and). Figure 2 shows an example graph. Nodes `a1`, `a2`, `a3` satisfy *Attribute*, node `a2` satisfies *Colour*, and node `c1` satisfies *Coord*.

$$\begin{aligned}
\mathit{Coord} &\to \langle\, x\ @T_{float}\ ;\ y\ @T_{float}\,\rangle\\
\mathit{Attribute} &\to \mathsf{extends}\ _\,\langle\, \mathit{name}\ @T_{str}\ ;\ \mathit{value}\ @T_{any}\,\rangle\\
\mathit{Colour} &\to \mathsf{extends}\ \mathit{Attribute}\,\langle\, \mathit{scope}\ @T_{str}\,\rangle\ \ \mathsf{and}\ \ \langle\, \mathit{name}\ @T_{\text{"colour"}}\,\rangle\\
\mathsf{abstract}\ \mathit{Figure} &\to \mathsf{extends}\ _\,\langle\, \mathit{coord}\ @\mathit{Coord}\,\rangle\\
\mathit{Circle} &\to \mathsf{extends}\ \mathit{Figure}\,\langle\, \mathit{attr}\ @\mathit{Radius}\,\rangle\\
\mathit{Radius} &\to \mathsf{extends}\ \mathit{Attribute}\,\langle\, \varepsilon\,\rangle\ \mathsf{and}\ \langle\, \mathit{name}\ @T_{\text{"radius"}}\ ;\ \mathit{value}\ @T_{float}\,\rangle\\
\mathit{ColouredFigure} &\to \mathsf{extends}\ \mathit{Figure}\,\langle\, \mathit{attr}\ @\mathit{Colour}\,\rangle\\
\mathit{ColouredCircle} &\to \mathsf{extends}\ \mathit{Circle}, \mathit{ColouredFigure}\,\langle\, \varepsilon\,\rangle
\end{aligned}$$

Fig. 1. A ShEx schema with inheritance.

A *Figure* has coordinates associated with the property *coord*. The shape expression *Figure* is marked as abstract, which indicates that it cannot be directly satisfied, rather one of its non-abstract descendants will be satisfied. A *Circle* is a *Figure* that has a *Radius* attribute reachable through the property *attr*, which in turn is an *Attribute* having *name* “radius” and a float *value*. Note that the

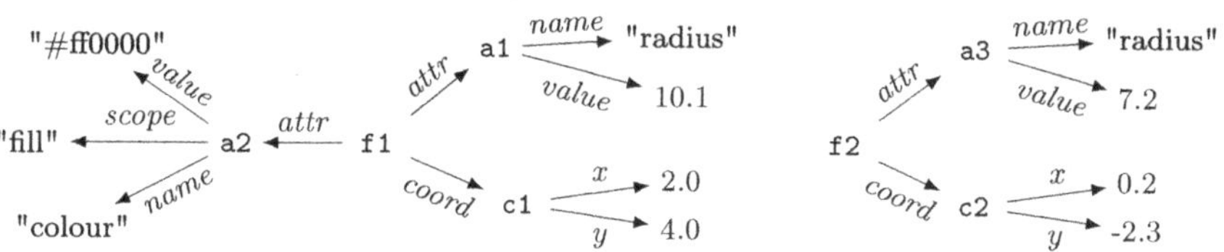

Fig. 2. Example graph with nodes `f1`, `f2`, `a1`, `a2`, `a3` and strings and float literals.

definition of *Circle* does not have a restriction after the and; in fact the restriction is optional. On the other hand, the extending portion of *Radius* is ε, meaning that a *Radius* does not require additional properties beyond those specified in its parent *Attribute*. In the example graph, `f2` is a *Circle*, `a1`, `a3` satisfy *Radius*.

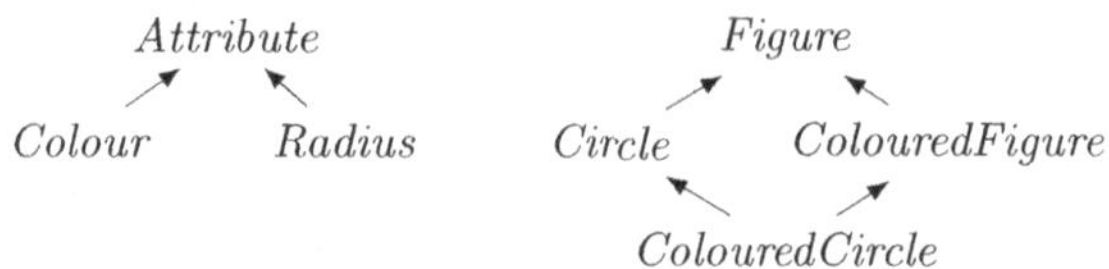

Fig. 3. Extension hierarchy graph for the example.

The final part of the example illustrates multiple inheritance. A *ColouredCircle* extends both on *ColouredFigure* and *Circle* which in turn both extend from *Figure*. Recall that a *Figure* has one *coord* property, which is inherited twice by *ColouredCircle*. The inheritance mechanism ensures that multiply-inherited properties are not used more than once. In this particular example, a *ColouredCircle* can have only one *coord*. In the example graph, `f1` is a *ColouredCircle*.

As in object-oriented programming languages, inheritance yields an extension hierarchy which is a directed acyclic graph. Figure 3 depicts the inheritance graph of the example schema, with arrows going from shape expressions to their direct ancestors. The inheritance mechanism requires that if a node satisfies some of the descendants of a shape expression, then it satisfies the shape expression itself. Therefore, `f1` is a *ColouredCircle*, a *ColouredFigure*, a *Circle*, and a *Figure*.

Background. IRI, Blank and Literal are three countable mutually disjoint sets of IRIs, blank nodes, and literals, respectively. Elements of $(\mathsf{IRI} \cup \mathsf{Blank}) \times \mathsf{IRI} \times (\mathsf{IRI} \cup \mathsf{Literal} \cup \mathsf{Blank})$ are called triples. If (n, p, o) is a triple, then n is its subject, p its property, and o its object. An RDF graph $\mathbf{G}$ is a finite set of triples. The set of nodes of $\mathbf{G}$ is the set of all subjects and objects of its triples and is denoted $nodes(\mathbf{G})$. Given a graph $\mathbf{G}$ and a node $n \in nodes(\mathbf{G})$, the *neighbourhood* of n in $\mathbf{G}$ is the set $neigh_{\mathbf{G}}(n)$ of triples having n as subject. Formally, $neigh_{\mathbf{G}}(n) = \{(n, p, o) \mid (n, p, o) \in \mathbf{G}\}$. A set of triples $M \subseteq \mathsf{Triples}$ is called a *neighbourhood set of triples* if all its elements have the same subject denoted $subject(M)$.

We use the symbol $\uplus$ for disjoint union, i.e. $M = M_1 \uplus \cdots \uplus M_k$ (for some $k \in \mathbb{N}$) means that $M = M_1 \cup \cdots \cup M_k$ and $M_i \cap M_j = \emptyset$ whenever $i \neq j$. The disjoint union of zero sets is the empty set, i.e., if $k = 0$, then $M = \emptyset$. In grammars or expressions, square brackets $[\,]$ surround optional elements.

3 Syntax

A ShEx schema consists of a set of labeled (named) shape expressions to be checked for conformance with the nodes of an RDF graph. We start by defining shape expressions, then give a formal definition of ShEx schemas with inheritance and the corresponding extension hierarchy.

Definition 1 (shape expressions). *Let* $\mathbf{Y}$ *be a finite set of* simple labels *and* $\mathbf{X}$ *be a finite set of* extendable labels *disjoint from* $\mathbf{Y}$. *A* shape expression *over* $\mathbf{Y}$ *and* $\mathbf{X}$ *is either a* reference *of the form* $@z$, *or a* plain shape expression *derivable from the non-terminal* u *in the following grammar:*

$u ::= h \mid t \mid c \mid u \text{ or } u \mid u \text{ and } u \mid \text{not } u\ .$	*(plain shape expression)*
$h ::= [\text{closed}]\ [\text{extra } P]\ \langle\, e \,\rangle\ .$	*(shape)*
$t ::= \text{extends}\, X\ h\ .$	*(shape with extends)*
$c ::=$ *a boolean function over* $\mathit{IRI} \cup \mathit{Blank} \cup \mathit{Literal}$.	*(node constraint)*
$e ::= \varepsilon \mid p\ @z \mid e \,\vert\, e \mid e\ ; e \mid e{*}\ .$	*(triple expression)*

where $z \in \mathbf{Y} \cup \mathbf{X}$, $X \subseteq \mathbf{X}$, $p \in \mathit{IRI}$, $P \subseteq \mathit{IRI}$, *the symbol '*ε*' is a constant, and the symbols '@', '|', '*' and ';' are operators. An expression derivable by the non-terminal* h, *resp.* t, *resp.* c, *resp.* e, *is called a* shape, *resp. a* shape with extends, *resp. a* node constraint, *resp. a* triple expression. *An expression of the form* $p\ @z$ *is called a* triple constraint.

A shape with extends is a new kind of shape expression. It is constructed from a (possibly empty) set X of extendable labels and a shape h. The labels in X indicate which shape expression. are being extended. All other ingredients of shape expression. are present in ShEx 2.1.

Definition 2 (ShEx schema). *A* ShEx schema (with inheritance) *is a tuple* $\mathbf{S} = (\mathbf{Y}, \mathbf{X}, \mathbf{def}, \mathbf{A})$, *where* $\mathbf{Y}$ *is a finite set of* simple labels, $\mathbf{X}$ *is a finite set of* extendable labels *disjoint from* $\mathbf{Y}$, $\mathbf{A} \subseteq \mathbf{X}$ *are the* abstract *labels, and* $\mathbf{def}$ *is a function in which every label from* $\mathbf{Y} \cup \mathbf{X}$ *associates its definition, with:*

- *if* $y \in \mathbf{Y}$, *then* $\mathbf{def}(y)$ *is a plain shape expression;*
- *if* $x \in \mathbf{X}$, *then* $\mathbf{def}(x)$ *is a shape expression of the form* extends $X\ h\ [\text{and } u]$, *also called an* extendable shape expression, *where* h *is a shape and* u *is a plain shape expression.*

Example 1. In Sect. 2 we presented a schema – denote it $\mathbf{S}_1 = (\mathbf{Y}_1, \mathbf{X}_1, \mathbf{def}_1, \mathbf{A}_1)$ – as a set of rules of the form $z \to s$, to be understood as $\mathbf{def}_1(z) = s$. The simple labels are $\mathbf{Y}_1 = \{Coord, T_{any}, T_{str}, T_{float}, T_{\text{“colour”}}, T_{\text{“radius”}}\}$. The definitions of the sub-scripted T's correspond to the node constraints described in the text, e.g., $\mathbf{def}_1(T_{str})$ is the function that is true for string literals and false for all other nodes. The extendable labels are those that appear on the left-hand sides of the rules, except for *Coord*, i.e.$\mathbf{X}_1 = \{Attribute, Colour, \ldots, ColouredCircle\}$. Their definitions are extendable shape expressions, in which we omit the curly braces around the elements of the set X and write _ if it is empty. Finally, $\mathbf{A}_1 = \{Figure\}$ is indicated by the `abstract` keyword.

In the sequel, all examples of schemas will use the same notation as the one used in Sect. 2 and explained in Example 1.

Definition 3 (extension hierarchy). *The* extension hierarchy *of the schema* $\mathbf{S} = (\mathbf{Y}, \mathbf{X}, \mathbf{def}, \mathbf{A})$ *is a directed graph denoted* $H_{\mathbf{S}}$ *whose set of nodes is* $\mathbf{X}$ *and that has an edge from* x *to* x' *if and only if* $\mathbf{def}(x) = \mathsf{extends}\, X\ h\ [\,\mathsf{and}\ u\,]$ *and* $x' \in X$. *If there exists a (possibly empty) path from* x *to* x' *in* $H_{\mathbf{S}}$, *we say that* x' *is an* ancestor *of* x *or, equivalently, that* x *is a* descendant *of* x'. *We denote* $anc(x)$, *resp.* $desc(x)$, *the set of ancestors, resp. descendants, of the label* x.

Example 2. Figure 3 shows $H_{\mathbf{S}_1}$, the extension hierarchy of the schema $\mathbf{S}_1$ from Example 1. Note that the extension hierarchy is determined only by the first shape with extends in the definition of an extendable shape label. For instance, let $\mathbf{def}(x) = \mathsf{extends}\ x_1, x_2 \langle \varepsilon \rangle\ \mathsf{and}\ \mathsf{extends}\ x_3, x_4 \langle \varepsilon \rangle$. According to Definition 3, x_1 and x_2 are ancestors of x, but x_3 and x_4 are not ancestors of x.

4 Semantics

Let $\mathbf{G}$ be a graph and $\mathbf{S} = (\mathbf{Y}, \mathbf{X}, \mathbf{def}, \mathbf{A})$ be a ShEx schema. A *typing of* $\mathbf{G}$ *by* $\mathbf{S}$ is a subset of $nodes(\mathbf{G}) \times (\mathbf{Y} \cup \mathbf{X})$. We use typings to define the semantics of triple expressions and shape expressions. Intuitively, a typing can be understood as a set of assertions about which nodes of the graph satisfy which shape expressions. It can be used to derive other such assertions. In Sect. 4.1 we define the semantics of expressions under a given typing. A typing τ is *correct* (w.r.t. a graph and a schema) if each of the assertions it contains can be derived from τ itself. In Sect. 4.2 we introduce well-defined schemas by forbidding some circular dependencies between shape expressions. Finally, in Sect. 4.3 we show that if a schema is well-defined, then there exists a unique maximal correct typing $\tau_{\max}$ which defines the semantics of ShEx schemas, in the sense that it contains all correct assertions about which nodes satisfy which shape expressions.

4.1 Satisfying an Expression Under Given Typing

The semantics of shape expressions and triple expressions are given by the means of three mutually recursive satisfiability relations. For triple expressions, the satisfiability relation is of the form $\mathbf{G}, M, \tau \models_{\mathbf{S}} e$ and for shape expressions we use two satisfiability relations, one of the form $\mathbf{G}, n, \tau \models_{\mathbf{S}} s$ and another one of the form $\mathbf{G}, M, \tau \models_{\mathbf{S}} s$ (where $\mathbf{S}$ is a schema, $\mathbf{G}$ is a graph, n is a node of $\mathbf{G}$, τ is a typing over $\mathbf{G}$ and $\mathbf{S}$, M is a set of neighbourhood triples from $\mathbf{G}$, e is a triple expression, and s is a shape expression). The different signatures of the three relations will always allow distinguishing them. We will omit $\mathbf{G}$ and $\mathbf{S}$ whenever they are clear from the context. The satisfiability relation of the form $\mathbf{G}, M, \tau \models_{\mathbf{S}} s$ is new for ShEx with inheritance. It is necessary for the definition of the semantics of shapes with extends.

For the remainder of the section, consider given a graph $\mathbf{G}$ and a ShEx schema $\mathbf{S} = (\mathbf{Y}, \mathbf{X}, \mathbf{def}, \mathbf{A})$ that will be implicit (omitted) in all satisfiability relation

Table 1. Definition of the satisfiability relation $M, \tau \models e$

e	$M, \tau \models e$
1. ε	$M = \emptyset$
2. p @z	$M = \{(n, p, o)\} \quad \wedge \quad o, \tau \models @z$
3. $e_1 \mid e_2$	$M, \tau \models e_1 \quad \vee \quad M, \tau \models e_2$
4. $e_1 ; e_2$	$\exists M_1, M_2.\ M = M_1 \uplus M_2 \quad \wedge \quad M_1, \tau \models e_1 \quad \wedge \quad M_2, \tau \models e_2$
5. e_1*	$\exists k \in 0..\lvert M\rvert.\ \exists M_1, \ldots, M_k.$ $M = M_1 \uplus \cdots \uplus M_k \quad \wedge \quad \forall i \in 1..k.\ M_i, \tau \models e_1$

Table 2. Definition of the satisfiability relation $n, \tau \models s$.

s	$n, \tau \models s$
6. c	$c(n)$
7. s_1 or s_2	$n, \tau \models s_1 \quad \vee \quad n, \tau \models s_2$
8. s_1 and s_2	$n, \tau \models s_1 \quad \wedge \quad n, \tau \models s_2$
9. not s_1	$n, \tau \neg \models s_1$
10. @z	$(n, z) \in \tau$
11. [closed] [extra P] $\langle e \rangle$	$neigh_{\mathbf{G}}(n), \tau \models$ [closed] [extra P] $\langle e \rangle$
12. extends X h	$neigh_{\mathbf{G}}(n), \tau \models$ extends X h

statements. The semantics of the satisfiability relations are given in Tables 1, 2 and 3 inductively on the structure of expressions. The right-hand side column in each table is a boolean expression giving the truth value for the case from the left-hand side column. The symbols $\neg$, $\wedge$, $\vee$, $\exists$ and $\implies$ have their usual meaning from logic. The symbol $=$ is test of equality, while $:=$ is assignment. When $\mathbf{G}, M, \tau \models_{\mathbf{S}} e$, resp. $\mathbf{G}, n, \tau \models_{\mathbf{S}} s$, resp. $\mathbf{G}, M, \tau \models_{\mathbf{S}} s$ holds, we say that M *satisfies* e, resp. n *satisfies* s, resp. M *satisfies* s under typing τ. In the sequel of this section we explain the satisfiability relations.

Triple Expressions (Table 1). The semantics of triple expressions are the same as those of ShEx 2.1. Only the empty set of triples satisfies the empty expression ε. A triple constraint p @z is satisfied by a singleton set which unique triple has p as property and an object that satisfies @z. The expression $e_1 \mid e_2$ is satisfied if one of e_1 or e_2 is satisfied. As for $e_1 ; e_2$, it is satisfied by M if M can be split in two parts M_1 and M_2 that satisfy the sub-expressions e_1 and e_2, respectively. Finally, the repeated expression $e*$ is satisfied by M if M can be split into k parts, for some k less than or equal to the number of elements in M, and each of them satisfies e. Note that the empty set always satisfies $e*$.

Example 3. Let τ be the typing and $M_{24}, M_{246}, M_{2\mathrm{a}}, M_{24\mathrm{a}}$ be the sets of triples defined below. Consider the triple expression $e = p\ @T_{even} ; (p\ @T_{<5}{}^* \mid p\ @T_{str})$. Then $M_{24}, \tau \models e$, $M_{246}, \tau \models e$, $M_{2\mathrm{a}}, \tau \models e$, but $M_{24\mathrm{a}}, \tau \neg \models e$.

$$M_{24} = \{(n, p, 2), (n, p, 4)\} \qquad M_{246} = \{(n, p, 2), (n, p, 4), (n, p, 6)\}$$
$$M_{2a} = \{(n, p, 2), (n, p, "a")\} \qquad M_{24a} = \{(n, p, 2), (n, p, 4), (n, p, "a")\}$$
$$\tau = \{(2, T_{even}), (2, T_{<5}), (4, T_{even}), (4, T_{<5}), (6, T_{even}), (6, T_{>5}), ("a", T_{str})\}$$

Shape Expression Satisfied by a Node (Table 2). A node constraint c is satisfied by n if $c(n)$ holds. The semantics of the boolean operators or, and and not are as expected. A reference $@z$ is satisfied by node n if the typing contains the pair (n, z). Both a shape and a shape with extends are satisfied by a node if the node's neighbourhood satisfies the shape.

Table 3. Definition of the satisfiability relation $M, \tau \models s$.

s	$M, \tau \models s$
13. c	$c(subject(M))$
14. s_1 or s_2	$M, \tau \models s_1 \quad \vee \quad M, \tau \models s_2$
15. s_1 and s_2	$M, \tau \models s_1 \quad \wedge \quad M, \tau \models s_2$
16. not s_1	$M, \tau \neg \models s_1$
17. [closed][extra P]$\langle e \rangle$	$M^{e}, \tau \models e \quad \wedge \quad props(M^{\not e}) \subseteq P' \quad \wedge$ closed is present $\Longrightarrow props(M) \subseteq props(e) \cup P'$ with $P' := P$ if extra is present, $P' := \emptyset$ otherwise $M^{e} := \{(n, p, o) \in M \mid \exists p\ @z \in tcs(e).\ o, \tau \models @z\}$ $M^{\not e} := \{(n, p, o) \in M \mid p \in props(e)\} \setminus M^{e}$
18. extends X h	$\exists M'$ and $\exists M_x$ for every $x \in anc(X)$ such that $M = M' \uplus \biguplus_{x \in anc(X)} M_x \quad \wedge$ $M', \tau \models h \quad \wedge$ for every $x \in anc(X)$. $M_x, \tau \models ext\text{-}te(x) \quad \wedge$ $restr(x)$ present $\Longrightarrow \left(\bigcup_{z \in anc(x)} M_z\right), \tau \models restr(x)$

Shape Expression Satisfied by a Set of Triples (Table 3). A node constraint is satisfied by a set M of neighbourhood triples if their common subject satisfies the constraint. The definitions of boolean operators are as expected. For a shape (line 17.), we use the following notations: $props(M) = \{p \mid (n, p, o) \in M\}$ is the set of properties of the triples in M, $tcs(e)$ is the set of triple constraints sub-expressions of the triple expression e, and $props(e) = \{p \mid p\ @s \in tcs(e)\}$ is the set of properties in the triple constraints of e. We identify two significant subsets of M:

- M^{e} are the triples from M that satisfy some triple constraint in e under τ,
- $M^{\not e}$ are the triples from M whose property appears in e but satisfy none of the triple constraints in e.

We require that M^e satisfies the triple expression e, while the triples in $M^{\not e}$ may use only extra properties from P. In particular, $M^{\not e}$ must be empty if $\mathsf{extra}\,P$ is not present. Additionally, if the shape is closed, then M contains only triples whose properties appear in e or are in the set P of extra properties.

Example 4. Let τ and $M_{24}, M_{246}, M_{2a}, M_{24a}$ be as in Example 3 and define the triple expression $e = p\ @T_{even}\ ;\ p\ @T_{<5}$. Then, for the sets used in line 17. we have e.g.: $M^e_{24a} = \{(n_2, p, 2), (n_2, p, 4)\}$, $M^{\not e}_{24a} = \{(n_2, p, "a")\}$, $M^e_{246} = M_{246}$ and $M^{\not e}_{246} = \emptyset$. For example, the following hold:

$$
\begin{array}{lll}
M_{24}, \tau \models \langle\, e \,\rangle & M_{24a}, \tau \models \mathsf{extra}\,\{p\}\,\langle\, e \,\rangle & M_{246}, \tau \neg\models \mathsf{extra}\,\{p\}\,\langle\, e \,\rangle \\
M_{2a}, \tau \neg\models \langle\, e \,\rangle & M_{24a} \cup \{(n, q, 2)\}, \tau \models \mathsf{extra}\,\{p\}\,\langle\, e \,\rangle & \\
M_{246}, \tau \neg\models \langle\, e \,\rangle & M_{24a} \cup \{(n, q, 2)\}, \tau \neg\models \mathsf{closed}\ \mathsf{extra}\,\{p\}\,\langle\, e \,\rangle &
\end{array}
$$

For the semantics of shapes with extends (line 18.), we use the following notations. If $X \in \mathbf{X}$, then $anc(X) = \bigcup_{x \in X} anc(x)$ is the union of the ancestor sets of the labels in X; remark that $X \subseteq anc(X)$. For an extendable label $x \in \mathbf{X}$ and assuming $\mathbf{def}(x) = \mathsf{extends}\,X\ h\ [\mathsf{and}\ \ s]$, we denote by $ext\text{-}te(x)$ the triple expression of h, and by $restr(x)$ the shape expression s whenever it exists.

For a shape with extends $t = \mathsf{extends}\,X\ h$, different parts of M will be used to satisfy h and the shape expressions extended in t. First, M must be partitioned as $M' \uplus M_{x_1} \uplus \cdots \uplus M_{x_k}$ where the x_i are the strict ancestors of t, i.e. $\{x_1, \ldots, x_k\} = anc(X)$ (for some $k \in \mathbb{N}$). Assume also that $\mathbf{def}(x_i) = \mathsf{extends}\,X_i\ h_i\ [\mathsf{and}\ \ s_i]$ for every $i \in 1..k$. Then we require that:

- M' satisfies h and for every $i \in 1..k$, M_{x_i} satisfies the triple expression $ext\text{-}te(x_i)$ which is the triple expression of the shape h_i;
- if present, the restriction s_i must be satisfied not by M_{x_i} alone, but by the union of all the M_z such that z is ancestor of x_i (recall that $x_i \in anc(x_i)$).

Example 5. Let $\mathbf{Y} = \{T_{even}, T_{<5}, T_{>5}\}$ and let τ, $M_{24}, M_{2a}, M_{246}, M_{24a}$ be as in Example 3. Consider the schema $\mathbf{S} = (\mathbf{Y}, \{x_1, \ldots, x_6\}, \mathbf{def}, \emptyset)$, where $\mathbf{def}$ is given by:

$$
\begin{array}{ll}
x_0 \rightarrow \mathsf{extends}\ \_\,\langle\, p\ @T_{even}\,\rangle & \\
x_1 \rightarrow \mathsf{extends}\ x_0\,\langle\, p\ @T_{even}\,\rangle & x_2 \rightarrow \mathsf{extends}\ x_1\,\langle\, p\ @T_{<5}\,\rangle \\
x_3 \rightarrow \mathsf{extends}\ x_0\,\langle\, p\ @T_{even}\,\rangle\ \mathsf{and}\ \langle\, p\ @T_{>5}{}^*\,\rangle & x_4 \rightarrow \mathsf{extends}\ x_6\,\langle\, p\ @T_{<5}\,\rangle \\
x_5 \rightarrow \mathsf{extends}\ x_0\ \mathsf{extra}\,\{p\}\,\langle\, \varepsilon\,\rangle & x_6 \rightarrow \mathsf{extends}\ x_3\,\langle\, p\ @T_{even}\,\rangle
\end{array}
$$

That is, $\mathbf{def}(x_1)$ requires exactly two p-triples with even values, and $\mathbf{def}(x_2)$ requires an additional p-triple which value is less than five. Then, e.g., $M_{24}, \tau \models \mathbf{def}(x_1)$ and $M_{246}, \tau \models \mathbf{def}(x_2)$, but $M_{24}, \tau \neg\models \mathbf{def}(x_2)$ and $M_{24a}, \tau \neg\models \mathbf{def}(x_1)$. In comparison, $\mathbf{def}(x_3)$ requires exactly two p-triples with even values greater than five, the latter being expressed by the restriction after the and. Therefore, none of the sets from Example 3 satisfies neither $\mathbf{def}(x_3)$, nor $\mathbf{def}(x_4)$, under τ. Finally, $M_{2a}, \tau \models \mathbf{def}(x_5)$ because $\mathbf{def}(x_5)$ allows the extra triple $(n, p, "a")$.

However, $M_{24a}, \tau \neg \models \mathbf{def}(x_6)$, because extra$\{p\}$ in the extended shape expression $\mathbf{def}(x_5)$ is not considered for satisfying $\mathbf{def}(x_6)$, thus nothing allows the triple $(n, p, "a")$.

We will be interested in correct typings only. They are coherent w.r.t. the satisfiability relation, and they ensure that an abstract label can be satisfied only through one of its non-abstract descendants.

Definition 4 (correct typing). *Let* $\mathbf{G}$ *be a graph and* $\mathbf{S}$ *be a schema. A typing* τ *of* $\mathbf{G}$ *by* $\mathbf{S}$ *is* correct *if for every* $(n, z) \in \tau$*:*

- *if* $z \notin \mathbf{A}$*, then* $\mathbf{G}, n, \tau \models_{\mathbf{S}} \mathbf{def}(z)$*,*
- *if* $z \in \mathbf{A}$*, then there exists* $x \in desc(z) \setminus \mathbf{A}$ *such that* $\mathbf{G}, n, \tau \models_{\mathbf{S}} \mathbf{def}(x)$*.*

We denote $\mathcal{C}(\mathbf{G}, \mathbf{S})$ *the set of correct typings of* $\mathbf{G}$ *by* $\mathbf{S}$*.*

4.2 Well-Defined Schemas

Shape expressions can refer to each other or extend one another and this could result in circular definitions. Well-defined schemas forbid some circular definitions. We start by defining a dependency graph that captures how shape expressions depend on one another, then use it to define well-defined schemas.

For every shape expression or triple expression s, let *sub-expr*(s) be the set of its sub-expressions, viewed as syntactic objects. Additionally, *neg-sub-expr*(s) is the set of sub-expressions of s that appear in s under an odd number of occurrences of the negation operator not. Let $\mathbf{S} = (\mathbf{Y}, \mathbf{X}, \mathbf{def}, \mathbf{A})$ be a schema. We define the following four binary relations for labels $z, z' \in \mathbf{X} \cup \mathbf{Y}$:

- *dep-extends*$_{\mathbf{S}}(z, z')$ iff $H_{\mathbf{S}}$ has an edge from z to z' or from z' to z;
- *dep*$_{\mathbf{S}}(z, z')$ iff z' appears as a reference in $\mathbf{def}(z)$, i.e. $@z' \in$ *sub-expr*$(\mathbf{def}(z))$;
- *dep-shape-neg*$_{\mathbf{S}}(z, z')$ iff [closed][extra P]$\langle e \rangle \in$ *neg-sub-expr*$(\mathbf{def}(z))$ and $@z' \in$ *sub-expr*(e);
- *dep-extra-neg*$_{\mathbf{S}}(z, z')$ iff [closed][extra P]$\langle e \rangle \in$ *sub-expr*$(\mathbf{def}(z))$, and $p\,@z' \in tcs(e)$, and $p \in P$.

The dependencies *dep-shape-neg* and *dep-extra-neg* are called *negative* dependencies. Obviously, *dep-extends* is new for ShEx with inheritance. Observe that *dep* subsumes *dep-shape-neg* and *dep-extra-neg*. We say that z depends on z' in schema $\mathbf{S}$ if *dep*$_{\mathbf{S}}(z, z')$ or *dep-extends*$_{\mathbf{S}}(z, z')$. The schema subscript is omitted whenever the schema is clear from the context.

Example 6. Let $\mathbf{S}_1$, $\mathbf{S}_2$ and $\mathbf{S}_3$ be the schemas below, from left to right, with respective sets of simple labels $\mathbf{Y}_1 = \{y_1, y_2, y_3\}$, $\mathbf{Y}_2 = \{y_4, y_5, y_6\}$ and $\mathbf{Y}_3 = \{y_7, y_8\}$ and with extendable labels $\mathbf{X}_3 = \{x_1, x_2\}$ for $\mathbf{S}_3$. Below each schema are enumerated the facts of its dependency relations.

$y_1 \to \langle\, p\ @y_2 \,\rangle$ and $\langle\, q\ @y_3 \,\rangle$ $y_2 \to \langle\, q\ @y_1 \,\rangle$ $y_3 \to$ extra$\{r\}$ $\langle\, p\ @y_1 \,\rangle$	$y_4 \to \langle\, p\ @y_5 \,\rangle$ or $\langle\, q\ @y_6 \,\rangle$ $y_5 \to$ not$\langle\, q\ @y_4 \,\rangle$ $y_6 \to$ extra$\{p\}$ $\langle\, p\ @y_4 \,\rangle$	$x_1 \to$ extends _ $\langle\, p\ @y_7 \,\rangle$ $x_2 \to$ extends x_1 $\langle\, p\ @y_8 \,\rangle$ $y_7 \to$ not $\langle\, q\ @x_2 \,\rangle$ $y_8 \to c$
$dep_{\mathbf{S}_1}(y_1, y_2)$ $dep_{\mathbf{S}_1}(y_1, y_3)$ $dep_{\mathbf{S}_1}(y_2, y_1)$ $dep_{\mathbf{S}_1}(y_3, y_1)$	$dep_{\mathbf{S}_2}(y_4, y_5)$ $dep_{\mathbf{S}_2}(y_4, y_6)$ $\textit{dep-shape-neg}_{\mathbf{S}_2}(y_5, y_4)$ $\textit{dep-extra-neg}_{\mathbf{S}_2}(y_6, y_4)$	$dep_{\mathbf{S}_3}(x_1, y_7)$ $dep_{\mathbf{S}_3}(x_2, y_8)$ $\textit{dep-shape-neg}_{\mathbf{S}_3}(y_7, x_2)$ $\textit{dep-extends}_{\mathbf{S}_3}(x_1, x_2)$ $\textit{dep-extends}_{\mathbf{S}_3}(x_2, x_1)$

The dependency graph of a schema $\mathbf{S}$ regroups all these dependency relations and, together with the extension hierarchy, is used to define well-defined schemas.

Definition 5 (dependency graph). *Let* $\mathbf{S} = (\mathbf{Y}, \mathbf{X}, \mathbf{def}, \mathbf{A})$ *be a schema. The* dependency graph *of* $\mathbf{S}$ *is a directed labelled graph denoted* $D_{\mathbf{S}}$ *whose set of vertices is* $\mathbf{Y} \cup \mathbf{X}$ *and that has an edge from* z *to* z' *labelled* d *if* $d(z, z')$ *holds, where* d *is one of the dependency relations* $\textit{dep-extends}_{\mathbf{S}}$, $dep_{\mathbf{S}}$, $\textit{dep-shape-neg}_{\mathbf{S}}$, $\textit{dep-extra-neg}_{\mathbf{S}}$.

Definition 6 (well-defined schema, stratified negation). *A schema* $\mathbf{S}$ *is* well-defined *if it satisfies these conditions:*

- $H_{\mathbf{S}}$, *the extension hierarchy graph of a* $\mathbf{S}$, *is acyclic;*
- *in* $D_{\mathbf{S}}$, *no cycle contains a negative dependency edge labelled dep-shape-neg or dep-extra-neg. We say in this case that* $\mathbf{S}$ *is* with stratified negation.

Example 7. Considering the schemas from Example 6: $\mathbf{S}_1$ is well-defined, as the cycles along y_1, y_2 and along y_1, y_3 are not forbidden; $\mathbf{S}_2$ is not well-defined as its dependency graph contains two cycles – one along y_4, y_5, and another one along y_4, y_6 – each containing a negative dependency; $\mathbf{S}_3$ is not well-defined as it has a forbidden cycle $\textit{dep-shape-neg}_{\mathbf{S}_3}(y_7, x_2)$, $\textit{dep-extends}_{\mathbf{S}_3}(x_2, x_1)$, $dep_{\mathbf{S}_3}(x_1, y_7)$. Note that the cycle won't be present without the *dep-extends* relation; this illustrates that extension and negation cannot be arbitrarily interleaved.

From here we will discuss only well-defined schemas. For well-defined schemas, satisfiability relations can be effectively computed.

Lemma 1. *The evaluation of* $\mathbf{G}, n, \tau \models_{\mathbf{S}} \mathbf{def}(z)$ *always terminates, for every schema* $\mathbf{S} = (\mathbf{Y}, \mathbf{X}, \mathbf{def}, \mathbf{A})$, *graph* $\mathbf{G}$, *node* n *of* $\mathbf{G}$, *typing* τ *of* $\mathbf{G}$ *by* $\mathbf{S}$, *and label* $z \in \mathbf{Y} \cup \mathbf{X}$.

4.3 Semantics of ShEx Schemas

The remaining of the section is devoted to the construction of the maximal correct typing. It is defined w.r.t. a stratification of the schema, which defines stratums (layers) of labels such that circular dependencies can happen only between labels on the same stratum, while negative dependencies can only go from upper to lower stratums.

Definition 7 (stratification). *Let* $\mathbf{S} = (\mathbf{Y}, \mathbf{def}, \mathbf{X}, \mathbf{A})$ *be a well-defined schema. A* stratification *of* $\mathbf{S}$ *is a function* $\sigma : \mathbf{Y} \cup \mathbf{X} \to \mathbb{N}$ *such that:*

- *if there exists a non-empty path from* z *to* z' *in* $D_{\mathbf{S}}$*, then* $\sigma(z) \geq \sigma(z')$*;*
- *if there exists in* $D_{\mathbf{S}}$ *a non-empty path from* z *to* z' *that contains an edge corresponding to a negative dependency, then* $\sigma(z) > \sigma(z')$*.*

It is well known that stratified negation, i.e. no cycles containing negative dependencies, guarantees the existence of a stratification. Note also that w.l.o.g. we can consider that the active range of a stratification is an interval of the form $[1..k]$ for some $k \geq 1$. In order to define $\tau_{\max}$, we introduce a series of typings $\tau_1, \ldots, \tau_k$ such that for every $i \in 1..k$, the typing τ_i contains only shape labels on stratums j such that $j \leq i$. Each of the intermediate typings is maximal w.r.t. set inclusion, in a sense to be made precise shortly. Then $\tau_{\max} = \tau_k$. We start by introducing some notations. Given a schema $\mathbf{S} = (\mathbf{Y}, \mathbf{def}, \mathbf{X}, \mathbf{A})$ and a stratification σ of $\mathbf{S}$ with active domain $[1..k]$, we denote $\mathbf{Y}_i^\sigma$, resp. $\mathbf{X}_i^\sigma$, resp. $\mathbf{A}_i^\sigma$ the subsets of labels from $\mathbf{Y}$, resp. $\mathbf{X}$, resp. $\mathbf{A}$ whose stratum is at most i. Formally, $\mathbf{Y}_i^\sigma = \{y \in \mathbf{Y} \mid \sigma(y) \leq i\}$, $\mathbf{X}_i^\sigma = \{x \in \mathbf{X} \mid \sigma(x) \leq i\}$ and $\mathbf{A}_i^\sigma = \mathbf{A} \cap \mathbf{X}_i^\sigma$. Additionally, we let $\mathbf{S}_i^\sigma = (\mathbf{Y}_i^\sigma, \mathbf{X}_i^\sigma, \mathbf{def}_i^\sigma, \mathbf{A}_i^\sigma)$ be the schema $\mathbf{S}$ restricted to the labels on stratum at most i. For a typing τ and a set of labels Z we denote by τ_Z the restriction of τ on Z, that is, $\tau_Z = \{(n, z) \in \tau \mid z \in Z\}$.

Definition 8 (maximal typing). *Let* $\mathbf{S} = (\mathbf{Y}, \mathbf{X}, \mathbf{def}, \mathbf{A})$ *be a schema,* $\sigma : \mathbf{Y} \cup \mathbf{X} \to [1..k]$ *be a stratification of* $\mathbf{S}$ *for some* $k \geq 1$*, and* $\mathbf{G}$ *be a graph. For every* $1 \leq i \leq k$*, let* τ_i *be the typing of* $\mathbf{G}$ *by* $\mathbf{S}_i^\sigma$ *defined inductively by:*

- τ_1 *is the union of all correct typings of* $\mathbf{G}$ *by* $\mathbf{S}_1^\sigma$*,*
- *for every* $1 \leq i < k$*,* τ_{i+1} *is the union of all correct typings* τ' *which restriction on* $\mathbf{Y}_i^\sigma \cup \mathbf{X}_i^\sigma$ *is* τ_i*. Formally,* $\tau_{i+1} = \bigcup \left\{ \tau' \subseteq \mathcal{C}(\mathbf{G}, \mathbf{S}_{i+1}^\sigma) \mid \tau'_{\mathbf{Y}_i^\sigma \cup \mathbf{X}_i^\sigma} = \tau_i \right\}$.

The typing τ_k *is called the* maximal typing *of* $\mathbf{G}$ *by* $\mathbf{S}$ *with* σ *and is denoted* $\tau_{\max}(\mathbf{S}, \sigma, \mathbf{G})$.

Proposition 1. *For every schema* $\mathbf{S}$*, stratification* σ *of* $\mathbf{S}$*, and graph* $\mathbf{G}$*, the typing* $\tau_{\max}(\mathbf{S}, \sigma, \mathbf{G})$ *is a correct typing.*

The maximal typing can be defined independently on a particular stratification, thanks to the lemma below. Therefore, we denote by $\tau_{\max}(\mathbf{S}, \mathbf{G})$ the unique maximal typing of $\mathbf{G}$ by $\mathbf{S}$.

Lemma 2. *For every schema* $\mathbf{S}$*, every graph* $\mathbf{G}$ *and all two stratifications* σ_1 *and* σ_2 *of* $\mathbf{S}$*, it holds that* $\tau_{\max}(\mathbf{S}, \sigma_1, \mathbf{G}) = \tau_{\max}(\mathbf{S}, \sigma_2, \mathbf{G})$.

5 Validation Algorithm

The validation problem for ShEx is, given a graph $\mathbf{G}$, a schema $\mathbf{S}$ and a typing τ of $\mathbf{G}$ by $\mathbf{S}$, determine whether $\tau \subseteq \tau_{\max}(\mathbf{S}, \mathbf{G})$. Intuitively, the latter means that for every $(n, z) \in \tau$, the node n conforms to the shape expression named z. A validation algorithm for ShEx with inheritance can be based on computing the maximal typing, using a standard *refinement* algorithm such as that presented in [4]. It goes as follows. Compute a stratification σ for $\mathbf{S}$, then construct the series of typings τ_i from Definition 8. Each of the τ_i is obtained by a refinement that starts with τ_i^c, a *complete typing for stratum* i, and successively removes from it the pairs that are not conformant. More precisely, $\tau_i^c = \tau_{i-1} \cup \{(n, z) \mid nodes(\mathbf{G}) \times (\mathbf{Y} \cup \mathbf{X}) \mid \sigma(z) = i\}$. Then, if τ is the current version of the progressively refined typing, we repeatedly remove from it pairs $(n, z) \in \tau$ such that $\mathbf{G}, n, \tau \neg \models_{\mathbf{S}} \mathbf{def}(z)$, until the typing becomes correct. This algorithm also works for ShEx with inheritance. As shown in Lemma 1, $\mathbf{G}, n, \tau \models_{\mathbf{S}} \mathbf{def}(z)$ can be effectively computed, so validation is decidable. In [4], the authors also present a recursive algorithm that computes only a portion of $\tau_{\max}(\mathbf{G}, \mathbf{S})$ that is relevant for the validation task. This algorithm also works for ShEx with inheritance. Because the validation algorithms for ShEx with inheritance are essentially the same as for ShEx 2.1, the effort required to adapt the existing ShEx validators is limited.

The inheritance mechanism does not increase the complexity of validation. As shown in [15], validation of ShEx limited to triple expressions only (i.e. the $M, \tau \models e$ relation) is NP-complete, while the refinement algorithm requires a polynomial number of steps. It is easy to see that computing the truth value of $\mathbf{G}, n, \tau \models_{\mathbf{S}} \mathbf{def}(z)$ is in NP. Therefore

Proposition 2. *Validation for ShEx with inheritance is NP-complete.*

6 Discussion

Comparison with the ShEx Specification. We use two simplifications. First, in the specification, references and shape expressions are interchangeable, e.g., it is possible to write $@y_1$ and $@y_2$ or $\langle\, p\ @(\mathsf{not}\ c_1\ \mathsf{or}\ @x)\,\rangle$. Following [4], we allow references only in triple constraints. This does not modify the expressive power of the language, yet it simplifies its presentation and formalization. Second, in the specification, the sets of labels $\mathbf{Y}$, $\mathbf{X}$ and $\mathbf{A}$ are not explicitly given. The set of extendable labels $\mathbf{X}$ is determined by the use of the extends keyword, and similarly the set of abstract labels $\mathbf{A}$ is determined by the use of abstract.

Inheritance-Like Features in ShEx 2.1. ShEx 2.1 allows for the expression of inheritance in some cases. If a shape expression needs to be extended with new properties only, then conjunction alone can be used, as in this example:

$$Figure \to \langle\, coord\ @Coord \,\rangle \qquad Circle \to @Figure\ \mathsf{and}\ \langle\, radius\ @T_{float} \,\rangle$$

A *Circle* is thus a *Figure* that has coordinates and a radius. This works because $\langle\, coord\ @Coord \,\rangle$ and $\langle\, radius\ @T_{float} \,\rangle$ use disjoint sets of properties and do not interact with each other. In a slightly different example, we have products with numeric reviews and want to extend them with text reviews:

$$Product \to \langle\, review\ @T_{int} * \,\rangle \qquad MyProduct \to @Product\ \mathsf{and}\ \langle\, review\ @T_{str} * \,\rangle$$

MyProduct cannot be satisfied because a value of a *review* property is required to be both integer and string. We actually want to be able to write something like $MyProduct \rightarrow @Product; \langle\, review\ @T_{str} * \,\rangle$, i.e. use the ';' operator to extend the contents of *Product* with additional properties. This would be possible in ShEx 2.1 if the definition of *Product* is a shape (non-terminal h in Definition 1), which is an important limitation. Furthermore, such a mechanism would not allow for easy restriction of the inherited shape, for instance if *MyProduct* required that numeric reviews have values between 1 and 5. Overall, ShEx 2.1 or mild extensions of it could provide inheritance-like features with the cost of modelling gymnastics.

Design Choices. We impose a syntactic restriction on extendable shape expressions. They must be conjunctions of a shape and a plain shape expression. That is, we disallow the extension of more general expressions, in particular expressions of the form s_1 or s_2. In what follows, we explain the rationale behind this design choice. On the one hand, the motivating use cases clearly demonstrated the need to extend shape expressions of that form, while the need to extend disjunctions was not supported by the use cases. On the other hand, extending on disjunctions together with multiple inheritance complicates the definition of the satisfaction relation, as illustrated by the following. In the example from Sect. 2, *ColouredCircle* inherits from *Figure* in two different ways. Suppose that the definition was $Figure \rightarrow \langle\, coord\ @CartesianCoord \rangle$ or $\langle\, coord\ @PolarCoord \rangle$. This raises the question whether we allow a *ColouredCircle* to have both Cartesian and polar coordinates, the one inherited from *ColouredFigure*, the other inherited from *Circle*. Allowing it would be undesirable. Disallowing it would complicate the satisfiability relation as it would require (1) recording which one of the disjuncts (here, Cartesian coordinate and polar coordinate) has been satisfied by each subtype (here, by *ColouredCircle* and by *ColouredFigure*), and (2) checking whether all the subtypes satisfy the same disjunct. We decided to take a conservative approach and limit the shape expressions that can be extended.

Limitations. The inheritance mechanism can be unintuitive in cases in which a node did not satisfy a shape expression before it was extended, but does conform after the expression is extended, as for instance node `f1` and shape expression *Circle* from Sect. 2. The language is backward compatible in the positive case, i.e. all nodes that satisfy a shape expression continue to satisfy it if it gets extended. Another possible drawback is the need to know all the descendants of a shape expression in order to validate it. This can limit the modularity of the validation.

Inheritance in SHACL? SHACL is an RDF validation language published as a W3C recommendation [7], and has been formalized in several scientific works [3,6,11], among others. We are not aware of the development of an inheritance mechanism for SHACL. We believe that it would require using conjunction, at least for some use cases. Using the syntax of SHACL from [6], the product example can be captured by:

$$Product \rightarrow \geq_0 review.T_{int} \qquad MyProduct \rightarrow Product \wedge \geq_0 review.T_{str}$$

Then every *MyProduct* would be a *Product*. The following example cannot be handled in the same way: a *Client* shape expression that has an email needs to be extended to *MyClient* that has an additional email. In ShEx with inheritance it is captured by $Client \rightarrow \mathsf{extends}\ \_\ \langle\ email\ @T_{str}\ \rangle$, $MyClient \rightarrow \mathsf{extends}\ Client\ \langle\ email\ @T_{str}\ \rangle$. A SHACL *MyClient* defined in a way similar to *MyProduct* above would have only one email.

Related Work. The semantics of ShEx with inheritance presented here is an extension of that of ShEx 2.1 from [4], e.g. based on the existence of a maximal typing. Overall, the proof follows the same steps, with additional technicality due to the new satisfiability relation $M, \tau \models s$ on the one hand, and to the fact that checking the satisfiability of a shape with extends requires dereferencing its ancestors (line 18.), on the other hand. The syntax of the language is slightly different from [4]. There, triple constraints use sets of properties instead of a single property, which allows encoding the closed and extra modifiers of a shape within its triple expression. We cannot rely on the same simplification as the closed and extra modifiers of the ancestors are ignored, only the triple expression *ext-te*(x) is used for satisfaction (line 12. in Table 2).

The notion of inheritance and its different incarnations has a long tradition in object-oriented programming [16], under various names such as sub-typing, generalization/specialization, etc. Many of these proposals date to the 80 s. The Liskov substitution principle was proposed in 1987 as a mechanism for conceptual inheritance [10] and the problem of name-collision in multiple-inheritance systems was already discussed at that time [8]. The notion of inheritance and inclusion polymorphism is a key feature of object-oriented languages [5]. We follow this tradition by allowing a hierarchy of shape expressions which is similar to sub-typing; we avoid the term subtype to prevent confusion with the notion of types identified by `rdf:type` in the RDF context.

Given that ShEx was inspired by XML Schema, the basic functionality of *extends* was also inspired by the extension behaviour in XML Schema [14], but we added multiple-inheritance and removed the requirement of type annotations in data. PG-Schema [2] has been proposed as a schema language for labelled property graphs and includes a notion of inheritance with support for multiple-inheritance. However, it checks that nodes conform exactly with their types, without considering their descendants as in our proposal. Extending ShEx to describe property graphs or RDF-Star has been explored in [9], although that paper does not include the inheritance mechanism presented here.

Conclusions. We presented an extension of ShEx with an inheritance mechanism. It is called for by use cases and inspired by inheritance in programming languages. Its semantics is well-founded, and builds upon the semantics of ShEx 2.1, which makes it easy to integrate into existing ShEx validators. The extension has several implementations [1]. The design of the inheritance mechanism required some decisions in order to ensure a good trade-off between keeping its conceptual complexity reasonable, while satisfying the motivating use cases. While our proposal has some limitations, we believe that it would allow for better reusability of ShEx schemas.

Acknowledgement. The work from Jose E. Labra has been partially funded by project ANGLIRU: Applying kNowledge Graphs to research data interoperabiLIty and ReUsability, code: PID2020-117912RB from the Spanish Research Agency and by the regional project SEK-25-GRU-GIC-24-089.

Disclosure of Interests. Nothing to report.

A Appendix

We view the satisfiability relations from Tables 1, 2 and 3 as boolean functions.

Lemma 1 (proof sketch). The proof goes by induction on the tree of recursive calls resulting from the evaluation of $n, \tau \models \mathbf{def}(z)$, let θ be that tree. In θ, every vertex corresponds to one of the lines 1.–18. of the satisfaction functions in Tables 1, 2 and 3 depending on the function being called in that vertex, and on the syntactic form of the shape or triple expression passed as parameter. Remark that almost all direct recursive calls are made on strict sub-expressions, except for lines 11., 12. and the two recursive calls $M_x, \tau \models \mathit{ext\text{-}te}(x)$ and $\left(\bigcup_{z \in anc(x)} M_z\right), \tau \models \mathit{restr}(x)$ on line 18.. Line 11. directly calls line 17. which in turns makes recursive calls on strict sub-expressions. Line 12. directly calls line 18. that we discuss below. The above two cases from line 18. are those that require attention. We show that in θ, if a vertex is a call $r = n, \tau \models \textsf{extends}\, X\, h[\textsf{ and } s_1]$ for line 12., then none of its direct recursive calls mentioned earlier has a descendant equal to r. This is due to the acyclic nature of the extension hierarchy. As a consequence, the tree θ of recursive calls is finite.

Proposition 1. The proof goes by induction on the number of stratums.
Base case There is a unique stratum and the dependency relations $\mathit{dep\text{-}extra\text{-}neg}_{\mathbf{S}}$ and $\mathit{dep\text{-}shape\text{-}neg}_{\mathbf{S}}$ are empty. As a consequence, (*) every sub-expression $\textsf{not}\, s_1$ that appears in the schema is s.t. s_1 contains no references, and every sub-expression $[\textsf{closed}][\textsf{extra}\, P]\langle\, e\,\rangle$ is s.t. for every $p\ @u \in tcs(e)$, if u is a reference, then $p \notin P$.

We show that (**) if τ' and τ'' are correct typings, their union $\tau = \tau' \cup \tau''$ is also a correct typing then, because there is a finite number of typings, the union of all correct typings (i.e. $\tau_{\max}$) is also correct. For (**), it is enough to show that if $n, \tau' \models \mathbf{def}(z)$[6] evaluates to true, then $n, \tau \models \mathbf{def}(z)$ evaluates to true, for every n and every $z \notin \mathbf{A}$ s.t. either $(n, z) \in \tau$, or there exists $x \in \mathbf{A}$ with $z \in desc(x) \setminus \mathbf{A}$ and $(n, x) \in \tau$. This goes by structural induction on the tree of recursive calls for the evaluation of $n, \tau' \models \mathbf{def}(z)$, let θ be that tree.

6 the case for τ'' is symmetric.

Once again, we distinguish the cases by the corresponding line in 1.–18. of the algorithms. Lines 1., 6., 10., 13. are base cases (leaves of θ). Lines 1., 6. and 13. do not rely on the typing, while for line 10., obviously $(n', z') \in \tau' \implies (n', z') \in \tau$ for all n', z'. For the induction case, we consider a vertex $r' = \frac{M}{n}, \tau' \models s$ of θ (where $\frac{M}{n}$ is the node or set of triples parameter of the call), and the following induction hypothesis: (ih) for every r'' strict descendant of r' in θ, if r'' evaluates to true, then r'' with same parameters except for the typing τ' being replaced by τ also evaluates to true. We show that then $\frac{M}{n}, \tau \models s$ evaluates to true.

Lines 2., 11., 12., 3., 7., 8., 14., 15. easily follow from (ih). For lines 4., 5., 18. we can take the same values for the sets M_i, M_x and M and then it again follows from (ih). For lines 9. and 16., the expression of the call is $\mathsf{not}\, s_1$ and then by (*), s_1 does not contain references, so its satisfiability does not depend on the typing. Therefore, the only challenge is line 17.. So, consider a call $r = M, \tau \models [\mathsf{closed}][\mathsf{extends}\, P]\langle\, e\,\rangle$, let $r' = M, \tau' \models [\mathsf{closed}][\mathsf{extends}\, P]\langle\, e\,\rangle$, and let:

- $M^e_{\tau'} = \{(n, p, o) \in M \mid \exists p\ @s \in tcs(e).\ o, \tau' \models s\}$, and
- $M^e_{\tau} = \{(n, p, o) \in M \mid \exists p\ @s \in tcs(e).\ o, \tau \models s\}$

be the corresponding sets computed in the calls r' and r, respectively. We show that if r' evaluates to true, then $M^e_{\tau'} = M^e_{\tau}$, which using (ih) directly allows deducing that r evaluates to true.

So, assume that r evaluates to true. Remark first that by (ih) we have $M^e_{\tau'} \subseteq M^e_{\tau}$. Suppose by contradiction that $(n_1, p, n_2) \in M$ and $p\ @s \in tcs(e)$ are s.t. $n_2, \tau \models s$ but $n_2, \tau' \neg \models s$; thus $(n_1, p, n_2) \in M^e_{\tau} \setminus M^e_{\tau'}$. Then by (*), we deduce that s is of the form $@z'$, so $p \notin P$. Then (n_1, p, n_2) belongs to the set $M^{\not e}_{\tau'} = \{(n_1, q, n') \in M \mid q \in props(e)\} \setminus M^e_{\tau'}$ computed during the evaluation of r, so $props(M^{\not e}_{\tau'}) \not\subseteq P$, which contradicts the fact that r evaluates to true.

Induction Case. Let $i > 1$ be in the range of σ, the induction hypothesis is (IH) for every $j < i$, τ_j is a correct typing. We again show that if τ', τ'' are correct typings for $\mathbf{S}^{\sigma}_i$, their union $\tau = \tau' \cup \tau''$ is also a correct typing. Thus, we need to show that if $n, \tau' \models \mathbf{def}(z)$ evaluates to true, then $n, \tau \models \mathbf{def}(z)$ evaluates to true, for every n and every $z \notin \mathbf{A}$ s.t. either $(n, z) \in \tau$, or there exists $x \in \mathbf{A}$ with $z \in desc(x) \setminus \mathbf{A}$ and $(n, x) \in \tau$. We assume that $\sigma(z) = i$, otherwise (IH) applies directly. Similarly to the base case above, the proof goes by structural induction on the tree of recursive calls θ for $n, \tau' \models \mathbf{def}(z)$. The challenging cases are lines 9., 16. and 17.. We do not have any more the property (*) from above allowing to easily deal with lines 9. 16.. However, we can show that if θ contains a node with expression $\mathsf{not}\, s_1$, then its evaluation depends only on labels lying on stratums strictly less than i. Moreover, by definition of stratification, all labels that appear in s_1 are necessarily on some stratum that is strictly less than i. Note however that, because of the extension mechanism and line 10., the evaluation of s_1 might depend on labels that are descendants of a label appearing in s_1. Because of line 18., it can also depend on labels that appear in the definitions of ancestors of z. Thanks to the *dep-extends* dependency relation, such labels are also on stratums strictly less than i. We thus show that $\frac{M}{n}, \tau' \models \mathsf{not}\, s_1$

evaluates the same as $\frac{M}{n}, \tau'_{|i-1} \models \mathsf{not}\, s_1$, where $\tau'_{|i-1}$ is the restriction of τ' to labels on stratum at most $i-1$. Similarly, $\frac{M}{n}, \tau \models \mathsf{not}\, s_1$ evaluates the same as $\frac{M}{n}, \tau_{|i-1} \models \mathsf{not}\, s_1$. By definition we know that $\tau'_{|i-1} = \tau_{|i-1} = \tau_{i-1}$, then $\frac{M}{n}, \tau' \models \mathsf{not}\, s_1$ and $\frac{M}{n}, \tau \models \mathsf{not}\, s_1$ have the same value.

As for line 12., the proof uses the arguments from above regarding negative dependencies and stratification, and the arguments from the same line in the base case.

Lemma 2 (proof sketch). Let σ' be a most refined stratification obtained by taking every strongly connected component of $H_{\mathbf{S}}$ as a separate stratum. Then for every stratification σ, every stratum of σ is a union of stratums of σ'. Using this property, we can show that for every stratification σ, $\tau_{\max}(\mathbf{S}, \sigma, \mathbf{G}) = \tau_{\max}(\mathbf{S}, \sigma', \mathbf{G})$. It then immediately follows that $\tau_{\max}(\mathbf{S}, \sigma_1, \mathbf{G}) = \tau_{\max}(\mathbf{S}, \sigma_2, \mathbf{G})$.

References

1. ShEx with inheritance: companion webpage. https://github.com/weso/shex_extends_paper_companion
2. Angles, R., et al.: PG-schema: schemas for property graphs. Proc. ACM Manag. Data **1**(2), 198:1–198:25 (2023). https://doi.org/10.1145/3589778
3. Bogaerts, B., Jakubowski, M., den Bussche, J.V.: SHACL: a description logic in disguise. In: Logic Programming and Nonmonotonic Reasoning - 16th International Conference, LPNMR 2022, Genova, Italy, September 5-9, 2022, Proceedings. Lecture Notes in Computer Science, vol. 13416, pp. 75–88. Springer, Cham (2022). https://doi.org/10.1007/978-3-031-15707-3_7
4. Boneva, I., Labra Gayo, J.E., Prud'hommeaux, E.G.: Semantics and validation of shapes schemas for RDF. In: d'Amato, C., et al. (eds.) ISWC 2017. LNCS, vol. 10587, pp. 104–120. Springer, Cham (2017). https://doi.org/10.1007/978-3-319-68288-4_7
5. Cardelli, L., Wegner, P.: On understanding types, data abstraction, and polymorphism. ACM Comput. Surv. **17**(4), 471–522 (1985). https://doi.org/10.1145/6041.6042
6. Corman, J., Reutter, J.L., Savković, O.: Semantics and validation of recursive SHACL. In: Vrandečić, D., et al. (eds.) ISWC 2018. LNCS, vol. 11136, pp. 318–336. Springer, Cham (2018). https://doi.org/10.1007/978-3-030-00671-6_19
7. Knublauch, H., Kontokostas, D.: Shapes Constraint Language (SHACL) (2017). https://www.w3.org/TR/shacl/. W3C Recommendation
8. Knudsen, J.L.: Name collision in multiple classification hierarchies. In: Gjessing, S., Nygaard, K. (eds.) ECOOP 1988. LNCS, vol. 322, pp. 93–109. Springer, Heidelberg (1988). https://doi.org/10.1007/3-540-45910-3_6
9. Labra Gayo, J.E.: Extending shape expressions for different types of knowledge graphs. In: Joint Proceedings of the 3rd International Workshop on Knowledge Graph Generation from Text (TEXT2KG) and Data Quality Meets Machine Learning and Knowledge Graphs, Co-located with the Extended Semantic Web Conference. CEUR-WS (2024)
10. Liskov, B.: Keynote address - data abstraction and hierarchy. In: Addendum to the Proceedings on Object-Oriented Programming Systems, Languages and Applications, OOPSLA 1987 Addendum, Orlando, Florida, USA, October 4-8, 1987, pp. 17–34. ACM (1987). https://doi.org/10.1145/62138.62141

11. Pareti, P., Konstantinidis, G., Mogavero, F., Norman, T.J.: SHACL satisfiability and containment. In: Pan, J.Z., et al. (eds.) ISWC 2020. LNCS, vol. 12506, pp. 474–493. Springer, Cham (2020). https://doi.org/10.1007/978-3-030-62419-4_27
12. Prud'hommeaux, E., Labra Gayo, J.E., Solbrig, H.R.: Shape expressions: an RDF validation and transformation language. In: Proceedings of the 10th International Conference on Semantic Systems, SEMANTiCS 2014, Leipzig, Germany, September 4-5, 2014, pp. 32–40. ACM (2014).https://doi.org/10.1145/2660517.2660523
13. Sharma, D.K., Prud'hommeaux, E., Booth, D., Nanjo, C., Jiang, G.: Shape expressions (ShEx) schemas for the FHIR R5 specification. J. Biomed. Inform. **148**, 104534 (2023). https://doi.org/10.1016/J.JBI.2023.104534
14. Siméon, J., Wadler, P.: The essence of XML. In: Conference Record of POPL 2003: The 30th SIGPLAN-SIGACT Symposium on Principles of Programming Languages, New Orleans, Louisisana, USA, January 15-17, 2003, pp. 1–13. ACM (2003). https://doi.org/10.1145/604131.604132
15. Staworko, S., Boneva, I., Labra Gayo, J.E., Hym, S., Prud'hommeaux, E.G., Solbrig, H.R.: Complexity and expressiveness of ShEx for RDF. In: 18th International Conference on Database Theory, ICDT 2015, March 23-27, 2015, Brussels, Belgium. LIPIcs, vol. 31, pp. 195–211. Schloss Dagstuhl - Leibniz-Zentrum für Informatik (2015). https://doi.org/10.4230/LIPICS.ICDT.2015.195
16. Taivalsaari, A.: On the notion of inheritance. ACM Comput. Surv. **28**(3), 438–479 (1996). https://doi.org/10.1145/243439.243441
17. Thornton, K., et al.: Using shape expressions (ShEx) to share RDF data models and to guide curation with rigorous validation. In: The Semantic Web - 16th International Conference, ESWC 2019, Portorož, Slovenia, June 2-6, 2019, Proceedings. Lecture Notes in Computer Science, vol. 11503, pp. 606–620. Springer, Cham (2019). https://doi.org/10.1007/978-3-030-21348-0_39

Author Index

E. Curry et al. (Eds.): ESWC 2025, LNCS 15718, pp. 501–503, 2025.
https://doi.org/10.1007/978-3-031-94575-5

Made in the USA
Monee, IL
03 May 2026

49438643R00295